CAMBRIDGE MONOGRAPHS ON MATHEMATICAL PHYSICS

LIQUID METALS
Concepts and Theory

LIQUID METALS

Concepts and Theory

N. H. MARCH

Coulson Professor

Theoretical Chemistry Department, University of Oxford

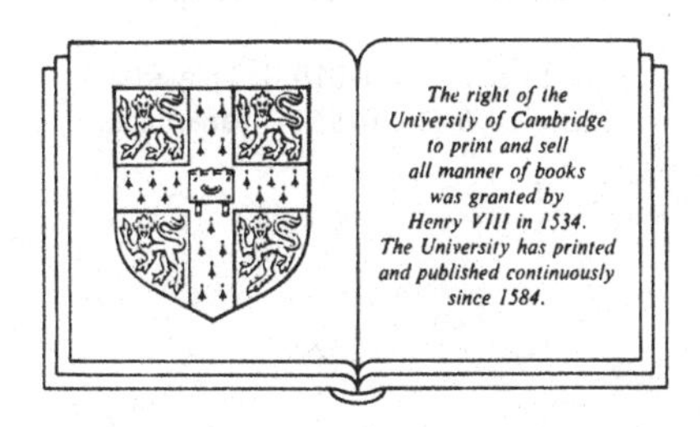

CAMBRIDGE UNIVERSITY PRESS

Cambridge

New York Port Chester Melbourne Sydney

CAMBRIDGE UNIVERSITY PRESS
Cambridge, New York, Melbourne, Madrid, Cape Town, Singapore, São Paulo

Cambridge University Press
The Edinburgh Building, Cambridge CB2 2RU, UK

Published in the United States of America by Cambridge University Press, New York

www.cambridge.org
Information on this title: www.cambridge.org/9780521302791

First published 1990
This digitally printed first paperback version 2005

A catalogue record for this publication is available from the British Library

Library of Congress Cataloguing in Publication data

March, Norman H. (Norman Henry), 1927–
Liquid metals / N. H. March.
p. cm. – (Cambridge monographs on mathematical physics)
ISBN 0-521-30279-X
1. Liquid metals. I. Title. II. Series.
QD171.M28 1990
546′.3–dc20 89-33208
CIP

ISBN-13 978-0-521-30279-1 hardback
ISBN-10 0-521-30279-X hardback

ISBN-13 978-0-521-01961-3 paperback
ISBN-10 0-521-01961-3 paperback

Contents

Preface

The origin of this volume can be traced to a letter from Dr. P. V. Landshoff (PVL), inviting me to write on liquid metals for his series. By then, my earlier book on the subject, published in 1968, described correctly by Professor N. E. Cusack in his generous review as 'Bare bones of liquid metals', was almost 20 years old. Of course, Dr. T. E. Faber's 1972 book was much more extensive, and I immediately recognized that in one important area of the subject: namely, weak scattering theory of electrical transport in liquid metals and alloys (Ziman and Faber-Ziman theory, respectively; see also below), I could not possibly compete with the quality of that.

However, after a long inner debate, I accepted the iniviation and sent PVL a proposed outline which already made clear that it would be a large volume. Back came a reply from PVL and an adviser: could I extend it somewhat?! Perhaps this may have been partly motivated by my long-standing interests in matter under extreme conditions of temperature and pressure, but, nevertheless, these additional proposals led me into areas in which I had not contributed myself for more than a decade. In the end, Chapter 16 became the main response to this challenge, and this could not have been written without the help of the 1985 review by Dr. M. Ross of the Lawrence Livermore Laboratory. While mentioning that, I must also acknowledge the 1987 survey by Dr. R. N. Singh, which was used so extensively in Chapter 13. Dr. Singh worked closely with the late Professor A. B. Bhatia, with whom I was also fortunate enough to collaborate over a decade or so.

This leads to acknowledgments of my indebtedness to numerous other people. First, colleagues from my Sheffield University days, Drs. T. Gaskell, W. Jones, J. S. Rousseau and J. C. Stoddart, provided much stimulation and help, while my research students R. C. Brown, G. K. Corless and M. D. Johnson notably influenced my outlook on the subject, the latter two in the general area of effective interionic interactions mediated by conduction electrons. Later, at Imperial College, London University, S. Cusack, through his Ph.D. thesis studies, aided considerably my own understanding of electronic correlation functions; C. M. Sayers strongly influenced the work in the area of magnetic properties. After moving to Oxford in 1977, my research students J. A. Ascough, R. G. Chapman, A. Ferraz and J. S. McCaskill and my colleague D. P. J. Grout are to be thanked for much help, particularly with transport in strong scattering systems, which has been made the basis for the whole discussion of electrical conductivity in the present volume; leaving T. E. Faber's account of weak scattering theory as the right source of that topic, as has already been mentioned. Though the present book is in the Mathematical Physics Series, I must add that I have been extremely lucky in also having close contact with the experimental groups of Professors J. E. Enderby and P. A. Egelstaff and later with Drs. D. I. Page and M. W. Johnson at Harwell and Rutherford Laboratories.

I owe the greatest debt to Professors M. P. Tosi (MPT) and W. H. Young (WHY) for the strong interactions with them and their research groups over two and three decades, respectively. To both of them I offer my warmest thanks for all the stimulation our group has received from them and also for their readiness to allow me to draw extensively on their own writings (e.g., WHY's 1987 article in Chapters 3 and 13 and my two books with MPT). The influence of them and their colleagues in visiting our group frequently has been of the utmost importance for us. Drs. M. Parrinello and G. Senatore from MPT's group spent extended periods of time with us and, deriving from WHY's influence, Professor J. A. Alonso and Dr. M. Ginoza also paid most valuable visits. It has been a pleasure to be involved with so many fine scientists and to have forged numerous firm friendships in the process.

In a work on this scale and with, at times, considerable theoretical detail, as in Chapter 14, it is almost inevitable that some errors will have crept in. I am, of course, solely responsible for any such, and I trust these will be, at worst, of detail and not of principle. But I shall count it a favour if readers who find my book either useful or interesting, and who spot places where I ought to do better, will write and tell me.

Finally, it is a pleasure to thank Cambridge University Press through the person of Rufus Neal, who gave much friendly and wise advice in seeing this work through to fruition.

NORMAN H. MARCH

1
Outline

This book is about the theory of liquid metals. The interplay between electronic and ionic structure is a major feature of such systems. This should occasion no suprise, as even a pure liquid metal is a two-component system: positive ions and conduction electrons. Therefore, as in a binary liquid mixture such as argon and krypton, where three partial structure factors S_{ArAr}, S_{KrKr}, and S_{ArKr} are required to describe the short-range atomic order, so in liquid metal Na, for instance, one needs $S_{\mathrm{Na^+Na^+}}$, $S_{\mathrm{Na^+e}}$, and S_{ee} for a structural characterization.

For a very fundamental treatment, the preceding description would be the correct starting point to treat liquid metal Na. Indeed, the theory of liquid metals has been developed in this manner. However, it is still true that, for many important purposes, a simpler picture suffices. Thus, in the chapter following this outline, attention will be focused on the ion-ion structure factor, which will simply be written as $S(k)$; $k = 4\pi \sin\theta/\lambda$, with 2θ the angle of scattering of X rays or neutrons and λ the wavelength of the radiation. It will be emphasized that it is indeed $S(k)$ that is measured in suitable neutron-scattering experiments.

Then, in the following chapter, the use of this knowledge of structure will be considered in relation to the thermodynamics of liquid metals. Following this, electron screening of ions will be treated with the theme stressed above, the interplay between electronic and ionic structure, leading to a treatment of effective interionic forces. This theory will then be confronted with an approach based on the so-called inverse problem—namely, that of extracting an effective ion-ion interaction from the measured structure factor $S(k)$. Following this study of interatomic forces, in which the Ornstein-Zernike direct correlation function $c(r)$ plays an important role, this same tool will be employed to treat the theory of freezing, following pioneering work of Kirkwood and Monroe (1941) and of Ramakrishnan and Yussouff (1977, 1979). A full discussion of electronic and also atomic transport then follows. This is closely linked to the study of liquids by inelastic neutron scattering; essentially described by the dynamical generalization $S(k, \omega)$ of the static structure factor $S(k)$. The frequency-dependent dielectric function and its use in treating optical properties will also be briefly considered.

The treatment of critical phenomena in liquid metals follows, this leading into a more detailed study of electron states plus some discussion of magnetism. Then, inhomogeneous systems are considered: the liquid-vapor surface, followed by surface segregation in binary alloys, the bulk properties of which are briefly treated in the penultimate chapter, which also focuses on phase diagrams of binary liquid alloys. This leads into the final chapter, which is concerned with the relevance of liquid metal theory to the hydrogen-helium mixtures in the giant planets, Jupiter and Saturn.

2
Pair correlation function and structure factor of ions

The most important single tool for dealing with the structural problem of liquid short-range order is the pair correlation function $g(r)$. If the bulk liquid number density is denoted by $\rho = N/\Omega$, for N atoms in volume Ω, then $g(r)$ is defined such that if one sits on an atom at the origin $r = 0$, then the probability of finding a second atom at distance between r and $r + dr$ is given by $g(r)4\pi r^2\,dr$ and the density of atoms is found by multiplying this by ρ.

2.1. Liquid structure factor

The pair function is accessible via diffraction experiments, as will now be outlined. Let $I(k)$ be the intensity of, say, X rays of wavelength λ, incident on a liquid sample and scattered through an angle 2θ, with

$$k = \frac{4\pi \sin\theta}{\lambda}. \tag{2.1}$$

Then if N is the number of atoms in the sample, the intensity $I(k)$ is related to the liquid structure factor $S(k)$ by

$$I(k) = Nf^2(k)S(k) \tag{2.2}$$

where $f(k)$ is the atomic scattering factor, given in terms of the electron density $\rho(r)$ in the atom (say, argon) by

$$f(k) = \int \rho(r)\exp(i\mathbf{k}\cdot\mathbf{r})\,d\mathbf{r}. \tag{2.3}$$

In turn, $S(k)$ is related to the pair function $g(r)$ by

$$S(k) = 1 + \rho\int [g(r) - 1]\exp(i\mathbf{k}\cdot\mathbf{r})\,d\mathbf{r}. \tag{2.4}$$

Figure 2.1 shows measured data on liquid Na, K, and Cs at the melting point from the work of Huijben and van der Lugt (1979; see also Greenfield, Wellendorf, and Wiser, 1971).

It is worth considering a number of features of this curve. First, the long wavelength limit of $S(k)$, i.e., $S(0)$, can be related to fluctuations $\langle \Delta N^2 \rangle$ in the number of particles and thereby to the isothermal compressibility K_T, the relation being

$$S(0) = \rho k_B T K_T, \tag{2.5}$$

which is proved in Appendix 2.1. For liquid argon near its triple point, $S(0) \sim 0.06$, whereas for the liquid metal K shown in Figure 2.1, $S(0) \sim 0.02$. This is in sharp contrast to a gas, where $S(0) \sim 1$ except near the critical point (see Section 9.2), where $S(0)$ diverges. In dense liquids the message is that $S(0) \ll 1$, which will later be seen to have important implications.

The second point to be made is that, as is clear from Figure 2.1, $S(k)$ has pronounced oscillations at large k. These come primarily from the

Figure 2.1. Measured structure factor $S(k)$ for liquid metals Na, K and Cs just above the melting point against scaled wave number (Huijben and van der Lugt, 1979; see also March and Tosi, 1984; Greenfield, Wellendorf and Wiser, 1971).

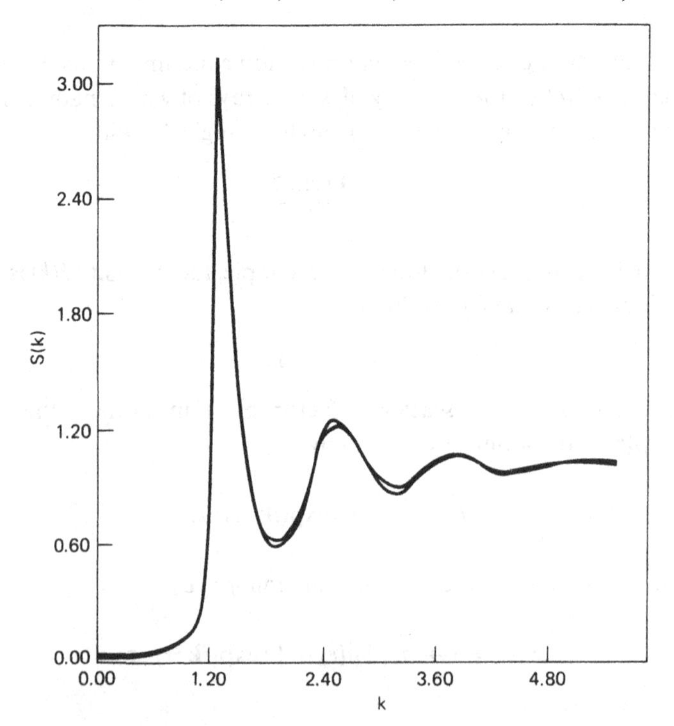

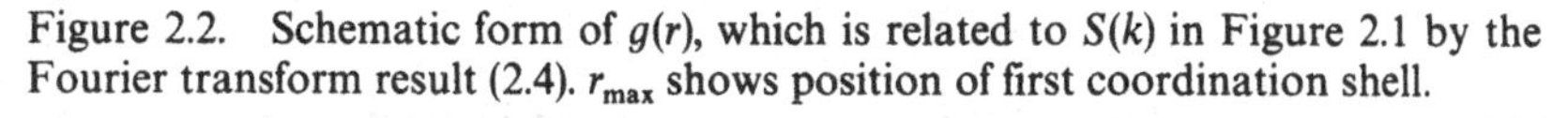

Figure 2.2. Schematic form of $g(r)$, which is related to $S(k)$ in Figure 2.1 by the Fourier transform result (2.4). r_{max} shows position of first coordination shell.

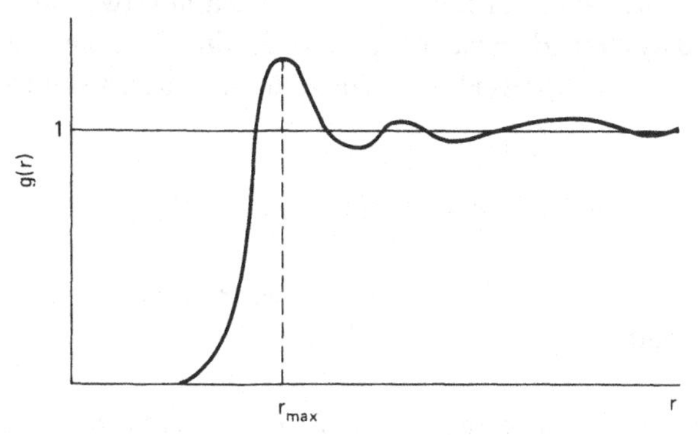

"hardness" of the core. In Figure 2.2 $g(r)$ is depicted schematically, and the important point is that there is a region inside the core diameter σ for which $g(r)$ is practically zero; then $g(r)$ rises steeply to its first peak. It is this steep rise that leads to the oscillations at large k in $S(k)$, from a basic property of Fourier transforms. The third point to be noted at this stage is that $S(k)$ reaches a first peak height of about 3.0 for the case plotted in Figure 2.1, where the temperature T is quite near the melting temperature T_m. This point will be taken up in the discussion of the theory of freezing in Chapter 6 (see also Ferraz and March, 1980).

Of course, a complete theory will eventually have to reproduce the structure factor $S(k)$ as measured by diffraction experiments (for neutrons, $f(k)$ in (2.2) is replaced by an appropriate scattering length) from a force field: a route that allows this, at least in principle, is set out in Chapters 3 and 4.

Before going on to deal with the relation between structure and forces, there is a further correlation function of central importance for liquid-state theory, namely, the Ornstein-Zernike correlation function $c(r)$.

2.2. Ornstein-Zernike direct correlation function

At this point, it will be useful to give the formal definition of $c(r)$; the definition will then be reexamined later in order to gain deeper insight into the reasons for its important status in liquid-state theories. The pair function $g(r)$ minus its asymptotic limiting value 1 at large r is called the total correlation function $h(r)$, i.e.,

$$h(r) = g(r) - 1. \tag{2.6}$$

$h(r)$, the total correlation function, is now divided into two parts, a direct part described by $c(r)$ and an indirect part. Following Ornstein and Zernike, the indirect part is expressed as a convolution; their definition of $c(r)$ being

$$h(r) = c(r) + \rho \int h(|\mathbf{r} - \mathbf{r}'|)c(r')\,d\mathbf{r}'. \tag{2.7}$$

The convolution is readily removed by Fourier transform, and (2.7) is then simply equivalent to

Figure 2.3. Schematic form of the Ornstein-Zernike direct correlation function in k-space, denoted by $\tilde{c}(k)$. This is related to $S(k)$ in Figure 2.1 by $\tilde{c}(k) = [S(k) - 1]/S(k)$ (Curve 1 resembles Ar; curve 2, Pb).

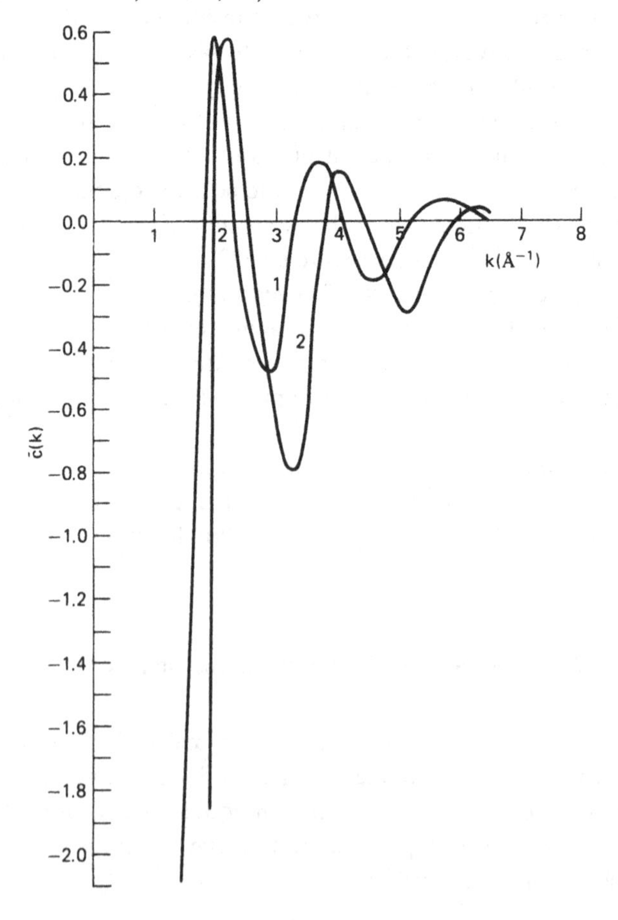

$$\tilde{c}(k) = \frac{S(k) - 1}{S(k)}, \tag{2.8}$$

where $\tilde{c}(k)$ is the Fourier transform of $c(r)$. One consequence of (2.8) follows from the small value of $S(0)$ for argon near its triple point or for liquid metals just above their freezing points. It is that for argon, $\tilde{c}(0) \sim -16$, whereas for liquid metals $\tilde{c}(0) \sim -40$ as a typical value. On the other hand, at the peak of $S(k)$, having the value from Figure 2.1 of about 2.9, $\tilde{c}(k) \sim 0.6$. Thus, $\tilde{c}(k)$ has the schematic form shown in Figure 2.3: it is seen to put much greater emphasis on the region of small k, i.e., small-angle scattering, than does Figure 2.1, which is largely dominated by the first peak of $S(k)$. Since, as will be seen shortly, $c(r)$ is more closely related to the microscopic force law than $g(r)$, one must expect small-angle scattering data to be of considerable importance when one wishes to draw conclusions about the force field from liquid structure factor measurements. However, before enquiring as to the precise form of this relation between liquid structure and interatomic forces, let us turn to consider in some detail the thermodynamics of liquid metals.

3
Thermodynamics

The thermodynamic properties of simple s-p bonded liquid metals under normal conditions will be discussed in this chapter. By restricting the range of systems in this way, one is permitted to focus only on those electronic aspects that are describable by nearly free electron (NFE) methods. Some of the theoretical methods and techniques used will by reviewed, and a selection of the results obtained will be presented.

The work of Stroud and Ashcroft (1972) on the melting of sodium indicated how the techniques of classical liquid theory might be combined with the pseudopotential method of calculating interionic forces in metals to produce a successful description of the thermodynamics. It is probably true to say that this problem is understood, albeit in a semiquantitative way; this "solution" is reviewed in this chapter, which follows rather closely the work of Young (1987).

3.1. Simple monatomic fluids

In this section, by way of an introduction, let us leave the problems specific to metals and, instead, consider the statistical mechanics appropriate to a monatomic fluid such as Ar. Many of the results that will be obtained can be applied to simple metals with little or no modification. Such changes as are necessary are dealt with in Section 3.3.

3.1.1. Basic features

Consider a simple classical monatomic fluid of N particles in a volume V, so that the number density is $\rho = N/V$. The Hamiltonian is taken to be

$$H = T + \tfrac{1}{2}\sum v(|\mathbf{r}_i - \mathbf{r}_j|) \tag{3.1}$$

where T is the kinetic energy and v is a potential energy function independent of density ρ.

Central to the statistical mechanical description afforded by (3.1) is the radial distribution function $g(r)$, which has been defined in Chapter 2 so that $\rho g(r)\, d\mathbf{r}$ is the probability of finding an atomic nucleus in the volume element $d\mathbf{r}$ around $\mathbf{r}$ given that there is one at the origin. In terms of this function, application of general theory leads to the expression

$$E = \tfrac{3}{2}k_B T + \tfrac{1}{2}\rho \int v(r)g(r)\, d\mathbf{r} \tag{3.2}$$

for the internal energy per atom and

$$P = \rho k_B T - \frac{1}{6}\rho^2 \int r \frac{d}{dr} v(r)g(r)\, d\mathbf{r} \tag{3.3}$$

for the pressure. Other results follow from these, of course, by the application of the various thermodynamic identities.

The structure factor $S(q)$* may be defined in terms of $g(r)$ by (2.4), which by Fourier inversion yields

$$g(r) = 1 + \frac{1}{(2\pi)^3\rho} \int \{S(q) - 1\} e^{-i\mathbf{q}\cdot\mathbf{r}}\, d\mathbf{q}. \tag{3.4}$$

For example, (3.2) and (3.3) may be rewritten in terms of $S(q)$ and the (assumed) Fourier transform $\tilde{v}(q)$ of $v(r)$ in (3.1) as

$$E = \frac{3}{2}k_B T + \frac{1}{2}\rho\tilde{v}(0) + \frac{1}{2(2\pi)^3} \int \tilde{v}(q)\{S(q) - 1\}\, d\mathbf{q} \tag{3.5}$$

and

$$P = \rho k_B T + \frac{1}{2}\rho^2\tilde{v}(0) + \frac{\rho}{2(2\pi)^3} \int \left\{\tilde{v}(q) + \frac{q}{3}\frac{\partial \tilde{v}(q)}{\partial q}\right\}\{S(q) - 1\}\, d\mathbf{q} \tag{3.6}$$

where

$$\tilde{v}(q) = \int v(r) e^{i\mathbf{q}\cdot\mathbf{r}}\, d\mathbf{r}. \tag{3.7}$$

As discussed in the previous chapters, $S(q)$ is measurable using neutron and other spectroscopies, and such data can be used in (3.5) and (3.6), for example, if $v(r)$ is specified. Such a procedure is of limited value, however, primarily because one wishes to investigate such expressions for varying thermodynamic states T, V, etc. Under such circumstances sufficient experi-

* k and q will be used interchangeably for wave numbers throughout this volume.

mental information is seldom available. What one needs, and what theory provides, are $S(q)$'s and $g(r)$'s compatible, in some approximation, with a given $v(r)$.

An important area of uncertainty with $S(q)$ measurements remains at low q, where, typically, there may be significant error below about 0.5 Å^{-1}. It is, therefore, worth stressing again that the limit $S(0)$ is known from density fluctuation theory (of Appendix 2.1) to be determined through (2.5), where the isothermal compressibility K_T is given by

$$K_T^{-1} = -V\left(\frac{\mathrm{d}P}{\mathrm{d}V}\right)_T. \tag{3.8}$$

$S(0)$ is thus accessible from thermodynamics and provides a "target" for the low-argument radiation measurements of $S(q)$.

Integration of (3.8) yields

$$P = k_B T \int_0^\rho \frac{\mathrm{d}\rho}{S(0)} \tag{3.9}$$

which is the so-called compressibility equation of state. This is to be compared with the virial pressure equation of state given by (3.3). In an exact theory of $g(r)$ and $S(q)$, these equations of state will be completely equivalent. In practice, however, one has to use approximate theories; thus discrepancies arise (see Appendix 5.2).

3.1.2. Structure factor S(q) as response function

Physical insight into the character of $S(q)$ is obtained by noting its role as a response function (see Young, 1987, whose account is followed closely below). Suppose the fluid atoms are exposed to a small external potential $\delta\varphi(q)\cos\mathbf{q}\cdot\mathbf{r}$ (i.e., a perturbation $\delta\varphi(q)\sum\cos\mathbf{q}\cdot\mathbf{r}_i$ is added to (3.1)). Then, to first order, the density becomes $\rho + \delta\rho(q)\cos\mathbf{q}\cdot\mathbf{r}$, where

$$\delta\rho(q) = -\frac{\rho}{k_B T} S(q)\delta\varphi(q). \tag{3.10}$$

At long wavelengths (small q) this tells us about the ability of the particles to clump, and this process will be assisted by attractions between the particles and hindered by repulsions. This discussion is compatible with the well-known general result (see (2.5)) that K_T diverges at a critical point, for the existence of which attractive interactions are essential (see Chapter 9 for details).

Returning briefly to the direct correlation function $c(r)$ introduced in Chapter 2, it is to be stressed that $c(r)$ is a particularly suitable starting point for theories because of its direct link with the interatomic forces. In fact, statistical mechanics yields the exact result (far from the critical point)

$$c(r) \sim \frac{-v(r)}{k_B T} \qquad \text{(large } r\text{)}. \tag{3.11}$$

This relation will be of considerable importance in what follows.

3.2. Approximate methods

As has been seen, the problem is to estimate $g(r)$, $S(q)$ in some way for use, for example, in such formulae as (3.3) and (3.9). It is, therefore, appropriate to examine next the character of typical interatomic potentials that establish these functions. Possible forms are shown schematically in Figure 3.1, and

Figure 3.1. Possible schematic forms of effective interionic potentials $v(r)$ in liquid metals. It is to be noted that the r axis is broken; there is generally an impenetrable core. The parameters r_0 and $v_{\min}$ play a role in the Weeks-Chandler-Andersen approximation to structure (see Appendix 3.2). In the third part of the figure, to the right, $v_{\min}$ and r_0 are measured at the (assumed) point of inflection (after Young, 1987).

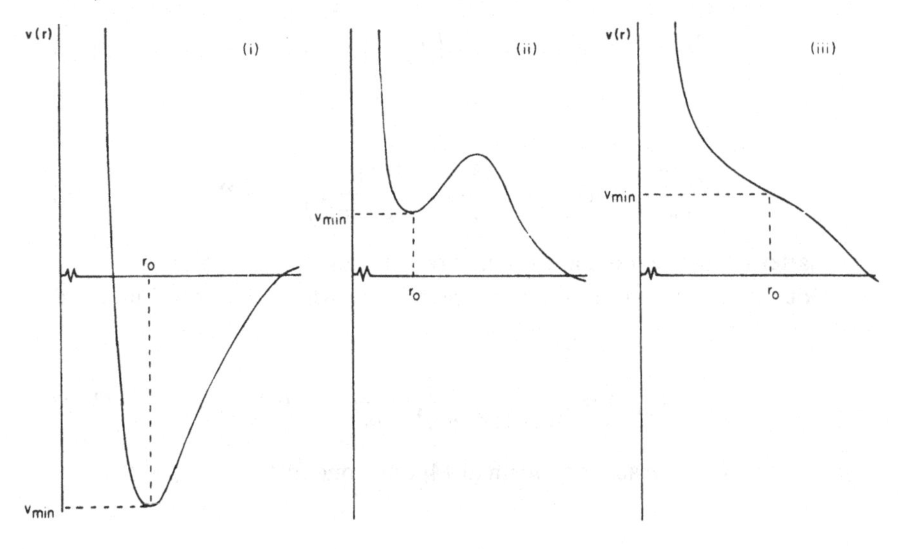

it is evident that in a zeroth approximation the atoms behave as though there is a quite well defined hard sphere diameter. One approach, therefore, is to try to establish a suitable hard sphere reference system and then to incorporate further details of $v(r)$ by suitable "corrections."

3.2.1. *Hard sphere fluids*

A good deal is known about hard sphere fluids because of their relative mathematical simplicity. Computer simulation studies (Alder and Wainwright, 1957) have established that they freeze at a packing fraction $\eta \equiv (\pi/6)\rho\sigma^3$ (σ = diameter) of about 0.45–0.46. In the liquid state, their structural and thermodynamic properties have been similarly investigated and found to be quite well reproduced in the Percus-Yevick (1958; PY) approximation, for which analytical results are available (see Appendix 3.1). For example, the Helmholtz free energy per particle of a hard sphere fluid is $\frac{3}{2}k_BT - TS_{hs}$ where machine results are well fitted by (Carnahan and Starling, 1969) an excess entropy (relative to the ideal gas value S_{gas}) of

$$\frac{S_\eta}{k_B} \equiv \frac{S_{hs} - S_{gas}}{k_B} = 3 - \frac{2}{(1-\eta)} - \frac{1}{(1-\eta)^2} \qquad \text{(CS).} \qquad (3.12)$$

This is plotted in Figure 3.2. The Percus-Yevick entropy expressions (differing according to whether the virial or compressibility route is used) are

$$\frac{S_\eta^{\text{vir}}}{k_B} = -2\ln(1-\eta) + 6\left(1 - \frac{1}{1-\eta}\right) \qquad \text{(PY)} \qquad (3.13)$$

or

$$\frac{S_\eta^{\text{comp}}}{k_B} = \ln(1-\eta) + \frac{3}{2}\left(1 - \frac{1}{(1-\eta)^2}\right) \qquad \text{(PY)} \qquad (3.14)$$

the latter being the more accurate, as direct substitution shows.

Double differentiation of these expressions with respect to volume and use of (2.5) yields

$$S_{\text{hs}}(0) = \frac{(1-\eta)^4}{(1+2\eta)^2 + \eta^4 - 4\eta^3} \qquad \text{(CS)} \qquad (3.15)$$

and using the more accurate form (3.14) only, one finds

$$S_{hs}(0) = \frac{(1-\eta)^4}{(1+2\eta)^2} \qquad \text{(PY)}. \tag{3.16}$$

Relationship (3.15) is also plotted in Figure 3.2.

Actually, the whole of $S_{hs}(q)$ is known analytically in the PY approximation (Thiele, 1963; Wertheim, 1963), and this is a major reason for its popularity as a substitute for "exact" machine-generated numerical output. The direct correlation function is obtained in the form

$$c_{hs}(r) = \begin{cases} a + b(r/\sigma) + d(r/\sigma)^3 & (r < \sigma) \\ 0 & (r > \sigma) \end{cases} \qquad \text{(PY)} \tag{3.17}$$

where the coefficients are recorded in terms of η in Appendix 3.1. The corresponding structure factor follows using (2.8). The height of the principal peak of this structure factor is also shown in Figure 3.2 as a function of η. The derivation of (3.17) is such that the corresponding structure factor at $q = 0$ agrees with the form (3.16).

Figure 3.2. Three different quantities (given in (3.12), (3.16), (2.8) and (3.14)). These are, respectively, (i) the excess entropy per particle S_η, (ii) the principal peak height S_{peak} of the structure factor, and (iii) $S(k)$ at $k = 0$. All are calculated from (approximate) hard sphere theory and are plotted as functions of the packing fraction η defined by $\eta = \frac{\pi}{6}\rho\sigma^3$, where ρ is the number density and σ the hard sphere diameter.

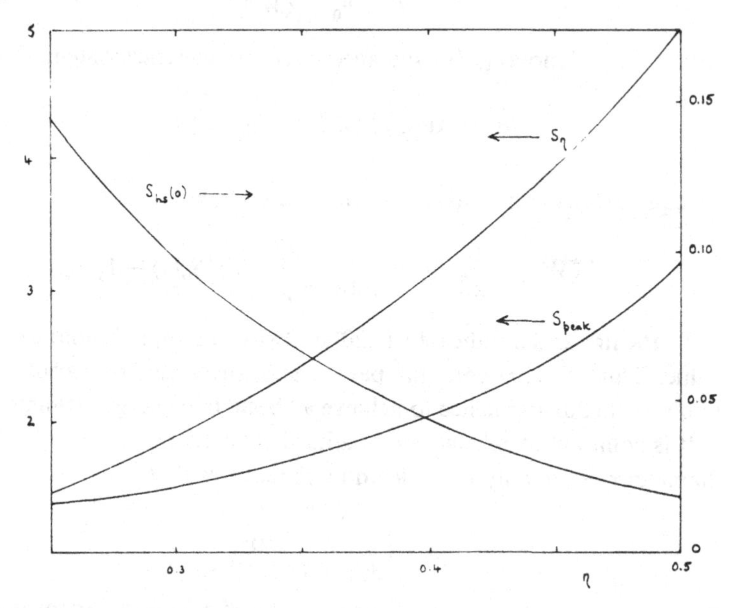

Two approximate methods have proved particularly useful for proceeding beyond the hard sphere picture to a description of fluids represented by potentials such as are sketched in Figure 3.1. Each of these is reviewed next. The first takes the virial route to the thermodynamics; the second, the compressibility alternative.

Gibbs-Bogoliubov (GB) Method

The method outlined below is commonly called the Gibbs-Bogoliubov method (Isihara, 1968; Lukes and Jones, 1968) but is alternatively known as the variational, or high-temperature, method, for reasons that will emerge in the discussion.

Let us begin (see Young, 1987) by writing the interatomic potential in the form

$$v(r) = v_0(r) + v_1(r) \tag{3.18}$$

where $v_0(r)$ is a core part and the remaining contribution $v_1(r)$ is "small." Then the system with interactions $v_0(r)$ becomes the unperturbed, reference system on which the perturbation

$$W = \tfrac{1}{2} \sum v_1(|\mathbf{r}_i - \mathbf{r}_j|) \tag{3.19}$$

acts. To first order in $W/k_B T$, one obtains a Helmholtz free energy

$$F = F_0 + \langle W \rangle_0 \tag{3.20}$$

where $\langle W \rangle_0$ denotes (3.19) averaged over the reference system. Explicitly,

$$\langle W \rangle_0 = \tfrac{1}{2}\rho \int v_1(r) g_0(r) \, d\mathbf{r} \tag{3.21}$$

or, equivalently, when $v(r)$ is Fourier transformable,

$$\langle W \rangle_0 = \frac{1}{2} \tilde{v}_1(0) + \frac{1}{2(2\pi)^3 \rho} \int \tilde{v}_1(q) \{S_0(q) - 1\} \, d\mathbf{q}. \tag{3.22}$$

To the first order indicated, (3.20) is always an upper bound to the exact value. Thus, if $v_0(r)$ contains parameters, these can be varied so as to minimize (3.20) and hence to achieve a "best" free energy estimate.

It is common to take a core of a hard sphere form ($v_0 = v_{\text{hs}}$). Then, the diameter σ is the only variable and one requires that

$$\left(\frac{\partial F}{\partial \sigma} \right)_{V,T} = 0. \tag{3.23}$$

When (3.23) applies, one finds (Edwards and Jarzynski, 1972) that (3.20) yields

$$S = S_{\text{hs}}(\sigma), \tag{3.24}$$

i.e., the entropy estimate is given by the hard sphere expression (e.g., (3.12) or (3.13)) with the optimizing diameter inserted. Furthermore (Watabe and Young, 1974) (3.2) and (3.3) apply with g replaced by $g_{\text{hs}}(\sigma)$ (or, equivalently, (3.5) and (3.6) apply with S replaced by $S_{\text{hs}}(\sigma)$).

For hard spheres, $g_{\text{hs}}v_{\text{hs}} = 0$, since the first factor vanishes when $r < \sigma$ and the second, when $r > \sigma$. Thus, (3.21) can be rewritten, in this case, as

$$\langle W \rangle_{\text{hs}} = \tfrac{1}{2}\rho \int_{\sigma}^{\infty} v(r) g_{\text{hs}}(r) 4\pi r^2 \, dr \tag{3.25}$$

and solutions using (3.25) then conform quite well to the empirical rule (see Young, 1987)

$$v(\sigma) - v_{\text{min}} \approx \tfrac{3}{2} k_B T \tag{3.26}$$

where v_{min} is the principal minimum of $v(r)$ (Figure 3.1).

A limitation of this method (at least when used in conjunction with a hard sphere reference system) is its relatively crude treatment of the tail. The latter affects the size of the diameter through its contribution to (3.25) but not the detailed shape of the radial distribution function, which always remains of hard sphere form. A method that rectifies this deficiency is discussed fully by Young (1987). Instead, the approximation

$$S^{-1}(0) = S_{\text{hs}}^{-1}(0) + \frac{\rho \tilde{v}_{\text{tail}}(\sigma)}{k_B T} \tag{3.27}$$

will be employed. This provides an approach to the thermodynamics by the compressibility route (cf. (3.9)).

3.3. Simple liquid metals

A simple liquid metal such as Na is more complicated than, say, Ar, to which the theory so far can be immediately applied. As in Chapter 1, one should really consider liquid Na to be a mixture of Na^+ ions and electrons and proceed by a two-component formalism (see Chapter 14) similar, in many general respects, to that to be given in Section 13.1. Under such circumstances, the $g(r)$ used so far still describes the ion-ion correlations, and (2.4) continues to define the corresponding structure factor, $S(q)$. Now, however, these quantities need to be calculated in the presence of the free electrons, which must be treated quantum-mechanically. It is interesting to note one simple general result (see Young, 1987) namely, that the requirement of charge neutrality leads to $S(0)$ given by (2.5), where ρ is, as before, the ion number density and K_T is the isothermal compressibility of the entire system of ions and electrons. It follows that observed thermodynamic data can continue to augment radiation measurements of $S(q)$.

3.3.1. Generalities

Notwithstanding the preceding remarks, a fruitful treatment (see also Chapter 1) is to convert the problem into one in which the ions move in an effective force field established by the electron gas (the Born-Oppenheimer approximation). Such a procedure is possible because the electron-ion coupling (the pseudopotential) is weak and can be regarded as merely perturbing the electron gas, assumed to be uniform in zeroth order. To second order, one calculates the energy for a given "frozen" configuration of ions and this acts as the potential energy function for the N ions alone. The resulting Hamiltonian has the form (Hasegawa and Watabe, 1972; see also Corless and March, 1961; Worster and March, 1964)

$$H = T + Nv_0(n) + \tfrac{1}{2}\sum v(|\mathbf{r}_i - \mathbf{r}_j|, n) \tag{3.28}$$

where n is the mean electron density ($= z\rho$, where z is the valence). In this equation $v_0(n)$, the volume term, is independent of the ionic positions $\{\mathbf{r}_i\}$, and the pair interaction v is density-dependent. Many ($\geqslant 3$) body potential energy terms in (3.28) are absent in this approximation.

On comparing (3.28) with (3.1) it is clear that the previous formalism is easily modified, (3.2) and (3.3) becoming, respectively,

$$E = \tfrac{3}{2}k_B T + v_0(n) + \tfrac{1}{2}\rho \int v(r, n) g(r)\, d\mathbf{r} \tag{3.29}$$

and

$$P = \rho k_B T + \rho n \frac{dv_0(n)}{dn} - \frac{1}{6}\rho^2 \int \left\{ r\frac{\partial v}{\partial r} - 3n\frac{\partial v}{\partial n} \right\} g(r)\, d\mathbf{r}. \tag{3.30}$$

Further differentiation leads, via (3.8), to a rather complicated expression for the compressibility, K_T. This, as has been noted, is the virial equation route to the thermodynamics.

Alternatively, the compressibility route, as one has seen, requires that $S(0)$ be inserted into (2.5). At constant volume, any standard method of calculating $g(r)$ requires knowledge of $v(r, n)$ but not of the volume term, $v_0(n)$. Thus, at first sight, (2.4), with $k = 0$ inserted, appears to be incompatible with (3.30), which explicitly depends on the volume term. A resolution of this point is to note (see Young, 1987) that structure-independent $0(\Omega^{-1})$ terms in $g(r)$, invisible when graphed, can be integrated in (2.4) to produce $0(1)$ contributions to $S(0)$, thus restoring compatibility.

The pseudopotential theory of v_0 and v will not be presented in detail at this point. Nevertheless, it will be useful to indicate the form the results take for a local electron-ion pseudopotential interaction, which, in free space, can be written

$$v_{ps}(r) = v_{ps}^{core}(r) - \frac{ze^2}{r}. \tag{3.31}$$

The core part represents a weak short-range-effective interaction between the electron and ion core; outside the core region, only the electrostatic attraction remains. Particularly simple is the Ashcroft (1966) empty core form

$$\begin{aligned} v_{ps}(r) &= \begin{cases} 0 & (r < r_c) \\ -ze^2/r & (r > r_c) \end{cases} \\ \tilde{v}_{ps}(q) &= -\frac{4\pi z^2 e^2}{q^2} \cos qr_c \qquad (q \neq 0) \end{aligned} \tag{3.32}$$

where r_c is a measure of the ionic radius.

For the ionic array in situ in the electron gas, linear response theory yields a pairwise interaction that can be written in inverse space as: (4.20)

$$\tilde{v}(q, n) = \frac{4\pi z^2 e^2}{q^2} + \frac{q^2}{4\pi e^2}\left\{\frac{1}{\varepsilon(q, n)} - 1\right\}\tilde{v}_{ps}^2(q). \tag{3.33}$$

The first term is clearly the direct coulomb interaction between the ions, whereas the second is the indirect electron-mediated term. In the latter, the dielectric screening function $\varepsilon(q, n)$ appears. This function has been much investigated over three decades, and a variety of forms for it have appeared in the literature (see Chapter 4). Asymptotically, it behaves like

$$\varepsilon(q, n) = \begin{cases} (K/K_f)^2(q_{TF}/q)^2 + 1 \\ 1 + 16\pi n(1 + G)/q^2 \end{cases} \tag{3.34}$$

where q_{TF}^{-1} is the Thomas-Fermi screening length,* K/K_f is the ratio of the compressibilities of the associated uniform interacting and noninteracting electron gases, and $1 - G$ is the radial distribution function of the former, evaluated at contact.

* See Section 4.1 below.

The volume term corresponding to the preceding description is

$$v_0(n) = z\left(1 - \frac{1}{2}V^2\frac{\mathrm{d}^2}{\mathrm{d}V^2}\right)v_{\mathrm{el}} + \frac{1}{2}\int\frac{q^2}{4\pi e^2}\left\{\frac{1}{\varepsilon(q,n)} - 1\right\}\tilde{v}_{\mathrm{ps}}^2(q)\,\mathrm{d}\mathbf{q} \quad (3.35)$$

where the final term is the self-energy of an ion in the electron gas. The first term arises from the (perturbed) electron gas only, the expression v_{el} being the energy per electron of the uniform gas in a neutralizing background.

The results (3.33) and (3.35) for the pair and volume terms are valid to second order in pseudopotential perturbation theory. Not only would higher-order terms introduce many ($\geqslant 3$)-body forces (as indicated earlier), but they would also lead to revised forms of $v_0(n)$ and $v(r, n)$. There has been little investigation of this problem at the time of this writing, the expressions above being almost invariably employed.

3.4. Specifics

In this section, the way in which the above formalism is able to explain some of the properties of liquid metals will be briefly reviewed. The matter largely revolves around the roles and character of the volume and pairwise terms. The former is, in fact, large and is responsible for most of the cohesive energy, as the results of Finnis (1974), shown in Table 3.1, indicate. These data are for the solids but make the point quite satisfactorily, since v_0 is phase-independent and dominantly large. Finnis noted that the first (electron gas) term of (3.35) is quite small compared with the second (self-energy) part and that the latter can be roughly evaluated as

$$\frac{1}{2}\int\frac{q^2}{4\pi e^2}\left\{\frac{1}{\varepsilon(q,n)} - 1\right\}\tilde{v}_{\mathrm{ps}}^2(q)\,\mathrm{d}\mathbf{q} \approx -\frac{1}{4}\frac{z^2e^2}{r_c} \quad (3.36)$$

where r_c is the empty core radius of (3.32). In fact, the results of Table 3.1 were calculated by neglecting the electron gas term and applying the self-energy part in the approximate form (3.36).

Since the largest term in $v_0(n)$ has little density-dependence (see (3.36)), one finds that the pairwise term plays a more prominent role in determining the pressure and compressibility derivatives. Any calculation of these quantities requires knowledge not only of $\tilde{v}_{\mathrm{ps}}^{\mathrm{core}}(q)$ in (3.31) for $q \neq 0$, but also of $q = 0$. One might plausibly invoke continuity for this purpose, e.g., if the empty core potential (3.32) is employed, one would write

$$\tilde{v}_{\mathrm{ps}}^{\mathrm{core}}(0) = \lim_{q\to 0}\left\{\frac{4\pi z e^2}{q^2} - \frac{4\pi z e^2}{q^2}\cos qr_c\right\} = 2\pi z r_c^2. \quad (3.37)$$

Table 3.1. *Cohesive energies [after Finnis (1974); see also Young, (1987)].*

1A	2A	2B	3	4	5	6
Li	Be					
0.512	1.13					
0.33	0.90					
Na	Mg		Al	Si		
0.460	0.892		1.38	1.96		
0.31	0.78		1.39	2.11		
K	Ca	Zn	Ga	Ge	As	Se
0.388	0.733	1.05	1.47	1.97	2.55	3.23
0.24	0.60	1.01	1.52	2.20	2.79	3.57
		Cd	In	Sn	Sb	Te
		0.993	1.36	1.77	2.24	2.73
		0.92	1.38	1.87	2.5	3.04
	Ba	Hg	Tl	Pb	Bi	
	0.617	1.10	1.43	1.81	2.21	
	0.47	1.07	1.40	2.15	2.34	

Note: First entry for each element is the observed value. The second is result of (3.35), calculated using approximation (3.36). The units are Rydbergs per z, where z is the number of free electrons per atom.

The results so obtained correlate with the experimental data, but it is usual to regard $\tilde{v}_{ps}^{core}(0)$ as an independent parameter to be adjusted to obtain the observed density at zero pressure.

Hasegawa and Young (1981) used the GB hard sphere method (Section 3.2.1) in conjunction with the Ashcroft pseudopotential description to calculate the compressibilities at melting by differentiation of (3.30). They chose r_c to yield a packing of $\eta = 0.45$ and $v_{ps}^{core}(0)$ to give zero pressure; the results are shown in Figure 3.3. Apart from one or two clear failures, these writers found a rough overall correlation between theory and observation. The results are quite sensitive to the pseudopotentials and other features involved, and it is believed (Hasegawa and Young, 1981) that sufficient refinement of these in any given case can bring theory into agreement with experiment.

Table 3.2 shows contributions from the various parts of (3.30) to the total bulk moduli, K_T^{-1}, calculated for Figure 3.3. These and similar calculations for the other metals suggest that volume and pairwise contributions are of comparable magnitude but of opposite sign. Furthermore, the volume dependence of $v(r, n)$ gives rise to only a small fraction of the total pairwise contribution; but for valence 4 and 5 metals, its effect on the final complete sum is important.

The alternative method, using (3.27) to calculate $S(0)$, is more qualitative but provides some further insight. In this case, one relies on the Weeks-Chandler-Andersen (WCA) formalism set out in Appendix 3.2 to obtain $S_{hs}(0)$. The result is about 0.018. This alone does not explain the variety of observed results for $S(0)$, but it is a convenient reference number. For as one knows quite generally ((3.10) and the discussion following) and as

Figure 3.3. Results of Hasegawa and Young (1981) for compressibilities of liquid metals at melting. Calculated versus observed values are plotted. The Gibbs-Bogoliubov hard sphere method, plus the Ashcroft pseudopotential description, was the basis of the calculation.

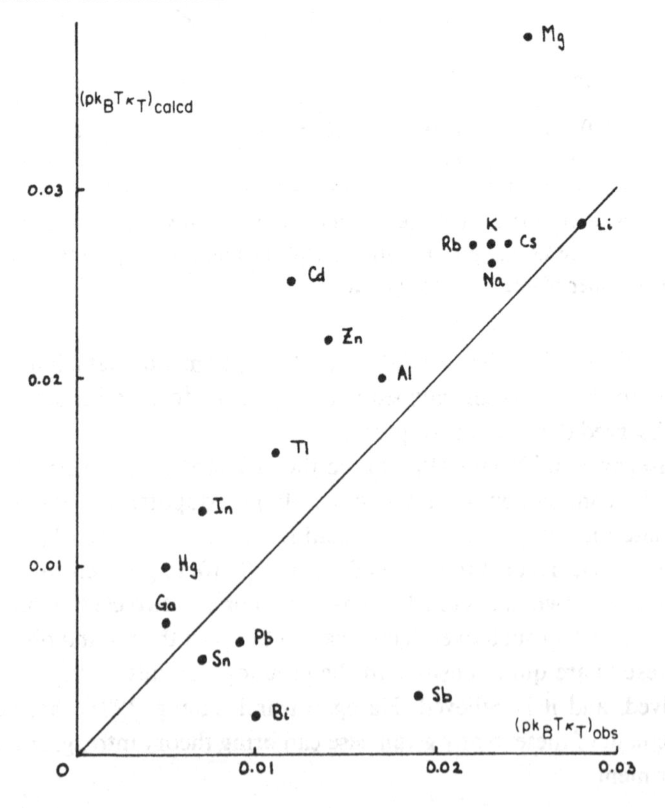

(3.27) shows specifically, attractive (on average) tails will raise this value and repulsive (on average) ones will lower it.

This brings us to the interatomic potentials of Hafner and Heine (1983) shown in Figure 3.4. These were calculated to explain the structures of the

Table 3.2. *Contributions to calculated bulk moduli near melting temperature [after Young (1987)].*

	Kinetic	Volume	Pair (1)	Pair (2)	Calculated	Experimental
Na	1	−28	60	4	37	43
Al	1	−88	133	5	50	60
Sn	1	−498	598	87	188	150

Note: The kinetic and volume terms arise, respectively, from the application of $\rho d/d\rho$ to the first two terms of (3.30). Similarly, the sum pair (1) + pair (2) comes from the third term, the second contribution being due to the explicit density dependence of $v(r, n)$.

Figure 3.4. A selection of potentials constructed for solid metals by Hafner and Heine (1983). The energy unit is z^2e^2/R_a, with z the number of free electrons per atom and $R_a = (3/4\pi)^{1/3}\rho^{-1/3}$. The length unit is π/k_f, where $2k_f$ is the diameter of the Fermi sphere. d_{cp} marked on the figure is the interatomic distance in the close-packed solid, while the arrows shown denote liquid-state diameters for 45 and 50% packing (scarcely distinguishable on the scale of the figure) (after Young, 1987).

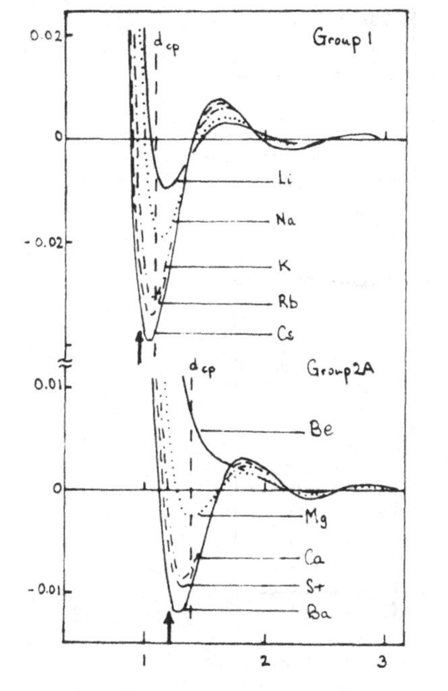

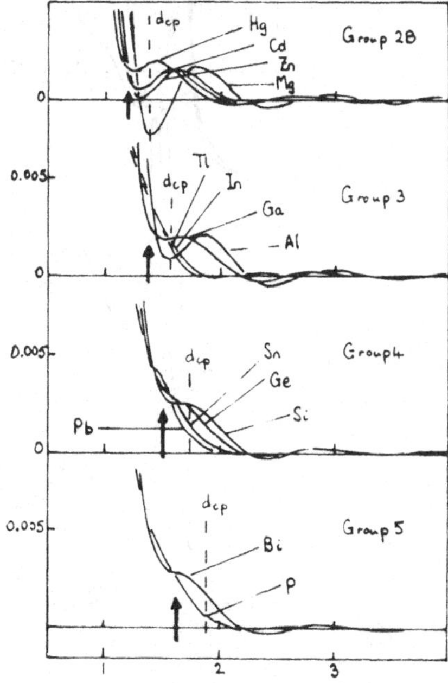

solids. The essential point here is that when a pronounced minimum exists, many atoms take advantage of it, and a fairly close-packed structure is obtained. On the other hand, when the minimum is lost, closer packing is no longer an advantage, and only detailed further enquiry can establish the lowest energy configuration.

These potentials vary a little (with density) when the liquid is considered (Hafner and Kahl, 1984), but the general characteristics are not altered. In the liquid state, the corresponding structural phenomena to be explained are the three $S(q)$ types, according to Waseda. These are defined through the shape of the principal peaks and are illustrated in Figure 3.5; every liquid metal seems to fit into one such category, and the distribution throughout the periodic table is indicated in Table 3.3. Liquids with well-defined principal minima (Lennard-Jones cases being the prototypes) belong to category (a) but repulsion in the tail (Silbert and Young, 1976) can yield type (c) character; all this is consistent with Figure 3.5.

Figure 3.5. Waseda's attempt to classify liquid-structure factors (after Young 1987). The three types shown can be classified as (a) hard sphere (b) skew peak, and (c) shouldered peak.

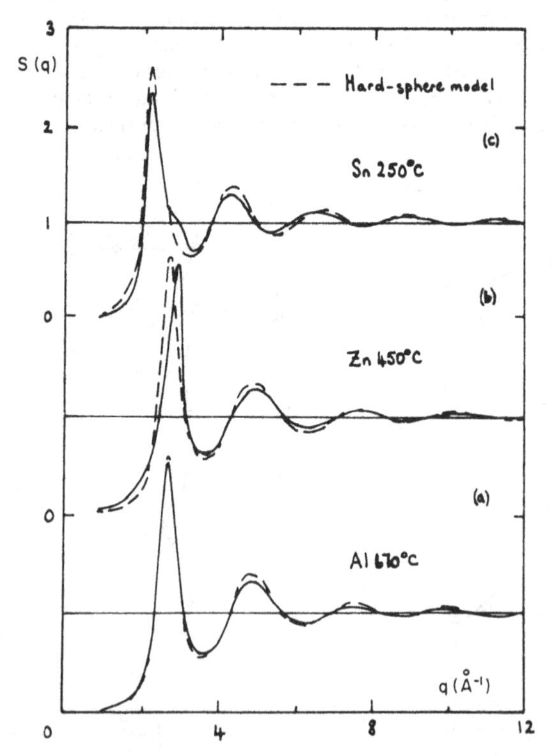

Table 3.3. *Solid (room temperature) and liquid (Waseda) structures. $S(0)$ at melting as deduced from thermodynamic data is also given [after Young (1987)].*

1A		2A		2B		3		4		5	
Li	BCC[a]	Be	HCP[b]			B					
[a]	0.028										
Na	BCC	Mg	HCP			Al	FCC[c]	Si	DIA[d]	P	
[a]	0.023	[a]	0.025			[a]	0.017	[c]			
K	BCC	Ca	FCC	Zn	HCP	Ga	ORC[e]	Ge	DIA	As	RHL[f]
[a]	0.023	[a]	0.035	[b]	0.014	[c]	0.005	[c]			
Rb	BCC	Sr	FCC	Cd	HCP	In	TET[g]	Sn	TET	Sb	RHL
[a]	0.022	[a]	0.031	[b]	0.012	[a]	0.007	[c]	0.007	[c]	0.019
Cs	BCC	Ba	BCC	Hg	RHL	Tl	HCP	Pb	FCC	Bi	RHL
[a]	0.024	[a]	0.036	[b]	0.005	[a]	0.011	[a]	0.009	[c]	0.010

Note: Some structures in solid change with temperature, but distinction between closer and looser packing persists.
[a] BCC: body centred cubic. [b] HCP: hexagonal close packed.
[c] FCC: face centred cubic. [d] DIA: diamond.
[e] ORC: orthorhombic. [f] RHL: rhombohedral. [g] TET: tetragonal.

In the present context, however, one is mainly interested in the variations in $S(0)$ from metal to metal and, for this purpose, Figure 3.5 continues to be useful. For, as is evident from the diagrams and as is required to explain the details, $S(0)$ lies above $S_{hs}(0) \approx 0.018$ when the tails are (on average) attractive and below when they are (on average) repulsive.

This account of $S(0)$ bears closer analysis (Young, 1987). For example, the increasing attractiveness of the tail (Figure 3.4) through the sequence Hg, Cd, Zn, Mg, Ca, Sr, Ba is reflected in the $S(0)$ trend shown in Table 3.3 (Sr being, perhaps, an exception). Other correct trends are discernable, but one should not attempt to read too much into results that might be sensitive to pseudopotential and screening characterisation.

It is to be noted that the preceding discussion depends on the use of a WCA diameter. If one had relied on the GB packing fraction of 0.45, (3.5) would have given 0.028, which is too high to be compatible with Figure 3.4. However, as discussed by Young (1987), there are good reasons to take an **apparently** larger packing for describing $S(q)$ as $q \to 0$.

The above discussion has been confined to liquids near their triple points. Before leaving this topic, note that a number of successful calculations have been made at elevated temperatures and pressures. Details may be found in the original papers (see McLaughlin and Young, 1984).

Finally, there are some properties that are mainly dependent on core size. For example, the GB method of Section 3.2.1, when used with a hard sphere reference system, leads to the conclusion that the structure factor is approximately given by a hard sphere form. It has been seen that there are departures from this result at $q = 0$ and there are others elsewhere, but, overall, a hard sphere shape is quite a fair approximation (with $\eta \approx 0.45$ near the triple point). The corresponding entropy is also given by that appropriate to hard spheres ((3.24) and (3.12)), and it is the case that there is reasonable agreement between the packings deduced from the observed principal peak heights of the structure factors and the measured entropies.

This proposition is illustrated in Figure 3.6(a), where experimentally inferred entropies are compared with those calculated by obtaining packing fractions from the measured first peak heights (Figure 3.2) and using them in (3.12) (Figure 3.2). The calculated values are underestimates presumably

Figure 3.6. (a) Entropies per atom, observed values plotted against the results of calculations. (b) Heat capacities per atom, again observed versus calculated (after Young 1987).

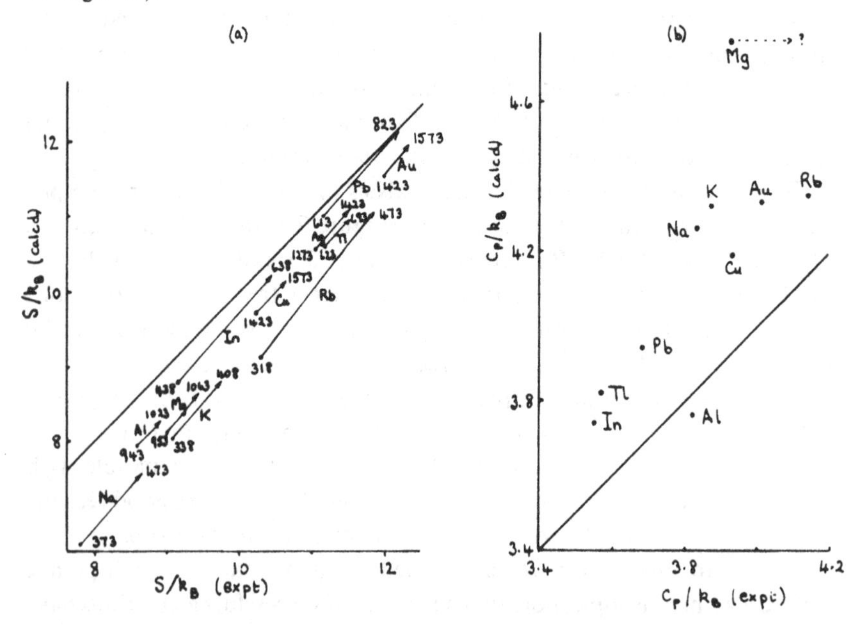

because of inadequate treatment of core softness and tail attraction. The error here diminishes at higher T, since (1) cores become effectively harder (noting (3.26) and the increasing steepness of $v(r)$ as r falls below r_0 in Figure 3.1)) and (2) the influence of $v_{\text{tail}}/k_B T$ on the first peak height decreases (since, although $|v_{\text{min}}|$, for instance, may increase, T increases faster). This is consistent with the behaviour evident in Figure 3.6(a) of the trajectories approaching a 45° asymptote. This figure also carries implications for estimates of the specific heat $C_p = T(\partial S/\partial T)_p$, and these are shown in Figure 3.6(b). As expected from the preceding discussion, the calculated values are overestimates, but there is an overall correlation (Young, 1977). Further discussion of specific heats occurs later (Chapter 8; see also Appendix 5.4).

3.5. One-component plasma as reference liquid

Having discussed the basis of a variational approach to the Helmholtz free energy F afforded by the Gibbs-Bogoliubov inequality, with a detailed example of the use of hard spheres as a reference liquid, it must be added that in the work of Ross et al. (1981; see also Mon et al., 1981), a comparison was made between such results and ones using the one-component plasma (OCP) as a reference liquid. Their conclusion was that for the alkali metals, the OCP is the favourable choice of reference liquid because it yields a lower free energy and better agreement with the Monte Carlo pressures. However, for polyvalent metals (in particular Al), the situation just described is reversed. It is also of interest to mention in concluding this section the calculations of Stroud and Ashcroft (1972) on the melting curve of Na (see also Stishov et al., 1973).

4

Electron screening and effective ion-ion interactions

In this chapter the framework of weak electron-ion coupling theory will be employed to

1. Complete the derivation of effective interactions in liquid metals, discussed in Chapter 3, and
2. Give a simple theory of the structure factor $S(k)$ for the alkali metals.

A useful starting point is the single-centre problem of a charged impurity in an electron liquid. In the following section, one takes a static point charge Ze, with Z sufficiently small (or, alternatively, the electron liquid density high enough) to use first-order perturbation theory in the screened potential round the point charge.

4.1. Screening of impurity centre in electron liquid

If one treats the problem by the linearized Thomas-Fermi approximation, then the self-consistent potential energy $V(r)$ of an electron at distance r from the test charge is obtained from the equation (Mott, 1936)

$$\nabla^2 V(r) = 4\pi Z e^2 \delta(\mathbf{r}) + q^2 V(r) \tag{4.1}$$

where q^{-1} is the Thomas-Fermi screening length l, given by multiplying a characteristic velocity, which in the degenerate electron liquid in molten metals is the Fermi velocity v_f, by a characteristic time τ. This is the period of the plasma oscillations of the electron liquid, $2\pi/\omega_p$, with ω_p the (Langmuir) plasma frequency: i.e.,

$$q^{-1} \doteqdot l = v_f \frac{2\pi}{\omega_p} = \left\{\frac{\pi a_0}{4k_f}\right\}^{1/2} \tag{4.2}$$

with k_f the Fermi wavenumber. Equation (4.1) has a solution satisfying the physical boundary conditions (1) $V(r) \to -Ze^2/r$ as $r \to 0$ and (2) $V(r) \to 0$ faster than $1/r$ as $r \to \infty$,

$$V(r) = -\frac{Ze^2}{r}\exp(-qr). \tag{4.3}$$

But the Thomas-Fermi approximation is valid for slowly varying potentials, and, as was shown by March and Murray (1960, 1961), one must treat the diffraction of the electron waves off the test charge. This is correctly accomplished by writing the nonlocal generalization of (4.1) as

$$\nabla^2 V(r) = 4\pi Ze^2\delta(\mathbf{r}) + \int F(\mathbf{rr}')V(\mathbf{r}')\,d\mathbf{r}' \tag{4.4}$$

which evidently reduces to (4.1) when the nonlocal response function $F(\mathbf{rr}') \equiv F(|\mathbf{r}-\mathbf{r}'|)$ in a homogeneous electron liquid is replaced by $q^2\delta(\mathbf{r}-\mathbf{r}')$. It can be shown (March and Tosi, 1984) that F is in fact given by

$$F(|\mathbf{r}-\mathbf{r}'|) = \frac{2mk_f^2e^2}{\pi^2\hbar^2}\frac{j_1(2k_f|\mathbf{r}-\mathbf{r}'|)}{|\mathbf{r}-\mathbf{r}'|^2} \tag{4.5}$$

with $j_1(x) = (\sin x - x\cos x)/x^2$ being the first-order spherical Bessel function.

Evidently, it is clear from (4.4) and (4.5) that the charge $\delta n(\mathbf{r})$ displaced by the introduction of the test charge is given in terms of the scattering potential $V(\mathbf{r})$ by

$$\delta n(r) = \frac{mk_f^2}{2\pi^3\hbar^2}\int\frac{j_1(2k_f|\mathbf{r}-\mathbf{r}'|)}{|\mathbf{r}-\mathbf{r}'|^2}V(\mathbf{r}')\,d\mathbf{r}'. \tag{4.6}$$

To study the form of the displaced charge far from the scattering centre, consider the example of a very localized scattering potential, i.e., $V(\mathbf{r}') = \lambda\delta(\mathbf{r}')$. Inserting this model form into (4.6), one finds

$$\delta n(r) \sim \frac{\cos 2k_f r}{r^3}; \qquad r \to \infty \tag{4.7}$$

when one makes use of the asymptotic form of the spherical Bessel function. Though one needs to obtain the scattering potential self-consistently from (4.4), it turns out that form (4.7) for the asymptotic displaced charge is again recovered for sufficiently large r.

4.2. Lindhard dielectric function

As anticipated, the behavior (4.7) reflects the diffraction of the electron de Broglie waves at the Fermi surface off the test charge, leading to these so-called Friedel oscillations. Equation (4.4), with F given by (4.5), has been solved numerically in $\mathbf{r}$ space for $V(\mathbf{r})$ for various electron densities (March and Murray, 1961), but if one Fourier transforms it, then the solution is obtained immediately in terms of the so-called Lindhard (1954) dielectric function $\varepsilon_L(k)$ given by

$$\varepsilon_L(k) = 1 + \frac{2mk_f e^2}{\pi\hbar^2 k^2}\left[1 + \frac{k_f}{k}\left(\frac{k^2}{4k_f^2} - 1\right)\ln\left|\frac{k - 2k_f}{k + 2k_f}\right|\right], \tag{4.8}$$

the Fourier transform $V(k)$ of $V(r)$ then being explicitly

$$V(k) = \frac{-4\pi Ze^2}{k^2\varepsilon_L(k)}. \tag{4.9}$$

Form (4.8) is established essentially from the form of the Fourier transform of the response function F in (4.5). In $\mathbf{k}$ space, it is the "kink" in $\varepsilon_L(k)$ at k equal to the diameter $2k_f$ of the Fermi sphere that is responsible for the long-range oscillations (4.7) in $\mathbf{r}$ space. (See the further discussion in Section 10.6.)

Quite briefly, it should be said that the r space oscillations have been detected in nuclear magnetic resonance experiments on field gradients due to charged impurities in metals (Rowland, 1960; Kohn and Vosko, 1960). Direct evidence for the kink in $\mathbf{k}$ space is provided by the observation of the so-called Kohn anomaly in phonon dispersion relations (Kohn, 1959).

It will be seen later, when strong scattering (for example, off a proton in metallic hydrogen) is treated, that such oscillatory behaviour of the displaced charge persists. The changes are that $\delta n(r)$ has now the asymptotic form $A\cos(2k_f r + \theta)/r^3$ and the amplitude A is changed from the first-order value. But as is evident, the most important point is that a phase factor θ is now introduced into the asymptotic form of $\delta n(r)$. In fact A and θ can be written explicitly in terms of the phase shifts δ_l for scattering of the Fermi surface electrons off the proton.

To discuss the two-centre problem, leading up to the effective pair interaction between ions in simple liquid metals, let us consider the model of two test charges, $Z_1 e$ and $Z_2 e$, separated by a distance R. It turns out that the most elementary electrostatic model is valid in this case. Thus, if one adopts the (oversimplified) form (4.3) of the screened potential, then the

Figure 4.1. Effective ion-ion interaction, with oscillations arising from singularity in the Lindhard dielectric function (4.8) at the Fermi sphere diameter $k = 2k_f$.

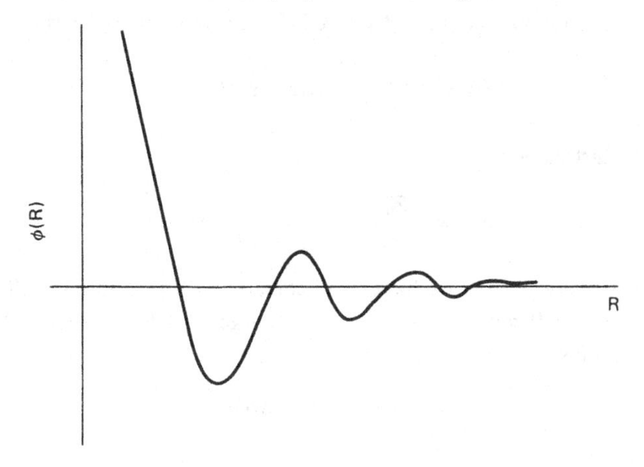

pair potential ϕ follows immediately as

$$\phi(R) = \frac{Z_1 Z_2 e^2}{R} \exp(-qR), \tag{4.10}$$

since we can view the charge $Z_2 e$ as sitting in the screened potential of charge $Z_1 e$ (or vice versa). Equation (4.10) is, of course, disappointing, as it tells us that like charges repel at all separations R, even in an electron liquid!

But to correct this (Corless and March, 1961), let us return to the Friedel oscillations. Provided R is taken to be sufficiently large to validate the asymptotic form (4.7) of the displaced charge, then $\phi(R)$ sketched in Figure 4.1 results. This form is, of course, simply the Fourier transform of

$$\phi(k) = \frac{4\pi Z_1 Z_2 e^2}{k^2 \varepsilon_L(k)} \tag{4.11}$$

in the Lindhard ($\equiv$Hartree) approximation.

4.3. Introduction of exchange and correlation

In the presence of exchange and correlation interactions, three modifications of the preceding argument are required, even at linear response level:

1. A one-body potential $V(\mathbf{r})$ exists, which generates the correct displaced charge through (4.6). But (4.4) is restricted to the Hartree approximation.
2. $V(\mathbf{r})$ becomes a functional of the displaced charge, and one may write

$$V(\mathbf{r}) = V_H(\mathbf{r}) + \int U(|\mathbf{r} - \mathbf{r}'|)\delta n(\mathbf{r}')\,d\mathbf{r}' \tag{4.12}$$

where the Hartree potential is

$$V_H(\mathbf{r}) = \frac{-Z_1 e^2}{r} + \int \frac{e^2}{|\mathbf{r} - \mathbf{r}'|}\delta n(\mathbf{r}')\,d\mathbf{r}', \tag{4.13}$$

U having subsumed into it exchange and correlation (Jones and March, 1973).
3. The potential $\phi(R)$ felt by a second test charge is determined by Poisson's equation and hence is given by

$$\phi(R) = -Z_2 V_H(\mathbf{R}). \tag{4.14}$$

From (1)–(3) and writing, for convenience,

$$\delta n(\mathbf{r}) = \int \chi_0(|\mathbf{r} - \mathbf{r}'|) V(\mathbf{r}')\,d\mathbf{r}' \tag{4.15}$$

then one finds

$$\phi(k) = \frac{4\pi Z_1 Z_2 e^2}{k^2 \varepsilon(k)} \tag{4.16}$$

and

$$V(k) = \frac{-4\pi Z_1 e^2}{k^2 \varepsilon_p(k)}. \tag{4.17}$$

Here

$$\varepsilon(k) = 1 - \frac{4\pi e^2}{k^2}\chi_0(k)/[1 - U(k)\chi_0(k)] \tag{4.18}$$

is the usual dielectric function but now corrected, via $U(k)$, for exchange and correlation interactions, whereas

$$\varepsilon_p(k) = 1 - \left[\frac{4\pi e^2}{k^2} + U(k)\right]\chi_0(k). \tag{4.19}$$

This latter screening function enters, through (4.6), the potential energy $V(k)$ of an electron round the test charge and therefore differs from $\varepsilon(k)$ because of exchange and correlation between such an electron and the electrons in the screening cloud. The two screening functions coincide, of course, in the Lindhard approximation, in which $V(\mathbf{r})$ is taken to satisfy Poisson's equation (4.4).

Equations (4.16) and (4.17) determine the effective ion-ion interaction and the effective ion-electron interaction, respectively, in the present case in which the ionic point charges are treated by linear response theory.

4.4. Effective ion-ion interactions in simple (s-p) metals

At this stage, one needs to introduce modifications to treat simple s-p liquid metals with ions having core electrons. The procedure adopted is to introduce pseudopotentials to describe the ion cores, and this is best implemented in **k** rather than **r** space.

Figure 4.2. Components of the effective ion-ion potential in liquid potassium (in units of e^2 Å^{-1}). The dot-dash curve is the negative of the direct Coulomb term e^2/R and the dashed curve is the electron-screening contribution. The total potential $\phi(R)$ near its minimum is shown in the inset on an enlarged scale.

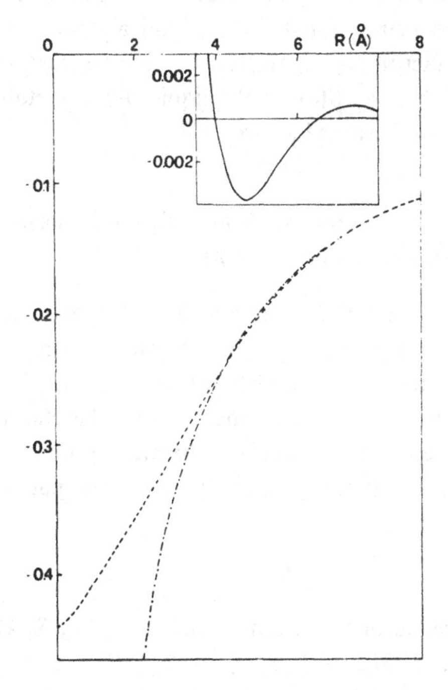

To this end, one introduces bare core potentials $v_i(k)$ (cf Chapter 3), which for point ions are simply $-4\pi Z_i e^2/k^2$, to describe the interaction between a bare ion and an electron. The result of (4.16) is then modified to read (see March and Tosi, 1984)

$$\phi(k) = \frac{4\pi Z_1 Z_2 e^2}{k^2} + \frac{v_1(k)v_2(k)}{(4\pi e^2/k^2)}\left[\frac{1}{\varepsilon(k)} - 1\right] \tag{4.20}$$

where the first term has been separated out to represent the direct ion-ion interaction, whereas in the remainder we have replaced the bare coulomb electron-ion interactions by the core potentials, which give the contribution due to electron screening. It may be noted that, in principle, the direct ion-ion term should also be supplemented by ion-core terms, but that these turn out to be normally negligible in practice for metals like Na or K.

The k^{-2} divergence in the direct term in (4.20) is exactly canceled in the limit $k \to 0$, where $v_i(k) \to -4\pi Z_i e^2/k^2$. The two components of the effective ion-ion potential in **R** space are illustrated for molten K near freezing in Figure 4.2. This shows the cancellation arising at large R, the oscillations at intermediate R, and the dominance of the direct term at small R. These predictions will be confronted in the following chapter with effective interionic forces extracted using statistical mechanical theory from the measured ionic structure factor $S(k)$. However, this chapter will conclude with a brief discussion of the way $S(k)$ for the liquid alkali metals can be modelled, utilizing the ideas presented above.

4.5. Structure factor of alkali metals modelled in terms of one-component plasma

As will be discussed in Chapter 6, Ferraz and March (1980) noted the relevance of the one-component plasma model to the freezing of liquid Na and K. This model considers classical point ions in a uniform non-responsive neutralizing background of electrons. The model is characterized by a single parameter Γ, measuring the ratio of the mean potential energy e^2/r_s to the thermal energy $k_B T$ (with r_s the mean interionic spacing)

$$\Gamma = \frac{e^2/r_s}{k_B T}. \tag{4.21}$$

Let us denote the structure factor of the ions in this model by $S_0(k)$, which clearly depends on the value of Γ.

Then the work of Tosi and March (1973) (see also Galam and Hansen, 1976; March and Tosi, 1984) allows the liquid alkali metal structure factors $S(k)$ to be modelled in terms of (1) $S_0(k)$; (2) a model potential representing the bare electron-ion interaction; and (3) the static dielectric function $\varepsilon(k)$ of the homogeneous electron fluid, already discussed in Section 4.3. Specifically the result takes the form, with n_i denoting the ionic density:

$$S(k) = \frac{S_0(k)}{[1 + n_i\tilde{v}(k)S_0(k)/k_B T]}. \tag{4.22}$$

Here $\tilde{v}(k)$ is related to the bare electron-ion potential $v(k)$ and $\varepsilon(k)$ by

$$\tilde{v}(k) = \frac{v^2(k)}{(4\pi e^2/k^2)}\left[\frac{1}{\varepsilon(k)} - 1\right]. \tag{4.23}$$

It is worth noting that (4.22) corresponds to the so-called random phase approximation of the electron screening, referred to in some detail in Section 4.2. A comparison of the long wavelength limit of (4.22) with experiment is worth making as follows (see also Chapter 3). One returns to the fluctuation theory result (2.5), to find (see Chaturvedi et al., 1981; March and Tosi, 1984)

$$\lim_{k\to 0} S(k) = n_i k_B T K_T. \tag{4.24}$$

This limit can then be taken in (4.22) as follows. For the classical one-component plasma, it can be shown (see March and Tosi, 1984) that

$$\lim_{k\to 0} S_0(k) = k^2 l_D^2 (1 + k^2/k_s^2)^{-1} \tag{4.25}$$

where $k_s^2 = 4\pi e^2/(\partial\mu/\partial n)_T$ with μ the chemical potential and l_D the Debye length. The other result one needs is the expansion

$$\lim_{k\to 0} v(k) = \frac{-4\pi Z e^2}{k^2}\left(1 - \frac{1}{2}k^2 r_c^2 + \cdots\right) \tag{4.26}$$

Table 4.1. *Isothermal compressibility for liquid alkali metals near freezing.*

		Na	K	Rb	Cs
	Theory	0.0215	0.0231	0.0236	0.0235
$n_i k_B T K_T$	Expt.	0.0236	0.0236	0.0220	0.0237

Adapted from March and Tosi, 1984.

for $v(k)$, introducing the parameter r_c which is representative of the core radius. The result from (4.24) is then (March and Tosi, 1984):

$$n_i k_B T K_T = \left(\frac{k_D^2}{k_{se}^2} + \frac{k_D^2}{k_{si}^2} + k_D^2 r_c^2 \right)^{-1}, \qquad (4.27)$$

where $k_D^2 = 4\pi n_i Z^2 e^2 / k_B T$, while k_{se}^2 and k_{si}^2 describe the "compressibilities" of the bare electronic and ionic components (March and Tosi, 1984). The first term in this expression (4.27) is essentially the Bohm-Staver result for the velocity of sound but now k_{se}^2 refers to the interacting electron fluid.

The numerical results obtained by Chaturvedi et al. (1981) with values of r_c, the core radius, taken from an analysis of phonon dispersion curves in solid alkalis are recorded in Table 4.1. The values of k_{se}^2 and k_{si}^2 were taken from electron-correlation theory (Vashishta and Singwi, 1972; Singwi and Tosi, 1981) and from computer simulation data on the free energy of the classical one-component plasma (Hansen et al., 1977). The quite good agreement with experiment shows that this simple treatment based on (4.22) is giving a consistent account of sound waves in the liquid.

5

Interionic forces and structural theories

Consider a situation such as depicted in Figure 5.1, in which attention is focused on atom 1 at position $\mathbf{r}_1$ in an environment in which there is a second atom 2 at distance $r_{12} = |\mathbf{r}_2 - \mathbf{r}_1|$. In a classical liquid, one next writes the pair function $g(r)$ as a Boltzmann factor:

$$g(r) = \exp\left[\frac{-U(r)}{k_B T}\right]. \tag{5.1}$$

This can be viewed as the definition of the potential of mean force $U(r)$, which in general will depend also on the temperature T. Returning to Figure 5.1, one can now write the total force acting on atom 1 as $-\partial U(r_{12})/\partial \mathbf{r}_1$; this can be separated into a direct part $-\partial\phi(r_{12})/\partial \mathbf{r}_1$ due to the assumed pair potential energy $\phi(r)$ acting between two atoms at separation r_{12} and that due to the rest of the atoms. If, as in Figure 5.1, one considers a third atom at $\mathbf{r}_3$, then clearly one must introduce into the theory a three-atom correlation function $g_3(\mathbf{r}_1\mathbf{r}_2\mathbf{r}_3)$ that measures the probability of finding three atoms simultaneously at $\mathbf{r}_1$, $\mathbf{r}_2$, and $\mathbf{r}_3$.

Figure 5.1. Geometry of three-atom configuration used in setting up the force equation (5.2). The quantity g_3 in that equation measures the probability of finding three atoms simultaneously at $\mathbf{r}_1$, $\mathbf{r}_2$, and $\mathbf{r}_3$.

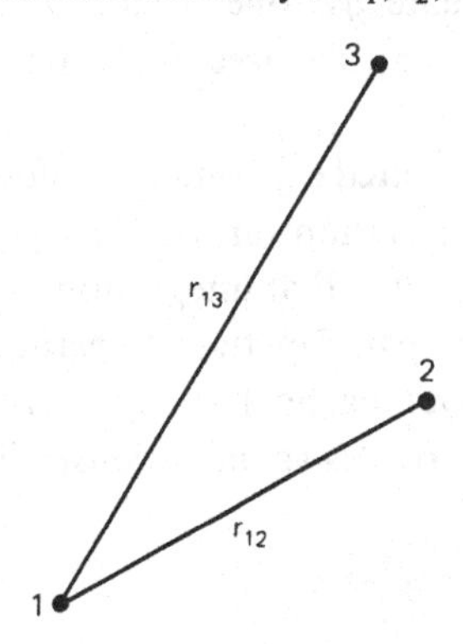

5.1. Force equation

One can write the so-called force equation:

$$\frac{-\partial U(r_{12})}{\partial \mathbf{r}_1} = -\frac{\partial \phi(r_{12})}{\partial \mathbf{r}_1} - \rho \int \frac{\partial \phi(r_{13})}{\partial \mathbf{r}_1} \frac{g_3(\mathbf{r}_1\mathbf{r}_2\mathbf{r}_3)}{g(r_{12})} \mathrm{d}\mathbf{r}_3. \tag{5.2}$$

The last term has the form shown in (5.2), since one has asserted that there are certainly atoms at $\mathbf{r}_1$ and $\mathbf{r}_2$ and hence the three-body correlation function g_3 must be divided by $g(r_{12})$ to take account of this. This conditional probability must be multiplied by the force $-\partial\phi(r_{13})/\partial\mathbf{r}_1$ on atom 1 due to atom 3 at $\mathbf{r}_3$, and the result must then be integrated over all positions $\mathbf{r}_3$. In (5.2), the total force is evidently expressed as a sum of a direct part, involving $\phi(\mathbf{r}_{12})$ and an indirect part. Equation (2.7) is a corresponding decomposition for correlation functions, $c(r)$ evidently reflecting somehow three-body correlations also.

Although the above construction of (5.2) is intuitive, (5.2) is, in fact, an exact consequence of classical statistical mechanics, given that the total potential energy $\Phi(\mathbf{r}_1, \ldots, \mathbf{r}_N)$ of the liquid can be expressed solely in terms of pair potentials by

$$\Phi = \sum_{i<j} \phi(r_{ij}). \tag{5.3}$$

Here then, in (5.2), is the desired link between structure and forces. However, it is plain that, since $U(r)$ is related directly to $g(r)$ by (5.1), one can forge an explicit relation between $g(r)$ and the pair potential $\phi(r)$ only if one has knowledge of the three-body correlation function g_3. This is a characteristic of many-body theories; they lead to hierarchical equations relating correlation functions. To calculate g_3, one needs to write an equation involving four-body correlations g_4, etc. (see, for example, Hill, 1956, and Appendix 5.4).

So far, while $g(r)$ can be measured, only limited experimental information can be obtained on the three-particle correlation function $g_3(\mathbf{r}_1\mathbf{r}_2\mathbf{r}_3)$. As shown in Appendix 5.1, one important result is that integration on g_3 can be related to the density derivative of the pair function. Experiments on this density derivative have been carried out, particularly by Egelstaff and his coworkers (1971; 1980). This already provides valuable constraints that any acceptable theory of g_3 must satisfy.

5.2. Simple structural theories

Following the ideas of Kirkwood, the simplest assumption to make for g_3 is that it is given by the product of pair functions, i.e.,

$$g_3(\mathbf{r}_1\mathbf{r}_2\mathbf{r}_3) \cong g(r_{12})g(r_{23})g(r_{31}). \tag{5.4}$$

Inserting this approximation (5.4) into the exact form (5.2) leads to the so-called Born-Green theory (see Green, 1952) of liquid structure. Though one must be careful to specify the range of the forces before deciding how to decouple g_3 ((5.4) being a simple example of decoupling), it is now known that the Born-Green theory is not sufficiently refined to yield a quantitative theory of liquid structure from a given pair potential $\phi(r)$. However, a refinement of it leads to the so-called hypernetted chain (HNC) theory, which, as the name implies, had its origins in diagrammatic methods (for a review, see McDonald and O'Gorman, 1978) of handling the classical many-body problem posed by a liquid such as argon.

To see how to reach the HNC theory, one notes that, following Rushbrooke (1960), the Born-Green theory leads to

$$\frac{U(r)}{k_B T} = \frac{\phi(r)}{k_B T} - \rho \int E(\mathbf{r} - \mathbf{r}')h(\mathbf{r}')\,d\mathbf{r}' \tag{5.5}$$

where

$$E(r) = \frac{1}{k_B T}\int_r^\infty g(s)\frac{\partial \phi(s)}{\partial s}\,ds. \tag{5.6}$$

In fact, there is a class of structural theories that have a convolution representation of $(U - \phi)/k_B T$, as in (5.5). Since $g(r) \to 1$ far from the critical point, if one calculates $E(r)$ at large r from (5.6), one has $c(r) \to -\phi(r)k_B T$,* $E(r) \to c(r)$. If, in embracing this class of theories, one replaces E in (5.5) by G, then the different members of this class are characterized by different forms of G, as collected in Table 5.1. Included there, in addition to HNC($G \to c$) and Born-Green theories ($G \to E$), are the treatments of de Angelis and March (1976) ($G \to h$) and a combination of a structural proposal of Liboff (1986) with Born-Green theory. It must be stressed that

* Though no general proof appears to exist to date, this asymptotic result for large r is the basis of the so-called mean spherical approximation (see for instance the book by March and Tosi, 1984), as well as being incorporated in the HNC theory. It ceases though to be valid on approaching the critical point (see (3.11)).

Table 5.1. *Structural theories related by G(r), which replaces E(r) in (5.5).*

Theory	Liboff (1986) + Born-Green (5.5)	Hypernetted chain	Born-Green	de Angelis and March	Gaskell
Form for $G(r)$	$h(r) + \frac{\rho}{k_B T} F(r)$: $F(r) = \int dr' \phi(r + r') g(r')$	$c(r)$	$E(r)$	$h(r)$	$E(r)$ + term with $\frac{\partial g}{\partial \rho}$

it has not been demonstrated that a convolution form such as (5.5) follows directly from the force equation (5.2). For a more general discussion of (5.2), reference may be made to examples of Ar and Na treated by Ebbsjö et al. (1983), which use computer simulation data. Also Appendix 5.2 considers the conditions to be satisfied by thermodynamically consistent structural theories.

5.3. Refined structural theories: factorization of triplet correlations

Following the philosophy of Kirkwood, structural theories transcending that of Born-Green can be constructed. Thus, it is of interest to record that Barrat et al. (1987) advocate the factorization of the triplet direct correlation function $c^{(3)}$:

$$c^{(3)}(r_1, r_2, r_3) = t(12)t(23)t(31) \tag{5.7}$$

where $12 \equiv r_{12}$, etc. In this form, t is now to be fixed by requiring that the exact relation between $\partial c(r)/\partial \rho$ and $c^{(3)}$ be satisfied (compare Appendix 5.1). Specifically, knowledge of the density derivative $\partial c/\partial \rho$ of the Ornstein-Zernike direct correlation function $c(r)$ would be sufficient to permit the unambiguous determination of the function $t(r)$ in (5.7) through (Barrat et al., 1987):

$$c(r) = t(r) \int t(r')t(|\mathbf{r} - \mathbf{r}'|)\, d\mathbf{r}'. \tag{5.8}$$

Given approximation (5.7), with $t(r)$ determined from equation (5.8), it remains only to utilize the force equation (5.2) together with the relation between $g^{(3)}$ and $c^{(3)}$ discussed by Hernando (1986) and by Senatore and

March (1986). To illustrate the procedure, one notes here that $\partial g/\partial \rho$, related to $g^{(3)}$ as in Appendix 5.1, is given by

$$\frac{\partial g(12)}{\partial \rho} = \int \mathrm{d}3\, h(13)h(32) + \frac{\partial c(12)}{\partial \rho} + 2\rho \int \mathrm{d}3\, h(13)\frac{\partial c(32)}{\partial \rho} + \rho^2 \int \mathrm{d}3 \int \mathrm{d}4\, h(13)h(24)\frac{\partial c(34)}{\partial \rho}, \tag{5.9}$$

which illustrates one aspect, admittedly limited, of the relationship between $g^{(3)}$ and $c^{(3)}$, with h as usual denoting the total correlation function $g - 1$. A more complete discussion is given by March and Senatore (1988), to which the interested reader is referred.

Returning to Kirkwood, being motivated by the factorization (5.7) of $c^{(3)}$ proposed by Barrat et al. (1987), it has been noted by March and Senatore (1988), from the work of Root et al. (1988) based on the Gaussian core model (see Appendix 5.3), in the completely aggregated state arising with purely attractive interactions in that model, that

$$g^{(3)}(\mathbf{r}_1, \mathbf{r}_2, \mathbf{r}_3) = s(12)s(23)s(31) \tag{5.10}$$

where again $s(r)$ is to be found from the $g^{(3)} - \partial g/\partial \rho$ relation (see Appendix 5.1). So far, except for this special model, this fact has not been utilized in explicit calculations on real liquids, although it is clearly promising and easier to work with than the $c^{(3)}$ factorization (5.7).

What is quite clear is that the density dependence of the pair function is going to play an increasingly important role in structural theories of liquids in the future.

5.4. Inversion of structure factor to yield effective interionic pair potentials

Johnson and March (1963; see also Johnson et al., 1964) made the proposal to invert a measured structure factor $S(k)$ (see Section 2.1) to extract an (assumed) pair potential $\phi(r)$. They presented an oscillatory potential for liquid Na near melting, which, although not quantitative because of (1) the approximate liquid structure theories used that are now known to need refinement for this purpose and (2) the use of experimental data of limited accuracy, has been refined into a fully quantitative form by Levesque et al. (1985) and Reatto et al., (1986). Such pair potentials for liquid metals, as anticipated by Johnson and March (1963), are mediated by the conduction or valence electrons and have qualitatively different form from that appropriate to the dense liquid phase of argon, for instance.

It is worth giving a concrete example of the way an effective pair potential can be derived from experimental data, plus an approximate liquid state theory of structure. If one adopts as the example the HNC theory defined by (5.5) with E replaced by the direct correlation function $c(r)$ (see Section 2.2), i.e.,

$$\frac{U - \phi}{k_B T} = -\rho \int c(\mathbf{r} - \mathbf{r}')h(\mathbf{r}')\,d\mathbf{r}' \tag{5.11}$$

then the definition of c in (2.7) immediately yields*

$$\frac{\phi_{\mathrm{HNC}}(r)}{k_B T} = \frac{U(r)}{k_B T} + h(r) - c(r). \tag{5.12}$$

Evidently the right-hand side (see (5.1) for definition of U) of (5.12) is determined solely by empirical $S(k)$ data, at least in principle, and hence $\phi_{\mathrm{HNC}}(r)$ can be calculated. A refined method, the modified HNC (MHNC) treatment, adds a so-called bridge correction $B(r)$ to (5.12) to yield

$$h - c = \frac{\phi - U}{k_B T} + B(r). \tag{5.13}$$

Rosenfeld and Ashcroft (1979) have argued that $B(r)$ is not sensitive to long-range forces and hence can be approximated by a hard sphere model (see, however, March, 1986). For liquid Na, as Figure 5.2 makes quite clear, the MHNC equation (5.13), with $B(r)$ approximated as just described, certainly requires further refinement before being quantitatively adequate for this inversion, since it leads to the solid line in Figure 5.2, which is substantially different from the lower curves giving the final potential obtained by Reatto et al. (1986) for this liquid metal. Suffice it to say that the "correct" potential $\phi(r)$ in Figure 5.2 was obtained by these workers using what is termed an iterative predictor-corrector method

* The so-called Percus-Yevick collective coordinate theory yields

$$\frac{\phi_{\mathrm{PY}}(r)}{k_B T} = \frac{U(r)}{k_B T} + \ln(1 + h(r) - c(r))$$

which expands to lowest order in $(h - c)$ to yield (5.12).

in which the predictor is the MHNC method, whereas the corrector is simulation.

March and Senatore (1988) have set out the way in which the decoupling of the triplet direct correlation function $c^{(3)}$ given in (5.7) can be made the basis of an inversion procedure. Alternatively, the use of $g^{(3)}$ in (5.10) would lead to a theory transcending the Born-Green equation (5.5). Neither of these approaches, at the time of writing, has been pushed through to yield a potential for Na to be compared with the lowest curves in Figure 5.2; that,

Figure 5.2. Potential $\phi(r)$ for liquid Na obtained from inversion of the measured diffraction data of Greenfield, Wellendorf, and Wiser (1971) by Levesque et al. (1985) and Reatto et al. (1986); see also Reatto, 1988). This is the result shown in the lowest curves. The solid line is the modified hypernetted chain result, whereas the middle curve is a result of an early iteration in their procedure.

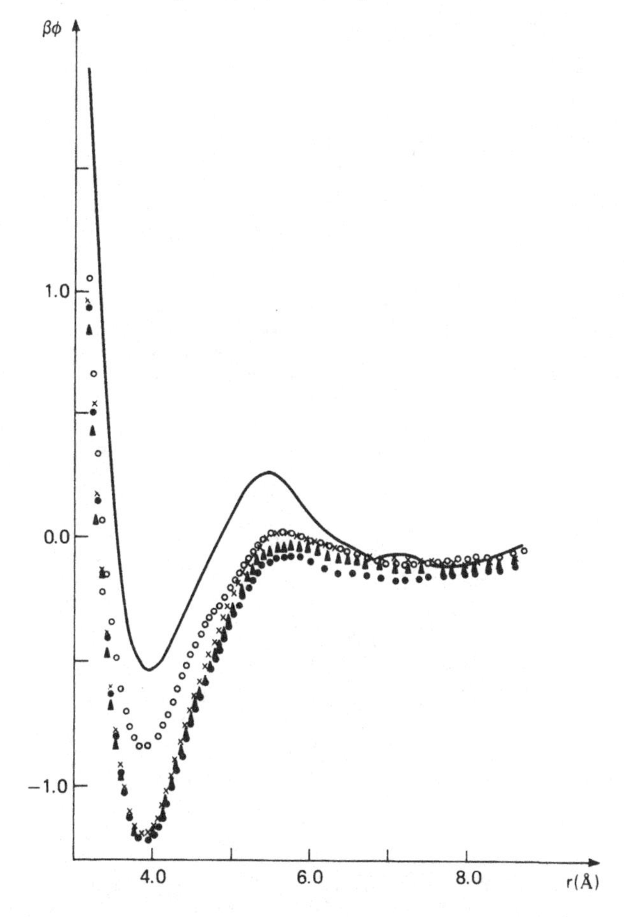

however, is of obvious interest for the future. For a fuller discussion of the inverse problem, reference may be made to the review by the writer (March, 1987d) and to the work of Reatto (1988), who also considers the quantum-mechanical liquid helium four in the same context. The work of Gaskell (1988) is also relevant; his approach fits into the framework of Table 5.1 and his result is therefore entered in that table. Appendix 5.5 briefly summarizes how the theory of metals can be utilized to extract potentials in liquid metals from $S(k)$, where, as emphasized before, the effective ion-ion interactions are mediated by the itinerant conduction electrons.

6
Statistical mechanics of inhomogeneous systems and freezing theory

In Chapter 5, the statistical mechanics of homogeneous bulk liquids was treated, mainly within a pair potential framework. The generalization of this theory will next be set up to deal with inhomogeneous systems. One important application will be to the study of the liquid-vapour interface, developed in Chapter 12. However, the immediate application will be to the theory of freezing of liquid metals.

Specifically, the focal point of this chapter is the idea that correlation functions in a bulk liquid near to its freezing point already contain valuable information pertaining to the properties of its solid near melting. An example that can be cited here is the theory of the liquid-solid transition due to Kirkwood and Monroe (1941). This theory was based on the so-called hierarchy of statistical mechanical equations for the various order distribution functions (see, for example, Hill, 1956), generalized to apply to inhomogeneous systems (e.g., a solid with a periodic rather than constant single-particle density). The Kirkwood-Monroe theory certainly requires as basic knowledge the pair potential $\phi(r)$ and the homogeneous pair function $g(r)$ discussed at length in the preceding chapters.

6.1. Single-particle density related to direct correlation function

Lovett and Buff (1980) have revived interest in the question of whether classical statistical-mechanical equations, such as the first member of the Born-Green-Yvon hierarchy, which connects the singlet density $\rho(\mathbf{r})$ and the liquid pair correlation function $g(r)$, can admit more than one solution for ρ for a given g (see Kirkwood and Monroe, 1941). Actually, these workers focused on the equation relating ρ and the Ornstein-Zernike direct correlation function $c(r)$ of a liquid. This latter equation has been derived by a number of workers and has the advantage over the equation relating $g(r)$ and $\rho(\mathbf{r})$ that no assumption of pairwise interactions need be invoked.

This equation relating $\rho(\mathbf{r})$ and $c(\mathbf{r})$ (see Appendix 13.3 for multi-component case), namely,

$$\ln \rho(\mathbf{r}_1) = \int \mathrm{d}\tau_2\, c(r_{12})\rho(\mathbf{r}_2) + \text{constant} \tag{6.1}$$

is the central tool employed in the present discussion. The integrated form (6.1) has been discussed, for example, by Lovett (1977). It will be demonstrated next, by direct solution of this equation, that for a given liquid direct correlation function, (6.1) admits not only a solution for which the singlet density is uniform, with a value $\rho(\mathbf{r}) = \rho_l$, but also a coexisting periodic solution $\rho_{\mathrm{p}}(\mathbf{r})$. If one then linearizes the equation determining $\rho_{\mathrm{p}}(\mathbf{r})$, one regains the bifurcation condition of Lovett and Buff (1980). The free-energy difference corresponding to the two types of singlet density will then be set up. Somewhat surprisingly, by using the equation for $\rho_{\mathrm{p}}(\mathbf{r})$ in this free energy, one finds a result obtained earlier in the theory of freezing of Ramakrishnan and Yussouff (1977, 1979), which was derived by making use of the hypernetted-chain approximation. The following presentation (March and Tosi, 1981a, b, c, see also Haymet and Oxtoby, 1981) thereby shows that this approximation is inessential in their theory of freezing.

6.2. Free-energy difference between homogeneous and periodic phases

Evidently there is, for a given liquid direct correlation function $c(r)$, a solution of (6.1) for which $\rho(\mathbf{r})$ is constant with a value ρ_l, for instance. What is more important for present purposes is to prove that for a given liquid $c(r)$, a periodic solution $\rho_p(\mathbf{r})$ for the singlet density also exists.

To do this, one applies (6.1) to both the singlet densities ρ_l and $\rho_p(\mathbf{r})$ and then subtracts to find

$$\ln\left(\frac{\rho_{\mathrm{p}}(\mathbf{r}_1)}{\rho_l}\right) - \int \mathrm{d}\tau_2\, c(|\mathbf{r}_1 - \mathbf{r}_2|)[\rho_{\mathrm{p}}(\mathbf{r}_2) - \rho_l] = 0, \tag{6.2}$$

where it is assumed that the constants in (6.1) are equal for the two phases in coexistence. Since, by assertion, ρ_{p} is periodic, one expands in a Fourier series using the reciprocal lattice vectors $\mathbf{G}$ to obtain

$$\rho_{\mathrm{p}}(\mathbf{r}) = \rho_0 + V^{-1} \sum_{G \neq 0} \rho_{\mathbf{G}} \exp(i\mathbf{G}\cdot\mathbf{r}) \tag{6.3}$$

where V is the total volume. Inserting (6.3) into (6.2) and integrating over

$\mathbf{r}_2$ yields

$$\ln\left(\frac{\rho_p(\mathbf{r})}{\rho_l}\right) = \frac{(\rho_0 - \rho_l)}{\rho_l}\tilde{c}(0) + (\rho_l V)^{-1} \sum_{G \neq 0} \rho_G \tilde{c}(\mathbf{G}) \exp(i\mathbf{G}\cdot\mathbf{r}) \quad (6.4)$$

where $\tilde{c}(k)$ is the Fourier transform of $c(r)$. Equation (6.4) is, for a given set of Fourier components of the liquid direct correlation function, to be solved for the Fourier components ρ_G of the singlet density.

Without seeking a specific solution of (6.4), a matter that will be referred to again later, it will be useful at this point to regard (6.4) as the Euler equation of a minimum free-energy principle. Of course, the thermodynamic requirement for the two phases to be in equilibrium is that this free-energy difference be zero. Actually, one works with the thermodynamic potential Ω, related to the Helmholtz free-energy F and chemical potential μ by

$$\Omega = F - N\mu. \quad (6.5)$$

The Helmholtz free energy can be conveniently divided into two parts, one corresponding to free particles and the other taking into account the interparticle interactions via the direct correlation function $c(r)$. The first part is well known for uniform density, and one merely takes the free-energy density over into the local density $\rho(\mathbf{r})$. The second part is also available in essence (see, for example, Kubo, 1965), and thus one can write

$$\frac{\Delta\Omega}{k_B T} = \int d\tau \left\{\rho_p(\mathbf{r}) \ln\left(\frac{\rho_p(\mathbf{r})}{\rho_l}\right) - [\rho_p(\mathbf{r}) - \rho_l]\right\}$$
$$- \frac{1}{2}\iint d\tau_1\, d\tau_2 [\rho_p(\mathbf{r}_1) - \rho_l] c(|\mathbf{r}_1 - \mathbf{r}_2|) [\rho_p(\mathbf{r}_2) - \rho_l]. \quad (6.6)$$

Performing the variation of $\Delta\Omega$ with respect to $\rho_p(\mathbf{r})$ is readily verified to lead back to (6.2).

At this stage, one inserts the Fourier expansions (6.3) and (6.4) into (6.6) to find, with $N = \rho_l V$,

$$\frac{\Delta\Omega}{k_B T} = \frac{1}{2N}\sum_{\mathbf{G}} \tilde{c}(\mathbf{G})|\rho_{\mathbf{G}}|^2 - \frac{N(\rho_0 - \rho_l)}{\rho_l} + \frac{1}{2}N\tilde{c}(0)\frac{(\rho_0^2 - \rho_l^2)}{\rho_l^2}. \quad (6.7)$$

This is the desired expression for the free-energy difference in terms of the Fourier components ρ_G of the periodic density and the volume change reflected in the difference between ρ_l and ρ_0.

The possibility of coexistence of homogeneous liquid and periodic phases is clear from (6.7) because of the balance between positive contributions from the first and third terms on the right-hand side and the negative term from the volume change, provided the periodic phase has the higher density. That these terms are strongly coupled is clear from the highly nonlinear nature of the Euler equation (6.4). The actual coexistence point is evidently determined by the properties of $\tilde{c}(\mathbf{G})$, including $\mathbf{G} = 0$, linking the liquid structure intimately with the appearance of the periodic phase.

6.3. Relation to the hypernetted-chain method

The preceding discussion of (6.7) and (6.4) has, of course, been focused on what is, in principle, possible from the structure of these two equations. It is remarkable that the work of Ramakrishnan and Yussouff (1977; 1979), using as it does the hypernetted-chain (HNC) approximation, nevertheless leads to the same Euler equation (6.4). This work is based directly on (6.2) and shows clearly that this use of HNC is inessential to their theory of freezing (see also March and Tosi, 1981a,b,c). However, a significant difference between this treatment and their approach resides in the variational principle from which the Euler equation is derived. The basic variation here is of $\Delta\Omega$ in (6.6). Their variation is on an equation resembling, but not identical to, (6.7). This latter equation, in this treatment, already embodies the result for the periodic density determined by the nonlinear Euler equation (6.4). The specific difference between (6.7) and their form for $\Delta\Omega$ is that they have an additional term of order $(\rho_0 - \rho_1)$, giving a result of

$$\frac{\Delta\Omega}{k_B T} = \frac{1}{2N}\sum_{\mathbf{G}} \tilde{c}(\mathbf{G})|\rho_{\mathbf{G}}|^2 - \frac{1}{2}N[1 - \tilde{c}(0)]\frac{(\rho_0^2 - \rho_l^2)}{\rho_l^2}. \qquad (6.8)$$

This form is appealing because $(1 - \tilde{c}(0))$ is essentially the inverse compressibility of the liquid. The work of Ramakrishnan and Yussouff, in fact, shows how an explicit periodic solution for a given liquid direct correlation function $c(r)$ can be derived from (6.4).

The final comment concerns the relation of the nonlinear (6.4) to the work of Lovett and Buff (1980) on bifurcation already mentioned. Whereas the preceding treatment is evidently describing a first-order transition, with a volume change and finite Fourier components $\rho_{\mathbf{G}}$ actually at the freezing point, their work explores the condition under which the $\rho_{\mathbf{G}}$ develop continuously from the homogeneous phase. The procedure they use corresponds to linearizing (6.4), which then has a solution of periodic form provided the condition $[1 - \rho_l\tilde{c}(G)] = 0$ is satisfied. This corresponds, in fact, to the structure factor $S(\mathbf{G}) = 1[1 - \rho_l \tilde{c}(\mathbf{G})]$ becoming infinite. This is an instability of the liquid phase.

It is not intended here to enter into a full discussion of the not inconsiderable technical aspects of actually calculating the freezing properties of some liquid metals from this theory (see, however, Sections 6.6 and 6.7). Suffice it here to say that successful applications have been made, although the convergence of the reciprocal lattice sum is not as fast as the success of Verlet's (1968) rule (see Section 6.4 below), which depends essentially only on the first such vector, might suggest at first sight. Nevertheless, one can have confidence that a first-principles statistical-mechanical theory of freezing is now laid down via the work of Kirkwood and Monroe (1941), Ramakrishnan and Yussouff, and other authors cited in the introduction to this chapter.

6.4. Verlet's rule related to Lindemann's law of melting

For Lennard-Jones liquids such as argon, Verlet (1968) drew attention to the fact that the liquids freeze when the peak height of the structure factor reaches a value of about 2.8 as the temperature is lowered. For liquid Na and K, Ferraz and March (1980) pointed out that the classical one-component plasma (see Section 4.5) would freeze under about the same condition and compared this with experiment for these two alkali metals. Later work (Chaturvedi et al., 1981; see also Tosi and March, 1973a) has elaborated upon the relation of the one-component plasma model to the structure factors of the alkali metals, as discussed briefly in Section 4.5.

Thus, for freezing there is the criterion that $S(q)$ at its maximum, say at position q_m, is about 2.8, whereas Lindemann's law of melting states that crystals melt when the root-mean-square vibrational amplitude becomes a fixed fraction of the lattice spacing.

It is worth elaborating here on the relation between these criteria for freezing and melting. The work of Bhatia and March (1984b), though designed to relate more generally the principal peak height, position, and width of the structure factor $S(q)$ of dense monatomic liquids, supplies such a relation. This is not to say that, in practice, Verlet's rule is not a somewhat better criterion for characterizing the phase transition than Lindemann's; this is a matter of detailed quantitative study, whereas the point here is to show a definite correlation between the two different rules.

6.4.1. Relation between principal peak height, position, and width of $S(q)$

In dense classical liquids, the pair function $g(r)$ must satisfy $g(r = 0)$. Using the Fourier-transform relation between the total correlation function $g(r) - 1$ and $S(q) - 1$, this condition $g(0) = 0$ reads as follows for N atoms in volume V:

$$-2\pi^2 \frac{N}{V} = -2\pi^2 \rho = \int_0^\infty (S(q) - 1) q^2 \, dq. \tag{6.9}$$

An approximate evaluation of the integral in (6.9) in such dense liquids satisfying $g(0) = 0$ can be carried out as follows: With q_m, as before, denoting the position of the principal peak of $S(q)$, let the peak width, denoted by $2\,\Delta q$, be measured by the distance between the two adjacent nodes of $S(q) - 1$, which embrace q_m. Furthermore, let us assume that any asymmetry of the peak about q_m can be neglected. If (6.9) is now written as

$$-2\pi^2 \rho = \int_0^{q_m - \Delta q} [S(q) - 1] q^2 \, dq + \int_{q_m - \Delta q}^{q_m + \Delta q} [S(q) - 1] q^2 \, dq + \int_{q_m + \Delta q}^\infty [S(q) - 1] q^2 \, dq \tag{6.10}$$

then for an $S(q)$ appropriate to dense fluids, such as argon near the triple point or Na and K near the melting point, the following approximations prove useful:

1. To replace $S(q) - 1$ by -1 over the range of the first integral in (6.10).
2. To neglect the third integral in (6.10) because of the oscillations of $S(q) - 1$ about zero.
3. To estimate the second integral by the triangular area

$$[[S(q_m) - 1] q_m^2 \, \Delta q]$$

Using these simplifications and introducing the mean interatomic separation R_A through $\rho = (3/4\pi R_A^3)$, it is readily shown that

$$S(q_m) q_m^2 \, \Delta q \doteqdot \frac{1}{3} q_m^3 \left(1 - \frac{9\pi}{2} \frac{1}{(R_A q_m)^3} \right). \tag{6.11}$$

Empirically, in dense liquids, $R_A q_m \simeq 4.4$, and the second term in the bracket in (6.11) contributes 0.15 compared with unity. Thus, one is left with the result

$$S(q_m) \sim \frac{0.3q_m}{\Delta q}. \tag{6.12}$$

It is instructive to confront the approximate prediction (6.12) with the accurate diffraction data of Yarnell et al. (1973) on liquid argon at 85 K. One finds from this data that $S(q_m) = 2.70$, $q_m = 2.00$, and $\Delta q = 0.275$ (q being in Å^{-1}), yielding $S(q_m)/(q_m/\Delta q) = 0.37$, which is nearer to $\frac{3}{8}$ than the predicted 0.3 in (6.12). It is satisfactory that the data of Greenfield et al. (1971) on liquid potassium at 65°C yields $S(q_m) = 2.73$, $q_m = 1.62$, $\Delta q = 0.225$, and hence a constant of 0.38, whereas for the experiment at 135°C, the constant in (6.12) is 0.37. Similarly for Na at 100°C and 200°C, the measured data yield 0.37 and 0.36, respectively. Thus for the five experiments referred to earlier, it turns out that (6.12) is quantitative when 0.3 is replaced by $\frac{3}{8}$. The fact that this is greater than 0.3 seems to indicate that the third integral in (6.10) actually has a nonzero negative value.

Turning to $g(r)$, it is worth noting that an argument in which $S(0)$ is evaluated via $[\int_0^\infty [g(r) - 1]r^2\,dr]$ can be carried through with assumptions paralleling (1) to (3) for calculating $g(0)$. The result is

$$g(r_m)r_m^2\,\Delta r = \tfrac{1}{3}r_m^3 - \tfrac{1}{3}R_A^3[1 - S(0)], \tag{6.13}$$

with definitions precisely paralleling those for $S(q)$. Since $g(r)$ is less readily accessible than $S(q)$, no comparison of (6.13) with experimental data will be attempted as it stands. However, for the data of Yarnell et al. (1973) on liquid argon at 85 K, one finds

$$g(r_m) = 3.05, \qquad \Delta r = 0.545\ \text{Å}, \qquad r_m = 3.68\ \text{Å} \tag{6.14}$$

and the value of $r_m/\Delta r$ is 6.7, as compared with $q_m/\Delta q = 7.2$. Thus, very approximately,

$$\frac{q_m}{\Delta q} \simeq \frac{r_m}{\Delta r}. \tag{6.15}$$

6.4.2. *Melting and freezing criteria related*

To complete this section, let us return to the criterion (Verlet 1968) that simple liquids like argon freeze when $S(q_m) \simeq 2.8$, this value also being applicable to Na and K. In the cases of Ar, Na, and K, where freezing involves only minor changes in local coordination, the use of $S(q_m)_{T_m} = 2.8$ at the melting temperature T_m yields the estimate $(\Delta r/r_m)_{T_m} \sim 0.11$ from (6.11) and (6.15). Lindemann's law of melting, according to Faber gives $(\Delta r/R_A)_{T_m} \sim 0.2$ if one identifies here Δr as the root-mean-square displacement of the atoms. Since $r_m \sim 1.8R_A$, these results are seen to be roughly consistent. Thus, there is no conflict between freezing criteria based on $S(q_m)_{T_m} = 2.8$ on the one hand and Lindemann's law on the other.

6.5. Cluster expansion and density functional theory

In the cluster expansion the free energy functional $F_e[n(\mathbf{r})]$ is expanded to infinite order around a homogeneous fluid, the coefficients of the expansion being the 2-body and higher direct correlation functions (DCF) of the fluid (see Rovere and Tosi, 1988). Truncation of the expansion at the lowest order yields the equilibrium equation for the density profile in the crystal at coexistence with the liquid, at the same temperature and chemical potential, with n now denoting the density, as

$$\ln[n(\mathbf{r})/n_l] = \int d\mathbf{r}'\, c_l(|\mathbf{r} - \mathbf{r}'|)[n(\mathbf{r}') - n_l], \tag{6.16}$$

$c_l(r)$ being the two-body DCF of the liquid. Equation (6.16), which can also be obtained directly from A6.2 through replacement of $c(\mathbf{r}, \mathbf{r}')$ by $c_l(|\mathbf{r} - \mathbf{r}'|)$, is equivalent to the hypernetted-chain theory of bulk liquid structure (HNC closure) as discussed in Section 6.3. It is evident that at this level the theory involves only the Fourier transform of $c_l(r)$ evaluated at the reciprocal lattice vectors (RLV) ($\hat{c}_l(G)$, say, and at $k = 0(c_0^l)$, this being determined by the isothermal compressibility of the liquid). For an illustration of the higher-order terms, let us consider them at the 3-body level. The corrections to the HNC closure ("bridge diagrams") then involve the following nonlinear physical properties:

1. $\hat{c}_l^{(3)}(0, 0)$, i.e., the density dependence of the compressibility;
2. $\hat{c}_l^{(3)}(\mathbf{G}, 0)$, i.e., the density dependence of the liquid structure factor at the RLV;
3. $\hat{c}_l^{(3)}(\mathbf{G}, \mathbf{G}')$, i.e., couplings between microscopic order parameters, whose angular dependence may be expected to start playing a role in determining the stable crystal structure. Some calculations (see Rovere and Tosi, 1988) have included (1) and sometimes also (2). The evaluation of (3) involves instead a large and rather delicate numerical effort.

6.6. Other approximations

6.6.1. Reference liquid at effective mean density (EMD)

The help of a reference liquid has been invoked by Baus and Colot (1985) and by Iglói and Hafner (1986). In brief, recourse is still made only to the homogeneous two-body DCF but for a reference liquid at density n_r, which is either treated as a variational parameter or chosen so that the liquid pair structure scales with that of the solid. The nonlinear effects (1) and (2) are clearly being approximately included to third and higher order.

6.6.2. Weighted density approximation (WDA)

Curtin and Ashcroft (1985, 1986), following earlier work by Tarazona (1984; 1985), have proposed another approach of the free-energy functional to the crystal, which parallels developments in the local density approximation for the inhomogeneous electron gas (see, for example, Lundqvist and March, 1983). In brief, F_e is expressed in terms of the excess free energy density $f_l(n)$ of a fluid as

$$F_e[n(\mathbf{r})] = \int d\mathbf{r}\, n(\mathbf{r}) f_l(\bar{n}(\mathbf{r})) \tag{6.17}$$

$\bar{n}(\mathbf{r})$ being a smoothed density profile. Smoothing of $n(\mathbf{r})$ is effected through a weighting function that depends self-consistently on $\bar{n}(\mathbf{r})$ and is determined so as to recover the HNC result for small deviations from homogeneity. All nonlinear effects of inhomogeneity are thus included approximately to infinite order.

6.6.3. Parametrization of the density profile

The Fourier series for the right-hand side of (6.16) converges rather slowly, thus requiring explicit calculation of a rather large number of order parameters $n_{\mathbf{G}}$ (of course, these still "drive" the higher ones through the same infinite set of equilibrium conditions). Haymet (1989) was the first to show by such a calculation that in the hard sphere system, excepting the first few order parameters, a harmonic formula is approximately correct. The implication is that, although the theory formally involves an infinite set of order parameters, the first few ones dominate and "enslave" the rest. Later workers have often assumed this relationship between the first-order parameter and all the others, by representing $n(\mathbf{r})$ as a superposition of Gaussians centred at lattice sites and with a width that is evaluated variationally. Further discussion of the evaluation of $n(\mathbf{r})$ has been given by Harrowell et al. (1989).

6.7. Applications

Table 6.1 (Rovere and Tosi, 1988) collects measured values for relevant parameters of atmospheric-pressure freezing in metals, classified according to their high-temperature crystal structure. These are the melting temperature T_m, the relative volume change $\Delta V/V$, and the entropy change ΔS; the value of $\hat{c}_1(k)$ at the main peak $k = k_p$ of the liquid structure factor $S(k)$, at the temperature T indicated in the next column, and the values of c_0^l and c_0^s from the compressibility of the liquid and of the solid at T_m.

The Lasocka (1975) and Tallon (1982) relationship between ΔS and $\Delta V/V$ is fairly well satisfied by all these systems. One notices from the next two columns the approximate constancy of $\hat{c}_1(k_p)$ at freezing: thus the empirical criterion of freezing that Hansen and Verlet (1970, see also Verlet, 1968) deduced from the study of the hard sphere system ($S(k_p) \simeq 2.85$, or $\hat{c}_1(k_p) \simeq 0.65$ at freezing) is well satisfied by real metals (compare Section 6.4). Comparison of the last two columns illustrates the magnitude of nonlinear effects due to the change of compressibility across melting.

The simplest microscopic theory that one can give for these data uses the HNC closure in the incompressible-liquid limit ($\Delta V/V \to 0$ and $c_0^l \to -\infty$ with $c_0^l\,\Delta V/V$ finite) with one microscopic order parameter corresponding to all the stars of reciprocal lattice vectors underlying the main peak of $S(k)$ (two stars for freezing into fcc). A complete solution of the equations for freezing can be given without invoking models for liquid structure and yields

$$\hat{c}_l(k_p) \approx 0.6 \div 0.7 \quad \text{and} \quad -c_0^l \frac{\Delta V}{V} \approx 1 \div 2, \tag{6.18}$$

differences between the three lattice structures being not significant.

Table 6.1. *Measured parameters of the liquid-solid transition for metals at atmospheric pressure.*

		T_m (C)	$\Delta V/V$ (%)	ΔS (Nk_B)	$\hat{c}_l(k_p)$	$(T - T_m)/T_m$ (%)	$-c_0^l \frac{\Delta V}{V}$	$-c_0^s \frac{\Delta V}{V}$
fcc:	Cu	1083	4.2	1.15	0.61	6.2	1.9	2.5
	Ag	962	3.8	1.16	0.60	4.0	2.0	2.7
	Au	1064	5.1	1.13	0.61	8.0	2.6	5.9
	Al	660	6.5	1.39	0.60	1.4	3.8	—
	Pb	328	3.6	0.96	0.60	3.8	3.9	4.7
	Ni	1453	5.4	1.23	0.59	3.2	1.9	—
	Co	1495	3.5	1.10	0.59	3.7	1.3	—
	Pd	1554	5.9	1.13	0.62	1.7	—	—
	Pt	1772	6.6	1.26	0.63	0.45	—	—
hcp:	Mg	649	4.1	1.17	0.61	4.8	1.4	—
	Zn	420	4.3	1.16	0.60	7.2	2.9	3.8
	Cd	321	3.8	1.24	0.60	9.1	3.2	4.3
	Sc	1539	—	1.21	0.56	1.4	—	—
bcc:	Li	180	1.65	0.80	0.63	5.3	0.5	—
	Na	98	2.5	0.85	0.63	7.4	1.06	1.14
	K	63	2.5	0.83	0.63	2.8	1.08	1.16
	Rb	39	2.5	0.84	0.63	2.8	1.1	—
	Cs	28	2.6	0.84	0.64	5.6	1.1	—
	Ca	839	4.7	0.93	0.62	1.3	1.3	—
	Sr	769	—	1.06	0.62	1.4	—	—
	Ba	725	(4.0)	0.93	0.62	0.7	(1.1)	—
	Tl	304	2.2	0.87	0.60	3.8	2.2	—
	Cr	1857	(4.3)	0.80	0.60	2.3	—	—
	Mn	1244	1.7	1.16	0.60	1.3	—	—
	Fe	1535	3.5	0.92	0.58	1.6	1.5	—
	Zr	1852	(5.3)	1.16	0.58	2.6	—	—
	La	921	0.6	0.62	0.53	5.3	—	—
	Pr	931	0.02	0.68	0.54	2.0	—	—
	Eu	822	4.8	1.02	0.55	6.8	—	—
	Gd	1313	2.0	0.76	0.54	6.4	—	—
	Er	1529	—	1.33	0.58	0.1	—	—

6.8. Correlation of melting temperature with vacancy formation energy

Following the preceding discussion of freezing, based on the statistical mechanics of inhomogeneous systems (and, in particular, on the bulk liquid direct correlation function in relation to the periodic crystal), this section will be concerned with another property, the vacancy formation energy E_v, which can also be characterized, in the hot crystal, by liquid structure. The theory, which originates from the **k**-space formulation of Faber (1972), is most conveniently developed here in the **r**-space formulation of Minchin et al. (1974). The following argument rests on two basic assumptions:

1. Pair potentials characteristic of the bulk liquid provide a useful description of the force field. Although this may be so for monovalent metals, the work of Gillan (1989) casts doubt on its quantitative validity for calculating the vacancy formation energy in the polyvalent metal Al.
2. Ionic relaxation around the vacant site in the crystal can be neglected. Although this seems to be a useful starting point for close-packed metals, it is inappropriate for, for example, the body-centred cubic alkalis (see Flores and March, 1981; Rashid and March, 1989).

6.8.1. *Vacancy formation energy in terms of liquid structure at melting point*

Pair potential calculations of the vacancy formation energy were mentioned earlier. Next, specific consideration will be given to the work of Faber (1972), as posed in **r** space by Minchin et al. (1974).

Consider a metal of volume $\mathscr{V}$ containing N ions. Then, with the ions at positions $\{\mathbf{r}_i\}$, one first introduces the structural description

$$a(q) = N^{-1} \sum_{i,j} \exp(i\mathbf{q}\cdot\{\mathbf{r}_i - \mathbf{r}_j\}). \tag{6.19}$$

Now suppose the vacancy is formed in two stages. First, there is a uniform expansion, all linear dimensions being increased by a factor $1 + \varepsilon$, where $\varepsilon \ll 1$. The new total volume is chosen to be the same as that of the final defective system so that, if no dilatation is permitted, $3N\varepsilon = 1$. Use of (6.19) shows that the new structure factor, to order ε, is

$$a(\mathbf{q}(1 + \varepsilon)) = a(\mathbf{q}) + \varepsilon\mathbf{q}\cdot\nabla a(\mathbf{q}). \tag{6.20}$$

Then, from this new configuration, a vacancy is produced by rearranging the ions at constant volume so that in the body of the metal, the mean ionic density has its unperturbed value. Then, Faber writes the structure factor as

$$a(\mathbf{q}) + N^{-1}[1 - a(\mathbf{q})]. \tag{6.21}$$

The simplifying feature of the present procedure is that, in the first step, the energy change should be quadratic in the strain ε and, therefore, entirely negligible. Thus the energy change one requires all comes from the second step, where the volume-independent part of the energy remains constant. Therefore, in second-order perturbation theory, only pairwise forces contribute; from the difference of (6.20) and (6.21), the result one requires is

$$E_v = \tfrac{1}{2}(N/\mathscr{V}) \sum_{\mathbf{q}} N^{-1}\left[1 - a(\mathbf{q}) - \frac{1}{3}\mathbf{q}\cdot\nabla a(\mathbf{q})\right] w(\mathbf{q}) \tag{6.22}$$

where $w(\mathbf{q})$ is the Fourier-transformed interatomic potential. Integration by parts yields

$$E_v = (1/6\mathscr{V}) \sum_{\mathbf{q}} (a(\mathbf{q}) - 1)\mathbf{q}\cdot\nabla w(\mathbf{q}), \tag{6.23}$$

which is the basic formula.

Minchin et al. (1974) give the direct-space formulation that is adopted below. Their result for E_v at arbitrary pressure is

$$E_v + \frac{p}{\rho} = -\frac{\rho}{2}\int\left(w + \frac{r}{3}\frac{\partial w}{\partial r}\right) g\, d\mathbf{r} \tag{6.24}$$

where $w(r)$ is the $\mathbf{r}$-space transform of $w(q)$. To see that it is equivalent to (6.22), Minchin et al. invoke the general identity

$$\rho \int f(r)g(r)\, d\mathbf{r} = \frac{1}{(2\pi)^3}\int f(q)[a(q) - 1]\, d\mathbf{q} \tag{6.25}$$

where f is arbitrary. If one takes

$$f(q) = \frac{qw'(q)}{6}, \tag{6.26}$$

then it follows after a short calculation that

$$f(r) = -\frac{1}{2}\left[w(r) + \frac{r}{3}w'(r)\right]. \tag{6.27}$$

The equivalence of the **r**- and **q**-space forms then follows after a brief calculation.

Even though Faber's work was done with the vacancy formation energy E_v in simple metals specifically in mind, he has expressed E_v for a pair force model and in the absence of relaxation, in terms of the pair potential, assumed to possess a Fourier transform as in the pseudopotential description of simple metals, and the structure factor. Bhatia and March (1984a) have used the alternative **r**-space description of Minchin et al. (1974), their formula for E_v being

$$E_v + p\Omega = -\frac{\rho}{2}\int g\phi\,d\mathbf{r} - \frac{\rho}{6}\int r\frac{\partial\phi}{\partial r}g\,d\mathbf{r} \qquad (6.28)$$

where the atomic number density ρ is the reciprocal of the atomic volume Ω.

Because of known empirical correlations between E_v and T_m, the melting temperature, and with $B\Omega$, bulk modulus times volume, Bhatia and March focused on the approximate evaluation of the result (6.28) in the hot solid near T_m. Of course, for any crystal, the determination of the pair function g in this case is a major problem necessitating a full treatment of large effects due to anharmonicity. To make progress in understanding further the status of the preceding empirical correlations, Bhatia and March considered specifically the case of condensed phases of rare gases, where there are known to be no major changes of local coordination on melting, in sharp contrast to, say, tetrahedrally bonded covalent solids like Ge and Si, which melt into a dense metallic phase. These workers argue therefore that one can usefully evaluate E_v in the hot solid by inserting the appropriate liquid structure, described by the pair function $g(r)$ at T_m.

Making this assumption, one can approximately evaluate the right-hand side of (6.28) when one inserts $g(r)$ at T_m. Invoking the virial equation for the pressure p, namely,

$$p = \rho k_B T - \frac{\rho^2}{6}\int r\frac{\partial\phi}{\partial r}g(r)\,d\mathbf{r} \qquad (6.29)$$

with pair potential ϕ, one notes that putting $p = 0$ yields the second term of (6.28) as $-k_B T_m$ and, since E_v is known to be much larger than this quantity, it is apparent that the term involving $\partial\phi/\partial r$ can be neglected.

In Appendix 6.1, by invoking a structural theory proposed by Woodhead-Galloway et al. (1968; compare Appendix 3.2) for liquid argon, the ratio $E_v/k_B T_m$ is written solely in terms of the direct correlation function $c(r)$ at the melting temperature T_m (see Chapter 13 for a generalization of this argument to binary liquid metal alloys). As shown by Bernasconi et al. (1986), a refined liquid structure theory is required, however, for metals.

In view of the result mentioned above, namely, that pair potential theory is not adequate to relate E_v and $k_B T_m$ in a polyvalent metal like Al (see Gillan, 1989), this topic will be concluded by relating this ratio $E_v/k_B T_m$ to departures from Joule's law in the liquid. To avoid invoking pair potentials at the outset, the writer has used an intuitive argument, which will be given next. Then, for pair forces, it will be shown how the result relates closely to the above discussion.

Relation of $E_v/k_B T_m$ to Departures from Joule's Law

In the hot solid, the vacancy energy can be obtained by taking out an atom and placing it on the surface (compare Appendix 6.2 for simple metals). This leaves a "localized hole," where the atom has been removed. Now consider that, at melting, the localized hole no longer exists, and therefore compare the change in internal energy E of the liquid at volume $\mathscr{V}$ and at volume $\mathscr{V} + \Omega$, with Ω the atomic volume. Again one is assuming no ionic relaxation, and therefore the argument is still appropriate only to close-packed solids. Thus one equates

$$E_v = E(\mathscr{V} + \Omega) - E(\mathscr{V}) \tag{6.30}$$

and, since Ω is $0(1/N)$ of $\mathscr{V}$, one can Taylor-expand to find

$$E_v \simeq \Omega \frac{\partial E}{\partial \mathscr{V}}. \tag{6.31}$$

One can now calculate the "departure from Joule's law," $(\partial E/\partial \mathscr{V})$, purely thermodynamically; after a little manipulation, one finds (March, 1987b)

$$\frac{E_v}{k_B T_m} = \left\{ \frac{(\gamma - 1)(c_v/k_B)}{S(0)} \right\}^{1/2} \tag{6.32}$$

where $\gamma = c_p/c_v$ is the ratio of the specific heats in the liquid metal at melting, while $S(0)$ is, as usual, the long wavelength limit of the structure factor at T_m, $S(k)_{T_m}$. Rashid and March (1989; see also Alonso and March, 1989) have considered (6.32) in relation to a number of liquid metals. It is not fully quantitative, as evidenced by the fact that the average value of $E_v/k_B T_m$ from experiment for the metals they consider is about 11, whereas (6.32) predicts a number of about 8 for close-packed metals.

Two other points should be made here. Equation (6.32) is quite inappropriate for the body-centred cubic metals, in agreement with the conclusions of Flores and March (1981) on Na and K, where ionic relaxation is a major component in calculating E_v. Secondly, if one attempts to relate (6.32) to the preceding theory based on the assumption of pair potentials, then a term has to be added (for a simple case like liquid argon) that involves the density derivative of the pair correlation function $g(r)$. For Ar and Kr, Rashid and March (1989) calculate this term explicitly; it is quite negligible compared with the contribution of (6.32), though this is no longer true for Ne. It remains to be seen what will be the main corrections to be added to (6.32) for close-packed metals. However, this formula already shows that in such metals, $E_v/k_B T_m$ is a large number, ~ 10, in general agreement with the empirical correlations that have been evident for a long time.

6.9. Supercooling of liquid Rb

The supercooling of liquid Rb has been studied by Mountain (1982) using computer simulation, with an effective interionic potential due to Price et al. (1976) of the kind discussed in Chapters 3 and 4. In particular, the energy-temperature relationship is presented for one density for liquid, amorphous solid, and body-centred cubic (bcc) crystal phases, together with the pair function $g(r)$ in the liquid phase as a function of temperature. Because of inherent difficulties of performing experiments on simple supercooled liquids, molecular dynamics is an attractive way to explore this subject, since both structure and dynamics (see Chapter 8) can be explored in detail.

Molecular Dynamics Calculations

The classical equations of motion for 432 interacting Rb particles were solved using the effective interaction of Price et al. (already referred to) for a lattice parameter of 5.739 Å. This leads to a reduced density $n^* = n\sigma^3 = 0.95$, where n is the number density and $\sigma = 4.48$ Å is the smallest distance at which the pair potential $\phi(r)$ is zero. The well depth is $\varepsilon = 393k_B$. The potential was truncated at 2.03σ. The units of length, energy, and time were chosen to be σ, ε, and

$$\tau = \left(\frac{m\sigma}{\varepsilon}\right)^{1/2} = 2.29 \times 10^{-12}\ \text{s}, \tag{6.33}$$

m denoting the mass of an Rb ion.

Figure 6.1. Energy temperature curve as given by Mountain (1982). Upper branch represents fluid states to right of break; amorphous solid states are to the left. Lower branch corresponds to body-centred cubic crystalline states.

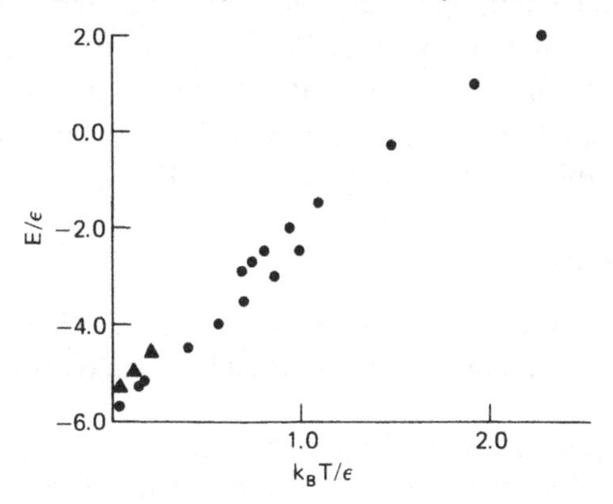

The explicit thermodynamic states considered by Mountain (1982) are indicated by the energy-temperature curve in Figure 6.1. The upper branch of the curve represents fluid states to the right of the break and amorphous solid states to the left. The lower branch corresponds to bcc crystalline states. For this density, the fluid becomes thermodynamically metastable below $T^* = 0.81$.

The empirical criterion (see Raveche et al., 1974; Abraham, 1980) that the ratio denoted by R, of the amplitude of $g(r)$ at the first minimum to that at the first maximum equals 0.20 ± 0.01 at freezing places the transition in the interval $0.80 < T^* < 0.95$. The ratio R as a function of temperature is shown in Figure 6.2. The break in the upper branch of Figure 6.1 corresponds to a region where the system is unstable against nucleation into the crystalline phase. That is, if a fluid is quenched into the energy-temperature

Figure 6.2. The ratio, denoted by R, of the amplitude of the pair distribution function $g(r)$ at the first minimum to that at the first maximum as a function of temperature for Rb (after Mountain, 1982).

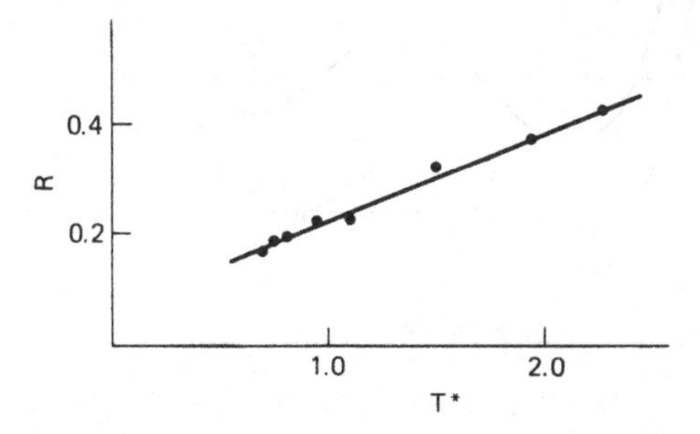

regime or if an amorphous solid is heated into that region, the system will spontaneously go over to the bcc phase (see Mountain (1982) for details).

The crystalline branch terminates at a temperature where the solid becomes unstable against thermal motion. This is not the melting point, as there is no free surface in the system, but instead is the limit to the metastability of the superheated crystalline state. For the density of the computer experiment, this process occurs at $T^* = 1.07$.

Pair Correlation Function $g(r)$

The variation with temperature of the liquid pair correlation function $g(r)$ is shown in Figure 6.3. This plot is for the normal liquid and also into the

Figure 6.3. Variation with temperature of the pair distribution function $g(r)$ as calculated for liquid Rb (after Mountain, 1982). Reduced temperatures are (a) 2.3, (b) 1.5, (c) 1.1, (d) 0.95, (e) 0.81, (f) 0.75 and (g) 0.69.

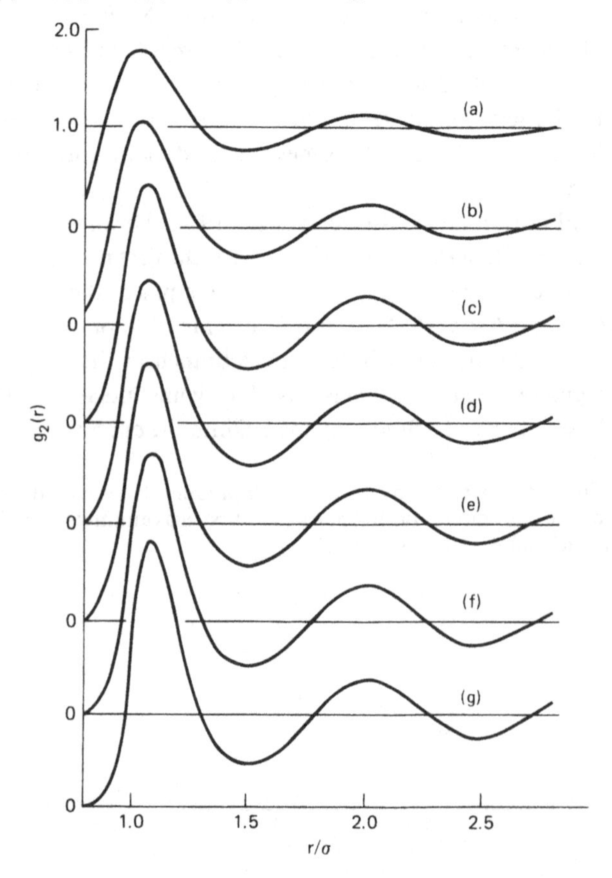

supercooled liquid region. There are two trends with temperature that Mountain stresses. The first is the shift in position and narrowing of the first peak, which occurs as the temperature T is reduced. The peak moves to larger values of r and becomes noticeably higher and sharper. The sharpening of the peak continues down to the instability region and is reflected in the temperature variation of the ratio R shown in Figure 6.2. The shift in the position of the first peak stops when $T^* = 0.81$ is reached.

The other feature is the evolution of the second peak in $g(r)$, which is characteristic of the changes that occur with decreasing temperature for $r/\sigma > 1.5$. The amplitude of this peak increases with decreasing T, whereas the r values where $g(r) = 1$ do not vary.

To summarize these findings, it is to be noted that the pair distribution function $g(r)$ represents the average environment of an atom. As Mountain emphasizes, in the supercooled region the near-neighbour environment becomes more sharply defined in space but does not shift its position as it does in the equilibrium liquid. That is to say, the local environment ceases to evolve, in the sense of a peak shift, once the freezing temperature is reached, and in the supercooled region this fully developed liquid structure can only become more sharply defined until crystallization intervenes.

Transport properties of supercooled liquid Rb will be considered in Chapter 8.

7
Electronic and atomic transport

In this chapter, an introduction to the theory of electronic and atomic transport in liquid metals is given. The theory is deepened and refined in subsequent chapters at the inevitable price of greater complexity and more formal results.

One pillar in the theory is the Wiedemann-Franz law, relating electrical and thermal conductivities. This is a convenient starting point for the present discussion.

7.1. Wiedemann-Franz law and Lorenz number

Let us denote the electrical conductivity by σ and the electronic contribution to the thermal conductivity of the liquid metal by K_e. Then the Wiedemann-Franz law asserts the constancy of $K_e/\sigma T$. Theory relates this precisely to the so-called Lorenz number, through

$$\frac{K_e}{\sigma T} = \frac{\pi^2 k_B^2}{3e^2} = L. \tag{7.1}$$

In the units employed in Table 7.1, where experimental data on a variety of liquid metals have been collected, L is 2.45.

Expression (7.1) requires solution of the Boltzmann equation, which is given in various places (see Ziman, 1960; or Faber, 1972). In the usual arguments, it is stressed that result (7.1) does not depend on specific assumptions about Fermi surfaces, density of electron states, nature of scattering potential, and the like. It does, however, involve the general assumption that the scattering is elastic.

It can be seen that while there are some 12 cases in Table 7.1 in which the Lorenz number lies in the (somewhat arbitrarily) chosen range 2.1 to 2.8, the cases of Zn and Tl lie outside these limits. To the writer's knowledge, no solution has been proposed to date for the spread of values

Table 7.1. ($K\rho/T$) *at melting point from experiment.*

Li	2.6	Al	2.4
Na	2.2	Ga	2.07
K	2.1	Tl	3.2
Cs	2.4	Sn	2.9
Cd	2.5	Pb	2.4
Hg	2.75	Sb	2.6
Zn	3.2	Bi	2.5

around the Lorenz number (see also Appendix 9.5 for further theory relating to the Wiedemann-Franz law).

Following this brief introduction to an outstanding question that remains unanswered in this area, let us turn to discuss the theory of electronic conductivity in liquid metals. It will be a useful starting point to find the scattering cross section for plane waves representing (approximately, of course) the conduction electron wave functions off a single ion of arbitrary strength. The ion can safely be assumed to be represented by a spherical potential of finite range; within an independent electron model, this problem can be solved exactly, as is discussed next.

7.2. Exact resistivity formula for finite-range spherical potential

Huang (1948) derived an exact expression in terms of phase shifts $\eta_l \equiv \eta_l(k_f)$ for the excess resistivity of a dilute metallic alloy in which independent free electrons are scattered by a spherical potential $V(r)$. Apart from constants, which are omitted here, the nub of the Huang formula is the sum

$$S = \sum_{l=0}^{\infty} [(2l+1)\sin^2\eta_l - 2l\sin\eta_l\sin\eta_{l-1}\cos(\eta_{l-1}-\eta_l)], \tag{7.2}$$

which is readily shown to be equivalent to

$$S = \sum_{l=1}^{\infty} l\sin^2(\eta_{l-1}-\eta_l). \tag{7.3}$$

Following March (1975), a result of Gerjuoy (1965) that was rediscovered by Gaspari and Gyorffy (1972) can now be inserted into (7.3). These workers show that the radial wave functions R_l for such scattering from a potential $V(r)$ of finite range but arbitrary strength satisfy the exact relation

$$\int_0^\infty \mathrm{d}r\, r^2 R_{l-1}(r)\frac{\partial V(r)}{\partial r} R_l(r) = \sin(\eta_l - \eta_{l-1}) \tag{7.4}$$

where, outside the range of V, R_l has the form

$$R_l(r) = j_l \cos\eta_l - n_l \sin\eta_l, \tag{7.5}$$

j_l and n_l being spherical Bessel and Neumann functions, respectively. Thus, from (7.3) and (7.4)

$$S = \sum_{l=1}^{\infty} l \int_0^\infty \mathrm{d}r_1\, r_1^2 R_{l-1}(r_1)\frac{\partial V(r_1)}{\partial r_1} R_l(r_1)$$
$$\times \int_0^\infty \mathrm{d}r_2\, r_2^2 R_{l-1}(r_2)\frac{\partial V(r_2)}{\partial r_2} R_l(r_2). \tag{7.6}$$

At this point, it is to be noted that an inverse transport theory proposed by Rousseau, Stoddart, and March (RSM) (1972; see also Appendix 7.1) has as its nub the force-force correlation function, F say, again stripped of unimportant multiplying factors:

$$F = \int \mathrm{d}\mathbf{r}_1\, \mathrm{d}\mathbf{r}_2 \frac{\partial V(\mathbf{r}_1)}{\partial \mathbf{r}_1}\cdot\frac{\partial V(\mathbf{r}_2)}{\partial \mathbf{r}_2}|\sigma(\mathbf{r}_1\mathbf{r}_2)|^2, \tag{7.7}$$

where $\sigma(\mathbf{r}_1\mathbf{r}_2)$ is the energy derivative of the Dirac density matrix for potential V evaluated at the Fermi energy $E_\mathrm{f} = k_\mathrm{f}^2/2$. Using the result of March and Murray (1960) that

$$\sigma(\mathbf{r}_1\mathbf{r}_2) = \sum_l (2l+1)\sigma_l(r_1 r_2)P_l(\cos\gamma) \tag{7.8}$$

(which is illustrated in Appendix 7.1 for nearly free electrons also), where γ is the angle between $\mathbf{r}_1$ and $\mathbf{r}_2$, the integrand, f, say, in (7.7) can be written as

$$f = \frac{\partial V(r_1)}{\partial \mathbf{r}_1}\cdot\frac{\partial V(r_2)}{\partial \mathbf{r}_2}|\sigma(\mathbf{r}_1\mathbf{r}_2)|^2$$
$$= \sum_l \sum_j (2l+1)\sigma_l(r_1 r_2)P_l(\cos\gamma)$$
$$\times (2j+1)\sigma_j(r_1 r_2)P_j(\cos\gamma)\frac{\partial V}{\partial r_1}\frac{\partial V}{\partial r_2}\cos\gamma, \tag{7.9}$$

the last step in (7.9) following because the potential is spherical. According to (7.7), on multiplying by $\sin\gamma\, d\gamma$ and integrating over γ from 0 to π:

$$\int_0^\pi f \sin\gamma\, d\gamma = \sum_l \sum_j (2l+1)\sigma(r_1 r_2)(2j+1)\sigma_j(r_1 r_2) \times \frac{\partial V(r_1)}{\partial r_1}\frac{\partial V(r_2)}{\partial r_2}\int_0^\pi \cos\gamma P_l(\cos\gamma)P_j(\cos\gamma)\sin\gamma\, d\gamma. \quad (7.10)$$

The integral in (7.10) is related to Clebsch-Gordon coefficients, but the answer can be written in elementary fashion by using the identity

$$(2l+1)\cos\gamma P_l = (l+1)P_{l+1} + lP_{l-1}. \quad (7.11)$$

On substituting (7.11) into (7.10) and performing the integration, a short calculation gives

$$\int_0^\pi f \sin\gamma\, d\gamma = 4\sum_{l=1}^{\infty} l\sigma_{l-1}(r_1 r_2)\sigma_l(r_1 r_2)\frac{\partial V(r_1)}{\partial r_1}\frac{\partial V(r_2)}{\partial r_2}. \quad (7.12)$$

But on writing (see March and Murray, 1960; Rousseau, 1971; Harris, 1972)

$$\sigma_l(r_1 r_2) \propto R_l(r_1)R_l(r_2), \quad (7.13)$$

it is readily seen from (7.13), (7.12) and (7.7) that, apart from constants,[†] the Huang (1948) formula, rewritten in the form (7.6), has exactly the same structure as the RSM formula. They are both force-force correlation functions of the type (7.7), as shown by March (1975; 1976).

7.2.1. *Extreme strong scattering limit*

This discussion, as already mentioned, is equivalent to "impurity" resistivity. It is the limit of "incoherent" scattering in which the liquid structure does not appear. It is to be contrasted with the weak scattering theory (Ziman, 1961; see Section 7.3) where the liquid structure factor $S(k)$ is a crucial ingredient.

At least under normal conditions, this incoherent theory is not good enough for $3d$ transition metals (e.g., liquid Fe); the scattering of conduction electrons of one Fe centre is not sufficiently strong to validate the approximation. It is almost good enough for rare earth metals, however.

[†] Inclusion of constants is elementary but tedious; the RSM formula is then found to be exactly equivalent to that of Huang for a spherical scatterer.

It will therefore be useful to start from this limit. Then one can construct, essentially from the same starting point, a nearly free electron theory appropriate to *sp* metals (alkalis, etc.), in which liquid structure $S(k)$ enters in a major manner.

7.2.2. *Incoherent scattering limit applied to liquid rare earth metals*

This section summarizes the work of Parrinello, March, and Tosi (PMT) (1977). Band structure calculations (Hill and Kmetko, 1975) indicate that a substantial amount of valence charge density in the rare earths arises from states of *d* character. The zeroth-order treatment corresponds to neglect of all η_l except for $l = 2$. This yields, with valence Z,

$$\rho_0 = \frac{20\pi\hbar}{Ze^2k_f}\sin^2\left(\frac{\pi Z}{10}\right). \tag{7.14}$$

A value of $Z = 2$ yields $\rho_0 \sim 400\mu\Omega$ cm. Including f and s phase shifts gives the curves (broken curve corresponds to $\eta_0 = 0$) shown in Figure 7.1.

The work of Beck and his colleagues extends somewhat the preceding treatment of PMT; see, for example, Delley and Beck (1979).

Figure 7.1. Resistivities of liquid rare earth metals; from the theory of Parrinello et al. (1977). Note that Eu and Yb are divalent.

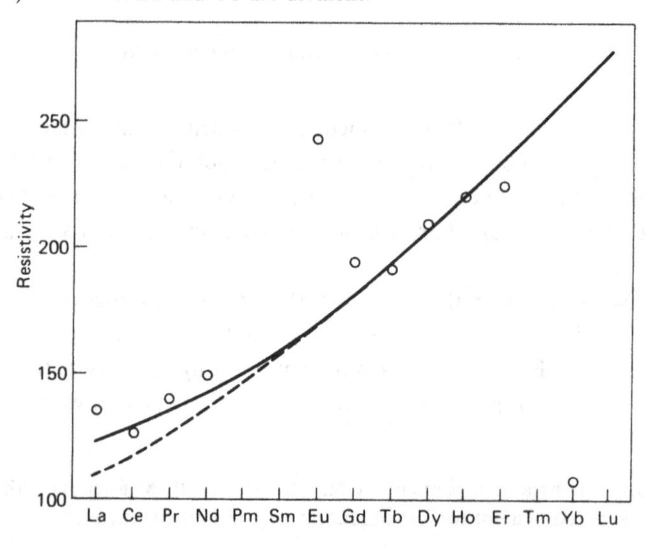

Let us turn next to the inverse transport theory, exemplified by (7.7), in the limit of weak scattering. In this way, one regains a formula that has its origins in the work of Bhatia and Krishnan (1945, 1948) and was brought to full fruition by Ziman (1961) using pseudopotentials (see Chapter 3).

7.3. Weak scattering theory of electrical resistivity

The idea behind weak scattering theory is to represent the total potential energy $V(\mathbf{r})$ scattering the conduction electrons by a sum of screened potentials $v(r)$ at the ionic sites $\mathbf{R}_i$, where one has taken a "snapshot" of the ions at a particular time:

$$V(\mathbf{r}) = \sum_i v(|\mathbf{r} - \mathbf{R}_i|). \tag{7.15}$$

Since the inverse transport theory expression for F in (7.7) already contains the potential energy V to second order explicitly, one can evaluate this force-force correlation function to second order in V by replacing the energy derivative σ of the density matrix by its free electron value. The density matrix ρ for free electrons is readily calculated for plane waves $\mathscr{V}^{-1/2}\exp(i\mathbf{k}\cdot\mathbf{r})$ normalized in a volume $\mathscr{V}$ as

$$\rho_0(\mathbf{r}_1\mathbf{r}_2) = \sum_{|\mathbf{k}|<k_f} \mathscr{V}^{-1}\exp(-i\mathbf{k}\cdot\mathbf{r}_1 - \mathbf{r}_2). \tag{7.16}$$

Replacing the summation of $\mathbf{k}$ by an integration with the usual constant density of states in $\mathbf{k}$ space as $(8\pi^3)^{-1}\mathscr{V}$, one finds

$$\rho_0(\mathbf{r}_1\mathbf{r}_2) = \frac{k_f^2}{\pi^2}\frac{j_1(k_f|\mathbf{r}_1 - \mathbf{r}_2|)}{|\mathbf{r}_1 - \mathbf{r}_2|}. \tag{7.17}$$

The energy derivative follows by using $k_f = (2E)^{1/2}$ and then forming $(\partial\rho_0/\partial E)_{E=E_f=k_f^2/2}$, say $\sigma_0(\mathbf{r}_1\mathbf{r}_2E_f)$. The resistivity ρ is then found by using this result for σ_0 in (7.7); clearly, as only pairs of sites $\mathbf{R}_i$ are now correlated, taking the liquid average one obtains a result in terms of the structure factor $S(k)$ and the Fourier transform of the localized potential, $\tilde{v}(k)$, say, in (7.15). The result, when one puts back all the numerical factors (see Section 14.14), is for weak scattering with a sharp Fermi surface of diameter $2k_f$:

$$\rho = \frac{3\pi}{\hbar e^2 v_f^2 \rho_i}\frac{1}{(2k_f)^4}\int_0^{2k_f} S(k)|\tilde{v}(k)|^2 4k^3\,dk. \tag{7.18}$$

This is the basic formula for the electrical resistivity of simple (s-p) nearly free electron metals such as Na and K. In (7.18), ρ_i is the ionic number density, and since $S(k)$ is measurable by diffraction experiments, the only quantity needed to determine ρ is the Fourier transform of the localized atomic-like screened potential energy $\tilde{v}(k)$. Some discussion of the way approximations may be set up for this quantity has already been given in Chapter 3. It is also relevant to note here that real liquid metals have blurred Fermi surfaces, in accord with the Heisenberg uncertainty principle

$$l\,\Delta k_{\mathrm{f}} \sim 1, \tag{7.19}$$

where Δk_{f} is the blurring of k_{f} and l is the electronic mean free path.

7.3.1. *Self-consistent treatment of Fermi surface blurring*

The weak scattering formula (7.18) is not self-consistent in the sense that on the right-hand side the integration is out to the Fermi sphere diameter $2k_{\mathrm{f}}$, which clearly implies a perfectly sharp Fermi surface, whereas on the left-hand side, there is a finite electrical resistivity; this, in turn, implies a finite mean free path through the elementary formula

$$\rho = \frac{\hbar k_{\mathrm{f}}}{ne^2 l}. \tag{7.20}$$

Ferraz and March (1979) make the assumption that the free electron density matrix $\sigma_0(|\mathbf{r}_1 - \mathbf{r}_2|)$ can be modified to take account of Fermi surface blurring by writing

$$\sigma(\mathbf{r}_1\mathbf{r}_2) = \sigma_0(|\mathbf{r}_1 - \mathbf{r}_2|)\exp\left(\frac{-|\mathbf{r}_1 - \mathbf{r}_2|}{2l}\right), \tag{7.21}$$

the arguments leading to a damping of the off-diagonal matrix as in (7.21) being quite analogous to those used to derive the probability of a mean free path of a given length l in classical kinetic theory. These arguments were first applied quantum mechanically by Bardeen (1956), who did not, however, derive a liquid metal transport theory. The result of Ferraz and March obtained by inserting (7.21) and (7.17) into (7.7) is given by

$$\rho = \frac{\hbar k_{\mathrm{f}}}{ne^2 l} = \int_0^\infty k^4 S(k)|\tilde{v}(k)|^2 \Gamma(k, k_{\mathrm{f}}, l)\,\mathrm{d}k \tag{7.22}$$

where

$$\Gamma(k, k_f, l) = \int \exp(i\mathbf{k}\cdot\mathbf{R})|\sigma_0(R)|^2 \exp(-R/l)\,d\mathbf{R}. \tag{7.23}$$

Figure 7.2 shows results thus obtained for Γ for Hg and for Li; the solid curve leads back to the weak scattering theory.

As noted by McCaskill and March (1982), the work of Götze (1979; 1981) for random systems constitutes an approach related to that of Ferraz and March. The detailed form of Γ is a little different in the two theories, but the essential features are the same. As shown by Chapman and March (1989) for liquid Cs along the coexistence curve, (7.22) and (7.18) give similar predictions within their region of validity.

Numerical results of this self-consistent approach to the mean free path, for liquid transition metals, have also been given by Brown (1981). No modification of the density of states $N(E)$ due to scattering is introduced. A method to incorporate such a modified density of states has been proposed by van Oosten and Geertsma (see van der Lugt and Geertsma, 1987), namely,

$$\rho = \rho_{\text{Ziman}} \times g^2 : g = N(E_f)/N_0(E_f) \tag{7.24}$$

Figure 7.2. Results for Γ defined in (7.23) for mean free paths l appropriate to Hg and to Li.

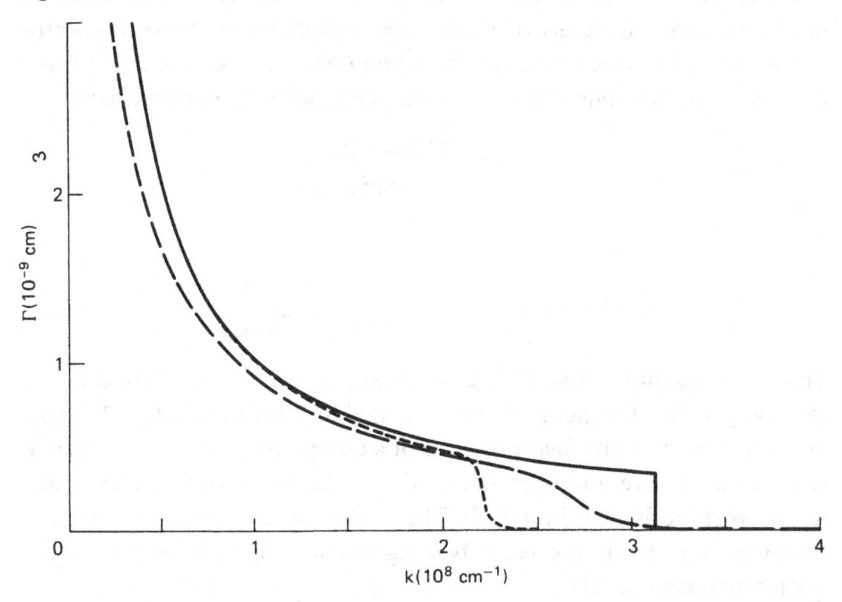

where $N_0(E)$ is the free electron density of states (the value of g is estimated to range from 1.13 in Li to 0.72 in Ba in metals they have studied so far).

7.3.2. Thermoelectric Power

In addition to the nearly free electron formula (7.18) for the electrical resistivity, it is worthwhile to consider briefly the results of the same theory for the thermoelectric power Q. The conventional representation of Q is through a dimensionless quantity ξ defined by

$$\xi = \frac{Q}{\dfrac{\pi^2 k_B^2 T}{3eE_f}}. \tag{7.25}$$

It can be shown (see, for example Ziman, 1960) that if the conductivity σ is known as a function of energy E in the region near to the Fermi energy E_f, then

$$\xi = E_f \left[\frac{\partial \ln \sigma(E)}{\partial E}\right]_{E=E_f}. \tag{7.26}$$

Using the theory developed in some detail above, it is a straightforward matter to find ξ in terms of the Fourier transform of the single-centre scattering potential, say $v(k)$, and the structure factor $S(k)$. The result takes the form (with the simplifying assumption of a local pseudopotential)

$$\xi \doteqdot 3 - \frac{2S(2k_f)|v(2k_f)|^2}{\langle Sv^2 \rangle} \tag{7.27}$$

where

$$\langle Sv^2 \rangle = 4 \int_0^1 S(k)|v(k)|^2 \left(\frac{k}{2k_f}\right)^3 \mathrm{d}\left(\frac{k}{2k_f}\right). \tag{7.28}$$

This latter quantity, from (7.18), essentially determines the liquid metal resistivity ρ; thus the thermopower is related to ρ, together with $v(2k_f)$ and the measured structure factor at $2k_f$. For a comparison between the theory summarized above and experiment, the reader is referred, for example, to the work of Sundström (1965). The agreement is quite reasonable for the simple s-p metals, especially bearing in mind the uncertainties in the scattering potential $v(k)$.

7.4. Hall coefficient

It is also of interest to note here that the Hall coefficient R_H in simple liquid metals is well described by the usual formula

$$R_H = \frac{1}{nec}. \tag{7.29}$$

Using this expression to interpret the measurements one finds that n comes out to agree well with the value obtained from the valence (e.g., four conduction electrons/atom for Sn). Although the result (7.29) is true for free electrons, it follows under less restrictive assumptions, as the following argument sets out (Szabo, 1972; March, 1989).

7.4.1. *Strong scattering regime*

Chapman and March (1989) have proposed a formula for the electrical resistivity R of liquid metals in the presence of strong electron-electron interactions. Their result is written in terms of the first-order density matrix describing the electronic structure, by generalizing the independent-electron formula (7.7).

The purpose of this section is to treat the Hall effect in the strong scattering regime (March, 1989). The following argument parallels that for R, in which the starting point was Huang's formula for impurity resistivity due to scattering from a finite-range spherical potential energy $V(r)$ as in Section 7.2. This is expressed in terms of the phase shifts η_1 of the partial waves. In the present treatment, use will be made of the single-centre scattering formula for the Hall coefficient R_H given by Szabo (1972) in terms of the same phase shifts. This has the form (Szabo's equation (11))

$$R_H = \frac{1}{nec}\left(1 + \frac{2}{3\pi n\Omega_0}\frac{\partial}{\partial\mu}\left\{\mu\sum_{l=0}^{\infty} 2(l+1)\sin^2(\eta_{l+1} - \eta_l)\right\}\right) \tag{7.30}$$

with n the electron density, Ω_0 the atomic volume, and μ the chemical potential. The sum appearing in (7.30) is simply related to that used in Section 7.2, which in turn was written in terms of the radial wave function R_l generated by $V(r)$. This was achieved using result (7.4) of Gerjuoy (1965).

Introducing the Dirac idempotent first-order density matrix

$$\rho(\mathbf{r}_1\mathbf{r}_2\mu) = \sum_i^{\mu} \psi_i^*(\mathbf{r}_1)\psi_i(\mathbf{r}_2) \tag{7.31}$$

where the ψ's are merely products of R_l and the spherical harmonics, one can show that the sum over l in (7.30) is simply related to the force-force correlation function $F(\mu)$, which in terms of V and ρ is given by (compare (7.7))

$$F(\mu) = \int \frac{\partial V}{\partial \mathbf{r}_1} \cdot \frac{\partial V}{\partial \mathbf{r}_2} \left| \frac{\partial \rho(\mathbf{r}_1\mathbf{r}_2\mu)}{\partial \mu} \right|^2 d\mathbf{r}_1\, d\mathbf{r}_2. \tag{7.32}$$

The burden of the remaining argument is to eliminate the one-body elements of (7.32) in favour of the density matrix, which has a ready many-electron generalization. This is done by Chapman and March (1989) by utilizing the equation of motion of the Dirac density matrix ρ, which in coordinate representation reads

$$\nabla^2_{\mathbf{r}_1}\rho - \nabla^2_{\mathbf{r}_2}\rho = \frac{2m}{\hbar^2}[V(\mathbf{r}_1) - V(\mathbf{r}_2)]\rho. \tag{7.33}$$

Dividing both sides by ρ and writing the difference in Laplacian operators as $\Delta L \equiv \nabla^2_{\mathbf{r}_1} - \nabla^2_{\mathbf{r}_2}$, one finds, with $\hbar^2/2m$ conveniently incorporated in the definition of L:

$$\text{grad}_1 V(\mathbf{r}_1) = \text{grad}_1 \left(\frac{\Delta L \rho}{\rho} \right). \tag{7.34}$$

Hence (7.32) becomes

$$F(\mu) = -\int \text{grad}_1 \left(\frac{\Delta L \rho}{\rho} \right) \cdot \text{grad}_2 \left(\frac{\Delta L \rho}{\rho} \right) \left| \frac{\partial \rho}{\partial \mu}(\mathbf{r}_1\mathbf{r}_2\mu) \right|^2 d\mathbf{r}_1\, d\mathbf{r}_2. \tag{7.35}$$

The final postulate in the present approach is that one can replace the one-body idempotent density matrix ρ in (7.35) by the many-electron first-order density matrix γ satisfying now the matrix inequality $\gamma^2 < \gamma$ to obtain $F_{\text{e-e}}$ in the presence of electron-electron interactions as

$$F_{\text{e-e}}(\mu) = -\int \text{grad}_1 \left(\frac{\Delta L \gamma}{\gamma} \right) \cdot \text{grad}_2 \left(\frac{\Delta L \gamma}{\gamma} \right) \left| \frac{\partial \gamma}{\partial \mu}(\mathbf{r}_1\mathbf{r}_2\mu) \right|^2 d\mathbf{r}_1\, d\mathbf{r}_2. \tag{7.36}$$

Since the sum in (7.30) is proportional to $F(\mu)$, it is now clear that the deviation ΔR_{H} of the Hall coefficient from $1/nec$ is characterized by the first-order density matrix γ.

This is then a formula for ΔR_{H} in liquid metals with strong electron-electron interactions, such as the expanded alkali metals discussed in

Chapters 10 and 11 in terms of the first-order density matrix. Given this formula, the following consequences arise:

1. Insertion of any translationally invariant $\gamma = \gamma(|\mathbf{r}_1 - \mathbf{r}_2|)$ whatsoever leads to $F_{\text{e-e}}(\mu) = 0$ and to the simple result $R_H = 1/nec$.
2. Assuming, as in the treatment of Chapman and March (1989) for R, that $\gamma(\mathbf{r}_1\mathbf{r}_2) \doteqdot f(\mathbf{r}_1 - \mathbf{r}_2)g(\mathbf{r}_1 + \mathbf{r}_2)$, one must expect ΔR_H to depend on the discontinuity, q, say, of the electronic momentum distribution of the liquid metal at the Fermi surface; see especially Chapman and March (1988) and also see Chapter 11.
3. Though further work is clearly required on the many-electron derivation of the formula for ΔR_H proposed here, this formula does motivate the suggestion that, in the presence of strong electron-electron interactions, ΔR_H may be closely correlated with measured conductivity and thermopower. However, quantitative numerical studies are required on this last point.

With this introduction to electronic transport, which will be deepened and extended in the discussion of Chapter 14, let us next consider the problem of atomic transport in liquid metals. Again the treatment of this chapter is introductory; a fuller understanding will come from the relation of hydrodynamic properties to neutron scattering and the dynamical structure factor $S(k, \omega)$ of a liquid metal (see Chapter 8).

7.5. Atomic transport: generalized Stokes-Einstein relation

When all the particles are identical, the random-walk formula $\langle r^2 \rangle = 6Dt$ for D, with $t \to \infty$, is equivalent to (see Appendix 8.6)

$$D = \left(\frac{1}{3N}\right) \int_0^\infty \mathrm{d}t \sum_j \langle v_j(t) \cdot v_j(0) \rangle. \tag{7.37}$$

If the displacements in a cell obey harmonic oscillator dynamics, then a normal mode transformation (Zwanzig, 1983) diagonalizes the force constant matrix, leading to the introduction of normal mode frequencies ω. The sum over coordinates in (7.37) can then be viewed as a sum over normal modes. Most of the normal modes may be treated as localized in some subvolumes, denoted as V^*; long wavelength modes cannot be localized in this way, but these make up only a small fraction of the entire spectrum. The time dependence of a single normal mode contribution in (7.37) varies as $\cos \omega t$, until a cell jump interrupts this motion (compare

Appendix A13.7). Normal modes in different subvolumes are interrupted at different times. This is accounted for (Zwanzig, 1983) by introducing a factor $\exp(-t/\tau)$, the waiting time distribution for cell jumps that destroy coherence of the oscillations in any V^*. The long wavelength modes are not substantially affected by cell jumps, but, as already stressed, these are but a small fraction of the entire spectrum. Then one obtains the following expression for the self-diffusion coefficient D:

$$D = \left(\frac{k_B T}{3MN}\right) \int_0^\infty \mathrm{d}t \sum_\omega \cos \omega t \exp\left(\frac{-t}{\tau}\right), \tag{7.38}$$

the sum being taken over all the $3N$ normal mode frequencies. The time integration is easily carried out to yield

$$D = \left(\frac{k_B T}{3MN}\right) \sum_\omega \frac{\tau}{1 + \omega^2 \tau^2}. \tag{7.39}$$

In the absence of much detailed information about the actual frequency distribution, Zwanzig now utilizes a Debye spectrum. Treating longitudinal and transverse oscillations separately, each with its own Debye cutoff q_0:

$$\omega_l(q) = v_l q, \qquad \omega_t = v_t q \tag{7.40}$$

for $0 < q < q_0$, v_l and v_t being longitudinal and transverse velocities of sound, respectively. The cutoff q_0 is chosen such that there are N modes in each branch of the spectrum, so that

$$\left(\frac{4\pi}{3}\right)\left(\frac{q_0}{2\pi}\right)^3 = \frac{N}{V} \tag{7.41}$$

where N/V is the number density of ions in the liquid metal. When the sum over frequencies in a single branch of the spectrum is replaced by an integral over q, the result is

$$\left(\frac{1}{N}\right) \sum_q \tau(1 + v^2 q^2 \tau^2) = \left(\frac{V}{N}\right)\left(\frac{1}{2\pi}\right)^3 (4\pi q_0/v^2\tau)\left[1 - \left(\frac{1}{q_0 v \tau}\right) \arctan(q_0 v \tau)\right]. \tag{7.42}$$

The preceding dynamical picture is physically meaningful only if the waiting time τ is much longer than a Debye period, or $q_0 v \tau \gg 1$. Then one can neglect the arctan term in (7.42). Introducing the mass density $d = MN/V$, one then arrives at the Zwanzig expression

$$D = \left(\frac{k_B T}{3\pi}\right)\left(\frac{3N}{4\pi V}\right)^{1/3}\left(\frac{1}{d v_l^2 \tau} + \frac{2}{d v_t^2 \tau}\right). \tag{7.43}$$

One next notes that in (7.43), dv_l^2 is an elastic bulk modulus, whereas dv_t^2 is a corresponding shear modulus. As Zwanzig notes, one expects that $dv_l^2\tau$ and $dv_t^2\tau$ are actually the longitudinal and shear viscosities η_l and η, respectively, which he interprets as follows. The viscosity coefficients are determined by time integrals of stress correlation functions. These correlation functions are short-ranged in space, so that one may focus on the local viscosity of the region V^*. As long as the system remains in the neighbourhood of a potential minimum, its motions are those of an elastic Debye continuum, and the local stress correlation function of the sub-volume V^* provides a time-independent local elastic modulus. But when the system leaves this cell, all correlations in V^* are lost, and the stress correlation function of this region vanishes. Thus the viscosity is the time integral of the constant elastic modulus multiplied by the exponential waiting time distribution, and the expected results follow. Then the self-diffusion coefficient becomes

$$D = \left(\frac{k_B T}{3\pi}\right)\left(\frac{3N}{4\pi V}\right)^{1/3}\left(\frac{1}{\eta_l} + \frac{2}{\eta}\right). \tag{7.44}$$

Equation (7.44) can then be rewritten in a form like the Stokes-Einstein relation (see Appendix 8.6 and Section 8.7) by defining the ionic volume $\Omega = V/N$. The quantity $\Omega^{1/3}$ is a length per particle and takes over the role of the molecular diameter σ. Then one can write

$$\left(\frac{D\eta}{k_B T}\right)\Omega^{1/3} = 0.0658\left(2 + \frac{\eta}{\eta_l}\right) = C'. \tag{7.45}$$

Although the actual value of C', as defined by this equation, clearly depends on the ratio of the shear viscosity to the longitudinal viscosity, which is still often not known quantitatively, Zwanzig (1983) notes that C' as defined can vary only between 0.13 and 0.18. Next (7.45) will be brought into direct contact with transport coefficients of liquid metals just above their melting points, following the study of Brown and March (1968; see also March, 1984).

7.6. Self-diffusion related to shear viscosity at the melting temperature

As with the argument of Zwanzig in Section 7.5, the Green-Kubo formulae for D and η were earlier used by Brown and March (1968). For liquid metals

Table 7.2. *Shear viscosities of liquid metals at freezing* (cp).

	Li	Na	K	Rb	Cs	Cu	Ag	Au	In	Sn
Experiment	0.60	0.69	0.54	0.67	0.69	4.1	3.9	5.4	1.9	2.1
Equation (7.47)	0.56	0.62	0.50	0.62	0.66	4.2	4.1	5.8	2.0	2.1

near T_m, these workers exploited the fact that the self-correlation function $S_s(k, \omega)$ and the dynamical structure factor (van Hove correlation function) $S(k, \omega)$, which enter the Green-Kubo formulae for diffusion and viscosity coefficients, have a rather well defined frequency range $0 < \omega < \omega_D$, the Debye frequency ω_D being analogous to that in a crystalline solid. Relating ω_D to the melting temperature T_m using Lindemann's law of melting, which has been shown in Chapter 6 to be consistent with the so-called Verlet rule (Section 6.4), Brown and March obtained the approximate relations for the atomic transport coefficients at T_m:

$$\frac{DM^{1/2}\rho^{1/3}}{T_m^{1/2}} = \text{constant} \tag{7.46}$$

and

$$\frac{\eta}{T_m^{1/2}M^{1/2}\rho^{2/3}} = \text{constant} \tag{7.47}$$

where ρ has been written for the atomic number density N/V. Forming the quantity $(D\eta/k_B T)/\rho^{1/3}$ appearing in (7.45), one immediately finds that this is a constant, in (approximate) accord with Zwanzig's formula.

The formulae (7.46) and (7.47) were compared directly with experiment in the work of Brown and March (1968). Provided the constant in (7.47) is chosen empirically, as in the early kinetic theory arguments of Andrade (1934), then η is reproduced quantitatively for liquid metals by this formula (see Table 7.2). Result (7.46) is less impressive, and when results of η/η_l eventually become available, it will be of obvious interest to test (7.45) for liquid metals at T_m in more detail than is possible at the time of writing. Transport properties will be taken up again in the following chapter as well as in Chapter 14.

8

Hydrodynamic limits of correlation functions and neutron scattering

This chapter is concerned with the generalization of the atomic transport coefficients discussed in Chapter 7 to relate to (1) hydrodynamic correlation functions and (2) more generally to neutron inelastic scattering. Before introducing a more general formalism for correlation functions (see also Chapter 14), let us begin by considering self-diffusion and its corresponding correlation function.

8.1. Self-motion in liquids and incoherent neutron scattering

Let us begin with a discussion of the motion of a single atom, which meanders in time in the liquid after starting out from the origin $\mathbf{r} = 0$ at time $t = 0$. Then the probability that the atom is at a later time t at position $\mathbf{r}$ is written $G_s(\mathbf{r}, t)$, the subscript s denoting self-motion (the following discussion is restricted to motion in classical liquids). In the hydrodynamic regime (actually large $\mathbf{r}$ and long time t), $G_s(\mathbf{r}, t)$ will satisfy the diffusion equation

$$\nabla^2 G_s(r, t) = \frac{1}{D}\frac{\partial G_s(r, t)}{\partial t}, \tag{8.1}$$

D being the self-diffusion coefficient. The Fourier transform on $\mathbf{r}(\mathbf{k})$ and on $t(\omega)$ of $G_s(r, t)$, namely, $S_s(k, \omega)$, is readily obtained from the solution of (8.1) satisfying the boundary condition already referred to, namely,

$$G_s(\mathbf{r}, t)_{t=0} = \delta(\mathbf{r}), \tag{8.2}$$

the solution being readily verified to be

$$G_s(\mathbf{r}, t) = \frac{1}{(4\pi D t)^{3/2}} \exp\left(\frac{-r^2}{4Dt}\right). \tag{8.3}$$

The double Fourier transform of (8.3) yields

$$S_s(k,\omega) = \frac{Dk^2}{\pi[\omega^2 + (Dk^2)^2]} \tag{8.4}$$

which, since (8.3) is valid for large $\mathbf{r}$ and t, applies for small $\mathbf{k}$ and ω. As will be mentioned again, $\hbar k$ and $\hbar\omega$ can be thought of as the momentum and energy transfer from a neutron to the liquid in incoherent neutron scattering experiments, S_s giving essentially the probability of such transfer.

8.2. Frequency spectrum (or spectral function) $g(\omega)$

The frequency spectrum $g(\omega)$ is now defined as

$$g(\omega) = \omega^2 \lim_{k\to 0} \frac{S_s(k,\omega)}{k^2}. \tag{8.5}$$

From (8.4) and (8.5), it is readily seen that one can obtain the self-diffusion coefficient D by either of the following procedures:

$$D = \pi \lim_{\omega\to 0} \omega^2 \lim_{k\to 0} \frac{S_s(k,\omega)}{k^2}, \tag{8.6}$$

which is a so-called Green-Kubo transport formula, or from

$$g(\omega = 0) = \frac{D}{\pi}. \tag{8.7}$$

The form of $g(\omega)$ for dense liquids is illustrated in Figure 8.1.

Figure 8.1. Schematic form of frequency spectrum (spectral function) $g(\omega)$ of a liquid metal. This quantity is defined in terms of the self-function $S_s(k,\omega)$, accessible by incoherent neutron scattering, through (8.5).

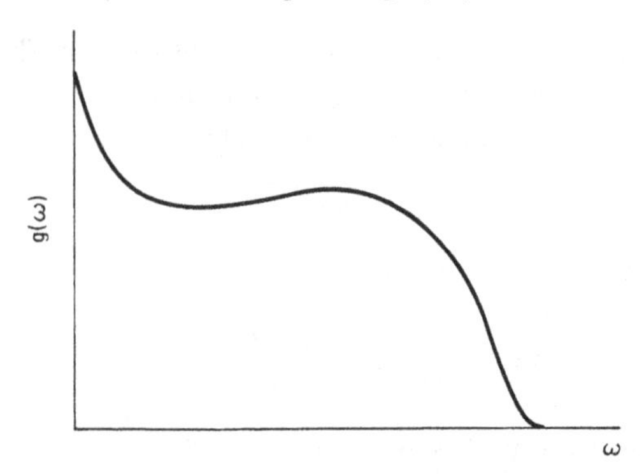

An equivalent form of D, going back essentially to Einstein, is obtained by considering the mean square displacement $\langle r^2 \rangle$ of an atom at long time t, when $\langle r^2 \rangle$ tends to $6Dt$. It is instructive to rewrite this result in terms of velocity $\mathbf{v}$ rather than displacement. One has evidently

$$\mathbf{r} = \int_0^t \mathbf{v}(s)\,ds \tag{8.8}$$

and hence (compare Chapter 7)

$$\langle r^2 \rangle = \int_0^t \int_0^t \langle \mathbf{v}(s_1) \cdot \mathbf{v}(s_2) \rangle \, ds_1 \, ds_2. \tag{8.9}$$

Here it has been assumed that one can interchange the order of the operations of time integration and ensemble averaging. Though this is usually permissible, pathological examples like hard spheres can arise in which this is not possible (Longuet-Higgins and Pople, 1956).

Since the velocity autocorrelation function cannot depend on the choice of the origin of time, (8.9) can be rewritten in terms of τ defined as $\tau = s_2 - s_1$. Then one finds, with change of variable and after some manipulation, that

$$\langle r^2 \rangle = 2t \int_0^t d\tau \langle \mathbf{v}(0) \cdot \mathbf{v}(\tau) \rangle \left(1 - \frac{\tau}{t}\right). \tag{8.10}$$

Provided the velocity autocorrelation function $\langle \mathbf{v}(0) \cdot \mathbf{v}(t) \rangle$ decays sufficiently rapidly at large t, then in the limit as t tends to infinity, one finds

$$D = \lim_{t \to \infty} \frac{r^2}{6t} = \frac{1}{3} \int_0^\infty dt \langle \mathbf{v}(0) \cdot \mathbf{v}(t) \rangle. \tag{8.11}$$

This is the desired alternative expression for D, already employed in Section 7.4.

From hydrodynamics, and in particular the Navier-Stokes equation (see Appendix 7.1), it can be shown (Ernst et al., 1970; 1971) that the velocity autocorrelation function falls off with a so-called long time tail (see Appendices 8.1 and 8.2) proportional to $t^{-3/2}$. In turn, the Fourier transform $g(\omega)$, as pointed out by Gaskell and March (1970), has therefore the small ω expansion.

$$g(\omega) = \frac{k_B T}{M\pi} \int_0^\infty \frac{\langle \mathbf{v}(0) \cdot \mathbf{v}(t) \rangle}{\langle v(0)^2 \rangle} \cos \omega t \, dt = \frac{D}{\pi} + a_1 \omega^{1/2} + O(\omega). \tag{8.12}$$

The form of a_1 is given by

$$a_1 = -(2\pi)^{1/2}\frac{2}{3\rho}\left[4\pi\left(D + \frac{\eta}{\rho M}\right)\right]^{-3/2}\frac{k_B T}{M\pi}. \tag{8.13}$$

It is, of course, of interest that the coefficient a_1 involves both D and the shear viscosity η, showing that a full microscopic theory of the self-function $S_s(k, \omega)$ would suffice to determine both transport coefficients. So far, it has not proved possible to exploit this route.

8.2.1. *Model of frequency spectrum with sharp Debye edge*

For liquid metals, the maximum or Debye frequency ω_d characterizing $g(\omega)$ is still reasonably well defined. One has also the sum rule, from small time expansion of the Fourier transform (see March, 1968, for a derivation; also see Appendix 8.4):

$$\int_{-\infty}^{\infty} \omega^2 S_s(k, \omega)\, d\omega = \frac{k^2 k_B T}{M} \tag{8.14}$$

for atoms of mass M composing the liquid, which yields readily

$$\int_{-\infty}^{\infty} g(\omega)\, d\omega = \frac{k_B T}{M}. \tag{8.15}$$

Noting that $g(\omega) = g(-\omega)$ in a classical liquid, and using area = range × height to gain orientation, one can write (Brown and March, 1968)

$$\frac{D}{\pi}\omega_d \simeq \frac{k_B T}{2M}. \tag{8.16}$$

Applying this order-of-magnitude estimate just at the melting point T_m of metals and invoking Lindemann's law of melting (see Faber, 1972) in the form $\omega_d \propto T_m^{1/2}$, one arrives at (7.46) for the self-diffusion coefficient D_m at T_m. A formula for shear viscosity, that has similar shape to (8.16) is discussed in Section 8.3 (see also (7.47)).

8.3. Van Hove dynamical structure factor $S(k, \omega)$

Having discussed the self-motion in liquids in relation to (1) the diffusion coefficient, (2) the so-called frequency spectrum, and (3) incoherent neutron scattering, we introduce here, following van Hove (1954), the dynamic generalization $S(k, \omega)$ of the static liquid structure factor $S(k)$ defined in Chapter 2. The first point to be made is that $S(k, \omega)$ has the physical interpretation (see $S_s(k, \omega)$ above) that it is the probability that a neutron incident on the liquid transfers momentum $\hbar k$ and energy $\hbar\omega$ to the liquid. The second point is that the integral of the dynamical structural factor $S(q, \omega)$ over all energy transfers $\hbar\omega$ leads back to $S(k)$; i.e.,

$$\int_{-\infty}^{\infty} S(k, \omega)\, d\omega = S(k). \tag{8.17}$$

In addition to this so-called zero moment theorem for the dynamical structure factor, it also follows from the small time behaviour of the intermediate scattering function, to be introduced later, that, for a classical liquid with atoms of mass M,

$$\int_{-\infty}^{\infty} \omega^2 S(k, \omega)\, d\omega = \frac{k^2 k_B T}{M}, \tag{8.18}$$

the so-called second-moment theorem. The fourth-order moment is also useful, being related to structure and forces (see March, 1968; also see Appendix 8.4):

$$\int_{-\infty}^{\infty} \omega^4 S(k, \omega)\, d\omega = \frac{3k^4 (k_B T)^2}{M^2} + \rho \frac{k^2 k_B T}{M^2} \int d\mathbf{r}\, g(r) \{1 - \cos kx\} \frac{\partial^2 \phi}{\partial x^2}, \tag{8.19}$$

with ρ, as usual, the atomic number density. In writing this form, it has been assumed that the total potential energy Φ can be decomposed into pair form $\sum_{i<j} \phi(r_{ij})$.

One direction that has proved somewhat fruitful for development of the theory of liquid dynamics has been that of relating self-motion, described by $S_s(k, \omega)$ in Section 8.1, to $S(k, \omega)$. Vineyard (1958) suggested the approximate relation (see Appendix 8.3)

$$S(k, \omega) \doteqdot S(k) S_s(k, \omega). \tag{8.20}$$

This is already useful for some purposes, but unfortunately, as discussed in Section 8.2, given that $S_s(k, \omega)$ has a hydrodynamic limit leading correctly to the diffusion constant D as in (8.6), $S(k, \omega)$ will not have the correct hydrodynamic form. This is because there is a corresponding (Green-Kubo) formula for $S(k, \omega)$ to that of (8.6), namely,

$$\lim_{\omega \to 0} \omega^4 \lim_{k \to 0} \frac{S(k, \omega)}{k^4} = \frac{k_B T}{\pi} \frac{(\frac{4}{3}\eta + \zeta)}{\rho M^2}, \tag{8.21}$$

where η as usual is the shear viscosity, whereas ζ is the bulk viscosity; the Vineyard approximation (8.20) does not possess the required analyticity properties to lead to such a limit as in (8.21). This is also true of other refinements, still based on the Vineyard philosophy of relating $S(k, \omega)$ to $S_s(k, \omega)$ plus static structure (e.g., Kerr, 1968). Gyorffy and March (1969, 1971) have considered various models, and the only case to date of a first-principles theory that possesses the Green-Kubo limit required in (8.21) is that of Hubbard and Beeby (1969); see also Section 8.6. Unfortunately, as emphasized in the work of March and Paranjape (1987a), this theory is deficient in the long-wavelength limit; in particular, it gives very poor numbers for the velocity of sound in a dense liquid. A possible modification has been proposed by Gyorffy and March (1969), based on the work of Sköld (1967), but this still lacks a truly fundamental basis. What has been established from first-principles, however, is the correctness of Vineyard's approach of trying to build in the dynamics of $S(k, \omega)$ through the self-function $S_s(k, \omega)$. This latter quantity, plus static structural knowledge, is indeed sufficient to determine $S(k, \omega)$ completely (Gyorffy and March, 1971; see also Appendix 8.3). The work of March and Paranjape points strongly, however, to the fact that a formula like (8.20) must be refined, at the very least, by inclusion of three-atom correlations in order to reproduce the velocity of sound in a dense liquid. This point will be referred to again in Section 8.6.

In concluding this relatively brief discussion of $S(k, \omega)$, it is to be noted that Bhatia, Thornton, and March (1974) have set up Green-Kubo-like formulae for transport coefficients in binary mixtures. Their results are summarized in Appendix 13.6.

8.4. Friction constant theory of transport

To amplify this discussion on transport in monatomic liquids, it will be a useful introduction to some of the material in Chapter 14 to return to the treatment of self-diffusion in a dense monatomic liquid near its melting point.

Many theories have indeed been proposed and tested (for a review, see Hynes, 1977). It is relevant to note here, in connection with the introductory discussion of this chapter, that Alder and Wainwright (1967) observed that if one chose the Enskog treatment as a standard of reference (see Chapman and Cowling, 1960), the diffusion constant of a dense fluid actually initially increases above the Enskog value as the packing fraction increases before decreasing rapidly. The discussions of Mazenko (1973, 1974), of Aliawadi, Rahman, and Zwanzig (1971), and of Hynes, Kapral, and Weinberg (1979) all shed light on the qualitative reasoning underlying the above trends. But interest remains in developing models of such behaviour of mass transport phenomena that quantitatively relate to the liquid structure in a fairly direct way.

Therefore, a summary will now be given of ideas based on Kirkwood's early work on friction coefficient theory (see Chapter 14) but developed especially by Sceats (1988a, b). As in the quantum problem of the electrical transport in liquid metals, replacement of random forces by total forces is of the essence of the treatment. In the electrical transport case, the work of Rousseau, Stoddart, and March (RSM) (1972) summarized in Appendix 7.1 can be regarded as a quantal parallel of the following classical argument.

Sceats has developed a theory of encounters based on an arbitrary attractive central potential of mean force $U(R)$ between two particles in the fluid (Sceats, 1986, 1988b); he applies the theory to the case in which both particles are solvent particles, and by combining this with a variant of Kirkwood's (1946) theory of friction, a model of self-diffusion is derived. Sceats has tested his model for the case of diffusion of a large particle in a gas. It is found to give a good description of the transition from the free-molecule, gas kinetic theory to the continuum theory of hydrodynamics with increase in the bath density. The application to self-diffusion in a dense monatomic liquid is more complex because of the influence of screening, caging (see Appendix 13.7), and collective effects (Sections 8.5 and 8.6), which are absent, for instance, in the case of aerosol diffusion (Sceats, 1988c). It is essential that models for these processes account for their simplest manifestation, namely, the density dependence of the self-diffusion coefficient.

Sceats (1988b), following numerous earlier workers, starts from the equations of motion of two atoms, A and B, say, in the form of generalized Langevin equations (see also Chapter 14):

$$M_j \ddot{\mathbf{R}}_j(t) = -\int_{-\infty}^{t} K_{\mathbf{R}}(t-t')\ddot{\mathbf{R}}_j(t')\,\mathrm{d}t' + \mathbf{F}_j(t) - \nabla_j U(R); \quad j = A, B. \tag{8.22}$$

Here a closed system of N particles has been assumed. Although one is interested in the motion of just one of these particles, say A, it is convenient to consider in (8.22) the relative motion of the two particles A and B. The projection techniques of Mori (1965a, 1965b) and of Zwanzig (1960) can be used formally to project out the degrees of freedom of the remaining $N-2$

particles. If the nonlocal spatial terms are then assumed to be only weakly dependent on the particle separation R, then (8.22) results.

The random force $\mathbf{F}_j(t)$ and the memory kernel $K_\mathbf{R}$ are related by the statistical properties

$$\langle \mathbf{F}_i(t) \cdot \mathbf{F}_j(t + t') \rangle = k_B T K_\mathbf{R}(t') \delta_{ij} \tag{8.23}$$

and

$$\langle \mathbf{F}_i(t) \rangle = 0. \tag{8.24}$$

The potential of mean force U in (8.22) is related, as usual (see (3.1)) to the pair function $g(r)$ by the Boltzmann factor, and one assumes from the outset that this static structural description is given. When R is sufficiently large, the two Langevin equations are uncoupled. The frequency-dependent friction of a single particle, $\xi(\omega)$, is the Fourier transform of $K_\infty(t)$, and the self-diffusion coefficient D is related to $\xi(0) \equiv \xi$ by the Einstein result $D = k_B T/\xi$.

The fluctuation dissipation theorem (see, for example the book by March and Tosi (1984)) relates ξ to the force correlation function by

$$\xi = \frac{1}{3k_B T} \int_0^\infty \langle \mathbf{F}(t) \cdot \mathbf{F}(0) \rangle \, dt. \tag{8.25}$$

The relationship of ξ to the total force on the test particle has not been established at the time of writing. Kirkwood (1946) developed a stochastic theory in which the random forces in (8.25) were replaced by the total force together with the change in the upper limit of the time integration being replaced by a "plateau time.' This was essential to his treatment, since, with total forces rather than random forces, the long time limit of the integral appearing in (8.25) vanishes. Unfortunately, difficulties in this approach were noted by Fisher and Watts (1972), whose computer simulation data did not reveal a plateau in the force correlation function at liquid densities.

Sceats (1988b) has therefore proposed an alternative to Kirkwood's plateau time approach. He manipulates (8.25) into what is an equivalent form by arguing that for a stationary stochastic process the order of the integration and the averaging can be interchanged and a time average taken for a system observation time τ, which is long compared to the force relaxation time scale. Thus, one can write (8.25) in the form

$$\xi = \lim_{\tau \to \infty} \frac{1}{6k_B T} \frac{1}{\tau} \left\langle \int_{-\tau/2}^{+\tau/2} \int_{-\tau/2}^{+\tau/2} \mathbf{F}(t) \cdot \mathbf{F}(t + t') \, dt' \, dt \right\rangle. \tag{8.26}$$

Still, the difficulty that the random force is required is not altered by this interchange.

Therefore, Sceats (1988b) introduced a cutoff into the preceding formula. His essential result is then expressed in product form, one factor being an encounter flux J^T, say, and the other the statistical average of the force "power spectrum" for an encounter, denoted by $\langle \hat{F}(0)^2 \rangle$ (see below).

The encounter flux J^T will be discussed qualitatively below and then a little further in Appendix 8.5. Sceats argues for the usefulness of the resulting form

$$\xi \doteqdot \frac{J^T}{6k_B T} \langle \hat{F}(0)^2 \rangle, \tag{8.27}$$

as it can be intuitively generalized away from zero frequency to yield $\xi(\omega) = J^T(\omega)\langle \hat{F}(\omega)^2 \rangle / 6k_B T$, which may eventually prove a valuable starting point to model chemical reactions (see Hynes, 1985). It is important, following Sceats (1988b), to note that J^T involves the concept that the total force on the test particle can be broken into pairwise contributions and the contribution to (8.26) is subject to a spatial constraint, namely, that a bath particle contributes to the force in (8.26) only when its separation R is less than some cutoff R_T to be sought. Obviously, the usefulness of (8.27) hinges on (1) whether a physically sensible choice of R_T exists and, if so, (2) whether J^T and $\langle \hat{F}(0)^2 \rangle$ can be modelled.

To do this, Sceats (1988b) returns to the Langevin equations (8.22). He uses Kramers' theory of one-dimensional barrier crossing, which is developed in Appendix 8.5, to show that the encounter flux can be usefully written as

$$J^T = \pi R_{\mathrm{T}}^2 \exp\left(\frac{-U(R_{\mathrm{T}})}{k_B T}\right)\left(\frac{8k_B T}{\pi\mu}\right)^{1/2} \rho \left\{\left[1 + \left(\frac{\xi_{\mathrm{R}}}{2\mu\omega_T}\right)^2\right]^{1/2} - \frac{\xi_{\mathrm{R}}}{2\mu\omega_T}\right\}. \tag{8.28}$$

Here $\mu = M/2$ is the reduced mass, ρ is the number density of the particles, and ω_T is the barrier frequency defined by the configurational integral

$$\omega_T = \frac{\left(\dfrac{2\pi k_B T}{\mu}\right)^{1/2}}{\displaystyle\int_{R_p}^{\infty} \exp\{[\tilde{U}(R) - \tilde{U}(R_{\mathrm{T}})]/k_B T\}\, \mathrm{d}R}. \tag{8.29}$$

The effective potential of mean force $\tilde{U}(R)$ must, Sceats (1988b) argues, have the form

$$\tilde{U}(R) = U(R) - 2k_B T \ln R \tag{8.30}$$

in order to preserve in a one-dimensional treatment the expansion of phase space as R increases. In (8.29), R_p denotes the inner minimum of (8.28) for

J_T. Equation (8.29) is valid provided a particular characteristic frequency ω'_T (see Sceats, 1988b, for its definition in precise terms) is sufficiently small so that it falls below the characteristic relaxation time scales of the bath.

The principal features of J^T can be summarized as follows:

(1) At low friction, where $\xi/\mu \ll \omega'_T$, the flux is that of canonical variational transition state theory, widely used in gas phase kinetics.
(2) At high friction, where $\xi/\mu \gg \omega'_T$, one regains Kramers' results (see Appendix 8.5).

Though the preceding approach appears promising, at the time of writing it has not been subjected to very direct contact between its theoretical predictions and experiment, both J^T and the force "power spectrum," entering the basic approximate formula (8.27) for the friction constant, deserving further study. It should be noted here that according to Sceats (1988b), the power spectrum $\langle|\tilde{F}(0)|^2\rangle$ can be written in terms of the well-depth $Q = \tilde{U}(R_\rho) - \tilde{U}(R_T)$ as $\langle|\tilde{F}(0)|^2\rangle = 8\mu k_B T \exp(Q/k_B T) + (2/9)8\mu k_B T(N_c - 1)\exp(2Q/k_B T)$, with N_c the number of bath molecules in the region $R < R_T$, but the reader should refer to the original paper for details of this very approximate treatment.

8.5. Observation of collective modes

To return to the discussion of transport in relation to neutron inelastic scattering, brief consideration will be given here to the use of such neutron techniques to reveal collective modes in liquids. The quite clear-cut case of collective modes is the work of Copley and Rowe (1974) on liquid metal rubidium. Therefore, most attention will be focused on this metal, both the measurements of collective modes and their interpretation. This latter work has been carried out at two levels: (1) the computer simulation of Rahman (1974a,b), using an appropriate pair potential mediated by the itinerant metal electrons, developed by Price et al. (1976) and (2) that based on simple collective models of almost independent density fluctuations (Eisenschitz and Wilford, 1962; Bratby, Gaskell and March, 1970; Gray, Yokoyama, and Young, 1980). The two approaches enable the origins of the collective modes to be understood. Evidence that is less decisive than for Rb has been presented by Johnson et al. (1977) for collective modes in liquid Ni just above its melting point: further work is called for here.

Table 8.1. *Specific heats of liquid metals at melting point [(from Kleppa (1950)].*

	Na	K	Rb	Zn	Cd	Ga	Tl	Sn	Pb	Bi
$\gamma = c_p/c_v$	1.12	1.11	1.15	1.25	1.23	1.08	1.21	1.11	1.20	1.15
c_v/Nk_B	3.4	3.5	3.4	3.1	3.1	3.2	3.0	3.0	2.9	3.1

As a little background, if one defines the density fluctuations in a classical liquid with atoms at positions $\mathbf{R}_i(t)$ as:

$$\rho_k(t) = \sum_{i=1}^{N} \exp(i\mathbf{k} \cdot \mathbf{R}_i(t)) \tag{8.31}$$

then independent density fluctuations, paralleling "phonons" in a crystal, would result in a time dependence of the form

$$\rho_k(t) = \rho_k(0) \exp(i\omega_k t). \tag{8.32}$$

Here ω_k would represent the dispersion relation of the collective mode. Such a model of independent density fluctuations receives almost immediate support for a variety of simple liquid metals near the melting point, since it would predict, as a purely harmonic model, that the ratio of the specific heats, $\gamma = c_p/c_v$, is near unity, whereas $c_v \geqslant 3R$, with R the gas constant. Indeed, for a number of simple liquid metals, this prediction is well borne out, as shown in Table 8.1. Development of such a theory based on density fluctuations leads to $C_v \geqslant 3R$, the correction being related to structure (Bratby, Gaskell, and March, 1970).

The dispersion relation for the collective modes in Rb is plotted in Figure 8.2 from the work of Copley and Rowe (1974). Matthai and March (1982) have argued that this is quite well fitted by the form

$$\omega^2(k) = \frac{k_B T k^2}{M S(k)} \tag{8.33}$$

and have given some theoretical arguments in support of this (see also Section 8.6). The computer simulation results of Rahman (1974a) are, of course, more accurate; there is excellent quantitative agreement between neutron and computer experiments.

McGreevy and Mitchell (1985; see also Egelstaff, 1987) have considered the question of collective modes in a wider context, also embracing insulating liquids such as argon. As they clearly recognized, no definitive experimental evidence has as yet been forthcoming in such insulating fluids. Certainly, independent density fluctuation models would need transcending

Figure 8.2. Dispersion relation $\omega(k)$ for collective modes in liquid Rb. (Observations from work of Copley and Rowe (1974) are shown: +). • is from (8.33). x is same theory but for Na; ∘ for K.

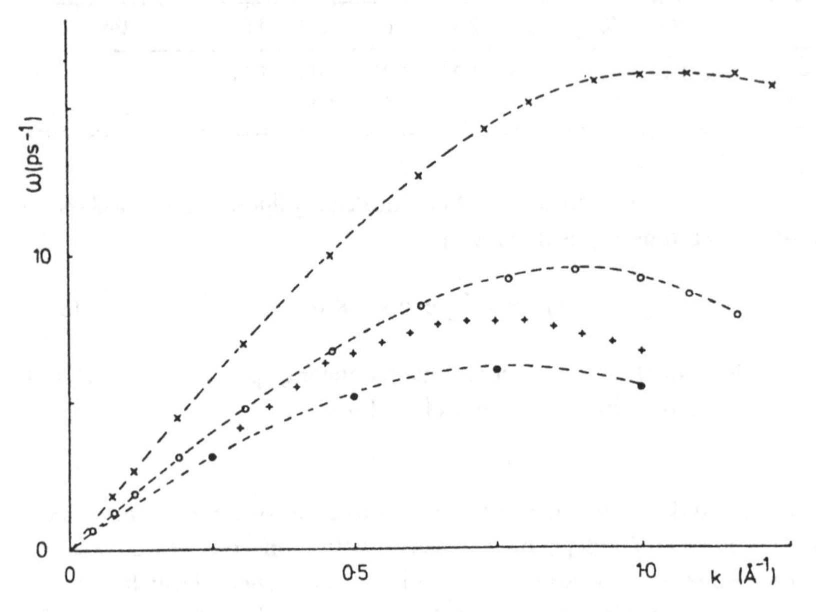

because, for instance, the ratio of the specific heats at the triple point of argon is $\gamma = 2.2$, in marked contrast to liquid Rb at its melting point, where $\gamma = 1.2$. However, it has been mentioned earlier that liquid Ni, with a much harder core than Rb, may also exhibit a collective mode, although more work is called for before that can be established with certainty. However, in the present context, γ for Ni at its melting temperature is 1.8, not so different from the argon value. Clearly, a great deal remains to be done on liquid dynamics as explored by inelastic neutron scattering.

8.6. Hubbard-Beeby theory

As a concrete example of the (formal) demonstration that the dynamics of a classical liquid is contained in the self-function $S_s(k, \omega)$, plus knowledge of static correlations, the theory of Hubbard and Beeby (1969) will be outlined. This was designed to treat collective motions in liquids. Though it is now known to have some rather severe quantitative defects, nevertheless it remains somewhat illuminating, and a correction can be applied to it to make its numerical predictions more reasonable.

Following their approach, one starts from a discussion of a cold amorphous solid, and then, as a second step, allows for particle motions (diffusion).

8.6.1. Cold amorphous solids

The method used by Hubbard and Beeby is to calculate the density-density response function. Suppose that a weak time-dependent potential field $V(\mathbf{r}, t)$ acts on the system. This will give rise to changes in the density, and for small V we have

$$\delta\rho(\mathbf{r}, t) = \int_{-\infty}^{\infty} \mathrm{d}t' \int \mathrm{d}\mathbf{r}' \, \chi(\mathbf{r} - \mathbf{r}', t - t') V(\mathbf{r}'t') \tag{8.34}$$

where χ is the density-density response function. If one takes the Fourier transform of $\chi(rt)$ into $\mathbf{k}\omega$ space and writes

$$\chi(\mathbf{k}\omega) = \chi'(\mathbf{k}\omega) + i\chi''(\mathbf{k}\omega), \tag{8.35}$$

then using the fluctuation-dissipation theorem (see March and Tosi, 1984) it can be shown that the van Hove scattering function or dynamical structure factor $S(\mathbf{k}\omega)$ is given by

$$S(\mathbf{k}\omega) = \frac{1}{\pi\rho} \frac{\hbar\omega}{1 - \exp(\beta\hbar\omega)} \frac{\chi''(\mathbf{k}\omega)}{\omega}. \tag{8.36}$$

One property of $\chi(\mathbf{r}t)$ is that it must vanish when $t < 0$, since $\delta\rho(\mathbf{r}t)$ cannot depend on $V(\mathbf{r}t)$ at later times. Hence $\chi(\mathbf{k}\omega)$ satisfies a dispersion relation of the form

$$\chi(\mathbf{k}\omega) = \lim_{\eta \to 0+} \int \frac{\mathrm{d}\omega' \, \chi''(\mathbf{k}\omega')}{\pi(\omega' - \omega - i\eta)}. \tag{8.37}$$

$\chi''(\mathbf{k}\omega)$ is an odd function of ω and its Fourier transform $\chi''(\mathbf{k}t)$ is correspondingly an odd function of t. Equation (8.37) implies that the transform $\chi(\mathbf{k}t)$ of $\chi(\mathbf{k}\omega)$ is given by

$$\chi(\mathbf{k}t) = 2i\theta(t)\chi''(\mathbf{k}t): \qquad \left.\begin{array}{ll} \theta(t) = 1, & t > 0 \\ \theta(t) = 0, & t < 0 \end{array}\right\}. \tag{8.38}$$

Next, the linear response of a disordered stationary array of atoms, as in a cold amorphous solid, will be studied. To do so, let us assume that the weak external potential energy has the explicit form

$$V(\mathbf{r}t) = V \exp(i\mathbf{k} \cdot \mathbf{r} - i\omega t). \tag{8.39}$$

Then the atoms will be displaced from their equilibrium sites by an amount $\mathbf{u}_i(t)$, $\mathbf{u}_i$ being infinitesimal if V is infinitesimal, as is assumed. Clearly, one can calculate the change in the force acting on the ith particle from a knowledge of the (assumed) pair interaction $\phi(r)$ and the external potential. This enables the equation of motion for $\mathbf{u}_i$ to be written as

$$M\ddot{\mathbf{u}}_i = -\sum_{j \neq i} \nabla\nabla\phi(\mathbf{r}_i - \mathbf{r}_j)\cdot(\mathbf{u}_i - \mathbf{u}_j) - iV\mathbf{k}\exp(i\mathbf{k}\cdot\mathbf{r}_i - i\omega t) \quad (8.40)$$

which is true to first order in V, and therefore in $\mathbf{u}$.

Secondly, since one wishes to relate $\delta\rho$ to V, one can calculate the change $\delta\rho_{\mathbf{k}}$ in $\rho_{\mathbf{k}}$ trivially, to the same order, to find

$$\delta\rho_{\mathbf{k}} = \sum_{i=1}^{N} i\mathbf{k}\cdot\mathbf{u}_i\exp(i\mathbf{k}\cdot\mathbf{r}_i). \quad (8.41)$$

In principle, one can now solve the equation of motion for $\mathbf{u}_i$, substitute in the above expression for $\delta\rho_{\mathbf{k}}$ and hence obtain $\chi(\mathbf{k}\omega)$. In practice, this cannot be done exactly, but it will be seen below that a rather natural approximation brings back a result that is well known for crystalline solids.

To tidy up a little, let us note that since (8.40) is linear in $\mathbf{u}_i(t)$, its time dependence must be $\exp(-i\omega t)$ and this suggests that one should work with a quantity $\bar{\mathbf{u}}_i$ defined by

$$\bar{\mathbf{u}}_i = \frac{\mathbf{u}_i(t)\exp(i\omega t - i\mathbf{k}\cdot\mathbf{r}_i)}{V}. \quad (8.42)$$

Then $\bar{\mathbf{u}}_i$ satisfies the equation

$$M\omega^2\bar{\mathbf{u}}_i = i\mathbf{k} - \sum_j \exp[i\mathbf{k}\cdot(\mathbf{r}_j - \mathbf{r}_i)]v_{ij}\bar{\mathbf{u}}_j \quad (8.43)$$

where v_{ij} is simply a shorthand notation for $\nabla\nabla\phi(\mathbf{r}_i - \mathbf{r}_j)$ for $i \neq j$ and $-v_{ii} = \sum_{j(\neq i)} v_{ij}$. Then one also has for the density change

$$\delta\rho_{\mathbf{k}}(t) = -iV\exp(-i\omega t)\mathbf{k}\cdot\sum_i \bar{\mathbf{u}}_i. \quad (8.44)$$

At this stage one iterates the equation of motion and finds

$$\bar{\mathbf{u}}_i = \frac{i\mathbf{k}}{M\omega^2} - \frac{i}{(M\omega^2)^2}\sum_i v_{ij}\exp(i\mathbf{k}\cdot(\mathbf{r}_j - \mathbf{r}_i))\mathbf{k}$$

$$+ \frac{i}{(M\omega^2)^3}\sum_{j=1} v_{ij}v_{j1}\exp(-i\mathbf{k}\cdot(\mathbf{r}_j - \mathbf{r}_i))\mathbf{k} + \cdots. \quad (8.45)$$

This equation has now to be averaged over all the configurations. The average of the linear term in v_{ij} in (8.45) is readily achieved and involves just the pair function $g(r)$.

In the higher terms, one does not know precisely what to take for the higher-order distribution functions. The Kirkwood (1935) approximation (see Chapter 5) might suggest itself but leads to considerable complications quite rapidly. The simplest procedure that comes to mind and that leads to (8.56) (Hubbard and Beeby, 1969) for $S(\mathbf{k}\omega)$ is to represent the average of any product of the v_{ij} by the corresponding product of the averages of the individual v_{ij}. Doing this, one relates the second-order term to the known linear term as follows. Thus the first-order terms may be obtained as

$$\sum_j v_{ij} \exp(i\mathbf{k} \cdot \mathbf{r}_i - \mathbf{r}_j) = -\rho\psi(\mathbf{k}) \tag{8.46}$$

where, explicitly in terms of the pair function $g(r)$, one has

$$\psi(\mathbf{k}) = \int d\mathbf{r}\, g(r) \nabla\nabla\phi(\mathbf{r})(1 - \exp(i\mathbf{k} \cdot \mathbf{r})). \tag{8.47}$$

Then in the approximation referred to above, the second-order term is $\rho^2[\psi(\mathbf{k})]^2$, and so on.

The result may be summed, the response function calculated from

$$\chi(\mathbf{k}\omega) = \frac{\langle \delta\rho_{\mathbf{k}} \rangle}{V \exp(-i\omega t)} \tag{8.48}$$

and one obtains

$$\chi(\mathbf{k}\omega) = \rho\mathbf{k} \cdot [M\omega^2 \mathbf{1} - \rho\psi(\mathbf{k})]^{-1} \cdot \mathbf{k}. \tag{8.49}$$

The singularities of the response function appear at the frequencies ω given by the roots of the secular equation

$$|M\omega^2 \mathbf{1} - \rho\psi(\mathbf{k})| = 0, \tag{8.50}$$

which determines the frequencies of the collective phonon modes.

These frequencies are real, and this means that the approximations made above do not include phonon lifetimes which is, of course, a rather severe limitation.

To make contact with well-established results for a cold crystalline solid, the pair function $g(\mathbf{r})$ can be written as

$$g(\mathbf{r}) = \frac{1}{\rho} \sum_{\mathbf{R}} \delta(\mathbf{r} - \mathbf{R}) \tag{8.51}$$

where the sum is now over lattice vectors. Then the result (8.50) reduces to

$$\left| M\omega^2 \mathbf{1} - \sum_{\mathbf{R}} \phi(\mathbf{R})(1 - \exp(-i\mathbf{k}\cdot\mathbf{R})) \right| = 0, \tag{8.52}$$

which is the usual result giving the phonon dispersion relations. Thus the above theory becomes exactly the same as the usual harmonic approximation in the crystalline case for pair potential ϕ.

8.6.2. *Implications of isotropy*

For isotropic systems, such as an amorphous solid or a liquid, the matrix $\psi(\mathbf{k})$ has a form, which, from symmetry considerations, must be

$$\frac{\rho}{M}\psi(\mathbf{k}) = \omega_{\mathbf{k}}^2 \hat{\mathbf{k}}\hat{\mathbf{k}} + \omega_{\mathbf{k}t}^2(\mathbf{1} - \hat{\mathbf{k}}\hat{\mathbf{k}}) \tag{8.53}$$

where $\hat{\mathbf{k}}$ is the unit vector $\mathbf{k}/k$. Equation (8.50) determining the frequencies of the collective modes then has one eigenvalue, $\omega_{\mathbf{k}}^2$, and a pair of degenerate eigenvalues, $\omega_{\mathbf{k}t}^2$, corresponding to one longitudinal mode and two transverse modes. Only the longitudinal mode contributes in (8.49), and one finds

$$\chi(\mathbf{k}, \omega) = \frac{\rho k^2}{M} \frac{1}{\omega^2 - \omega_{\mathbf{k}}^2}, \tag{8.54}$$

where

$$\begin{aligned} \omega_{\mathbf{k}}^2 &= \frac{\rho}{M} \hat{\mathbf{k}} \cdot \psi(\mathbf{k}) \cdot \hat{\mathbf{k}} \\ &= \frac{\rho}{M} \int \frac{\partial^2 \phi}{\partial z^2} g(r)(1 - \cos kz)\, d\mathbf{r}. \end{aligned} \tag{8.55}$$

For small $\mathbf{k}$ the right-hand side is proportional to k^2, and so $\omega_k = v_s k$, corresponding to a longitudinal sound velocity v_s. The scattering function has the form

$$S(\mathbf{k}\omega) = \frac{\hbar k^2}{2M\omega_{\mathbf{k}}} [(n_{\mathbf{k}} + 1)\delta(\omega - \omega_{\mathbf{k}}) + n_{\mathbf{k}}\delta(\omega + \omega_{\mathbf{k}})]$$

where

$$n_{\mathbf{k}} = [\exp(\hbar\beta\omega_{\mathbf{k}}) - 1]^{-1}. \tag{8.56}$$

Relation (8.52) predicts nonzero frequencies for the transverse modes even for small k. This reflects the fact that an amorphous solid has rigidity, and thus one can expect it to sustain transverse modes. The lack of rigidity of a liquid, in contrast, is a consequence of relaxation processes that result from the movement of the particles in a liquid. This will be allowed for in the following extension.

All that needs to be added at this point is that from (8.56), there will be a peak in the neutron scattering at the longitudinal collective mode frequency ω_k.

8.6.3. Generalization to liquid state

When the motion of particles is incorporated, one must consider not particular configurations $\{\mathbf{R}_i\}$ but particular histories $\{\mathbf{R}_i(t)\}$ of the system. The equation of motion (8.40) for a particular history $\{\mathbf{R}_i(t)\}$ can be derived in exactly the same way as before, the only change being that the force constants become time-dependent:

$$\left.\begin{aligned} \boldsymbol{\phi}_{ij}(t) &= \nabla\nabla v\{\mathbf{R}_i(t) - \mathbf{R}_j(t)\}, \qquad i \neq j \\ \boldsymbol{\phi}_{ii}(t) &= -\sum_{j \neq i} \boldsymbol{\phi}_{ij}(t) \end{aligned}\right\}. \tag{8.57}$$

Because the time variation of the $\mathbf{R}_i(t)$ produces only this one change, the procedure adopted in the previous case carries through to the liquid state readily.

If one integrates the equation of motion twice, one now obtains for atoms of mass m:

$$m\mathbf{u}_i(t) = -\int_{-\infty}^{t} dt' \int_{-\infty}^{t'} dt'' \left[\sum_j \boldsymbol{\phi}_{ij}(t'')\mathbf{u}_j(t'') + iV\mathbf{k}\exp\{i\mathbf{k}\cdot\mathbf{R}_i(t'') - i\omega t''\} \right]. \tag{8.58}$$

The quantities $\boldsymbol{\xi}_i(t)$ are defined according to (see (8.42))

$$\boldsymbol{\xi}_i(t) = \mathbf{u}_i(t)\frac{\exp\{i(\omega t - \mathbf{k}\cdot\mathbf{R}_i)\}}{V};$$

they are, as indicated, no longer independent of time but are still related to $\delta\rho_k(t)$ by (8.44). From (8.58) and (8.42) one finds

$$m\xi_i(t) = -\int_{-\infty}^{t} \mathrm{d}t' \int_{-\infty}^{t'} \mathrm{d}t'' \Big(i\mathbf{k} \exp[i\mathbf{k}\cdot\{\mathbf{R}_i(t'') - \mathbf{R}_i(t)\} + i\omega(t - t'')]$$

$$+ \sum_j \phi_{ij}(t'') \exp[i\mathbf{k}\cdot\{\mathbf{R}_j(t'') - \mathbf{R}_i(t)\} + i\omega(t - t'')]\cdot\xi_j(t'')\Big), \quad (8.59)$$

which one can now iterate to obtain the analogue of (8.45). Again, by averaging over all histories $\{\mathbf{R}_i(t)\}$ (with appropriate weights), the analogue of (8.43) is obtained. The nature of the resulting series for $\langle\xi_i(t)\rangle$, the average being independent of t although the individual $\xi_i(t)$ are not, can be understood by considering the first few terms. The leading term is

$$-\frac{i}{m}\mathbf{k}\left\langle \int_{-\infty}^{t} \mathrm{d}t' \int_{-\infty}^{t'} \mathrm{d}t'' \exp[i\mathbf{k}\cdot\{\mathbf{R}_i(t'') - \mathbf{R}_i(t)\} + i\omega(t - t'')]\right\rangle$$

$$\equiv -\frac{i}{m}\mathbf{k}Q(\mathbf{k},\omega), \quad (8.60)$$

which defines the function $Q(\mathbf{k},\omega)$. The average is independent of t because the average $\langle \exp[i\mathbf{k}\cdot\{\mathbf{R}_i(t) - \mathbf{R}_i(t'')\}]\rangle$ is a function of $t - t''$ only. The function $Q(\mathbf{k},\omega)$ is related to the self-correlation function $G_s(\mathbf{r}, t)$ of Section 8.1, which gives the probability that an atom of the liquid will move by an amount $\mathbf{r}$ in time t, the relation being

$$Q(\mathbf{k},\omega) = \int_0^{\infty} \mathrm{d}t \int \mathrm{d}\mathbf{r}\, t G_s(\mathbf{r}, t) \exp(-i\mathbf{k}\cdot\mathbf{r} + i\omega t). \quad (8.61)$$

The second term in the series takes the form

$$\frac{i}{m^2}\sum_j \int_{-\infty}^{t} \mathrm{d}t' \int_{-\infty}^{t'} \mathrm{d}t'' \Big\langle \boldsymbol{\phi}_{ij}(t'') \exp[i\mathbf{k}\cdot\{\mathbf{R}_j(t'') - \mathbf{R}_i(t)\} + i\omega(t - t'')]\cdot\mathbf{k}$$

$$\times \int_{-\infty}^{t''} \mathrm{d}t_1' \int_{-\infty}^{t_1'} \mathrm{d}t_1'' \exp[i\mathbf{k}\cdot\{\mathbf{R}_j(t_1'') - \mathbf{R}_j(t'')\} + i\omega(t'' - t_1'')]\Big\rangle. \quad (8.62)$$

As in the previous discussion, the averaging will be approximated, following Hubbard and Beeby (1969), by averaging separately the ϕ factor and its associated exponential factor. Thus the expression (8.62) becomes

$$\frac{i}{m^2}\sum_j \int_{-\infty}^{t} \mathrm{d}t' \int_{-\infty}^{t'} \mathrm{d}t'' \langle \boldsymbol{\phi}_{ij}(t'') \exp[i\mathbf{k}\cdot\{\mathbf{R}_j(t'') - \mathbf{R}_i(t)\} + i\omega(t - t'')]\rangle\cdot\mathbf{k}$$

$$\times \int_{-\infty}^{t''} \mathrm{d}t_1' \int_{-\infty}^{t_1'} \mathrm{d}t_1'' \langle \exp[i\mathbf{k}\cdot\{\mathbf{R}_j(t_1'') - \mathbf{R}_j(t'')\} + i\omega(t'' - t_1'')]\rangle. \quad (8.63)$$

Comparing this with (8.60), it can be seen that the second factor is proportional to $Q(\mathbf{k},\omega)$, and one may write (8.63) as $-(i/m)Q(\mathbf{k},\omega)(\rho/m)\boldsymbol{\psi}(\mathbf{k},\omega)\cdot\mathbf{k}$ if one now defines $\boldsymbol{\psi}(\mathbf{k},\omega)$ by

$$\rho\boldsymbol{\psi}(\mathbf{k},\omega) = \left\langle \sum_j \int_{-\infty}^{t} \mathrm{d}t' \int_{-\infty}^{t'} \mathrm{d}t''\, \boldsymbol{\phi}_{ij}(t'') \exp[i\mathbf{k}\cdot\{\mathbf{R}_j(t'') - \mathbf{R}_i(t)\} + i\omega(t - t'')] \right\rangle. \tag{8.64}$$

Similar approximations are made in the higher-order terms. When this is done, the nth term assumes the approximate form $-(i/m)Q(\mathbf{k},\omega) \times \{(\rho/m)\boldsymbol{\psi}(\mathbf{k},\omega)\}^{n-1}\cdot\mathbf{k}$. Summing the resulting series, one finds an approximation for $\langle\boldsymbol{\xi}_i(t)\rangle$ analogous to (8.49), namely,

$$\langle\boldsymbol{\xi}_i(t)\rangle \simeq -i\exp(-i\omega t)Q(\mathbf{k},\omega)\left\{\mathbf{1} - \frac{\rho}{m}\boldsymbol{\psi}(\mathbf{k},\omega)\right\}^{-1}\cdot\mathbf{k} \tag{8.65}$$

and a response function of the form

$$\chi(\mathbf{k},\omega) = -\frac{\rho}{m}Q(\mathbf{k},\omega)\mathbf{k}\cdot\left\{\mathbf{1} - \frac{\sigma}{m}\boldsymbol{\psi}(\mathbf{k},\omega)\right\}^{-1}\cdot\mathbf{k} \tag{8.66}$$

analogous to (8.49). In fact, if in (8.60) and (8.64) one assumes $\mathbf{R}_i(t)$ to be independent of t, one obtains $Q(k,\omega) = -1/\omega^2$ and $\boldsymbol{\psi}(\mathbf{k},\omega) = (\rho/\omega^2)\boldsymbol{\psi}(\mathbf{k})$, and (8.66) reduces to (8.49).

When the summation over j and the average in (8.64) have been carried out, one finds

$$\boldsymbol{\psi}(\mathbf{k},\omega) = \int \mathrm{d}\mathbf{r} \int \mathrm{d}\mathbf{r}' \int_0^\infty \mathrm{d}t\, \nabla\nabla v(\mathbf{r}')t\mathscr{G}(\mathbf{r},\mathbf{r}',t)\exp(i\mathbf{k}\cdot\mathbf{r}) \times [\exp(-i\mathbf{k}\cdot\mathbf{r}') - 1]\exp(i\omega t) \tag{8.67}$$

where $\rho\mathscr{G}(\mathbf{r},\mathbf{r}',t)$ is the correlation function that measures the joint probability that (1) at $\tau = 0$, there is a particle at the origin and another at the point $\mathbf{r}'$, and (2) that at $\tau = t$, the atom at the origin has moved to $\mathbf{r}$.

This correlation function, unlike the self-function G_s entering the formula (8.61) for $Q(k,\omega)$, is practically unknown. In order to make progress, it is obviously necessary to make some approximation for it. Hubbard and Beeby (1969) argue that the simplest approximation is to neglect the correlation between the motion of the particle at the origin and the position of the atom at $\mathbf{r}'$ and, therefore, to write

$$\mathscr{G}(\mathbf{r},\mathbf{r}',t) \simeq G_s(\mathbf{r},t)g(\mathbf{r}'), \tag{8.68}$$

which is somewhat like the approximation (8.20) of Vineyard (1958), though introduced in a slightly different context.

If this approximation is used, one obtains

$$\boldsymbol{\psi}(\mathbf{k},\omega) \simeq -Q(\mathbf{k},\omega)\boldsymbol{\psi}(\mathbf{k}) \tag{8.69}$$

and (8.66) becomes

$$\chi(\mathbf{k},\omega) = -\frac{\rho}{m}Q(\mathbf{k},\omega)\mathbf{k}\cdot\{1 + Q(\mathbf{k},\omega)\boldsymbol{\psi}(\mathbf{k})\}^{-1}\cdot\mathbf{k}. \tag{8.70}$$

In the case of a liquid, $\boldsymbol{\psi}(\mathbf{k})$ assumes the form (8.53) and (8.70) reduces to

$$\chi(\mathbf{k},\omega) = -\frac{\rho k^2}{m}\frac{Q(\mathbf{k},\omega)}{1 + \omega_{\mathbf{k}}^2 Q(\mathbf{k},\omega)} \tag{8.71}$$

which yields in the classical limit

$$\begin{aligned} S(\mathbf{k},\omega) &= \frac{k^2}{\pi m\beta\omega}\,\mathrm{Im}\left\{\frac{Q(\mathbf{k},\omega)}{1 + \omega_{\mathbf{k}}^2 Q(\mathbf{k},\omega)}\right\} \\ &= \frac{k^2}{\pi m}\frac{Q''(\mathbf{k},\omega)/\omega}{\{1 + \omega_{\mathbf{k}}^2 Q'(\mathbf{k},\omega)\}^2 + \{\omega_{\mathbf{k}}^2 Q''(\mathbf{k},\omega)\}^2} \end{aligned} \tag{8.72}$$

where $Q(\mathbf{k},\omega) = Q' + iQ''$. If Q'' is small, as it certainly is for small k, then $S(\mathbf{k},\omega)$ is sharply peaked at the frequency determined by the dispersion relation

$$1 + \omega_{\mathbf{k}}^2 Q'(\mathbf{k},\omega) = 0 \tag{8.73}$$

which gives the collective mode frequency. The width of the peak is proportional to Q'', which reflects the effect of particle motion in damping the collective mode (the analogue of the Landau damping of collective oscillations).

Whereas approximation (8.72) for $S(\mathbf{k},\omega)$ satisfies the ω^2 and ω^4 sum rules (see Appendix 8.4) exactly, it is not expected to satisfy the zeroth-order ω^0 sum rule at all well, as Hubbard and Beeby (1969) stress. This means it is not a suitable theory for the static structure of the liquid. But the structure of the theory has merit; for example, as discussed by Gyorffy and March (1971), the theory has a Green-Kubo limit, as shown in Appendix 8.3, though the quantitative form of the answer requires a modification set out there. Also March and Paranjape (1987a; see also Appendix 8.3) have examined the dispersion relation (8.55) for model liquids and pointed out a way in which this form of Hubbard and Beeby can be transcended.

Some discussion will now be given concerning a topic closely related to the preceding remarks on the Green-Kubo formula for transport, namely, the wave number-dependent shear viscosity. This has been examined by Balucani et al. (1987), and a brief summary of their results will be presented immediately below.

8.7. Generalized shear viscosity in liquid Rb

The transverse current correlation function (see 8.74) has been studied in liquid Rb by computer simulation (Balucani et al., 1987: see also 1988). The associated memory function has been directly determined from this data in their work for a number of values of the wave vector **q**. The usual phenomenological relaxation time approximation for the memory function proves to be inadequate in the wave vector range where well-defined shear waves are supported by the liquid. The related generalized q-dependent shear viscosity $\eta(q)$ is derived also, and its importance in the context of a microscopic Stokes-Einstein relation (see Section 7.5) is discussed.

The shear viscosity η is, in fact, the simplest transport property connected with the collective behaviour of the liquid. (See also Appendix 8.6.) The associated dynamical quantity is the transverse current correlation function referred to before. This is defined by

$$C_{\mathrm{T}}(q,t) = \left\langle \sum_i v_i^x(0)\exp[-iqz_i(0)] \sum_j v_j^x(t)\exp[iqz_j(t)] \right\rangle \tag{8.74}$$

with $\mathbf{v}_i$ and $\mathbf{r}_i$ denoting velocity and position, respectively, of particle i. The wave vector **q** is here taken along the z axis. In the hydrodynamic limit one finds that

$$[C_{\mathrm{T}}(q,t)]_{\mathrm{hydr}} = C_{\mathrm{T}}(q,0)\exp\left[-\left(\frac{\eta q^2}{\rho m}\right)t\right] \tag{8.75}$$

where, as usual, $\rho = N/V$ is the number density and m denotes the ionic mass.

Following Balucani et al. (1987), the normalized quantity $\rho_T(q,t) = C_T(q,t)/C_T(q,0)$ will be referred to later. Equation (8.75) is expected to be valid in a physical situation where length and time scales are substantially larger than the ones associated with microscopic processes involving the average range and duration of an atomic collision. When such conditions are not satisfied, the monotonic decay implied by (8.75) is not valid, and the system is expected to support shear-wave propagation.

On the other hand, this solidlike behaviour gives rise to well-defined oscillations of $C_T(q,t)$ only in a limited range of wave vectors. As q becomes much larger than the inverse mean free path, the particles can be considered to be essentially free. This yields the purely kinetic Gaussian decay

$$C_T(q,t) = C_T(q,0)\exp\left[-\left(\frac{k_B T}{2m}\right)q^2t^2\right], \tag{8.76}$$

where every vestige of collective behaviour is absent.

Let us recall briefly at this point the memory function formalism. In terms of Laplace transforms, defined as

$$\hat{f}(z) = \mathscr{L}f(t) = \int_0^\infty dt\exp(-zt)f(t), \tag{8.77}$$

one finds that

$$\hat{\rho}_T(q,z) = [z + \hat{K}_T(q,z)]^{-1} = \left[z + \left(\frac{q^2}{\rho m}\right)\hat{\eta}(q,z)\right]^{-1} \tag{8.78}$$

with

$$K_T(q,t) = \mathscr{L}^{-1}\hat{K}_T(q,z) \tag{8.79}$$

denoting the transverse current memory function whose initial value is given by

$$K_T(q,t=0) = \frac{q^2}{\rho m}\left[\rho k_B T + \frac{\rho^2}{q^2}\int d\mathbf{r}\frac{\partial^2\phi(r)}{\partial x^2}(1-\exp(-iqz))g(r)\right]. \tag{8.80}$$

In (8.80), the quantity in the square brackets denotes the wave vector rigidity modulus $G(q)$ expressed in terms of the pair potential $\phi(r)$ and the pair distribution function $g(r)$. An excellent approximation is

$$G(q) = \rho k_B T + \left(\frac{\rho m\omega_E^2}{q^2}\right)\left[1 - \frac{3j_1(q\sigma)}{q\sigma}\right] \tag{8.81}$$

where ω_E is the Einstein frequency ($\omega_E = 0.61 \times 10^{13}\ s^{-1}$ in liquid Rb at the temperature and density of the computer experiment of Balucani et al., 1987) (j_1 is the first-order spherical Bessel function: see (4.5) and following). Again, it will be convenient to deal with the normalized memory function defined by

$$n_T(q,t) = \frac{K_T(q,t)}{(q^2/\rho m)G(q)}. \tag{8.82}$$

In the final part of (8.78) for $\hat{\rho}_T$, the generalized wave vector and frequency-dependent viscosity $\hat{\eta}(q,z)$ has been introduced. In the hydrodynamic limit $q \to 0$, $z \to 0$, this is simply the shear viscosity coefficient η discussed at some length in Chapter 7. Beyond this regime, a somewhat simpler description of the nonhydrodynamic effects is provided by the quantity $\eta(q) = \hat{\eta}(q, z = 0)$. From (8.79) and (8.82), one has

$$\eta(q) = G(q)\hat{n}_T(q, z = 0) = G(q) \int_0^\infty \mathrm{d}t\, n_T(q,t). \tag{8.83}$$

If one assumes $n_T(q,t)$ to decay exponentially with a time constant $\tau(q)$, which is the so-called viscoelastic model, one then can deduce that $\eta(q) = G(q)\tau(q)$.

An equivalent expression for the q-dependent viscosity $\eta(q)$ can be written as

$$\eta(q) = \frac{\rho m}{q^2}\left[\int_0^\infty \mathrm{d}t\, \rho_T(q,t)\right]^{-1}. \tag{8.84}$$

Balucani et al. (1987) have used both these routes to calculate $\eta(q)$ from their simulation data for liquid Rb. They stress, as a result, that the gross dynamical features of the transverse current, e.g., the presence of shear excitations, have an important effect on $\eta(q)$. A rigorous limit for this quantity is known only in the large q, or free-particle, regime, namely,

$$\eta(q) \sim \left(\frac{2m\rho^2 k_B T}{\pi}\right)^{1/2} q^{-1}, \qquad q \to \infty. \tag{8.85}$$

8.7.1. *Simulation results*

Balucani et al. (1987) considered $N = 500$ Rb atoms in a box of length $L = 36.15$ Å, interacting by the effective potential of Price et al. (1976). They worked with the same reduced density $\rho^* = \rho\sigma^3 = 0.905$ used by Rahman (1974a,b) in his studies of collective modes in Rb to compare with the neutron experiments of Copley and Rowe (1974; see also Section 8.5). At the above density, the potential parameters used were $\sigma = 4.405$ Å and $\varepsilon/k_B = 402.8$ K. Balucani et al. (1987) display $\rho_T(q,t)$, $\rho_T(q,\omega)$ and the memory function $n_T(q,t)$ at several wave vectors.

The data for $\eta(q)$ will be only briefly considered here. This is shown in

Figure 8.3, where the data are compared, for example, with the predictions of the viscoelastic model.

With regard to the Stokes-Einstein expression, $D = k_B T/4\pi\eta R$, where R is the particle "radius" and slip boundary conditions have been used, this equation is expected to be rigorous for a Brownian particle immersed in a continuous medium with viscosity η. This is saying that it will work for a situation where the separation of length and times scales typical of hydrodynamics is certainly valid.

Balucani et al. (1985) propose instead

$$D = \frac{k_B T}{4\pi}\frac{4\rho}{3\pi}\int_0^\infty dq \frac{f(q)}{\eta(q)} \tag{8.86}$$

Figure 8.3. Generalized shear viscosity $\eta(q)$, as defined in (8.83) and (8.84) for liquid Rb (after Balucani et al., 1987). Dotted and solid lines come from versions of the viscoelastic model. Dashed line-two-exponential fit to memory function.

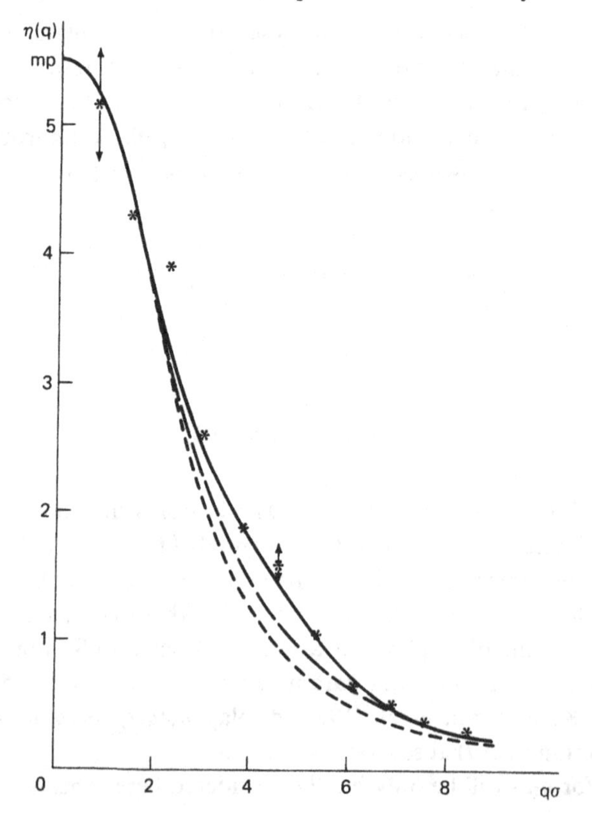

where $f(q) = 4\pi a^2 j_1(qa)/q$ and the length a is defined by $\frac{4}{3}\pi\rho a^3 = 1$. If the q dependence of $\eta(q)$ is neglected, the generalized viscosity then being replaced by its hydrodynamic value η, one recovers the Stokes-Einstein expression given above, with a playing the role of the particle radius. Although this leads to reasonable values for the diffusion coefficient, this is generally underestimated by around 25 to 30%. Since clearly the approximation $\eta(q) \sim \eta$ is not appropriate in the whole wave-vector range, a significant improvement is to be expected by taking into account the strong dependence of the generalized viscosity $\eta(q)$ with increasing q (see Figure 8.3).

8.8. Transport in supercooled liquid Rb

The structure of supercooled liquid Rb has already been studied in Section 6.9. In the equilibrium liquid, it was demonstrated there that the local environment, through which an ion diffuses, is evolving as well as becoming more sharply localized as the temperature decreases. In the supercooled liquid, only further sharpening occurs. The saturation of the structure means that self-diffusion occurs in similar environments for supercooled states at different temperatures. As a result, the temperature dependence of D is different in the two regimes (see Figure 8.4, where the difference from

Figure 8.4. Self-diffusion coefficient D as function of temperature for phases of Rb. For comparison, results for a Lennard-Jones fluid (e.g., argon) are also plotted (after Mountain, 1982; see correction, Mountain, 1983). Closed circles: Rb.

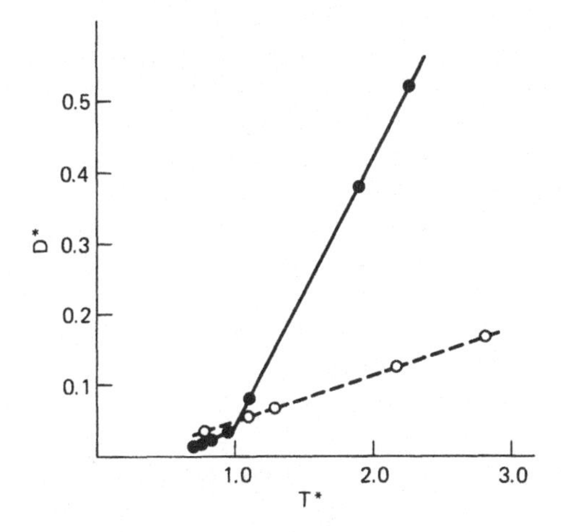

the behaviour of a Lennard-Jones fluid such as argon is seen to be marked; Mountain (1982; 1983)).

8.8.1. *Dynamic correlations*

The dynamical correlations considered in the previous section will now be summarized in the supercooled liquid states of Rb, again following the work of Mountain (1982). The transverse-current correlations (see Section 8.7) turn out to be significantly enhanced in the supercooled region. The dispersion relation for the propagating transverse mode was used to determine the lower limit for the wavelength of fluctuations whose decay can be described by linearized hydrodynamics. This limiting length was found to increase significantly with decreasing temperature and to be larger than the size of the cube to which periodic boundary conditions were applied in Mountain's study. This growth in the spatial extent of the correlations of transverse current fluctuations is reflected in the increase in the q- and ω-dependent shear viscosity $\tilde{\eta}(q, \omega)$ in the small q and ω regime. It also means that one would have to go to significantly larger systems before the long-wavelength low-frequency limit could be taken for supercooled liquid Rb. Though not specifically designed for liquid metals, in contrast to the preceding studies, the later computer experiments of Amar and Mountain (1987) on supercooling in soft-sphere fluids are nevertheless of some relevance in the present context.

9
Critical behaviour

In this chapter, an introduction will first be given to some progress in characterizing critical behaviour in terms of critical exponents. This will then be illustrated by calculations based on model equations of state. These will, in fact, give a route to the calculation of the liquid-vapour coexistence curves of liquid metals. Then some other nonuniversal properties, and especially the critical constants of the fluid alkalis, will be considered in relation to simple plasma models. Reference will also be made to spinodal curves.

In Chapter 6, when dealing with freezing theory, the order parameters were referred to; these were, in fact, the Fourier components of the periodic density in the crystalline phase. Again, in the treatment of the liquid-vapour critical behaviour, the concept of the order parameter is basic: Let us take this as a starting point.

9.1. Concept of order parameter

In the liquid-gas system, to be treated in detail below, the transition is from the high-temperature (or low-pressure) gas phase to the liquid phase, which has, as discussed quantitatively in the previous chapters, marked short-range order that is absent in the gaseous phase. Further lowering of temperature or increase of pressure normally produces a further transition to the solid phase, having crystalline long-range order (see Chapter 16). A typical phase diagram is shown in Figure 9.1. Having used the variables T and P, the third variable V is dependent, of course, and could have been used to distinguish the phases. For the liquid-gas transition, a better choice is the difference $\rho_l - \rho_g$ between the liquid and gas densities; this is zero in the high-temperature "disordered" gas phase and becomes nonzero in the "ordered" liquid phase. This is the behaviour typical of an order parameter, to be chosen always to become nonzero in the ordered phase.

Figure 9.1. Solid-liquid-gas phase diagram, showing state (T_c, P_c) corresponding to the critical point.

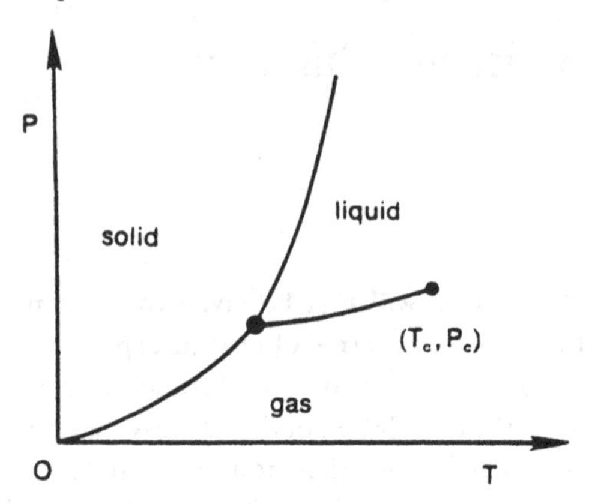

As mentioned above, an order parameter that is nonzero only in the solid phase could be constructed from an appropriate Fourier component of the periodic, spatially varying density. The gas-liquid-solid system of prime importance in this volume is one in which there are a series of ordered phases, characterized by different order parameters.

9.2. Liquid-vapour critical point: a second-order phase transition

Let us begin with an elementary introduction to the liquid-vapour critical point and then turn to a very brief summary of some of the results of the modern theory. A great deal of work remains to be done on this transition, in the context of liquid-state structural theory developed in the earlier chapters of this text.

The most obvious definition of this critical point is that point at which the isotherm has a point of inflexion (see Figure 9.2):

$$\left(\frac{\partial p}{\partial \rho}\right)_T = 0, \quad \left(\frac{\partial^2 p}{\partial \rho^2}\right)_T = 0. \tag{9.1}$$

From this it follows immediately that the isothermal compressibility K_T diverges at the critical point. Usually, the form of K_T is taken as

$$K_T \sim (T - T_c)^{-\gamma}, \tag{9.2}$$

measured at the critical density ρ_c, T_c being the critical temperature. For

Figure 9.2. Point of inflexion in critical isotherm (see (9.1)). $D_1 E_1$ is vapour region; $E_1 F_1$ is liquid-vapour while $F_1 G_1$ is liquid region, etc.

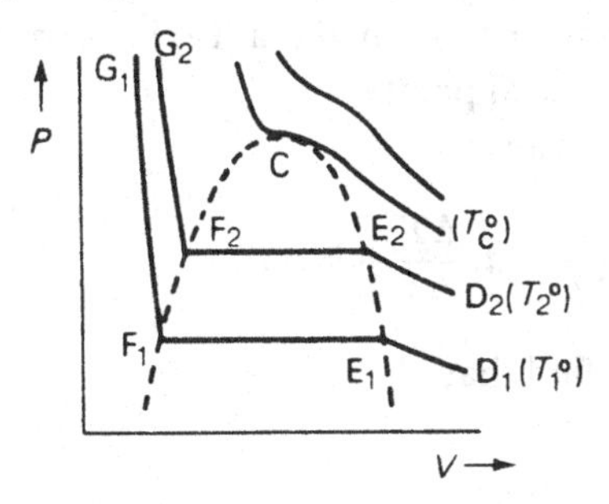

insulating liquids, experiment indicates that the exponent $\gamma \sim 1.1$. As will be discussed next, this is near the value for a fluid described by the equation of state of van der Waals.

Also, one can write for the difference between liquid (l) and gas (g) densities (the order parameter introduced in Section 9.1)

$$\rho_l - \rho_g \sim (T_c - T)^\beta \tag{9.3}$$

where the exponent β is found from experiments on insulating liquids to have a value of 0.35 ± 0.02.

9.2.1. *Fluid described by van der Waals equation of state*

Let us now examine how such critical behaviour can arise by studying the model represented by the van der Waals equation of state, namely,

$$\frac{p}{k_B T} = \frac{\rho}{1 - b\rho} - \frac{a\rho^2}{k_B T}. \tag{9.4}$$

The critical values of pressure, density, and temperature will be denoted by p_c, ρ_c, and T_c, respectively. We now find the critical exponents for K_T and $\rho_l - \rho_g$ in (9.2) and (9.3) by expanding about the critical point. Writing $\Delta p = p - p_c, \ldots$, we find

$$\Delta p = a_1 \Delta T + b_1 \Delta T \Delta\rho + d(\Delta\rho)^3 + \cdots \tag{9.5}$$

where the terms in $\Delta\rho$ and $(\Delta\rho)^2$ are absent because of conditions (9.1).

It is interesting that although the constants a_1, b_1, and d can, of course, be determined from the parameters appearing in the van der Waals equation of state, the exponents γ and β do not depend on them. These exponents depend only on the assumption (unfortunately, as discussed later, inaccurate, though still useful) that one can make a Taylor expansion of the equation of state about the critical point.

Dividing (9.5) by $\Delta\rho$, one can write

$$d(\Delta\rho)^2 = \frac{\Delta p}{\Delta\rho} - a_1\frac{\Delta T}{\Delta\rho} - b_1\Delta T. \tag{9.6}$$

In the limit $\Delta\rho$ tends to zero we find that

$$\frac{\Delta p}{\Delta\rho} \to \left(\frac{\partial p}{\partial\rho}\right)_{T=T_c} = 0 \tag{9.7}$$

and

$$\frac{\Delta T}{\Delta\rho} \to \left(\frac{\partial T}{\partial\rho}\right)_p = -\frac{(\partial p/\partial\rho)_{T=T_c}}{(\partial p/\partial T)_{\rho=\rho_c}} = 0, \tag{9.8}$$

since the ratio $\Delta p/\Delta T$ tends to a finite constant at the critical point. Thus (9.6) takes the form

$$\Delta\rho = \pm\text{const.}(\Delta T)^{1/2}. \tag{9.9}$$

It follows that $\beta = \frac{1}{2}$ in this van der Waals case. Equation (9.9) predicts further that the coexistence curve is symmetrical about $\rho = \rho_c$ in the neighbourhood of the critical point (see Section 9.4.1 for details, including the rule of equal areas).

The value $\beta = \frac{1}{2}$ is evidently not in quantitative agreement with the measured value of 0.35 quoted before for insulating fluids. The Taylor expansion assumed (which is certainly valid for the van der Waals equation of state) is, therefore, clearly an oversimplification for insulating fluids. Nevertheless, one finds from the van der Waals equation of state the correct general behaviour for $\rho_l - \rho_g$ as T tends to T_c, although the model is quantitatively in error.

One can also calculate $(\partial p/\partial\rho)_T$; one then finds

$$\left(\frac{\partial p}{\partial\rho}\right)_T \sim b_1\Delta T + 3d(\Delta\rho)^2. \tag{9.10}$$

Along the critical isochore, $\Delta\rho$ is zero and

$$K_T = \frac{1}{\rho}\left(\frac{\partial\rho}{\partial p}\right)_T = \frac{1}{a_1\rho_c}(\Delta T)^{-1}, \qquad T \to T_c^+. \tag{9.11}$$

Similarly, along the coexistence curve, (9.9) applies, and one has

$$K_T = \frac{1}{2a_1\rho_c}(\Delta T)^{-1}, \qquad T \to T_c^-. \tag{9.12}$$

One has that the critical exponent $\gamma = 1$, but it should be noted that the coefficients of $(\Delta T)^{-1}$ in (9.11) and (9.12) differ by a factor of 2. The exponent $\gamma = 1$ is in fair agreement with experiment.

A little further discussion of these critical exponents is given later. However, at this point it is important to summarize the way in which the structure factor $S(k)$ and the pair distribution function $g(r)$ of a fluid vary as the critical point is approached. A summary of the salient features of liquid structure as the critical point is approached will be presented; some further detail is to be found in Appendices 9.1 and 9.2.

9.3. Ornstein-Zernike theory of static structure

In (2.5), the long-wavelength limit $S(0)$ of the structure factor $S(k)$ was related to the isothermal compressibility K_T. But as has been seen, the quantity K_T diverges as T tends to T_c; therefore, from (2.5), $S(0)$ also diverges. But from the relation (2.4) between $S(k)$ and the radial distribution function $g(r)$, the divergence of $S(0)$ signals the fact that $g(r)$ must become so long-ranged as $T \to T_c$ that $\int [g(r) - 1]\,\mathrm{d}\mathbf{r}$ diverges.

In fact, in the argument of Ornstein and Zernike presented in Appendix 9.1, the large r behaviour of $g(r) - 1$ as the critical point is approached is given by

$$g(r) - 1 \sim \text{constant} \exp\frac{(-\kappa r)}{r}. \tag{9.13}$$

This equation leads to an important characterization of critical behaviour through a correlation length $\xi = \kappa^{-1}$, ξ becoming infinite as $T \to T_c$. This, in turn, implies from (9.13) that $g(r) - 1 \sim r^{-1}$ for large r as T tends to T_c, which does indeed cause the integral of $g(r) - 1$ through the liquid volume to diverge. The correlation length ξ is characterized by a third critical exponent ν through

$$\xi \sim |T - T_c|^{-\nu}. \tag{9.14}$$

Actually, as with the preceding van der Waals model, the Ornstein-Zernike assumptions are oversimplified. In reality, at $T = T_c$ and for large r,

$$g(r) - 1 \sim \frac{\text{constant}}{r^{1+\eta}} \tag{9.15}$$

where available experimental evidence points to the fact that η is small ($\sim$0.1). If $g(r) \sim r^{-1}$ at large r, then by Fourier transform, $S(k) \sim$ constant xk^{-2} for small k, showing how $S(k)$ diverges at T_c as k tends to zero. This latter result, of course, must be modified because in real fluids $\eta \neq 0$, the result for $S(k)$ being readily shown then to take the form

$$S(k) \sim \text{constant}/k^{2-\eta}; \qquad T = T_c, k \to 0. \tag{9.16}$$

Modern critical point theories (see Fisher, 1967; Widom, 1965; Kadanoff, 1966, also Appendix 9.2) lead to a relation between β, ν, and η, namely,

$$\frac{\beta}{\nu} = \frac{1+\eta}{2}. \tag{9.17}$$

With $\beta = \frac{1}{2}$, the van der Waals value, and $\eta = 0$ resulting from the Ornstein-Zernike treatment, one finds from (9.17) that $\nu = 1$, which is again the Ornstein-Zernike value. In practice, as we have seen, $\beta \sim \frac{1}{3}$ for insulating fluids; hence, with η small, $\nu \simeq 2\beta \simeq \frac{2}{3}$. A little more will be said on critical phenomena in this vein when we turn to dynamics in Section 9.7.

However, from these somewhat general considerations, it is of some interest to consider specific liquids and ask how these differ in detail as to their critical point properties and, in particular, as to their specific values of the critical constants p_c, V_c, and T_c. One particular combination that is usefully embraced by the discussion in Appendix 9.3 is the so-called compressibility ratio $Z_c = p_c V_c / k_B T_c$ (see Tables 9.1 and 9.2).

9.4. Predictions from more general equation of state

For the alkali metals, where the compressibility ratio Z_c just introduced varies markedly through the series (see Table 9.1 on page 112), Chapman and March (1987) have proposed a more general equation of state that allows for this variation. Of course, with van der Waals theory, Z_c is determined; it is simply the number $\frac{3}{8}$. Likewise, Dieterici's equation of state (see, for example, Guggenheim, 1949) yields a different number. Thus these equations of state are not general enough to allow for the variation of Z_c through the liquid alkali metal series. Therefore, Chapman and March, in the course of a study of experiments on thermal pressure coefficients, set up an equation of state (see (9.18)) that embraces both van der Waals and Dieterici models but into which one can put the value of the compressibility ratio.

The main purpose of this section is to use this equation of state (see Appendix 9.3 for its use in interpreting thermal pressure coefficients) to construct the liquid-vapour coexistence curve. By invoking the fact that β is known from experiment to be less than $\frac{1}{2}$, the van der Waals prediction, March, Tosi, and Chapman (1988) have proposed a way of getting round the difficulties associated with the van der Waals equation of state.

9.4.1. *Model equation of state: liquid-vapour coexistence curve*

Measuring pressure, temperature, and density in terms of the critical constants p_c, T_c, and ρ_c as units—i.e., $p^* = p/p_c$, and so on—one has the equation of state proposed by Chapman and March (1987) in the form

$$[p^* + f_1(T^*, \rho^*)][1 - \alpha\rho^*] = \frac{T^*\rho^*}{Z_c} f_2(T^*, \rho^*). \tag{9.18}$$

Using (9.18) together with the thermodynamic result

$$p = \rho^2 \left(\frac{\partial F}{\partial \rho}\right)_T, \tag{9.19}$$

with F the Helmholtz free energy per particle and the chemical potential

$$\mu = F + \frac{p}{\rho}, \tag{9.20}$$

one finds the following equilibrium equations. The first, namely,

$$\frac{T^*}{Z_c}\left\{\frac{\rho_l^* f_2(T^*, \rho_l^*)}{1 - \alpha\rho_l^*} - \frac{\rho_g^* f_2(T^*, \rho_g^*)}{1 - \alpha\rho_g^*}\right\} = f_1(T^*, \rho_l^*) - f_1(T^*, \rho_g^*) \tag{9.21}$$

comes from the equality of the pressures in the two phases. The second,

$$T^* \ln\frac{\rho_g^*}{\rho_l^*} = T^*[F_2(T^*, \rho_l^*) - F_2(T^*, \rho_g^*)] + Z_c[F_1(T^*, \rho_l^*) - F_1(T^*, \rho_g^*)], \tag{9.22}$$

is the condition for the equality of the chemical potentials. Here the quantities F_1 and F_2 are defined by integrals on the isotherms as

$$F_1(T^*, \rho^*) = -\int^{\rho^*} \frac{d\rho^*}{\rho^*} \frac{\partial f_1(T^*, \rho^*)}{\partial \rho^*} \tag{9.23}$$

and

$$F_2(T^*,\rho^*) = -\int^{\rho^*} \frac{d\rho^*}{\rho^*}\left\{1 - \frac{\partial}{\partial\rho^*}\left[\frac{\rho^* f_2(T^*,\rho^*)}{1-\alpha\rho^*}\right]\right\}. \tag{9.24}$$

EXAMPLE OF VAN DER WAALS EQUATION

Equation (9.18) has now the explicit form determined by

$$f_1(T^*,\rho^*) = 3\rho^{*2} \tag{9.25}$$

$$f_2(T^*,\rho^*) = 1 \tag{9.26}$$

together with $\alpha = \frac{1}{3}$ and the compressibility ratio $Z_c = \frac{3}{8}$.

The explicit form of the chemical potential μ is then

$$\frac{\mu}{k_B T_c} = \frac{\mu_{\text{ideal}}}{k_B T_c} - T^* \ln\left(1 - \frac{1}{3}\rho^*\right) + \frac{T^*}{(1-\frac{1}{3}\rho^*)} - \frac{9}{4}\rho^*. \tag{9.27}$$

The liquid-vapour equilibrium conditions μ = constant and p = constant, then lead to

$$T^* = \tfrac{1}{8}(\rho_l^* + \rho_g^*)(3 - \rho_l^*)(3 - \rho_g^*) \tag{9.28}$$

and

$$\ln\left[\frac{\rho_g^*}{\rho_l^*}\frac{3-\rho_l^*}{3-\rho_g^*}\right] = 3\frac{\rho_l^* - \rho_g^*}{(3-\rho_l^*)(3-\rho_g^*)}\left[1 - \frac{6}{\rho_l^* + \rho_g^*}\right]. \tag{9.29}$$

It is a straightforward matter from (9.29) to plot the liquid-vapour coexistence curves for this model. As can be seen, for example, from Figure 1 of March, Tosi, and Chapman (1988), such curves are not in agreement with data on, say, the insulating fluids neon or ethylene. Nor is the situation appreciably altered if one uses Dieterici's equation of state.

March, Tosi, and Chapman (1988) argue that to remove this disagreement, the function f_1 in (9.18) should be nonanalytic in the order parameter $\rho_l - \rho_g$. This constraint is shown to lead to a relation between average density of liquid and vapour and the temperature, transcending rectilinear diameters. The final point to be made in this context is that the above procedure involving equating chemical potentials is equivalent to Maxwell's equal areas construction invoked in Section 9.2. Though ample data exist on insulating liquids, some of which has been utilized previously, let us now turn to the case of the fluid alkali metals in more detail.

9.5. Critical constants of fluid alkali metals

Work by Chapman and March (1986) has drawn attention to some empirical relations between the critical constants of fluid alkali metals that are relevant in the present context. In particular, as will be discussed in some detail here, they show that $T_c V_c^n$ = constant for the critical temperature T_c and volume V_c of the alkali metals, where the exponent n is about 0.3. This is in marked contrast, as they emphasize, with the noble gases, where n is about -2. Furthermore, the so-called compressibility ratio $Z_c = p_c V_c / R T_c$ for the alkalis ranges from 0.06 for Li to 0.20 for Cs, whereas for the noble gases the value is 0.29 ± 0.01. Here, following the above workers, the importance of Coulomb forces in interpreting the results for the alkali metals will be emphasized.

As regards the regularities between T_c and V_c, it can be seen from Figure 9.3 that a ln-ln plot of T_c versus V_c leads, with useful accuracy, to a straight line with a slope of $n = 0.28$. The data for Rb and Cs seem appreciably better than for the other metals. The discussion of the interpretation of this exponent will be based on a model already referred to in Section 5.2, the one-component plasma, incorporating long-range Coulomb forces. By way of motivation for this model, Figure 9.4 shows a similar ln-ln plot of T_c versus V_c for the neutral noble-gas fluids. In contrast to the exponent $n \sim 0.3$ for the alkalis, an exponent of -2 is obtained for the noble gases. In addition, the compressibility ratio $Z_c = p_c V_c / R T_c$ is recorded along with the critical constants for the alkalis in Table 9.1,

Figure 9.3. A ln-ln plot of critical temperature T_c versus critical volume V_c using measured data for fluid alkalis. Straight line is well represented by $T_c V_c^n$ = constant, with $n = 0.28$ (Chapman and March, 1986).

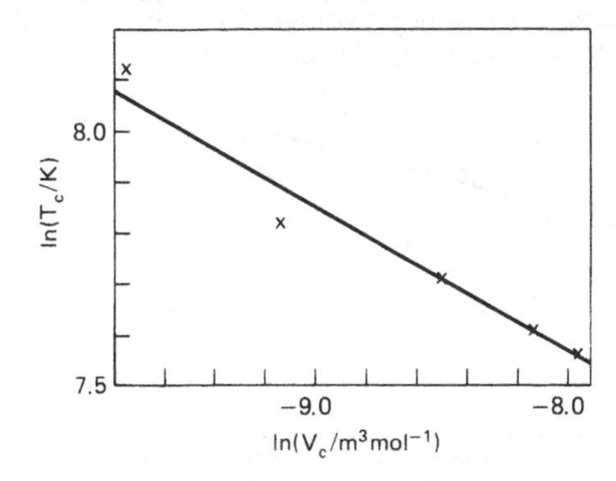

whereas Table 9.2 records similar data for the noble gases. These tables show that Z_c varies significantly through the five alkali metals, but in all cases the compressibility ratio is substantially lower than the almost constant value of 0.29 obtained for the noble gases. It is to be noted again

Table 9.1. *Observed critical constants of alkali metals.*

Element	p_c (MPa)	$10^4 V_c$ ($m^3 mol^{-1}$)	T_c (K)	Z_c
Li	30.4	0.588	3344	0.064
Na	25.22	1.09	2497	0.132
K	15.95	2.04	2239	0.175
Rb	12.45	2.93	2017	0.217
Cs	9.25	3.51	1924	0.203

Table 9.2. *Critical data for noble gases, for comparison with alkali metal data in Table 9.1.*

Element	p_c (MPa)	$10^5 V_c$ ($m^3 mol^{-1}$)	T_c (K)	Z_c
Ne	2.7262	4.17	44.45	0.308
Ar	4.8641	7.52	150.75	0.292
Kr	5.4917	9.22	209.35	0.291
Xe	5.8938	11.42	289.75	0.279

Figure 9.4. Similar to Figure 9.3 but for rare gases. The exponent n in $T_c V_c^n$ = constant required to fit the data is now negative and of magnitude 2. It should be noted that He behaves quite differently from the rest.

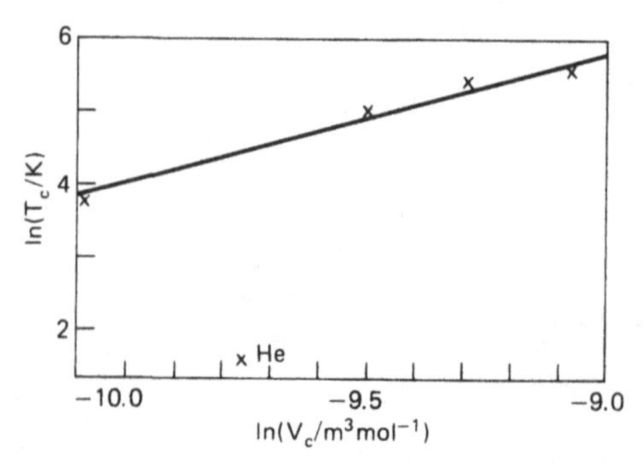

here that data for the lighter alkali metals involves some extrapolation from experiment and that only for Rb and Cs are all three critical quantities obtained simultaneously by direct measurement.

Following Chapman and March (1986), the relation of the empirical regularities for the alkalis discussed earlier to the predictions of the one-component plasma (OCP) model will be treated. The thermodynamic properties of this system are well established from the simulation work of Hansen et al. (1977) and may be written in terms of the OCP coupling parameter given by (see Chapter 4, eqn. (4.21))

$$\Gamma = \frac{e^2}{ak_B T} \tag{9.30}$$

where $\rho^{-1} = (4/3)\pi a^3$, if ρ denotes the particle density. One finds that the isothermal compressibility vanishes at $\Gamma = 3.1$ and that the pressure is nearly linear in Γ over a substantial range. For the pressure p in an OCP model, modified to account for core size, one therefore writes the expression

$$\frac{p}{\rho k_B T} = \frac{p_{\text{hs}}}{\rho k_B T} - \alpha\Gamma \tag{9.31}$$

where α is found to be 0.24 in the linear approximation from the fact that in the OCP, the value of Γ corresponding to $p = 0$ is 4.1.

Using expression (3.20) of Carnahan and Starling (1969) for the hard-sphere pressure p_{hs}, one readily obtains the form $T_c V_c^n =$ constant, with the exponent $n = \frac{1}{3}$. This is encouragingly close to the value 0.28 derived from the slope of the line in Figure 9.3, but it must be added that this model predicts only the order of magnitude of the quantity $T_c V_c^n$, the theoretical value of this quantity being a factor of four greater than the empirically determined value. It is also of interest that the Coulomb-force model predicts the compressibility ratio to be 0.14, which is considerably smaller than what would be obtained from classical equations of state, such as those of van der Waals or of Dieterici (see Guggenheim, 1949).

Clearly then, the merits of the simple model based on (9.31) are that it offers a simple explanation of the relation $T_c V_c^n =$ constant, with exponent n near to 0.3, and it also yields a low compressibility ratio Z_c, as required. One must caution, however, that in addition to the failure to predict the constant in the preceding equation to better than a factor of four, the model based on (9.31) fails to predict the coexistence curve at all correctly on the high-density side of the critical density, ρ_c, except very near T_c. These points indicate, of course, that important physical processes must be subsumed, in addition to the effects embodied already in writing (9.31). Two points that immediately come to mind are the existence of dimers in the fluid phases and the existence of a metal-insulator transition along the high-density side of the coexistence curve (compare Chapter 11). Nevertheless, the results of the work of Chapman and March (1986) can leave no doubt that the long-range Coulomb interaction e^2/r_{ij} between ions is playing an essential role in interpreting the empirical regularities for the critical constants of the fluid alkalis.

It is relevant here to point out the more formal, but also much more rigorous, theoretical discussion prompted by the results of Jüngst, Knuth and Hensel (1985) on the critical behaviour of Rb and Cs, given by Goldstein and Ashcroft (1985).

9.6. Spinodal curves

Having discussed the calculation of the coexistence curve of a monatomic fluid from a given equation of state, we now consider the theory of the spinodal curves. The treatment follows that given by March and Tosi (1988), the specific examples taken being the equations of state of van der Waals (W) and Dieterici (D) (see Guggenheim, 1949).

The spinodal curve is normally defined as corresponding to zero second derivative of the Helmholtz free energy F with respect to density ρ. Using the expression $p = \rho^2(\partial F/\partial \rho)_T$, where p is the pressure, this condition can be posed as $(\partial p/\partial \rho)_T = 0$. This is used in conjunction with the model equation of state (9.18). The functions f_1 and f_2 and the constant α will be made explicit in the examples W and D (see also Section 9.4).

From (9.18) $(\partial p^*/\partial \rho^*)_T = 0$ yields

$$\frac{T^*}{Z_c}\left[\frac{f_2(T^*,\rho^*)}{1-\alpha\rho^*} + \rho^*\left(\frac{\partial f_2(T^*,\rho^*)}{\partial \rho^*}\right)_{T^*}\right] = (1-\alpha\rho^*)\left(\frac{\partial f_1(T^*,\rho^*)}{\partial \rho^*}\right)_{T^*}. \tag{9.32}$$

The solution of (9.32) gives T^* as a function of ρ^*, and substituting in (9.18) one finds the equation of the spinodal curve in the (p^*, ρ^*) plane as

$$p^*(\rho^*) = \frac{T^*(\rho^*)}{Z_c}\rho^*\frac{f_2(T^*(\rho^*),\rho^*)}{1-\alpha\rho^*} - f_1(T^*(\rho^*),\rho^*). \tag{9.33}$$

At the critical density ρ_c one evidently has $(dp^*(\rho^*)/d\rho^*)_{\rho^*=1} = 0$, with similar conditions on p^*.

Although these equations define the spinodal curve for general functions f_1 and f_2, their use will be illustrated by taking in turn the models D and

W. For the first of these, one has explicitly

$$f_1(T^*,\rho^*) = 0, \qquad f_2(T^*,\rho^*) = \exp\left(-\frac{2\rho^*}{T^*}\right)$$
$$\alpha = \tfrac{1}{2}, \qquad Z_c = 2\exp(-2). \tag{9.34}$$

Equations (9.32) and (9.33) then become

$$T^*(\rho^*) = \rho^*(2-\rho^*) \tag{9.35}$$

and

$$p^*(\rho^*) = \rho^{*2}\exp\left[2\frac{1-\rho^*}{2-\rho^*}\right]. \tag{9.36}$$

From (9.35) the relation between density and temperature on the spinodal curve is

$$\rho^* = 1 \pm (1-T^*)^{1/2}. \tag{9.37}$$

For case W,

$$f_1(T^*,\rho^*) = 3\rho^{*2}, \qquad f_2(T^*,\rho^*) = 1, \qquad \alpha = \tfrac{1}{3}, \qquad Z_c = \tfrac{3}{8}. \tag{9.38}$$

Equations (9.32) and (9.33) now take the form

$$T^*(\rho^*) = \tfrac{1}{4}\rho^*(3-\rho^*)^2 \tag{9.39}$$

and

$$p^*(\rho^*) = \rho^{*2}(3-2\rho^*). \tag{9.40}$$

The two physically acceptable solutions of (9.39) for the density-temperature relation on the spinodal curve are

$$\rho^*(T^*) = [2-\cos(\tfrac{1}{3}\arctan g\tau)] \pm \sqrt{3}\sin(\tfrac{1}{3}\arctan g\tau) \tag{9.41}$$

where

$$\tau = \frac{[4T^*(1-T^*)]^{1/2}}{2T^*-1}. \tag{9.42}$$

On approaching the critical point ($T^* \to 1$), (9.41) and (9.42) yield

$$\rho^* \to 1 \pm \frac{2\sqrt{3}}{3}(1-T^*)^{1/2}. \tag{9.43}$$

As discussed in Section 9.2 for W, it is known from critical exponent theory that both W and D models give an incorrect description of the coexistence curve near the critical point. The relevance of this to the present

Figure 9.5. Shows schematic way in which spinodal (S) and coexistence (C) curves can be expected to relate to one another (March and Tosi, 1988). Dash–dot lines show results of van der Waals model.

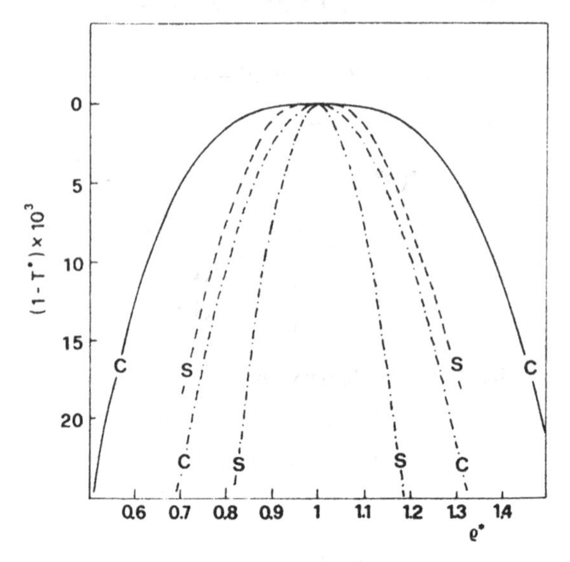

discussion is that, in the (T^*, ρ^*) plane, the spinodal forms (9.37) and (9.43) have the same exponents as for these same models of the coexistence curve. It is to be anticipated that such behaviour will be erroneous near T_c for spinodal properties also, but it is to be stressed that, at the time of writing, experiments on monatomic fluids that are decisive on this point do not seem to exist.

Nevertheless, it seemed of interest to show schematically in Figure 9.5 the way spinodal and coexistence curves might relate to one another in a more general model. In particular, a critical exponent β, known to be very near to $\frac{1}{3}$ for the coexistence curve, can be expected to have relevance for the spinodal behaviour near the critical point. As discussed in Section 9.4 on the coexistence curve using the model equation of state (9.18), the function f_1 is general enough to permit the inclusion of non-analytic behaviour near the critical point, leading to $\beta = \frac{1}{3}$.

9.7. Dynamics of critical behaviour

Having said something in Section 9.3 about static correlation functions $S(k)$ and $g(r)$ near the liquid-vapour critical point, it is relevant to the preceding discussion of dynamics to consider this aspect as the critical point is approached.

The oldest theory of critical dynamics goes back to Van Hove (1954). Consider some variable, $Q(\mathbf{r}, t)$ say, which relaxes to its equilibrium value via a single, nonpropagating mode. If Q is chosen such that it is conserved by the equations of motion of the system (i.e., $\int Q(\mathbf{r}, t)\,d\mathbf{r}$ is independent of time), then at any temperature $T \neq T_c$, the relaxation of Q in the long-wavelength limit k tends to zero is given by a diffusion equation. The characteristic frequency associated with the relaxation rate of Q can be written as

$$\bar{\omega} = \frac{\lambda_Q k^2}{\chi_Q} \tag{9.44}$$

where λ_Q is the transport coefficient for the variable Q, while χ_Q denotes the appropriate susceptibility describing the response of the system to an applied field. In the case of entropy fluctuations, for example, λ_E is the thermal conductivity, whereas χ_E is the heat capacity per unit volume at constant pressure, ρc_p, and therefore

$$\bar{\omega}_E = \frac{\lambda_E k^2}{\rho c_p}. \tag{9.45}$$

On the other hand, if the variable Q is not conserved, then the relaxation to equilibrium should take place at a finite rate even at $k = 0$, and hence

$$\bar{\omega} = \frac{\Lambda_Q}{\chi_Q} \tag{9.46}$$

where Λ_Q is the kinetic coefficient for Q.

In Van Hove's theory, the assumption made is that the transport coefficients λ_Q and also the kinetic coefficients Λ_Q are finite and nonzero at the critical point.

If the variable Q is, in fact, the order parameter (say ψ) of the assembly, then χ_Q diverges near T_c as the inverse coherence length κ, to the power $2 - \eta$:

$$\chi_Q \propto \kappa^{2-\eta}, \tag{9.47}$$

where η is a small positive quantity. If the relaxation of ψ is described by a single nonpropagating mode, then the Van Hove theory predicts a "critical slowing down" of the order parameter fluctuations as k tends to zero near T_c, with

$$\bar{\omega} \propto \begin{cases} \kappa^{2-\eta}k^2 & \text{if } \psi \text{ is conserved,} \\ \kappa^{2-\eta} & \text{if } \psi \text{ is not conserved.} \end{cases} \tag{9.48}$$

Unfortunately there are many situations in which both first-principles theories and experiment are not in agreement with the predictions of this simple theory.

For the specific case of the liquid-gas critical point, the order parameter is conserved (also true for, say, the critical point for phase separation) and is dominated by a nonpropagating diffusive mode at long wavelength, both above and below the transition. For instance, in the liquid-gas case, we saw in Section 9.2 that the order parameter may be taken as the density difference $\rho_l - \rho_g$, and long-wavelength density fluctuations near T_c are dominated by entropy fluctuations at constant pressure, whose relaxation is described by the thermal conductivity formula (A9.3.1). Nevertheless, a mode-mode analysis (see, for example, Kawasaki, 1976; Stanley, 1971) shows that the conventional theory is not correct.

Universality

Let us briefly motivate the ensuing discussion by referring here to the notion of universality. As with static critical phenomena, universality does not mean that all systems have the same critical exponents but rather that there exist large classes of systems with the same critical behaviour. However, it is known that a class of systems having the same static critical behaviour may be divided into several classes of dynamic behaviour. Also it needs to be said here that the dynamical characteristics of the liquid-vapour critical phenomena may have somewhat complex features. For a fuller discussion, the reader may consult Kadanoff (1970).

Finally, let us turn, in this context, to dynamic scaling. Following Halperin and Hohenberg (1967; see also Halperin, 1973) one says that the correlation function $Q(\mathbf{k}, \omega)$ obeys dynamic scaling if

1. At $T = T_c$, for sufficiently small $\mathbf{k}$,

$$Q(\mathbf{k}, \omega) \sim 2\pi\bar{\omega}_k^{-1}Q(\mathbf{k})f_Q\left(\frac{\omega}{\bar{\omega}_{\mathbf{k}}}\right) \tag{9.49}$$

with

$$\bar{\omega}_k = \text{constant} \times k^z \tag{9.50}$$

where $Q(\mathbf{k})$ is the Fourier transform of the time $t = 0$ form of Q, z is an as yet unspecified exponent, and f_Q is a function normalized to unity and dependent only on $\omega/\bar{\omega}_k$. The dynamic scaling hypothesis supposes also that at T_c, f_Q is a well-behaved function of its argument, with a characteristic width of order unity. Thus the spectrum of $Q(\mathbf{k}, \omega)$, at a fixed wave vector $\mathbf{k}$, has a single relaxation frequency $\bar{\omega}_k$.

2. For temperatures near T_c, the dynamic scaling hypothesis requires that the function $Q(\mathbf{k}, \omega)$ has a form similar to (9.49), but now the functions depend in an essential way on the ratio k/κ, where the inverse coherence length κ goes to zero at T_c. (Along the coexistence curve, or the "critical isochore", κ varies as $(T - T_c)^\nu$ (see (9.14)). For example, away from the critical point, $\bar{\omega}_k$ is assumed to take the form

$$\bar{\omega}_k \propto k^2 \Omega\left(\frac{k}{\kappa}\right) \tag{9.51}$$

where Ω depends on k only through the ratio k/κ. If the limit k/κ tends to zero, the characteristic frequency will have the form

$$\bar{\omega} \propto k^z \left(\frac{k}{\kappa}\right)^x \tag{9.52}$$

where x is determined by macroscopic or hydrodynamic considerations. Usually x will be 0, 1, or 2, depending on the variable Q under consideration.

As discussed in detail by Halperin (1973), the exponent z depends not only on whether or not the order parameter ψ is conserved, but also on whether or not the energy is conserved. When ψ is not conserved, conventional theory would yield $z = 2 - \eta$, whereas when ψ is conserved, $z = 4 - \eta$.

When neither ψ nor E is conserved, the renormalization methods sketched in Appendix 9.2 have been applied to some simple models. The exponent z in this case is found to be greater than the conventional results.

When ψ is conserved but E is not, evidence points to the applicability of conventional theory. When ψ is not conserved but E is, the conventional theory again appears to fail, as discussed more fully by Halperin et al. (1972) (see also Halperin, 1973).

In concluding this brief discussion of critical dynamics, reference should be made to the studies of Olchowy and Sengers (1988). These workers have solved mode-coupling equations (Kawasaki, 1976; see also Appendix 9.4) for dynamics of critical fluctuations that incorporate the crossover from the singular behaviour of transport properties of fluids asymptotically close to the critical point to the regular behaviour characteristic when one is far from the critical region. Their calculations show good agreement with experimental thermal diffusivity, thermal conductivity, and viscosity data for the insulating fluid carbon dioxide at all temperatures and pressures where critical effects in these transport properties are observed.

10

Electron states, including critical region

So far, one has assumed that the electron states in liquid metals can be described by the nearly free electron approximation. It seems unlikely that this can be true in the presence of strong electron-ion scattering, such as exists in (1) rare earth metals, (2) transition metals, and (3) even *sp* metals such as the heavy alkalis, as these are taken up along the coexistence curve toward the critical point (Chapman and March, 1988). Therefore, in this chapter, it is important to summarize a basic approach to the calculation of electron states in a framework that transcends the nearly free electron approximation.

10.1. Electron states in simple *s-p* metals

Following Edwards (1958, 1962; see also Cusack, 1963; 1987), Ballentine (1966; 1975) has employed the Green function formalism to make perturbation calculations of the density of states N(E) for Al and Zn, thereby allowing him to discuss the range of its validity and limitations and motivating more satisfactory calculations for Al and Bi.

As discussed more specifically later in this chapter, in order to study the electronic structure of a liquid metal, represented by a model of independent electrons interacting with a disordered array of ions, one should start from a fixed configuration of ions and then average over the ensemble of all possible arrangements of the ions. The Green function formalism as set up by Edwards (1958; 1962) allows the calculation of ensemble averages of physical quantities without ever calculating the wave functions for a specific ionic array. The summary of the Green function procedure below follows closely the account of Ballentine (1966; 1975).

The Green function operator is defined as

$$G(E) = (E - H)^{-1} \tag{10.1}$$

where $H = H_0 + V$ is the one-electron Hamiltonian and V is the potential due to N ions. In terms of the eigenfunctions ψ_n of the equation

$$H\psi_n = E_n\psi_n \tag{10.2}$$

$G(E)$ has the representation

$$G(E) = \sum_n \frac{|\psi_n\rangle\langle\psi_n|}{E - E_n}. \tag{10.3}$$

The diagonal matrix element of $G(E)$ in the momentum representation, namely,

$$G(\mathbf{k}, E) = \langle\mathbf{k}|\,G(E)\,|\mathbf{k}\rangle \tag{10.4}$$

will be used. It is convenient to introduce a spectral function

$$\begin{aligned}\rho(\mathbf{k}, E) &= (-1/\pi)\,\mathrm{Im}\{G(\mathbf{k}, E + i\varepsilon)\}\\ &= \sum_n |\langle\mathbf{k}|\psi_n\rangle|^2\delta(E - E_n),\end{aligned} \tag{10.5}$$

where $\varepsilon \to 0+$. This function, which measures the momentum distribution of the electrons with energy E, provides a quantitative measure of the extent to which the electronic states differ from free-particle states. The completeness of the set of states ψ_n implies the sum rule

$$\int_{-\infty}^{\infty} \rho(\mathbf{k}, E)\,\mathrm{d}E = 1. \tag{10.6}$$

The density of states (ignoring spin) per unit energy range is given by

$$N(E) = \sum_{\mathbf{k}} \rho(\mathbf{k}, E) = \frac{L^3}{(2\pi)^3}\int \mathrm{d}\mathbf{k}\,\rho(\mathbf{k}, E), \tag{10.7}$$

with L^3 the normalization volume.

In order to calculate the ensemble average Green function and hence the average spectral function and density of states, one makes use of the formal perturbation theory due to Edwards. The Green operator is formally expanded as

$$G(E) = G_0 + G_0VG_0 + G_0VG_0VG_0 + \cdots \tag{10.8}$$

where $G_0(E) = (E - H_0)^{-1}$ and H_0 is the kinetic energy operator. If one takes the total potential V to be the sum of (suitably screened) spherically symmetric potentials v centred at each of the N ions:

$$V(r) = \sum_\alpha v(\mathbf{r} - \mathbf{R}_\alpha), \tag{10.9}$$

then the dependence of the momentum representation matrix element of V

on the ionic positions can be factored out to yield

$$\langle \mathbf{k}|V|\mathbf{k}'\rangle = \langle \mathbf{k}|v|\mathbf{k}'\rangle \sum_{\alpha} \exp(i(\mathbf{k}-\mathbf{k}')\cdot\mathbf{R}_{\alpha}). \tag{10.10}$$

The nth-order term in the expansion of $G(\mathbf{k}, E)$ can be written, now employing units with $\hbar = 2m = 1$,

$$\frac{1}{E-k^2} \sum_{\mathbf{k}',\mathbf{k}''\ldots\mathbf{k}^{(n-1)}} \langle \mathbf{k}|v|\mathbf{k}'\rangle \frac{1}{E-(k')^2} \langle \mathbf{k}'|v|\mathbf{k}''\rangle \cdots$$

$$\frac{1}{E-(k^{(n-1)})^2} \langle \mathbf{k}^{(n-1)}|v|\mathbf{k}\rangle \frac{1}{E-k^2} C_n(\mathbf{k}'-\mathbf{k}, \mathbf{k}''-\mathbf{k}', \ldots, \mathbf{k}-\mathbf{k}^{(n-1)}), \tag{10.11}$$

where the exponentials from (10.10) have been grouped together into the function C_n defined by

$$C_n(\mathbf{p}_1, \mathbf{p}_2, \ldots, \mathbf{p}_n) = \sum_{\alpha\ldots\omega} \exp(i\mathbf{p}_1\cdot\mathbf{R}_{\alpha} + \cdots + \mathbf{p}_n\mathbf{R}_{\omega}). \tag{10.12}$$

To obtain the ensemble average of $G(\mathbf{k}, E)$, one replaces C_n by its average

$$\bar{C}_n(\mathbf{p}_1, \mathbf{p}_2, \ldots, \mathbf{p}_n) \equiv \langle C_n\rangle_{\text{av}}, \tag{10.13}$$

which involves the n-particle correlation function of the liquid.

Following Edwards, Ballentine then shows that this procedure allows $G(\mathbf{k}, E)$ to be approximately summed into the form*

$$G(\mathbf{k}, E) = [E - k^2 - \Sigma(\mathbf{k}, E)]^{-1}. \tag{10.14}$$

This procedure then evidently reduces the problem to approximating the self-energy $\Sigma(\mathbf{k}, E)$.

10.1.1. Second-order calculation of self-energy

The first-order contribution to $\Sigma(\mathbf{k}, E)$ is just the average potential energy, which can conveniently be chosen as zero. One is then led to evaluate Σ to second order, which yields

$$\Sigma_2(\mathbf{k}, E) = \sum_{\mathbf{k}'\neq\mathbf{k}} \frac{\langle \mathbf{k}|v|\mathbf{k}'\rangle^2}{E-k'^2} \bar{C}_2(\mathbf{k}, -\mathbf{k}). \tag{10.15}$$

Introducing the momentum transfer $\mathbf{K}$ by

* Ballentine shows, for the reader initiated in graph theory, that the function $\Sigma(\mathbf{k}, E)$ is equal to the sum of all irreducible graphs (see his Figure 1).

$$\mathbf{K} = \mathbf{k}' - \mathbf{k} \tag{10.16}$$

and

$$u(\mathbf{K}) = L^3\langle \mathbf{k}|v|\mathbf{k}'\rangle = \int \exp(i\mathbf{K}\cdot\mathbf{r})v(\mathbf{r})\,d\mathbf{r}, \tag{10.17}$$

with $n_a = N/L^3$ as the atomic density, one can show that

$$\Sigma_2(\mathbf{k}, E) = \frac{n_a}{(2\pi)^3}\int d\mathbf{k}'\frac{|u(\mathbf{K})|^2 S(\mathbf{K})}{E-(k')^2}, \tag{10.18}$$

with $S(\mathbf{K})$ as usual the liquid structure factor.

Ballentine (1966) has used suitable model potentials to calculate the density of states for Al and Zn; the reader is referred to his paper for details. It turns out, as he stresses, that the calculations thus made are unsatisfactory for $E \langle |\mathrm{Re}\,\Sigma_2|$. He proposes to remove this difficulty by replacing the free-particle Green function in the integral of (10.18) by the interacting Green function, to yield the self-consistent equation to be solved for the self-energy:

$$\Sigma(\mathbf{k}, E) = \frac{n_a}{(2\pi)^3}\int \frac{d\mathbf{k}'\,|u(\mathbf{k}'-\mathbf{k})|^2 S(\mathbf{k}'-\mathbf{k})}{E-(k')^2-\Sigma(\mathbf{k}', E)} \tag{10.19}$$

This, it turns out from these calculations, leads to physically plausible results for the cases of Al and Bi that Ballentine (1966) studied.

10.2. Electronic structure of nonsimple liquid metals

Considerable attention (Bansil et al., 1976) has been devoted to the calculation of the electronic structure of nonsimple liquid metals and in particular liquid transition and noble metals. Below, two approaches will be surveyed: first, the tight-binding approximation and second, the muffin-tin model.

10.2.1. Tight-binding approach

Extensive studies of the tight-binding approximation (TBA) for calculating the electronic density of states of liquid nonsimple metals have by now been made, following the work of Ishida and Yonezawa (1973) and of Roth

(1972, 1973). The presentation here draws heavily on the review by Watabe (1977).

Below, for simplicity of presentation, let us consider the TBA for the single s-band model. Here it is assumed that the one-electron wave function, and hence the one-electron Green function, $G(\mathbf{r},\mathbf{r}')$, can be expanded in terms of atomic s-orbitals, $\phi_i(\mathbf{r}) = \phi(|\mathbf{r} - \mathbf{R}_i|)$, centred on the various atomic sites:

$$G(\mathbf{r},\mathbf{r}') = \sum_{i,j} \phi_i(\mathbf{r}) G_{ij} \phi_j^*(\mathbf{r}'). \tag{10.20}$$

For G_{ij} one has the equation

$$\sum_l (zS_{il} - H_{il}) G_{lj} = \delta_{ij} \tag{10.21}$$

where

$$S_{il} = S(\mathbf{R}_{il}) = \int \phi_i^*(\mathbf{r}) \phi_l(\mathbf{r}) \, d\mathbf{r} \tag{10.22}$$

is the overlap integral, while H_{il} is the matrix element of the one-electron Hamiltonian:

$$H_{il} = \int \phi_i^*(\mathbf{r}) H \phi_l(\mathbf{r}) \, d\mathbf{r}. \tag{10.23}$$

In what follows, these two assumptions will be made for H_{ij}:

1. For $i \neq j$, H_{ij} is a function only of the distance between atom i and j, that is, $H_{ij} = H(R_{ij})$.
2. For $i = j$, H_{ii} is a constant independent of the distribution of neighbouring atoms, that is $H_{ii} = \varepsilon_0$.

The equation for G_{ij} is given by

$$(z - \varepsilon_0) G_{ij} - \sum_{l \neq i} H'(R_{il}) G_{lj} = \delta_{ij} \tag{10.24}$$

where

$$H'(R_{il}) = H(R_{il}) - zS(R_{il}). \tag{10.25}$$

In the usual two-centre approximation, we can write H' as

$$H'(R_{il}) = t(R_{il}) - (z - \varepsilon_0) S(R_{il}) \tag{10.26}$$

where $t_{il} = t(R_{il})$ is the transfer integral. It is frequently assumed further that the atomic orbitals have zero overlap, so that $S_{ij} = \delta_{ij}$ and $S(R_{il})$ in $H'(R_{il})$ is neglected.

DENSITY OF STATES

The ensemble-averaged density of states (per atom) is given in terms of the Green functions with $z = E + i0^+$ as

$$
\begin{aligned}
N(E) &= -\frac{1}{\pi N}\operatorname{Im}\int \langle G(\mathbf{r},\mathbf{r})\rangle \, d\mathbf{r} \\
&= -\frac{1}{\pi}\operatorname{Im}\left[\left\langle \sum_i G_{ii}\right\rangle + \left\langle \sum_{i\neq j} S_{ji}G_{ij}\right\rangle\right. \\
&= -\frac{1}{\pi}\operatorname{Im}\left[\langle G_{ii}\rangle_i + n_a \int S_{ji}\langle G_{ij}\rangle_{ij} g_2(R_{ij})\, d\mathbf{R}_{ij}\right.
\end{aligned} \tag{10.27}
$$

where $n_a = N/\mathscr{V}$ is the number density of atoms, $\mathscr{V}$ being the volume of the system. The notation is such that $\langle \ldots \rangle$ denotes the configuration average over all atomic sites, whereas $\langle \ldots \rangle_i$ and $\langle \ldots \rangle_{ij}$ denote, respectively, the conditional average with the position of the ith atom held fixed and that with the positions of both the ith and jth atoms held fixed. To derive the preceding equation, one has to use the fact that the system can be assumed to be statistically homogeneous so that $\langle G_{ii}\rangle_i$ is independent of i (or $\mathbf{R}_i$) and $\langle G_{ij}\rangle_{ij}$ is a function only of R_{ij}. As usual, $g_2(R_{ij})$ in (10.27) is the pair-distribution function for the ions. If one neglects overlap, then

$$
N(E) = -\frac{1}{\pi}\operatorname{Im}\langle G_{ii}\rangle_i. \tag{10.28}
$$

To calculate $\langle G_{ii}\rangle_i$ and $\langle G_{ij}\rangle_{ij}$, one expands G_{ij} in terms of H' as

$$
\begin{aligned}
G_{ij} &= G_0\delta_{ij} + G_0(1-\delta_{ij})H'_{ij}G_0 + \sum_l{}' G_0H'_{il}G_0H'_{lj}G_0 \\
&\quad + \sum_{l,m} G_0H'_{il}G_0H'_{lm}G_0H'_{mj}G_0 + \cdots
\end{aligned} \tag{10.29}
$$

where $G_0 = (z-\varepsilon_0)^{-1}$ and the prime on the summation indicates that any successive indices are different from each other. Taking the configuration average term by term, one finds

$$
\begin{aligned}
\langle G_{ii}\rangle_i &= G_0 + G_0 n_a \int d\mathbf{R}_1\, g_2(\mathbf{R}_i,\mathbf{R}_l)H'_{il}G_0H'_{li}G_0 \\
&\quad + G_0 n_a^2 \int d\mathbf{R}_1\, d\mathbf{R}_m\, g_3(\mathbf{R}_i,\mathbf{R}_l,\mathbf{R}_m)H'_{il}G_0H'_{lm}G_0H'_{mi}G_0 + \cdots
\end{aligned} \tag{10.30}
$$

and

$$\langle G_{ij}\rangle_{ij} = G_0 H'_{ij} G_0 + G_0 n_a \int d\mathbf{R}_l [g_2(\mathbf{R}_i, \mathbf{R}_j)]^{-1} g_3(\mathbf{R}_i, \mathbf{R}_j, \mathbf{R}_l)$$
$$\times H'_{il} G_0 H'_{lj} G_0 + \cdots, \tag{10.31}$$

where one needs the s-body distribution function, $g_s(\mathbf{R}_1, \ldots, \mathbf{R}_s)$, to express the average of the term containing s different atomic sites.

Of course, it is necessary to approximate g_s, usually in terms of g_2, and also to partially sum the preceding series. As to the latter, many existing tight-binding theories employ the single-site approximation (SSA) (though some discussion of theories beyond SSA is given by Katz and Rice (1972)), in which all multisite terms describing the repeated scattering of an electron between two or more sites are neglected. In addition, one often breaks g_3, for example, into product of g_2's, following Kirkwood. This is done by Roth (1974).

10.2.2. Muffin-tin potential model

The muffin-tin model has been studied extensively for describing the electronic band structure of crystalline metallic systems. Investigation of this model for a liquid metal led to Lloyd's (1967a) derivation of a formula for the density of states of a system of nonoverlapping spherical potentials that is valid for any configuration of scattering centres. The formalism is a natural generalization of the Korringa-Kohn-Rostoker (KKR) method of determining the band structure of crystalline systems and reduces to it for a regular lattice (see Jones and March, 1985).

Lloyd's formula for the integrated density of states $n(E)$ for this model is

$$n(E) = n_0(E) + \frac{1}{\pi N} \langle \mathrm{Im}\,\mathrm{Tr}\ln \mathbf{T}(E)\rangle. \tag{10.32}$$

Here $n_0(E)$ is the integrated density of states for free electrons, whereas $\mathbf{T}(E)$ is a generalized total t matrix (see Appendix 7.1):

$$\mathbf{T}^{-1}(E) = \|\delta_{ij}\delta_{LL'} t_l^{-1}(E) - G_{0,LL'}(\mathbf{R}_i - \mathbf{R}_j, E)\|. \tag{10.33}$$

The labelling of the matrix specifies the positions of the individual atoms as well as the angular momentum subscripts $L = (l, m)$.

$$t_l(E) = -E^{-1/2} \exp[i\delta_l(E)] \sin \delta_l(E) \tag{10.34}$$

is the t matrix of the individual atom, $\delta_l(E)$ being the scattering phase shift at the energy E and $G_{0,LL'}(\mathbf{R}_i - \mathbf{R}_j, E)$ is a quantity related to the free-particle propagator (see Lloyd, 1967b; Lloyd and Smith, 1972). From (10.32) the excess density of states is obtained as

$$\begin{aligned}
\delta N(E) &= N(E) - N_0(E) \\
&= \frac{-1}{\pi N}\left\langle \operatorname{Im} \operatorname{Tr} \mathbf{T}(E) \frac{\partial \mathbf{T}(E)^{-1}}{\partial E} \right\rangle \\
&= \frac{-1}{\pi N} \operatorname{Im}\left[\left\langle \sum_i \operatorname{Tr} \mathbf{T}_{ii}(E) \frac{\partial \mathbf{t}(E)^{-1}}{\partial E} \right\rangle \right. \\
&\qquad \left. - \sum_{i \neq j} \operatorname{Tr}\left\langle \mathbf{T}_{ij}(E) \frac{\partial \mathbf{G}_0(\mathbf{R}_j - \mathbf{R}_i, E)}{\partial E} \right\rangle \right]
\end{aligned} \tag{10.35}$$

where in the last expression the trace operation with respect to the atomic labels is explicitly written and the remaining trace denotes only that over the angular momentum subscripts. The matrices are defined as

$$\mathbf{t}(E) = \|\delta_{LL'} t_l(E)\| \tag{10.36}$$

$$\mathbf{G}_0(\mathbf{R}, E) = \|G_{0,LL'}(\mathbf{R}, E)\|, \tag{10.37}$$

and $\mathbf{T}_{ij}(E)$ is the (i, j) matrix element of $\mathbf{T}(E)$, which is given as a perturbation series expansion with respect to $\mathbf{t}(E)$ by

$$\begin{aligned}
\mathbf{T}_{ij} &= \mathbf{t}\delta_{ij} + \mathbf{t}(1 - \delta_{ij})\mathbf{G}_0(\mathbf{R}_i - \mathbf{R}_j)\mathbf{t} \\
&\quad + \sum_l{}' \mathbf{t}\mathbf{G}_0(\mathbf{R}_i - \mathbf{R}_l)\mathbf{t}\mathbf{G}_0(\mathbf{R}_l - \mathbf{R}_j)\mathbf{t} \\
&\quad + \sum_{l,m}{}' \mathbf{t}\mathbf{G}_0(\mathbf{R}_i - \mathbf{R}_l)\mathbf{t}\mathbf{G}_0(\mathbf{R}_l - \mathbf{R}_m)\mathbf{t}\mathbf{G}_0(\mathbf{R}_m - \mathbf{R}_j)\mathbf{t} + \cdots .
\end{aligned} \tag{10.38}$$

Comparison with Multiband Tight-Binding Theory

By comparing the preceding results with the equations in the multiband tight-binding theory, it can be shown that the two formalisms are equivalent if one notes the following correspondence:

Muffin-tin model		*Tight-binding model*
$\mathbf{t}(E)$	$\leftrightarrow$	$G_0(E)$
$\mathbf{G}_0(\mathbf{R}_{ij}, E)$	$\leftrightarrow$	$H'(\mathbf{R}_{ij})$
$\mathbf{T}_{ij}(E)$	$\leftrightarrow$	$G_{ij}(E)$
$\delta N(E)$	$\leftrightarrow$	$N(E)$

It is also worth noting that if one introduces a matrix

$$\begin{aligned}\mathbf{G}(E) &= \|\mathbf{G}_{ij}(E)\| \\ &= \|\delta_{ij}\delta_{\mu\nu}G_{0,\mu}(E)^{-1} - H'_{\mu\nu}(\mathbf{R}_{ij}, E)\|^{-1}\end{aligned} \tag{10.39}$$

where ν and μ are the suffixes specifying the relevant atomic levels, one can write the integrated density of states in the tight-binding model as

$$n_{<E}(E) = \frac{1}{\pi N}\langle \mathrm{Im}\,\mathrm{Tr}\ln \mathbf{G}(E)\rangle. \tag{10.40}$$

By noting the correspondence in (10.40), approximations proposed for the tight-binding model can readily be extended to the muffin-tin model (e.g., the Ishida-Yonezawa (1973) method is paralleled by Lloyd (1967a)). The modified quasi-crystalline approximation proposed by Schwartz et al. (1975), can be shown to be the non-self-consistent version of the Lloyd-Ishida-Yonezawa approximation.

Liquid noble and transition metals have been studied by the muffin-tin model (e.g., Peterson et al. (1976) in liquid Cu; Asano and Yonezawa (1980) for liquid Ni).

10.3. Partition function of a liquid metal

Rousseau, Stoddart, and March (1970) have proposed a method for calculating the partition function $Z(\beta)$ of a liquid metal. If $Z(\beta)$ were known precisely, then the density of states could be obtained by an inverse Laplace transform of essentially $Z(\beta)$. However, in practice, $Z(\beta)$ from this method is known only to finite numerical accuracy over a limited portion of the real $\beta(=(k_B T)^{-1})$ axis. Thus, the calculation of $N(E)$ from $Z(\beta)$ is itself a matter that needs careful investigation.

Nevertheless, the method of calculation of the partition function $Z(\beta)$ is of interest in its own right. The motivation for this procedure can be viewed as follows. Suppose the total scattering potential $V(\mathbf{r})$ felt by an electron when the ions are "frozen" in configuration $\{\mathbf{R}_i\}$ (the ith ion evidently being at position $\mathbf{R}_i$) can be represented by (see also (10.9))

$$V(\mathbf{r}) = \sum v(\mathbf{r} - \mathbf{R}_i) \tag{10.41}$$

where $v(\mathbf{r})$ is an atomic-like localized potential. In the Thomas-Fermi approximation the integrated local density of states $n(\mathbf{r}, E)$ is given (see

March, 1968) by

$$n(\mathbf{r}, E) = \frac{8\pi}{3h^3}(2m)^{3/2}[E - V(\mathbf{r})]^{3/2} \tag{10.42}$$

and because of the nonlinear character of this n-V relation, the superposition property cannot be exploited. However, the so-called Bloch, or canonical, density matrix, defined in terms of the (as yet unknown) eigenvalues ε_i and eigenfunctions ψ_i of the potential $V(\mathbf{r})$ by

$$C(\mathbf{r}\beta) = \sum \exp(-\beta\varepsilon_i)\psi_i^*(\mathbf{r})\psi_i(\mathbf{r}), \tag{10.43}$$

which is a generalized partition function such that

$$\int C(\mathbf{r}\beta)\,d\mathbf{r} = Z(\beta), \tag{10.44}$$

is simple in the Thomas-Fermi approximation, since all the one-electron levels can be regarded as "shifted" by $V(\mathbf{r})$ in a slowly varying V for which the Thomas-Fermi theory is valid. Then if $C_0(\beta)$ is the generalized partition function for free particles in atomic units

$$C_0(\beta) = (2\pi\beta)^{-3/2}, \tag{10.45}$$

one can write the approximate formula

$$C(\mathbf{r}\beta) = (2\pi\beta)^{-3/2}\exp(-\beta V(\mathbf{r})) \tag{10.46}$$

and hence, given the superposition property one can write

$$C(\mathbf{r}\beta) = (2\pi\beta)^{-3/2}\prod_i \exp(-\beta v(\mathbf{r} - \mathbf{R}_i)). \tag{10.47}$$

The fact that the "local" partition function $C(\mathbf{r}\beta)$ factorizes into a product of single-centre terms is the basic observation underlying the method of Rousseau et al.

Of course, realistic choices of v in liquid metals make the Thomas-Fermi approximation too crude. Rousseau, Stoddart, and March (1970) therefore used the method of Hilton, March, and Curtis (1967) for solving the one-centre problem: In essence this replaces $v(r)$, now taken as spherical, by an effective potential $U(\mathbf{r}\beta)$. Using this, their proposal is to write, by analogy with (10.47)

$$C(\mathbf{r}\beta) = (2\pi\beta)^{-3/2}\prod_i \exp(-\beta U(\mathbf{r} - \mathbf{R}_i)) \tag{10.48}$$

where, for notational convenience, the β dependence has been dropped from U. Of course, this feature is essential for the theory of Rousseau, Stoddart, and March (1970).

These workers show that the above approximation is valid in a variety of circumstances:

1. β small.
2. ∇U small.
3. Overlaps of U on different sites are small.

Thus, for the "frozen ion" configuration $\{\mathbf{R}_i\}$, (10.48) is the basic approximation used for the electronic states.

LIQUID AVERAGING

As in the previous sections, liquid averaging must now be carried out, this time to get the liquid-metal partition function $Z(\beta)$. Rousseau, Stoddart, and March (1970) use a modification of the Kirkwood approximation to express three-atom and higher-order correlation functions in terms of the pair function $g(r)$. In this way, they obtain $Z(\beta)$ solely in terms of the single-centre effective potential $U(r\beta)$ and the pair function $g(r)$. They carried out calculations of liquid metal Be that were encouraging in revealing a dip in the density of states characteristic of a divalent metal. In subsequent work, Rousseau, Stoddart, and March (1971) used idempotency of density matrices to refine the method, while Pant, Das, and Joshi (1971; 1973) have made further practical applications of the approach outlined in this section.

10.4. Electron states in expanded liquid Hg

A very different approach to those just described has been employed by Mattheiss and Warren (1977) to treat the electronic structure of expanded liquid Hg.

Just because this liquid exhibits a relatively low critical temperature and pressure ($T_c = 1750$ K, $p_c = 1670$ bar) it can be expanded to very low densities by heating it under pressure. This has permitted the measurement of electrical conductivity, thermopower, Hall coefficient, and Knight shift as functions of pressure. Each of these properties exhibits rather remarkable variation with density. As an example, the electrical conductivity decreases by eight orders of magnitude as the density ρ is reduced from 13.6 to 2 g/cm^3. A comparison of these various pieces of experimental data just referred to leads one to conclude that a rather gradual metal-semiconductor transition occurs in liquid Hg in the range 8 to 9 g/cm^3.

This type of metal-semiconductor transition is expected at reduced densities for a divalent metal such as Hg when the $6s$ and $6p$ conduction bands no longer overlap. In the case of a liquid, the sharpness of this transition would almost certainly be reduced by density fluctuations and the general loss of long and short-range order.

One approach to this problem of expanded liquid Hg, adopted by Mattheiss and Warren (1977), is to model the electronic structure of the liquid by means of band-structure calculations for appropriate crystalline structures. The validity of this method rests on the assumption that gross features of the electronic structure are determined by the density and local arrangement of atoms and are relatively insensitive to the degree of long-range order. The small changes of the electronic properties observed for most metals on melting provide support for this assumption. Even in cases where the bands are far from the free-electron form, there may still be little change in the density of states $N(E)$ on melting, as strikingly demonstrated by the X-ray photoemission study of Bi by Baer and Myers (1977).

10.4.1. Detailed band structure calculations for crystalline Hg

The first band-structure calculations carried out for hypothetical forms of crystalline Hg in the low-density limit were those of Devillers and Ross (1975). These workers used the pseudopotential approach to calculate the energy bands for crystalline Hg with the *bcc*, *fcc*, and rhombohedral structures. For each structure, they obtained a band gap at $\rho = 8.5$ g/cm^3 in general accord with the electronic transport data. Overhof et al. (1976) subsequently carried out relativistic Korringa-Kohn-Rostoker (KKR) calculations for *fcc* and simple cubic (*sc*) Hg. Their finding was that *fcc* Hg becomes semiconducting at $\rho = 9.3$ g/cm^3, whereas *sc* Hg remains a conductor until the density is lowered to around 5.5 g/cm^3. A similar dependence on crystal structure was obtained by Fritzson and Berggren (1976), who carried out again pseudopotential calculations for *fcc*, *bcc*, and *sc* Hg and found that band gaps open up at $\rho = 6.5$ g/cm^3 for *fcc*, 5.5 g/cm^3 for *bcc*, and 4 g/cm^3 for *sc* Hg.

It is relevant here to return briefly to the tight-binding approach to the calculation of the density of states $N(E)$ for a disordered liquid metal. Using this approach, in a self-consistent single-site approximation, Yonezawa et al. (1976; 1977) also exhibit the development of an energy gap but now in the range of densities 2 to 4 g/cm^3. This is significantly lower than the band-structure models, which do tend to produce results that are in better numerical agreement with experiment.

In the approach of Mattheiss and Warren (1977), a different quasi-crystalline model is applied to approximate the electronic properties of expanded liquid Hg. In their study, the basic assumption is that the nearest-neighbour bond distance is constant, so that the density variation in expanded liquid Hg is due entirely to changes in coordination number. Although Mattheiss and Warren stress that this assumption is something of an oversimplification, available data on liquid structure suggests it is a useful first approximation. For instance, X-ray studies by Waseda and Suzuki (1970) on liquid Hg at low temperatures ($\rho \simeq 13.6$ g/cm^3) suggest a temperature-independent nearest-neighbour bond distance $a = 3.07$ Å and a coordination number of 10 to 11. Less complete determinations of the structure factor for liquid Hg at higher temperatures by Waseda et al. (1974) suggest the likelihood that the nearest-neighbour bond distance remains essentially unchanged and that density reductions are due primarily to a decrease in coordination number.

The study of Mattheiss and Warren is based on a series of augmented plane-wave (APW) calculations for crystalline Hg with the *fcc*, *bcc*, *sc*, and diamond structures and fixed nearest-neighbour bond distance of 3.07 Å. In a perfectly ordered crystal, these calculations would correspond to densities of 16.3, 15.0, 11.5, and 7.5 g/cm^3, respectively. However, Mattheiss and Warren argue that these densities are reduced by such factors as disorder and fluctuations in the liquid state for the structures having coordination number $C_n < 12$, so that these quasi-crystalline results correspond to densities of 16.3, 10.9, 8.1, and 5.4 g/cm^3, respectively.

Results for 6*s* and 6*p* bands

The results of their calculations show that the Hg 6*s* and 6*p* bandwidths and their overlap are reduced systematically as the coordination number is decreased at a fixed near-neighbour distance. In particular, the 6*s*-6*p* band overlap is reduced to about 0.1 eV in the tetrahedrally coordinated diamond structure. A band gap is then opened up if the nearest-neighbour bond distance is increased by about 1% to 3.10 Å.

Density of States $N(E_F)$ at Fermi Level as Function of Density for Liquid Hg

The Slater-Koster (1954) linear-combination-of-atomic-orbitals (LCAO) interpolation method was applied by Mattheiss and Warren (1977) to construct accurate density-of-states curves for each structure, including their decomposition into s and p components (see Figure 4 of their paper).

In selecting a structural model for liquid Hg, Mattheiss and Warren assumed $a = 3.07$ Å over the entire liquid range. This serves to emphasize that the dominant effect of density reduction on the electronic properties derives from the reduced average coordination number rather than an increased near-neighbour separation.

The simplest structural model with constant a is one in which vacancies are introduced randomly on a close-packed lattice. Furukawa (1960) has described such a model for normal high-density liquid metals based on an *fcc* lattice. His expression for the average coordination number

$$C_n = 6\sqrt{2}\rho a^3 \tag{10.49}$$

yields $C_n = 10$ for Hg at the melting point, which is consistent with structural measurements. If one assumes with Mattheiss and Warren that C_n remains linearly related to the density in the expanded liquid, one finds from (10.49) that $C_n = 4$ near the critical density $\rho_c \simeq 5.5$ g/cm^3, which seems plausible.

10.5. Band model for electronic structure of expanded liquid caesium

Warren and Mattheiss (1984) have subsequently carried out band-structure calculations for four crystalline forms of Cs (*bcc*, *sc*, simple tetragonal and diamond structures) at a fixed nearest-neighbour bond distance. The results of these calculations are used to model the variation with density of several one-electron properties of expanded liquid caesium, including the total density of states $N(E_F)$, its 6s component $N_s(E_F)$, and the average Fermi-electron charge density at the nucleus.

The calculations do not explain an observed enhancement of the magnetic susceptibility at low densities. Also, in contrast to experimental evidence derived from the combined analysis of Knight shift and susceptibility data, the calculated probability density at the nucleus for Fermi-surface electrons tends to increase rather than decrease as a function of decreasing density. This all points to the fact that many-electron correlation effects play an essential role in determining the electronic properties of liquid Cs in the low-density limit where $\rho \lesssim 1.3$ g·cm^{-3} and the nearest-neighbour coordination number $C_n \lesssim 6$.

In spite of the importance of correlation, knowledge of one-electron band structure results is clearly useful. Again, as in Section 10.4, the principal justification for the application of a band model to a liquid metal is the fact that many important features of electronic structure are determined largely by local properties such as the number and distance of nearest neighbours. An experimental demonstration of this is the observation that most electronic properties of metals are only slightly affected by the loss of long-range order at the melting transition.

10.5.1. Structural model

Warren and Mattheiss have adopted for their quasi-crystalline treatment of expanded liquid Cs a structural model suggested by the neutron diffraction data on expanded liquid Rb (Franz et al., 1980). According to these results, a 50% density reduction in liquid Rb is achieved primarily by a nearly linear decrease in the nearest-neighbour coordination number C_n and involves only a modest ($\sim$4%) increase in the nearest-neighbour bond distance a. In their work, these small variations in a are neglected, and it is assumed that density changes are due solely to variations in the coordination number C_n.

The value $a = 5.31$ Å for the nearest-neighbour bond distance for liquid Cs has been determined from neutron-diffraction studies (Gingrich and Heaton, 1961) at a temperature ($\sim$30°C) just above the melting point. This study also gave an average coordination number $C_n = 9.0 \pm 0.5$ for the normal liquid. It is interesting to note that this value for a in the liquid is identical to that for the solid at -10°C, where the structure is *bcc* and the coordination number $C_n = 8$. The lattice parameter for *bcc* Cs decreases by about 1.4% at low ($\sim$5 K) temperatures.

Of the structures listed in Table 10.1, the simple cubic and diamond structures are natural choices for modeling the six- and four-fold coordinated forms of the liquid because of their high symmetry and the fact that these

Table 10.1. *Crystal structures used by Warren and Mattheiss (1984) in quasicrystalline treatment of expanded liquid* Cs.

	BCC	SC	Simple tetragonal (*st*)	Diamond
Coordination number z	8	6	4	4
a (Å)	6.13	5.31	5.31	12.26
c (Å)			7.965	
Atomic volume	115.26	149.72	224.58	230.51
Solid density ($g \cdot cm^{-3}$)	1.92	1.47	0.98	0.96
Liquid density ($g \cdot cm^{-3}$)	1.73	1.30	0.87	0.87

structures are fully determined by the nearest-neighbour bond length a. The desirability of an alternative structural model for $C_n = 4$ will emerge shortly. A reasonable alternative is provided by the simple-tetragonal (*st*) structure with $c/a > 1$. In this case, the nearest-neighbour coordination is planar rather than tetrahedral. The c/a ratio has been arbitrarily set equal to 1.5, since this choice yields an atomic volume Ω that is close to the one for the diamond structure and simplifies the density-of-states calculations. (Warren and Mattheiss, 1984).

Without going into much detail here, the electronic-structure calculations reported by Warren and Mattheiss (1984) for these various crystalline forms of Cs have been carried out with the use of a self-consistent, scalar-relativistic version of the linear augmented-plane-wave (LAPW) method.

10.5.2. *Conduction electron wave functions*

Let us write the Knight shift K in the form

$$K = \frac{8\pi}{3} \langle |\psi(0)|^2 \rangle_{E_F} \chi \tag{10.50}$$

where $\langle |\psi(0)|^2 \rangle_{E_F}$ is the average probability density at the nucleus averaged over states at the Fermi level. Using experimental values of K, plus a combination of theory and experiment for the spin susceptibility χ, Warren and Mattheiss have evaluated $\langle |\psi(0)|^2 \rangle_{E_F}$ from (10.50). These values, normalized to the experimental atomic value $|\psi(0)|^2_{\text{atom}} = 3.81(a \cdot u)^{-3}$ are plotted in Figure 10.1 for liquid Cs along the coexistence curve. The open circles are derived from experiment, whereas the one-electron band structure values for the different crystal lattices are shown by the solid circles. The one-electron model fails badly in the low-density range; this situation has been discussed from many-electron theory by Chapman and March (1988; see also Chapter 11).

In summary, the one-electron band structure model investigated by Warren and Mattheiss is a good model for high-density liquid Cs, yielding reasonable results for the density of states at the Fermi level and the electronic wave functions at the nucleus. However, this one-electron model begins to fail at intermediate densities. For low densities, it is quite clear that electron-electron interactions play a dominant role.

Now that the theory of electron states in liquid metals has been discussed in some detail, the following chapter will take up the magnetic behaviour of liquid metals, both under normal conditions and when expanded considerably. This will, in turn, throw some further light on the nature of the electronic structure of liquid alkalis especially. Before doing this, however, a summary on optical properties of liquid metals follows.

Figure 10.1. Shows $\langle\psi(0)^2\rangle_{F_f}$, normalized to experimental atomic value $|\psi(0)^2|^2_{\mathrm{atom}}$ plotted for liquid Cs along coexistence curve. (After Warren and Matthiess, 1984). Open circles are experimental values.

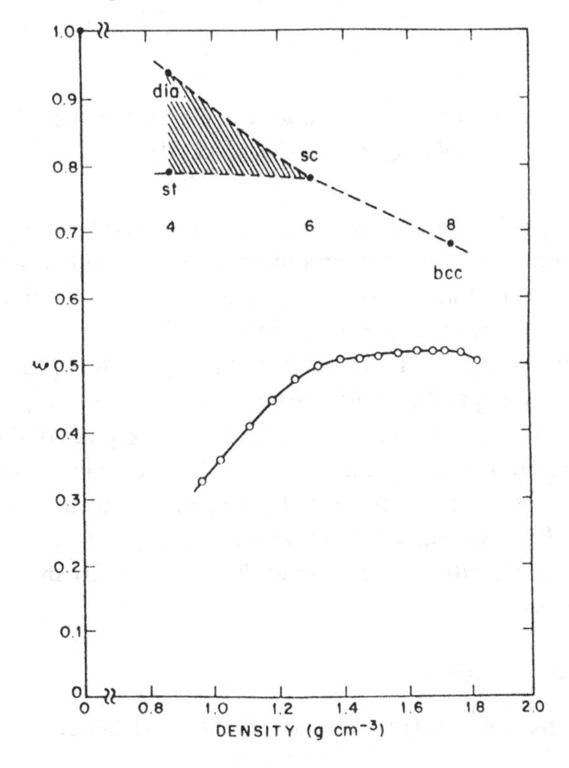

10.6. Optical properties

To conclude this chapter, something should be added concerning the optical properties of liquid metals. First, it is to be recalled that Chapter 7 included an account of the theory of the d.c. conductivity of liquid metals. Here, some discussion of the a.c. conductivity will be presented, in relation to the optical properties of liquid metals (see also Faber, 1966; 1972).

It is worthy of note that the optical properties of liquid metals are easier to measure than those of solids. Often, too, the a.c. conductivity $\sigma(\omega)$ is adequately represented by the Drude-Zener formula (see for example, Donovan (1967)):

$$\sigma(\omega) = \frac{\sigma(0)}{1 + \omega^2\tau^2} \tag{10.51}$$

and $\omega\tau \sim 1$ in the visible, compared with the case of solids, where $\omega\tau \sim 1$ lies in the far infrared. Equation (10.51) is, of course, the real part of the complex conductivity, the imaginary part following from this by using the Kramers-Krönig relations.

10.6.1. Frequency-dependent dielectric constant and optical properties of liquid metals

In liquid metals the electronic mean free path is finite, and this implies, via the uncertainty principle, that there is a blurring Δk_F of the Fermi surface given by (7.17). This fact was utilized by Ferraz and March (1979) in their treatment of electrical resistivity given in Section 7.3.

In this section, the effect of Fermi surface blurring on the frequency (ω) and wavenumber (q)-dependent dielectric function $\varepsilon(q, \omega)$ of a simple liquid metal will be studied. To this end, a summary will first be given of the main results of Leavens et al. (1981) for $\varepsilon(q, 0)$. Then the frequency dependence will be the focal point, the Lindhard formula being taken as starting point (Mermin, 1970; Götze, 1979). However the major departure from their methods is to introduce Fermi surface blurring directly through the Lindhard expression.

Static Dielectric Function

Following Leavens et al. (1981), the real part of the static dielectric function is

$$\varepsilon_1(q,l) = 1 + \frac{4k_F}{\pi a_0 q^2} f(q,l) \tag{10.52}$$

where $f(q,l)$ is given by

$$f(q,l) = \frac{1}{2}\left\{1 - \frac{1}{2k_F l}\left(\tan^{-1}\left[\frac{4k_F l}{1 + (ql)^2 - (2k_F l)^2}\right] + \pi(4k_F^2 - q^2 - l^{-2})\right)\right\}$$
$$+ \frac{[1 - (q/2k_F)^2 + (2k_F l)^{-2}]}{8(q/2k_F)}$$
$$\times \ln\left\{\frac{[(ql)^2 + 1 + 2k_F q l^2]^2 + (2k_F l)^2}{[(ql)^2 + 1 - 2k_F q l^2]^2 + (2k_F l)^2}\right\}, \tag{10.53}$$

from which one recovers the Lindhard (1954) dielectric function in the limit $l \to \infty$.

Approximate Inclusion of Exchange and Correlation

Leavens et al. (1981) include the effects of exchange and correlation in an approximate manner by replacing $f(q,l)$ in (10.52) by the function

$$F(q,l) = \frac{f(q,l)}{1 - \lambda(1 + 0.1534\lambda)f(q,l)}, \tag{10.54}$$

where $\lambda = (\pi a_0 k_F)^{-1}$. In the large l limit, Taylor (1978) has shown that this dielectric function is a good approximation to that of Geldart and Taylor

Figure 10.2. How finite mean free path effects influence the static dielectric constant of an electron gas (eqns. (10.52) and (10.53)). $k_f l = 1000$, 30 and 10 in curves I, II and III).

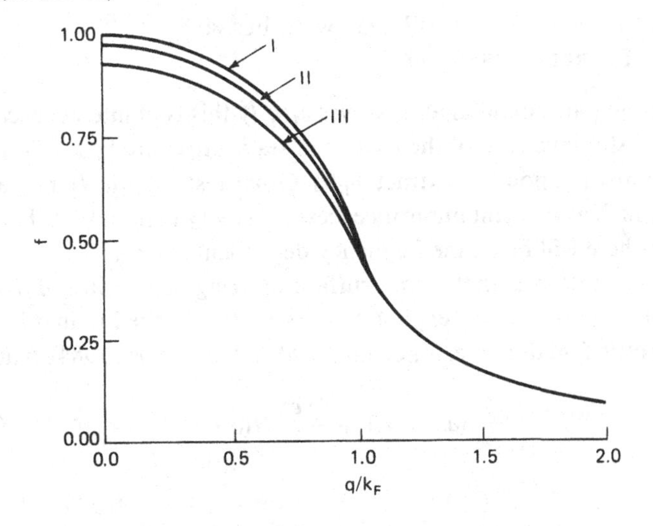

Figure 10.3. Includes exchange and correlation, but otherwise has similar purpose to that of Figure 10.2. Actual quantity plotted in this figure is F in (10.54). Abscissa is in units of q/k_f, with k_f the Fermi wave number. Different curves correspond to different mean free paths.

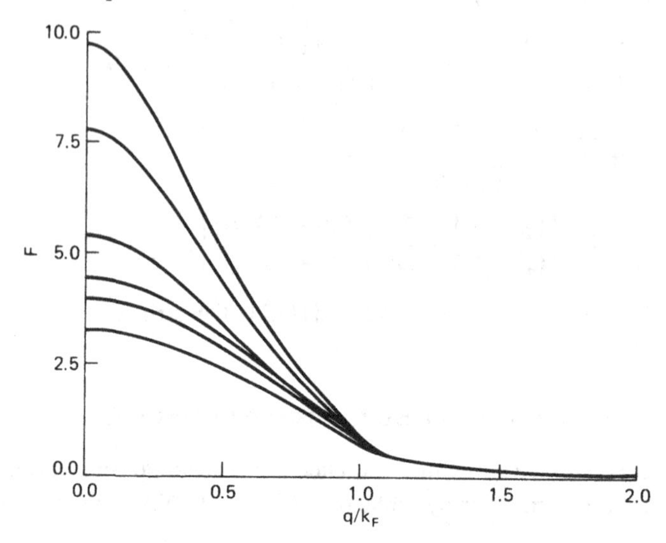

(1970; see also Geldart and Rasolt, 1987). With the seemingly reasonable assumption that the mean free path changes local field effects in a minor fashion, (10.54) represents an obvious generalization of (10.53). Figures 10.2 and 10.3 indicate how finite free path effects influence the static dielectric function $\varepsilon_1(q)$ without and with exchange and correlation.

Consequences of Lindhard Form, with Fermi Surface Blurring, of $\varepsilon(q, \omega)$

The ion-ion pair potential Φ_{ii} involves $\varepsilon_1(q, l)$; this is of interest because of the successful inversion of the measured static structure factor $S(q)$ of Na near the melting point to extract $\Phi_{ii}(r)$. Other tests of $\varepsilon_1(q, l)$ may also be feasible, such as divalent impurity excess resistivity in liquid Na. However, the focus here will be on the frequency-dependent $\varepsilon(q, \omega)$.

The assumption is that Fermi surface blurring can be introduced into $\varepsilon(q, \omega)$ of a liquid metal by exploiting the structure of the Lindhard form of the dielectric function (see, for example, March and Tosi, 1984), namely,

$$\varepsilon(q, \omega) = 1 + \frac{4\pi e^2}{q^2} H(q, \omega) \tag{10.55}$$

where (March and Paranjape, 1987):

$$H(q,\omega) = \sum \frac{f_{\mathbf{k}} - f_{\mathbf{k}+\mathbf{q}}}{\varepsilon_{\mathbf{k}+\mathbf{q}} - \varepsilon_{\mathbf{k}} - \hbar\left(\omega + \dfrac{i}{\tau}\right)}. \tag{10.56}$$

It is to be noted that if Fermi surface blurring is introduced via the occupation probabilities f_q, then it is consistent to have a finite relaxation time τ in the denominator, Boltzmann's transport equation yielding the appropriate modification as in (10.56); i.e., ω is replaced by $(\omega + i/\tau)$.

Forming the real and imaginary parts of H in (10.56), it can be shown that

$$\operatorname{Re} H = F(q,\omega) - \omega\tau \operatorname{Im} H \tag{10.57}$$

where $F(q,\omega)$ is given by

$$F(q,\omega) = \sum \frac{(f_{\mathbf{k}} - f_{\mathbf{k}+\mathbf{q}})(\varepsilon_{\mathbf{k}+\mathbf{q}} - \varepsilon_{\mathbf{k}})}{(\varepsilon_{\mathbf{k}+\mathbf{q}} - \varepsilon_{\mathbf{k}} - \hbar\omega)^2 + \dfrac{\hbar^2}{\tau^2}}. \tag{10.58}$$

Combining (10.55) and (10.57) with the general relation between ε and conductivity $\sigma = \sigma_1 + i\sigma_2$,

$$\varepsilon(q,\omega) = 1 + \frac{4\pi i\sigma(q,\omega)}{\omega}, \tag{10.59}$$

one readily finds

$$\varepsilon_1(q,\omega) = 1 + \frac{4\pi e^2}{q^2} \operatorname{Re} H(q,\omega) \tag{10.60}$$

and

$$\varepsilon_2(q,\omega) = \frac{4\pi e^2}{q^2} \operatorname{Im} H = \frac{4\pi\sigma_1(q,\omega)}{\omega}. \tag{10.61}$$

Substituting in (10.57) after multiplying by $4\pi e^2/q^2$ yields

$$\varepsilon_1(q,\omega) - 1 = \frac{4\pi e^2}{q^2} F(q,\omega) - 4\pi\tau\sigma_1(q,\omega). \tag{10.62}$$

Of course, further physical content will come from (10.62) when one inputs energy levels and Fermi surface blurring into (10.58). However, it will emerge in the following section that, in the small q regime, $F(q,\omega)$ must reflect the dynamical structure factor of the ions in the liquid metal. To introduce this limit, let us note that

$$\varepsilon_1(0,\omega) - 1 = \lim_{q\to 0} \frac{4\pi e^2}{q^2} F(q,\omega) + 4\pi\tau\sigma_1(0,\omega) = f_1 + 4\pi\tau\sigma_1(0,\omega), \tag{10.63}$$

this equation evidently defining $f_1(\omega)$. It is easy to show that the Drude-Zener model

$$\sigma(\omega) = \frac{\sigma_0}{(1 - i\omega\tau)} \tag{10.64}$$

with σ_0 the d.c. conductivity, corresponds to $f_1(\omega) = 0$, although not to $F(q,\omega) = 0$. In the following section, a form of $f_1(\omega)$ will emerge in terms of $f_{\mathbf{k}}$, including blurring.

Long Wavelength Evaluation of $\varepsilon(q,\omega)$

To evaluate $F(q,\omega)$ in (10.58), $\varepsilon_{\mathbf{k}+\mathbf{q}} - \varepsilon_{\mathbf{k}}$ can be put to zero in the denominator. Then, Taylor expansion in the numerator is made to lowest order in q. With $\mu = \cos\theta$, $\theta = \mathbf{k}\cdot\mathbf{q}/kq$, one finds, using $\varepsilon_{\mathbf{k}} = \varepsilon(k)$ in the liquid,

$$\lim_{q\to 0} F(q,\omega) = \frac{q^2}{\dfrac{\hbar^2}{\tau^2}(1+\omega^2\tau^2)} \iint_{-1}^{1} \frac{\partial f}{\partial \varepsilon}\left(\frac{\hbar^2}{2m}\right)^2 (2kq\mu)^2\, d\mu \frac{2\pi k^2}{4\pi^3}\, dk \tag{10.65}$$

from which $f_1(\omega)$ in (10.63) is found as

$$f_1(\omega) = \frac{4\pi e^2}{(1+\omega^2\tau^2)} \int \frac{\partial f}{\partial \varepsilon}\left(\frac{\hbar\tau}{2m}\right)^2 \frac{4}{3\pi^2} \frac{k^4}{(\partial\varepsilon/\partial k)}\, d\varepsilon. \tag{10.66}$$

This shows the frequency dependence explicitly and that the magnitude depends on the detailed nature of the Fermi surface blurring through $\partial f/\partial\varepsilon$, as well as on departures of the $\varepsilon(k)$ relation from $\hbar^2k^2/2m$ for free electrons. The conclusion is that, with n the electron density,

$$f_1(\omega) = \frac{\tau^2}{(1+\omega^2\tau^2)} h(\tau, n, T) \tag{10.67}$$

where the function h can be determined from (10.66) given $\partial f/\partial\varepsilon$ and $\varepsilon(k)$. With $f_1(\omega)$ given by (10.67), some test of (10.63) will be made using experimental data on $\varepsilon_1(0,\omega)$ and $\sigma_1(0,\omega)$ for liquid Na.

REAL PART OF AC CONDUCTIVITY IN TERMS OF DYNAMICAL STRUCTURE OF IONS

Tosi, Parrinello, and March (1974; see also Hinkelmann, 1970) show that (see Chapter 14 also)

$$\operatorname{Re}\sigma_e(0,\omega) = \frac{\pi e^2 M^2 \omega}{m^2}(1-\exp(-\beta\omega)) \lim_{q\to 0}\frac{1}{q^2}S_{ii}(q,\omega) - \frac{\pi n_i M e^2}{m^2}\delta(\omega) \tag{10.68}$$

where $M = m_i + Zm$ with Z the valency and

$$\sigma_e(0,\omega) \equiv \frac{\sigma(0,\omega)}{\varepsilon(0,\omega)}. \tag{10.69}$$

Writing the factor multiplying $\lim q^{-2}S_{ii} = T(\omega)^{-1}$ in (10.68) as $g(\omega)^{-1}$, one finds (March and Paranjape, 1987b):

$$\frac{16\pi^2}{\omega^2}\sigma_1^2 - gT\sigma_1 + \varepsilon_1^2 = 0, \tag{10.70}$$

with solution near the plasma frequency, where $\varepsilon_1 \to 0$:

$$\sigma_1(\omega \sim \omega_p) = \frac{\omega_p^2}{16\pi^2} gT|_{\omega\sim\omega_p}. \tag{10.71}$$

Equation (10.70) can, alternatively, be used to gain information on S_{ii} as reflected by $T(\omega)$, in frequency regimes away from $\omega = 0$, where $\sigma_1(\omega)$ and $\varepsilon_1(\omega)$ are known from experiment. This route is mentioned briefly immediately below.

SOME POINTS OF CONTACT WITH EXPERIMENT

Here the theory outlined above will be brought into some contact with experiment. Let us begin with the prediction in (10.63) and (10.67).

Relation between $\sigma_1(\omega)$ and $\varepsilon_1(\omega)$ in liquid Na. The data on the optical constants n and k of liquid Na (Inagaki et al., 1976) have been utilized to test (10.63), with $f_1(\omega)$ taking the form (10.67). Thus, Figure 10.4 shows nk versus $1 - n^2 + k^2$, the former being proportional to $\sigma_1(\omega)$ and the latter to $\varepsilon_1(\omega)$. The striking thing to note is that, at the high-frequency end of this plot, which is near the origin, one can draw a straight line (through the origin) passing though the observed points (March and Paranjape, 1987b).

Taking the relaxation time τ as (Inagaki et al., 1976) 1.6×10^{-14} s, it is easily verified that the slope of the $\sigma_1 - \varepsilon_1$ plot is near $4\pi\tau$, as predicted by (10.63) and (10.67) ($\omega^2\tau^2 \gg 1$). Since in the high-frequency limit, $f_1(\omega) \to 0$, one is recovering the Drude-Zener relation between σ_1 and ε_1. Since, according to (10.67), $f_1(\omega) \to$ constant for $\omega^2\tau^2 \ll 1$, at the low-frequency end the prediction is another parallel straight line, which, when extrapolated back to $\varepsilon_1 = 0$, cuts the σ_1 axis at the value $-h(\tau, n, T)$. The deviation from the Drude-Zener line drawn in Figure 10.4 is compatible with this prediction. However, as March and Paranjape (1987b) stress, the calculation of h from first principles is a task well beyond the theoretical framework outlined above.

Dispersion of plasmon. Here brief comment will be made on the way the plasmon dispersion relation $\omega_p(q)$ and—more importantly—plasmon damping are affected by mean free path—or, equivalently, finite lifetime—effects. The pointers are that the main effect in determining the dispersion

Figure 10.4. Product of optical constants nk versus $1 - n^2 - k^2$ for liquid Na, the former quantity being proportional to $\sigma_1(\omega)$ and the latter to $\varepsilon_1(\omega)$. $\triangle$ = experiment. Straight line = Drude-Zener.

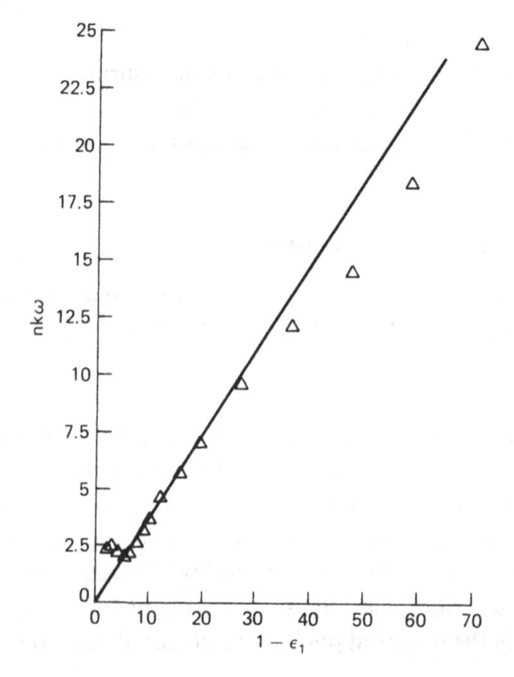

relation

$$\varepsilon_1(q, \omega_p(q)) = 0 \tag{10.72}$$

is the presence of τ in the denominator of (10.56). Thus, in determining $\omega_p(q)$, the present approach can be simplified by neglecting the blurring of the Fermi surface in the numerator of (10.56) when one is led back to the treatment of Mermin (1970; see Appendix 10.1). The dependence of $\omega_p(q)$ on τ thereby obtained in displayed in Figure A10.1.1.

DISCUSSION

The experimental results on n and k for liquid Na support (10.63) with $f_1(\omega)$ as in (10.67). The function h in (10.67) does, however, depend crucially on the liquid structure as well as on the Fermi surface blurring; this is not reflected sufficiently in (10.66).

It is also of interest to stress that, as is clear from (10.68), experimental measurement of the frequency dependence of both $\sigma_1(\omega)$ and $\varepsilon_1(\omega)$ will allow $\lim_{q\to 0} q^{-2} S_{ii}(q, \omega) \equiv s(\omega) = T(\omega)^{-1}$ to be extracted. Since one of the aims of liquid metals theory is to predict electronic properties from observed structure, both static $S(q)$ and dynamic $S(q, \omega)$, it is of interest to establish features of $S_{ii}(q, \omega)$ beyond those presently known from neutron inelastic scattering. For liquid Rb, for example, such neutron studies by Copley and Rowe (1974) reveal collective modes, which were subsequently found in the computer experiments of Rahman (1974a; see also Chapter 8).

The functional relation between $\sigma_1(\omega)$ and $\varepsilon_1(\omega)$ discussed above (see Figure 10.4) is quite different, of course, from the exact connection between nk and $n^2 - k^2$ given by the Kramers-Krönig relation. When n and k measurements become available for a particular liquid metal over a fuller frequency range, it would be of interest to insert $[\varepsilon_1(0, \omega) - f_1(\omega)]$ appearing in (10.63) with $f_1(\omega)$ given by (10.67) into such a Kramers-Krönig integral and hence to explore the dependence of $\sigma_1(\omega)$ thereby obtained on the choice of $h(\tau, n, T)$. As already mentioned, it is strongly suspected that h must turn out to be substantially smaller than implied by a formula such as (10.66).

To conclude this brief discussion of optical properties in relation to electron states, it should be noted that, in principle, photoemission provides information on the density of electronic states. Unfortunately, to date, however, there is no technique for a liquid metal that gives this quantity directly. As discussed earlier (March, 1977), the general trends of photoemission data support qualitative predictions of the theory. There is still a long way to go before one can say that one has a full theory of electron states in liquid metals, including strong scattering cases like rare earth (see Chapter 7) and, to a lesser extent, transition metals.

11
Magnetism of normal and especially of expanded liquid metals

Following the discussion of electron states in the previous chapter, the focus here will be the magnetic properties of liquid metals, both near to their melting points and along the coexistence curve towards the critical point.

In discussing the magnetism of normal metals, it has been traditional to separate the magnetic susceptibility into three contributions:

1. Core diamagnetism
2. Pauli spin susceptibility
3. Conduction electron orbital diamagnetism

Available evidence is that (1) can be dealt with adequately and thus it will be treated only in passing. The interest in (2) turns out, at least for simple *s-p* metals, to focus on the role of electron-electron interactions in enhancing the independent electron Pauli susceptibility. Contribution (3), as will be seen, is also of interest from the point of view of neutron scattering studies on a solid metal like Ni. The discussion will also embrace nuclear magnetic resonance and, in particular, the Knight shift in liquid metals.

Following this, which is centred on liquid metals just above their melting points, some consideration will also be given to the magnetic properties of expanded liquid metals. Here both phenomenology and also a treatment that leans on *heavy Fermion* theory (see Appendix 11.1) will be developed in order to interpret the properties of expanded liquid Cs, in particular, taken up toward the critical point along the coexistence curve.

11.1. Spin susceptibility of normal liquid metals

The most elementary theory of the spin susceptibility of a liquid metal follows that given in the early days of metal physics theory by Pauli. Here one considers a density-of-states curve $N(E)$ populated with electrons filling the levels two by two with opposed spins. Then one applies a magnetic field, which lowers the energy levels of one spin direction by $\mu_B B$, with μ_B the Bohr magneton, and raises the others by an equal amount. Of course, the Fermi level must be the same in the two subbands, and this leaves an imbalance of magnetic moment proportional to the applied field **B**. Thus the spin susceptibility is obtained and is given by the Pauli formula in terms of the density of states $N(E)$ at the Fermi level E_f as

$$\chi_{\text{Pauli}} = \text{const}\, N(E_f). \tag{11.1}$$

It has been known for a long time that this formula is sensitive to the introduction of electron-electron interactions. One experimental way of demonstrating this in simple (solid) metals is to compare (11.1) with that for the electronic contribution to the specific heat at low temperatures, usually written γT. The quantity γ is then also proportional to the density of states at the Fermi level. This formula turns out to be less sensitive than the Pauli value to electron-electron interactions and therefore is more useful for estimating $N(E_f)$ from experiment. The use of such a value in (11.1) then considerably underestimates the spin susceptibility, as measured, for instance, by electron spin resonance.

The work of Sampson and Seitz (1940) is classical in this area. These workers showed, by means of what is now called density functional theory, that electron-electron interaction enhances the spin susceptibility over and above the Pauli value. A useful way of parametrizing this enhancement is by means of Landau Fermi liquid theory (see Appendix 11.1), which is, therefore, briefly considered later. Needless to say, a full theory will eventually have to take complete account of both electron-electron and electron-ion interactions (see the following) and will no doubt require an extension of the treatment of Chapter 14. For present purposes, it will suffice to demonstrate later, admittedly by pseudopotential theory, that evidence presently available is that electron-ion interactions are much less important in determining the spin susceptibility of the simple *s*-*p* metals than are electron-electron correlations.

11.2. Relation between spin and orbital magnetism of simple liquid metals

It is natural enough to begin with band theory, at first neglecting electron-electron interaction; a neglect not justified, as will emerge, for quantitative work. If one starts from conduction electrons described by plane waves $\exp(i\mathbf{k}\cdot\mathbf{r})$, which is certainly a good approximation for liquid Na, say,

$$-\chi_{\text{diam}} = \frac{1}{3}\chi_{\text{Pauli}}; \qquad \chi_{\text{Pauli}} = \frac{3}{2} n_0 \mu_B^2 \frac{1}{E_F} \tag{11.2}$$

relates the Landau diamagnetism to the Pauli spin paramagnetism, n_0 being the conduction electron density (see, for example, Hebborn and March, 1970).* Since $n_0/E_F \propto E_F^{1/2}$, the rather obvious generalization of (11.2) for χ_{Pauli} to an energy band with density of states $N(E)$ is equivalent to (11.1).

The effect of electron-electron interactions on these formulae is important, χ_{Pauli} being relatively strongly enhanced by the interactions, in the range of electron density appropriate to the simple liquid metals. In contrast, for liquid Na the electron-ion contribution to χ_{Pauli}, according to second-order pseudopotential theory (Borchi and de Gennaro, 1972), appears to be small compared to the effect of electron-electron correlations. Later references to pure liquid metal studies are given by Dupree and Sholl (1975), though their main concern is with alloys, to be treated later.

11.2.1. Total and spin susceptibilities of liquid metals

Dupree and Seymour (1970) have plotted χ, the total susceptibility in cgs volume units versus $r_s(n_0 = 3/4\pi r_s^3)$ in atomic units (see Figure 11.1).

* These workers discuss orbital and spin susceptibility contributions to neutron scattering from Ni, concluding that the orbital contribution is rather small.

Figure 11.1. Total magnetic susceptibility versus mean interelectronic separation r_s for liquid metals (Dupree and Seymour, 1970; see March, 1977).

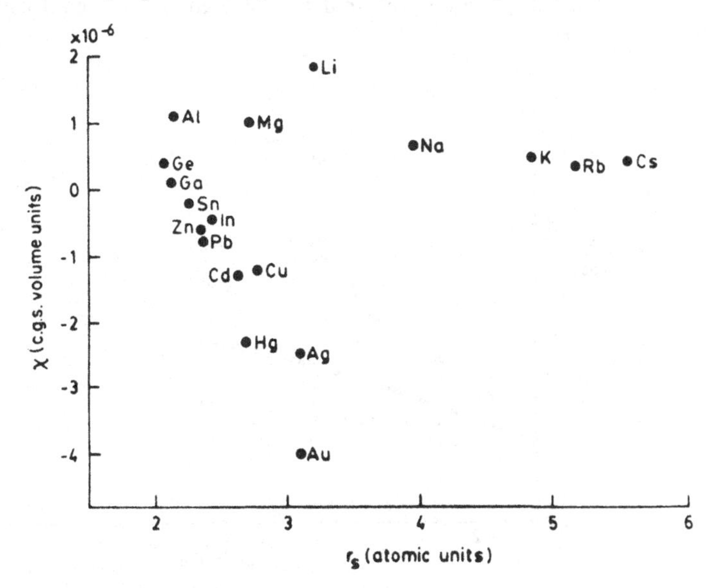

Writing $\chi = \chi_i + \chi_d + \chi_p$ as ion-core + orbital + spin pieces, they then take χ_i from the work of Angus (1932). For χ_d they use the interacting electron gas result (Kanazawa and Matsudaira, 1960; March and Donovan, 1954) which, in units of $\chi_{\mathrm{diam}} = \chi_{\mathrm{L}}$ in (11.2), is

$$\frac{\chi_d}{\chi_{\mathrm{L}}} = 1 + \frac{\alpha r_s}{6\pi}\left[\ln r_s + 4 + \ln\left(\frac{\alpha}{2\pi}\right)\right]; \qquad \alpha = \left(\frac{4}{9\pi}\right)^{1/3}. \tag{11.3}$$

The following points are noteworthy from the spin susceptibilities thereby extracted (with effective mass $m^* = m$) in Figure 11.2:

1. χ_p values lie in the range 0.9 to 2.1×10^{-6} (see Figure 11.2).
2. Values deduced from χ agree well with direct conduction electron spin resonance data for χ_p in Li (Hanabusa et al., 1976), Na, and K (crosses).
3. Figure 11.2, curves 1–3, depict

$$\frac{\chi_p}{\chi_{\mathrm{Pauli}}} = \frac{1 + A_1}{1 + B_0} \tag{11.4}$$

from Landau Fermi liquid theory of an interacting electron gas for different values of the Landau parameters A_1 and B_0 (see, e.g., Jones and March 1973); actually $A_1 \ll |B_0|$, B_0 being negative). More refined theories (Hamann and Overhauser, 1966; see also Dupree and Geldart, 1971) of $\chi_p/\chi_{\mathrm{Pauli}}$ reproduce the trends in Figure 11.2. To do better, obviously electron-ion interaction and the liquid structure (see Chapter 10) must be introduced, as is discussed briefly later.

Figure 11.2. Spin susceptibilities extracted from total susceptibilities in Figure 11.1. Crosses denote direct conduction electron spin resonance determination for Na and K. Lines 1, 2, and 3 correspond to different Fermi liquid parameters (see March, 1977).

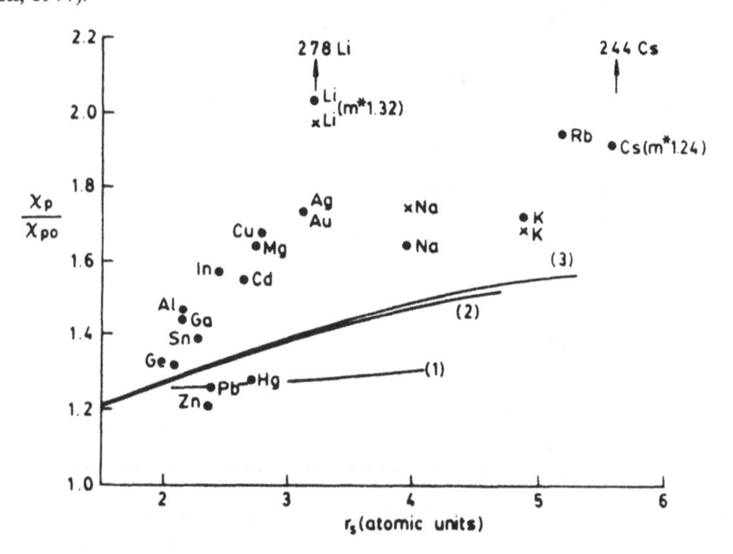

11.3. Effects of electron-ion interaction

Timbie and White (1970), following Baltensperger (1966), have studied χ_d with electron-ion interaction; its effect is relatively small and leaves intact the conclusion that electron-electron interactions play an important role in the theory of χ_d for simple liquid metals.

Takahashi and Shimizu (1973) and Timbie and White (1970) use identical expressions for χ_p in terms of liquid structure factor $S(q)$ and pseudo-potential, although their numerical estimates differ somewhat. Again, electron-ion interactions cause only small changes when calculated by such a nearly free electron approach.

The reader should be cautioned that for a metal like Li, which is not free-electron-like in the solid, it is not possible to treat the electron-ion interaction in the spin susceptibility in the above manner. Here, the similarity between liquid and solid results is a much better guideline.

11.4. Nuclear magnetic resonance: normal conditions

Nuclear magnetic resonance (NMR) has been observed in more than 20 liquid metals. Seymour (1974) has reviewed the data on Knight shifts at the melting points in liquid metals. The experimental findings are summarized in Table 11.1.

Table 11.1. *Experimental Knight shifts for liquid metals at their melting points.*

	K (%)		K (%)
Metal	Experimental	Metal	Experimental
Li	0.026	In	0.786
Na	0.116	Sn	0.73
Al	0.162	Sb	0.72
K	0.265	Te	0.38
Cu	0.264	Cs	1.44
Ga	0.449	Hg	2.42
As	0.32	Tl	1.48
Rb	0.662	Pb	1.49
Ag	0.575	Bi	1.41
Cd	0.795		

11.4.1 Knight shifts in liquid metals

Leaving out transition metals, one can write the Knight shift K in the usual way as

$$K = \left(\frac{8\pi}{3}\right)\Omega P_{\mathrm{F}}\chi_p. \tag{11.5}$$

Here Ω is the atomic volume, P_{F} is the probability density at the nucleus for conduction electrons at the Fermi surface, and χ_p is the spin susceptibility discussed earlier.

For the electron contact density factor ΩP_{F}, expressions have been given by Faber (1967), Cyrot-Lackmann (1964) and Watabe et al. (1965), with modifications given by Perdew and Wilkins (1970, 1973). Unfortunately, the Knight shifts remain difficult to interpret in a really convincing way.

Seymour (1974) also reviews the temperature coefficient of K and the change in Knight shift on melting, but the details are not pursued here.

11.5. NMR study of expanded liquid caesium

Warren et al. (1989) made a comprehensive study of the ^{133}Cs Knight shift and nuclear relaxation rates in liquid caesium from the vicinity of the melting point to the critical region of the liquid-gas transition. The data cover a temperature range 55 to 1590°C at pressures up to 90 bars and include a wide range of sample densities from 0.70 to 1.93 g/cm^3. Measurements extended to pressures of 900 bars in the lower part of the temperature range. The data yield the isobaric temperature dependence, the isothermal pressure dependence, and the isochoric temperature dependence of the Knight shift as well as the isobaric temperature dependence of the nuclear relaxation rate. At low densities, a strong enhancement of the Knight shift was observed, which is related to enhancement of the static susceptibility. In the range of increasing enhancement of the shift and susceptibility, they observed strong deviations of the relaxation rate from the Korringa relation (see Faber 1972; also eqn. (11.15)) signifying breakdown of the conventional (Stoner) model of exchange-correlation enhancement (see eqn. (11.18) later). The charge density at the nucleus exhibits a surprising decrease with decreasing density in the range 0.8 to 1.4 g/cm^3. These effects led them to propose a description of expanded Cs as a highly correlated metal with incipient antiferromagnetic spin correlations between electrons on neighbouring ions.

As has just been stressed, the magnetic properties of monovalent metals are particularly sensitive to the effects of electron-electron interactions. Even at ordinary densities, the uniform, static susceptibilities of the alkali metals are enhanced significantly by exchange-correlation effects. The enhancements lie in the range 1.5–2.0 and are usually described in terms of the Stoner model (see Jones and March, 1973; 1985; also eqn. (11.18) below). Close to the metal-nonmetal transition, where the ratio U/W* becomes large, Brinkman and Rice (1970), using a method due to Gutzwiller (1965), showed that correlation leads to a further enhancement. Their result is expressed as an enhancement of the effective mass; both the paramagnetic susceptibility and electronic specific heat being enhanced. The ground state of such a "highly correlated metal" is predicted to be antiferromagnetic for the Hubbard model with on-site repulsion energy U and bandwidth W. Similarly, several calculations using density functional theory have shown that the ground state of metallic, atomic hydrogen should be antiferromagnetic. The energies of the ferromagnetic and antiferromagnetic states are much closer for the alkali metals, and calculations differ as to which would form first in an expanded crystalline alkali metal.

Despite the importance for theory of the concept of an expanded monovalent crystal, such a system has not been realized experimentally in a single component material. However, as already discussed in Chapter 9, large reductions of density can be obtained for *liquid* alkali metals using thermal expansion as the metal is heated to the region of the liquid-gas critical point. The range of possible density variation can be seen, for example, from the temperature-density phase diagram (Stone et al., 1966; Franz, 1980; Jüngst et al., 1985) of Cs shown in Figure 11.3. Reduction of the density by about a factor of five is possible in the subcritical liquid, and arbitrarily low densities are possible in the supercritical fluid. Studies of the electrical conductivity of the alkali metals have established that a continuous metal-nonmetal transition occurs on passing through the critical region (Renkert, Hensel and Franck, 1969; 1970; Pfeifer, Freyland and Hensel, 1979; Franz et al., 1980). Because of this transition, the interparticle forces are strongly affected by changes in density. This has important consequences for the liquid-gas equilibrium. Early work by Landau and Zeldovich (1943) on the interplay between the metal-nonmetal and liquid-gas transitions, has taken explicit form in studies of singularities in the rectilinear diameters of metals. (Jüngst et al., 1985; Goldstein and Ashcroft, 1985; see also Chapter 9).

* Here, U is the energy cost of putting two electrons with opposed spins on the same site: the so-called Hubbard U, while W is the electronic bandwidth.

Figure 11.3. Temperature-density phase diagram of Cs (after Warren et al, 1989; fine shading shows regions they studied). Upper scale shows electron density for 1 electron per atom.

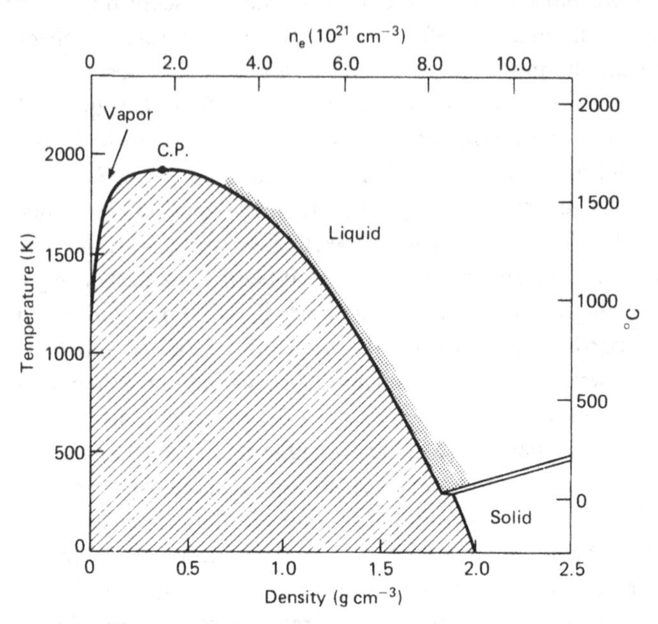

Measurements of the static magnetic susceptibility of Cs and Rb show strikingly increased enhancement in the low density metal (Freyland, 1979, 1980). Interpretation of this effect in terms of models of expanded monovalent crystals is obviously complicated by liquid-state disorder and the high temperatures at which these experiments are carried out. A particular consequence of the latter is limitation of the susceptibility below the Curie value at high temperatures. This is a clear demonstration that the low density enhancement is a mass enhancement of the type considered by Brinkman and Rice, rather than enhancement of the Stoner type (see Warren, 1984).

In the work of Warren et al. (1989), a comprehensive investigation is made of expanded liquid Cs using NMR measurements. These extend from the vicinity of the normal melting point (T_m) to the critical region of the liquid-gas transition ($T_c = 1651$°C, $P_c = 92.5$ bar). The density varies by roughly a factor 2.7 over experimental range and at the highest temperature reached, $(T_c - T)/T_c = 3 \times 10^{-2}$. These workers have measured Knight shifts, obtaining information about the behaviour of the uniform, static susceptibility and the electronic charge density at the nucleus, and the linewidth (free induction decay lifetime) related to the low-frequency limit of the integrated nonuniform susceptibility. Their experiments permitted measurement of the pressure dependence of these quantities at constant temperature at all but the highest temperatures attained. Although the measurements approached the critical temperature, all data reported by Warren et al. were obtained within the metallic range of expanded Cs. The data reveal pronounced changes in electronic character due to electron-electron interactions in the very low density metal. These are reflected in the real and imaginary parts of the frequency and wave-number dependent magnetic susceptibility (see (11.9); also Hebborn and March, 1970) and in the charge distribution of conduction electrons.

11.6. Experimental results: Knight shift for Cs

^{133}Cs Knight shifts were measured as a function of temperature under isobaric conditions. In Figure 11.4 data are shown for two pressures (90 and 120 bars) covering the full temperature range of this investigation. These results demonstrate an initial decrease in shift up to 600°C followed by a strong increase as the temperature is raised to the critical region. Their complete data for all isobars are collected in Table 11.2.

Figure 11.4. Knight shift (^{133}Cs) as function of temperature for two pressures (90 and 120 bars) (after Warren et al., 1989).

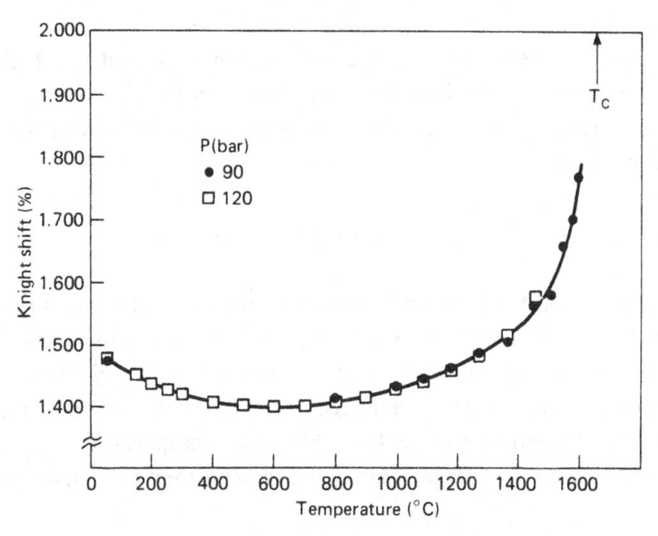

Table 11.2. ^{133}Cs *Knight shift isobars.*

	Knight shift (%)				
T (°C)	P = 30 bar	90	120	500	900
55	1.472	1.474	1.476	1.492	1.512
150	1.449	—	1.450	1.464	1.479
200	1.436	—	1.436	1.452	1.464
254	1.424	—	1.427	1.440	1.455
300	1.415	—	1.418	1.429	1.440
400	1.405	—	1.405	1.414	
500	1.400	—	1.400	1.403	
600	1.397	—	1.395	1.396	
700	1.398	—	1.399	1.393	
800	1.407	1.411	1.406	1.392	
895	1.420	—	1.414		
990	1.434	1.430	1.424		
1083	1.454	1.440	1.439		
1176	1.479	1.459	1.459		
1269		1.486	1.482		
1363		1.504	1.518		
1456		1.564	1.679		
1503		1.581			
1549		1.660			
1571		1.701			
1590		1.768			

The pressure dependence of the shift is shown explicitly in Figure 11.5. These results were obtained by isothermal compression of the liquid at various temperatures. At 55°C, Warren et al. (1989) measured a logarithmic pressure coefficient

$$\left(\frac{\partial \ln K}{\partial P}\right)_{55} = 3.2 \pm 0.2 \times 10^{-5}\,\text{bar}^{-1} \tag{11.6}$$

which agrees within experimental error with the value measured for the solid at 20°C by Benedek and Kushida (1958). With increasing temperature, however, the pressure coefficient gradually decreases and changes sign in roughly the same temperature range as the minimum in the temperature dependence shown in Figure 11.4. At higher temperatures, the pressure dependence becomes increasingly negative, and they obtain, for example,

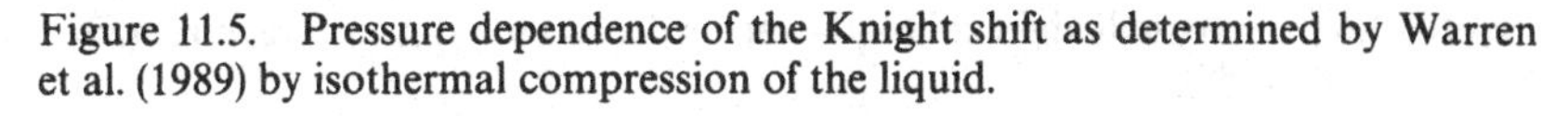

Figure 11.5. Pressure dependence of the Knight shift as determined by Warren et al. (1989) by isothermal compression of the liquid.

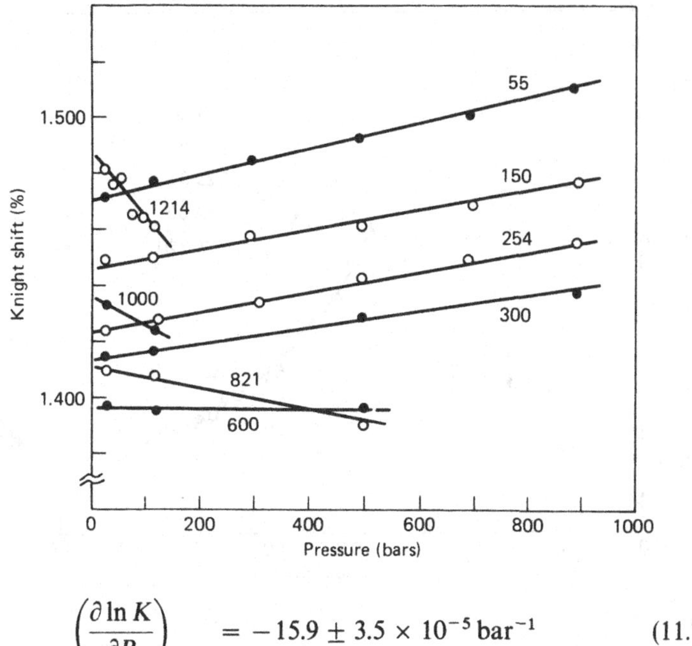

$$\left(\frac{\partial \ln K}{\partial P}\right)_{1190} = -15.9 \pm 3.5 \times 10^{-5}\,\text{bar}^{-1} \tag{11.7}$$

for the shift coefficient at 1190°C.

The isobaric Knight shift data are plotted against the density for various pressures in Figure 11.6. This plot shows the variation of the shift over nearly a factor of three. The minimum density, 0.7 g/cm^3, is slightly less than twice the critical density ($\rho_c = 0.379$ g/cm^3). The minimum in the density dependence of the shift near 1.5 g/cm^3 corresponds to the minimum in the temperature dependence around 600°C.

When the relationship between the temperature and density dependences of the shift is examined in more detail, a clear difference emerges between the low-temperature and high-temperature regions. On the high-density, low-temperature side of the minimum, isobars and isotherms are not coincident, and there is an explicit negative temperature dependence for the shift at constant volume. The constant-volume temperature coefficient, $(\partial \ln K/\partial T)_v$, is negative and about 65% greater than the corresponding value for solid caesium. At low densities, in contrast, the isobars and isotherms are coincident within experimental error. This has the implication that the increase in the Knight shift on approaching the critical

Figure 11.6. Isobaric Knight shift data versus density for various pressures (after Warren et al., 1989).

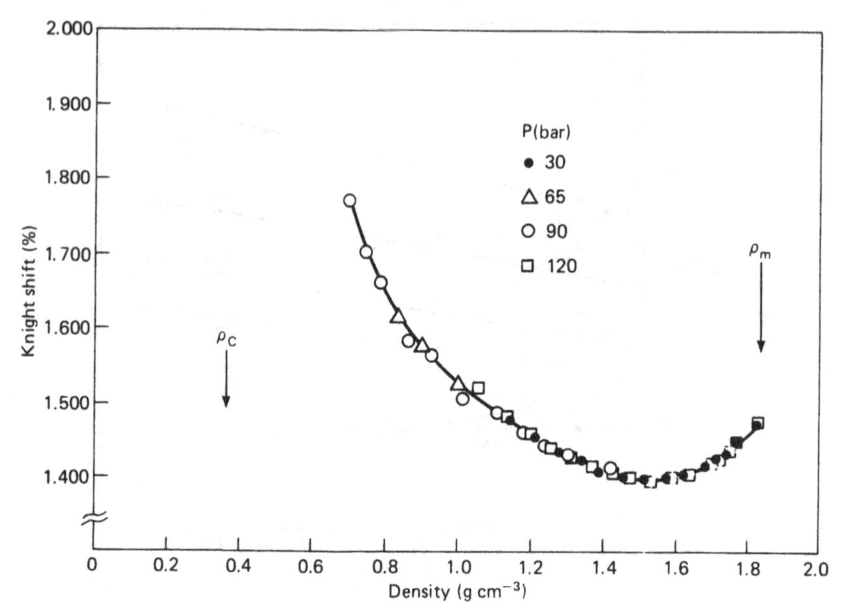

region is primarily due to the reduction of the density with the explicit temperature dependence playing a minor role.

11.6.1. Relaxation rates

^{133}Cs nuclear relaxation rates are presented in Figure 11.7 plotted against temperature for various pressures. As discussed in the previous section, these rates are the observed inverse free induction decay lifetimes corrected for inhomogeneous broadening. By direct measurement of the spin-lattice relaxation times (see Faber, 1972) $1/T_1$, it was confirmed by Warren et al. (1989) that the corrected value T_2 equals T_1 in the low temperature range. As shown in Figure 11.7, the quantity $1/T_2T$ is roughly constant up to 800°C and then rises sharply with further heating toward the critical region. Because of poor signal-to-noise ratios at the highest temperatures, these workers were able to obtain reproducible data for the relaxation rate only to 1500°C, roughly 100°C lower than the highest Knight shift measurement.

Figure 11.7. Temperature dependence of $1/T_2T$ for Cs as measured by Warren et al. (1989) at various pressures.

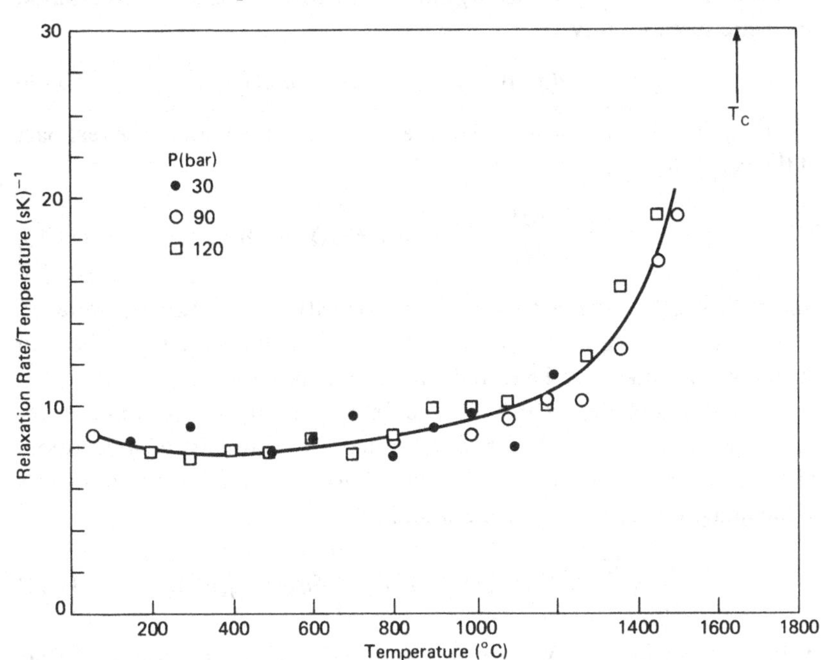

11.7. Theory for NMR shifts and relaxation in fluid metals

In this section a summary is given of the basic theoretical expressions for NMR shifts and relaxation in fluid metals in a form suitable for interpretation of the preceding results (Winter, 1971). Particular emphasis is placed on the effects of electron-electron interactions on the magnetic response of the electrons (compare Section 11.2).

The dominant coupling between the nuclear spin system and the conduction electrons is the magnetic hyperfine interaction

$$H_c = \frac{8\pi}{3}\gamma_n\gamma_e\hbar^2 \sum_j \mathbf{I}^j\cdot\mathbf{S}(\mathbf{R}_j), \tag{11.8}$$

where **I** and **S** are the nuclear and electronic spins, respectively, and γ_n and γ_e are the corresponding gyromagnetic ratios. The summation is taken over all nuclear sites $\mathbf{R}_j$ in the liquid. The consequences of the hyperfine interaction for NMR—namely, shifts of the resonance frequency and nuclear spin relaxation—are governed by different contributions to the generalized magnetic susceptibility

$$\chi(q,\omega) = \chi'(q,\omega) + i\chi''(q,\omega). \tag{11.9}$$

The Knight shift is proportional to the uniform, static limit of the real part of the generalized susceptibility $\chi'(0,0)$:

$$K \equiv \frac{\Delta H}{H_0} = \frac{8\pi}{3}\langle|\psi(0)|^2\rangle_F \Omega \chi'(0,0), \tag{11.10}$$

where $\langle|\psi(0)|^2\rangle$ is the electronic charge density at the nucleus, averaged over all states at the Fermi energy. The wave functions are normalized in the atomic volume Ω and $\chi'(0,0)$ is the susceptibility per unit volume.

Fluctuations of the local hyperfine field at the nuclear resonance frequency ω_0 are responsible for longitudinal, or "spin-lattice," relaxation. The relaxation rate depends on an integral over the transverse components of the imaginary part of the generalized susceptibility

$$\frac{1}{T_1} = \frac{32}{9}\langle|\psi(0)|^2\rangle_F^2 \gamma_n^2 kT\Omega^2\omega_0^{-1} \int dq\, q^2 \chi''(q,\omega_0)_\pm. \tag{11.11}$$

In metals, the fluctuations are fast compared with the inverse nuclear Larmor frequency $1/\omega_0$, so that "extreme narrowing" conditions apply and $1/T_1 = 1/T_2$.

11.7.1. Noninteracting electrons

For noninteracting electrons, the uniform, static susceptibility is just the Pauli susceptibility (compare (11.1) and (11.2)):

$$\chi_0'(0,0) = \tfrac{1}{2}(\gamma_e \hbar)^2 N(E_F), \tag{11.12}$$

where $N(E_F)$ is the density of states at the Fermi level for a single direction of spin. The integral in (11.11) over the dynamic susceptibility can be evaluated explicitly for noninteracting electrons in the low frequency limit to yield

$$\omega_0^{-1} \int dq\, q^2 \chi_0''(q,\omega_0)_\pm = 2\pi^3 \gamma_e^3 \hbar^3 [N(E_F)]^2. \tag{11.13}$$

Substitution of (11.13) in (11.11) yields the Korringa (1950) expression for the relaxation rate

$$\frac{1}{T_1} = \frac{64}{9}\pi^3\hbar^3\gamma_n^2\gamma_e^2\langle|\psi(0)|^2\rangle_F^2\Omega^2 kT[N(E_F)]^2 \tag{11.14}$$

and the combination of (11.10), (11.12), and (11.14) leads to the Korringa relation between the relaxation rate and the Knight shift (see (11.10)):

$$\left(\frac{1}{T_1}\right)_{\text{Korr}} = \left(\frac{4\pi k}{\hbar}\right)\left(\frac{\gamma_n}{\gamma_e}\right)^2 K^2 T. \tag{11.15}$$

It can be seen from (11.10) and (11.11) that the Korringa relation, (11.15), expresses a particular relationship between the integral of the q-dependent dynamic susceptibility and the static, uniform susceptibility for noninteracting electrons

$$\omega_0^{-1}\int \mathrm{d}q\, q^2\chi_0''(q,\omega_0)_\pm = \left(\frac{8\pi^3}{\gamma_e^2\hbar}\right)[\chi_0'(0,0)]^2. \tag{11.16}$$

Consideration will be given next to how this relationship is changed by interactions.

11.8. Interactions between electrons

The magnetic effects of electron-electron interactions are manifested as enhancements of the generalized susceptibility. These fall into two classes: a cooperative exchange-correlation enhancement of the local magnetic field and correlation enhancement of the effective mass or density of states at the Fermi level. The first is exemplified by the Stoner model and the second by the Brinkman-Rice model of a highly correlated metal.

11.8.1. Stoner enhancement

The effects of interactions in alkali metals under ordinary conditions are usually described in terms of the Stoner model of exchange-correlation enhancement. The generalized susceptibility is expressed in a molecular field or random phase approximation (RPA) as

$$\chi(q,\omega) = \frac{\chi_0(q,\omega)}{1 - V_{xc}(q)\chi_0(q,\omega)}, \tag{11.17}$$

in which $V_{xc}(q)$ is proportional to the q-dependent exchange-correlation potential. The enhanced static, uniform susceptibility then takes the simple form

$$\chi'(0,0) = \frac{\chi'_0(0,0)}{1 - \alpha} \tag{11.18}$$

where $\alpha = V_{xc}(0)\chi'_0(0,0)$. Enhancement of the imaginary part of the dynamic susceptibility is q-dependent and at low frequencies may be written

$$\chi''(q,\omega) = \frac{\chi''_0(q,\omega)}{[1 - \alpha f(q)]^2}, \tag{11.19}$$

where the q-dependence $f(q)$ is determined by the form of the exchange-correlation potential and the nonuniform susceptibility for noninteracting electrons

$$f(q) = \frac{V_{xc}(q)\chi'_0(q,0)}{V_{xc}(0)\chi'_0(0,0)}. \tag{11.20}$$

The q-dependent enhancement of $\chi''(\mathrm{q},\omega_0)$ leads to breakdown of the Korringa relation, (11.15) or (11.16). As shown first by Moriya (1963), substitution of specific functions for $V_{xc}(\mathrm{q})$ and $\chi'_0(0,0)$ gives a quantitative relation between the Korringa ratio

$$\eta \equiv \frac{(1/T_1)}{(1/T_1)_{\mathrm{Korr}}} = \frac{\omega_0^{-1} \int \mathrm{d}q\, q^2 \chi''(q,\omega_0)_{\pm}}{(8\pi^3/\gamma_e^2\hbar)[\chi'(0,0)]^2} \tag{11.21}$$

and the static enhancement parameter α. For Cs at normal densities, the experimental Korringa ratio interpreted with Shaw's exchange-correlation potential (Shaw and Warren, 1971) yields a value $\alpha = 0.44$ or, using (11.21), a uniform enhancement of about 1.8. This is a typical value for alkali metals.

The function $f(q)$ is a decreasing function of q in simple metals, and the enhancement in this model is therefore lower at finite q than at $q = 0$. Thus, the Stoner enhancement is ferromagnetic in the sense that the largest enhancement occurs for the uniform magnetization. Although the quantitative relation between η and α depends on the form of the exchange-correlation potential, inspection of (11.20), (11.19), and (11.18) shows that the Korringa ratio (11.21) is less than 1, and η decreases further the more the static enhancement α increases.

11.8.2. Correlation enhancement

The enhancement of the susceptibility by correlation was described by Brinkman and Rice (1970) using a variational method due to Gutzwiller (1965). Unlike the Stoner model, the Brinkman-Rice description explicitly introduces the role of the atoms. The majority of sites in the highly correlated metal are assumed to be instantaneously singly occupied, leaving only a small fraction f doubly occupied. Brinkman and Rice showed that the effective mass in the highly correlated metal is enhanced according to

$$\frac{m^*}{m_e} = \frac{1}{2f}. \tag{11.22}$$

The enhancement leads to a new degeneracy temperature T_d reduced relative to the Fermi temperature T_F of the noninteracting system:

$$T_d = 2fT_F. \tag{11.23}$$

Enhancement in the highly correlated metal differs in two important respects from Stoner enhancement. First, both the susceptibility and electronic specific heat are enhanced, whereas Stoner enhancement affects only the susceptibility. Although enhancement of the specific heat is unlikely to be observed in expanded liquid metals due to the dominance of the "ionic" specific heat, this enhancement is commonly seen in heavy Fermion metals (see Appendix 11.1). The second important difference is that the static, uniform susceptibility of the highly correlated metal saturates at the Curie value as the experimental temperature approaches the degeneracy temperature T_d. In contrast, the Stoner susceptibility, (11.18), can increase without limit as $\alpha \rightarrow 1$ and, in fact, this divergence indicates the onset of ferromagnetism. The limitation of the susceptibility to the Curie value is important for expanded Cs because of the high experimental temperatures and possible reduction of the degeneracy temperature by correlation enhancement (see also Section 11.12).

Calculations for the Hubbard model indicate that the ground state of the highly correlated metal is antiferromagnetic. Likewise, band calculations for crystalline atomic hydrogen show a progression from paramagnetic metal to antiferromagnetic metal to antiferromagnetic insulator as the crystal is expanded. The nonuniform susceptibility of an antiferromagnetic metal should have a maximum at a finite q-value corresponding to the ordering wavelength. Direct calculations of $\chi'(q, 0)$ by Kelly and Glötzel (1986) show, indeed, that the susceptibility of hydrogen at the zone boundary exceeds that at $q = 0$. The calculations for expanded alkali metals, however, are contradictory. Band calculations for expanded lithium by Callaway, Zou, and Bagayoko (1983) indicate that a ferromagnetic ground state is slightly favored; it seems probable that a similar situation will hold for the other alkali metals. Kelly and Glötzel (1986), in contrast, found that antiferromagnetic order sets in at a higher density than ferromagnetic order as various crystalline structures of Cs are expanded.

Development of a tendency toward antiferromagnetic ordering has an important consequence for nuclear relaxation. According to (11.13) and the arguments given in the preceding section, enhancement of $\chi''(q, \omega)$ at finite q in excess of the enhancement at $q = 0$ should lead to an increase of the Korringa ratio relative to the value $\eta = 1$ expected in the noninteracting limit. Since, as has been seen, the Korringa ratio is reduced for Stoner (ferromagnetic) enhancement, measurement of η provides a means of distinguishing between the two types of enhancement.

11.9. Characteristics of high-density liquid Cs

The electronic structure and magnetic properties of liquid Cs near the melting point are very similar to those of the solid just below the melting point. The Knight shift, for example, decreases by only 1.8% on melting and the spin-lattice relaxation rate changes by less than 10%. The Korringa enhancements are 0.58 and 0.61 in the solid (Narath and Weaver, 1968) and liquid, respectively, showing that the susceptibility enhancement is only slightly affected by melting. With the exception of the isochoric temperature coefficient, $(\partial \ln K/\partial T)_v$, the various temperature and pressure coefficients of the Knight shift in the liquid lie within 20% of their values for solid Cs (Table 11.3).

These small changes in magnetic properties at T_m correlate with strong indications that relatively little change in the local atomic arrangement occurs during melting. The molar volume increases by only 2.5%, for example, and the average near-neighbour distance in the liquid, 5.31 Å (Gingrich and Heaton, 1961), is identical to the near-neighbour distance in *bcc* Cs about 40°C below T_m (Pearson, 1964). The density of states at the Fermi level in solid Cs is strongly influenced by the *d*-states, which lie mostly above E_F. Band calculations (Ham, 1962) give a value $m^* = 1.76$ for the effective mass. The small changes in magnetic properties on melting show that this "band-structure" enhancement of the density of states per-

Table 11.3. ^{133}Cs *differential Knight shift coefficients in solid and liquid* Cs *near the melting point* (*after Warren et al.*, 1989).

	Solid	Liquid
$\left(\frac{\partial \ln K}{\partial T}\right)_P$	$-1.85 \times 10^{-4} K^{-1}$	$-1.66 \pm 0.10 \times 10^{-4} K^{-1}$
$\left(\frac{\partial \ln K}{\partial P}\right)_T$	3.4×10^{-5} bar^{-1}	$3.2 \pm 0.2 \times 10^{-5}$ bar^{-1}
$\left(\frac{\partial \ln K}{\partial \ln V}\right)_T$	-0.62	-0.48 ± 0.07
$\left(\frac{\partial \ln K}{\partial T}\right)_V$	-0.7×10^{-4}	$-1.16 \pm 0.09 \times 10^{-4}$

sists in the liquid. Most of the change in Knight shift on melting can be attributed to the small volume change at the phase transition viewed as extension of the thermal expansion of the solid. Accordingly, one calculates the change at the melting point, ΔK, using the volume coefficient of the Knight shift measured for the solid,

$$\frac{\Delta K}{K_{\text{solid}}} = \left(\frac{\Delta V}{V_{\text{solid}}}\right)\left(\frac{\partial \ln K}{\partial \ln V}\right)_{\text{solid}} = -1.6\%, \qquad (11.24)$$

which can be compared with the observed value -1.8%. This is strong evidence that the electric and magnetic properties are determined mainly by the volume and local distribution of near neighbours and are not strongly influenced by the loss of long-range order on melting. This basic idea motivates the use of crystalline models to calculate the electronic structure and properties of expanded liquid Cs (Kelly and Glötzel, 1986; Warren and Mattheiss, 1984; see also Chapter 10).

The volume dependence of the shift in the solid and dense liquid results mainly from the decrease in density of states as the Fermi level moves away from the *d*-states. Close to T_m, the isothermal volume dependence is very similar in liquid and solid. This is in agreement with the basic assumption of the *uniform fluid model* (UFM), according to which the structure is unaffected by pressure except for a simple scaling of distances by $V^{1/3}$ (Egelstaff et al., 1971, 1980). Neutron studies of liquid Cs (Winter et al., 1989) show that the UFM is valid close to T_m but that it breaks down dramatically at higher temperatures. This is consistent with the data shown in Figure 11.5, which display a marked change in the isothermal volume dependence of the shift between 300°C and 600°C.

Even near the melting point, however, the explicit temperature dependence of the Knight shift in liquid Cs is substantially larger than in the solid. Further, the explicit temperature dependence decreases rapidly with temperature in the liquid, whereas in the solid it is nearly the same at helium temperatures and close to the melting point (Kaeck, 1972; see also Muto et al., 1962). The origin of the explicit temperature dependence is not well understood in either liquid or solid, although it may be partially explained by effective thermal averaging of the structure factor and a resulting tendency toward a more free-electron-like density of states at high temperature. A theory based on this effect was quite successful in explaining the unusually strong temperature dependence of the Knight shift in solid cadmium (Kasowski, 1969).

The divergence with increasing temperature of liquid-state properties from those of the solid may reflect in part an essential difference in the nature of thermal expansion and thermal excitations in the two phases. Whereas thermal expansion of the crystal is due to anharmonic lattice vibrations that lead to increased average interatomic separations, neutron studies of liquid Cs (Winter et al., 1987) and Rb (Franz et al., 1980) show that expansion of the liquid is mainly due to reduced numbers of near neighbours. Only slight increases in the interatomic distances are observed. In other words, as the liquid expands with increasing temperature, it undergoes a continuous change of structure. Continuous structural evolution is, of course, forbidden by the requirements of crystalline symmetry in the solid, and it is therefore not surprising to find that significant differences develop in the temperature and pressure dependence of electronic structure and properties of solid and liquid, even with relatively modest increases in temperature.

11.9.1. Static, uniform susceptibility in expanded Cs

The behaviour of the static, uniform susceptibility reveals that significant changes in the electronic properties of Cs develop at low density. In Figure 11.8 the volume electronic susceptibility extracted from Freyland's (1979) measurements of the mass susceptibility along the coexistence curve is shown (see also Chapman and March, 1988). To obtain the electronic paramagnetic contribution from the total susceptibility, a correction for diamagnetism was made using a value -35×10^{-6} cm^3/mole for the ionic diamagnetism (Brindley and Hoare, 1937) and values of the electronic diamagnetic susceptibility calculated from the theory of Kanazawa and Matsudaira (1960; see also March and Donovan, 1954). The dominant features of the data shown in this figure are a gradual decrease of $\chi'(0,0)$ as the density decreases from the normal value at the melting point, a sharp rise as the density falls below about 1.5 g/cm^3, and a decrease in the lowest density range.

Since the Stoner model adequately explains the enhancement of $\chi'(0,0)$ in the solid and normal liquid, it is natural to ask whether increased enhancement at low density can be explained within the framework of this theory. There is substantial evidence that this is not the case (Warren et al., 1989). First, theoretical calculations (Kelly and Glötzel, 1986; Warren and Mattheiss, 1984) for various low-density Cs crystal structures offer no indication that the density of states (per unit volume) at the Fermi level increases as the density decreases. To take account of the structural evolution of the expanding liquid, band calculations (Kelly and Glötzel, 1986; Franz, 1984) were carried out for various structures having different near-neighbour coordination numbers with a constant value of the near-neighbour distance. This feature of the structural evolution was also exploited by Franz (1984) for a model of expanded metals viewed as an "alloy" of atoms and vacancies. These models predict either decreasing or constant densities of states at the Fermi level and, accordingly, (11.12) and

Figure 11.8. Spin susceptibility for liquid caesium along coexistence curve (Chapman and March 1988; see also Warren, 1984). (a) Measurement. (b) Curie limiting law.

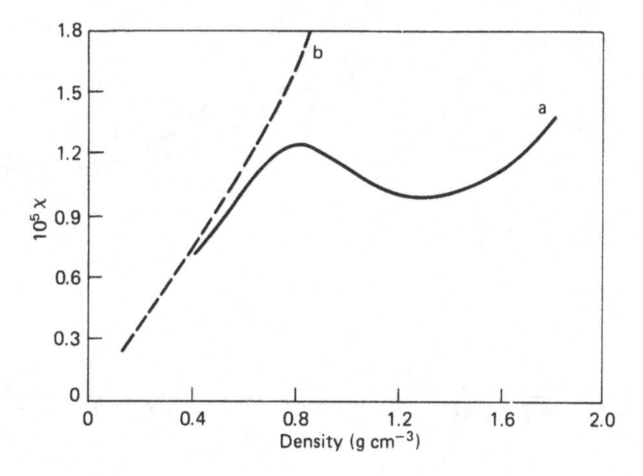

(11.18) do not predict increased enhancement of the volume susceptibility at low density unless the Stoner parameter α increases sharply. The latter possibility is precluded, however, by the behaviour of the Korringa enhancement shown in Figure 11.9. Warren et al. (1989) observed that η increases significantly in the low-density range and reaches values slightly greater than 1 at the lowest densities. The same trend was observed for liquid Na over a more limited density range by Bottyan et al. (1983). Since the Stoner picture predicts a *decrease* in η if α increases, an increase in the Korringa enhancement is inconsistent with increased enhancement of $\chi'(0,0)$ on the conventional model.

Finally, Warren et al. (1989) considered the decrease of the susceptibility when the density falls below 0.8 g/cm^3. As may be seen in Figure 11.8, $\chi'(0,0)$ is limited by the Curie value for densities below this value. This behaviour is not expected for Stoner enhancement and indicates that the low-density enhancement is due to reduced degeneracy temperature as in the Brinkman-Rice model (Warren, 1984). Chapman and March (1988; see Section 11.12) show how this crossover from Stoner to Brinkman-Rice enhancement follows from the finite temperature extension (Rice et al., 1985) of the theory of Brinkman and Rice. They confirm that caesium is highly correlated for densities below $\sim$0.8 g/cm^3.

Figure 11.9. Behaviour of Korringa enhancement versus density at the pressures indicated (after Warren et al., 1989).

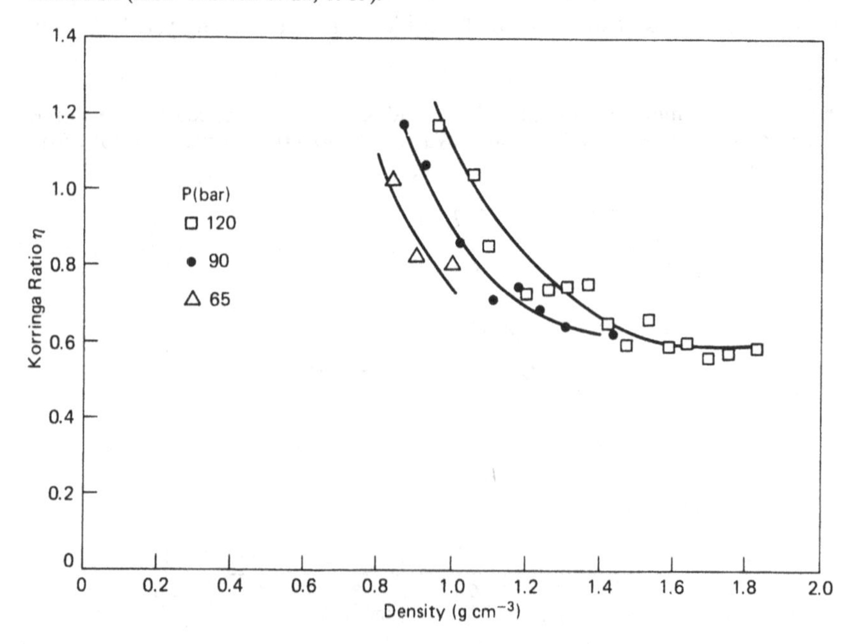

The preceding analysis has been developed largely using concepts familiar from the study of condensed metals at low temperatures. The influence of the high experimental temperatures has been introduced only through the Curie law limitation of the susceptibility. Warren et al. (1989) argue that this approach is justified by the essentially metallic character of caesium throughout the range covered in this investigation. For example, the minimum value of the electrical conductivity in their experimental range is about 700 (Ω cm^{-1}) and the Hall coefficient remains close to the free electron value for one electron per atom down to at least $\rho = 1.1$ g/cm^3, the lowest density measured. A complementary theoretical approach developed by Alekseev and Iakubov (1983) begins with a description of the low-density, high-temperature gas as a weakly ionized plasma. Hernandez (1986a, b) has adopted this point of view and calculated the thermodynamic equilibria for formation of various polyatomic species as the density of the gas is increased. This approach provides a consistent interpretation of the electrical and magnetic properties at densities up to the critical region and predicts, in particular, an important role for the diamagnetic dimers in this range. However, the high-density limit of this theory is a uniform metal ("jellium"), and thus the model does not include the important ion-ion and electron-ion correlations that characterize the developing condensed state. Chapman and March (1988; see Section 11.12) have pointed out that the magnetic susceptibility in this region shows a marked difference from that of jellium and they attribute this to the substantial influence of electron-ion interactions on the electronic structure in the highly expanded fluid metal.

11.10. Dynamic nonuniform susceptibility in expanded Cs

The breakdown of the Stoner enhancement description of expanded Cs has been stressed in the previous section. Let us now consider more explicitly the implications for $\chi''(q, \omega_0)_\pm$ of the observed increase in the Korringa ratio at low densities (Warren et al., 1989). Using the low-frequency expression for the dynamic, nonuniform susceptibility in the noninteracting system (Doniach, 1967)

$$\chi_0''(q, \omega_0)_\pm = \frac{8\pi^3}{\gamma_e^2 \hbar^2} [\chi_0'(0, 0)]^2 \frac{\hbar\omega_0}{2q_F^2} \frac{1}{q} \tag{11.25}$$

one can show from (11.21) that the Korringa ratio may be written

$$\eta = \frac{1}{2q_F^2} \int_0^{2q_F} dq\, q g(q). \tag{11.26}$$

The function $g(q)$ provides a general description of the q-dependence of $\chi''(q, \omega_0)_\pm$ relative to that of the noninteracting system:

$$\frac{\chi''(q, \omega_0)_\pm}{\chi_0''(q, \omega_0)_\pm} = \frac{\chi''(0, \omega_0)_\pm}{\chi_0''(0, \omega_0)_\pm} g(q). \tag{11.27}$$

In the conventional Stoner description, (11.19) gives

$$g_s(q) = \frac{(1-\alpha)^2}{[1-\alpha f(q)]^2}. \tag{11.28}$$

It is the monotonic decrease in $g_s(q)$ that leads to values $\eta < 1$ in cases of Stoner enhancement.

The data presented in Figure 11.9 show that η increases strongly as the density falls below about 1.4 g/cm^3 and reaches values in excess of 1 at the lowest densities. It is evident from (11.26) that values of $\eta > 1$ require that

Figure 11.10. Function g(q) defined in (11.29) plotted for three sets of the parameters.

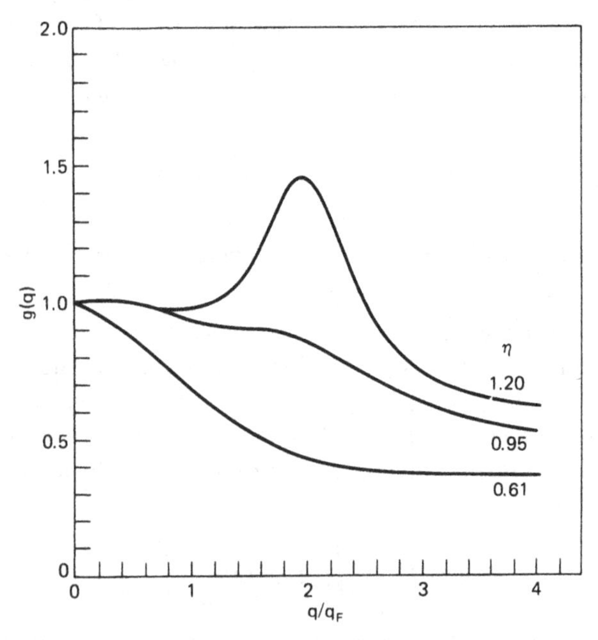

$g(q)$ increase and exceed $g(0)$ in some range of nonzero q. Such behaviour corresponds to a change in the enhancement of the dynamic, nonuniform susceptibility from ferromagnetic in the normal metal to antiferromagnetic enhancement in the expanded metal.

To get a semiquantitative description of the antiferromagnetic enhancement, Warren et al. (1989) have modeled the evolution of $g(q)$ using a function characterized by increased enhancement in the vicinity of $q = 2q_F$,

$$g(q) = g_s(q)\left[1 + \left(\frac{q}{q + q_F}\right)\frac{\lambda}{1 + (q - 2q_F)^2\lambda^2}\right]. \tag{11.29}$$

This function is plotted in Figure 11.10 for three sets of parameters that yield η values in the range 0.6–1.2 measured in the experiments of Warren et al. (1989). Since η depends only on the integral of $g(q)$, as shown by (11.26), it is obvious that the particular form chosen for (11.29) has no special significance. One could have chosen any function in which the high q enhancement occurs at a nonzero value of q in the range $0 < q < 2q_F$. The presence of a peak in $g(q)$ and η values greater than 1 implies a tendency for anticorrelation of spins or, in other words, formation of slightly underdamped spin density waves in the low-density metal. For a spin density wave of wavelength π/q_F, corresponding to (11.29) and Figure 11.10, a correlation length comparable with the wavelength and interatomic spacing yields the η values observed at the lowest densities.

11.10.1. Conduction electron wave functions

The average density of Fermi energy electrons at the nucleus, $\langle|\Psi(0)|^2\rangle_F$, can be obtained from the Knight shift and susceptibility results using (11.10). Susceptibility data (Freyland, 1979) were obtained only along the liquid-gas coexistence curve, so Warren et al. (1989) have extrapolated their Knight shift data to the coexistence curve using the density dependence exhibited in Figure 11.6. The paramagnetic, electronic susceptibility extracted from the data of Freyland is that shown in Figure 11.8. For convenience, the convention is followed of expressing the charge density at the nucleus in units of the atomic value $\xi = \langle|\Psi(0)|^2\rangle_F/|\Psi(0)|_a^2$, where $|\Psi(0)|_a^2 = 2.58 \times 10^{25}$ cm^{-1}. The result plotted in Figure 11.11 shows a density-independent value $\xi \sim 0.5$ at high densities, but ξ decreases markedly when ρ drops below about 1.4 g/cm^3. At the lowest densities, ξ reaches a minimum value of about 0.28 and then starts to increase.

Figure 11.11. Charge density at nucleus in units of atomic value (after Warren et al. 1989).

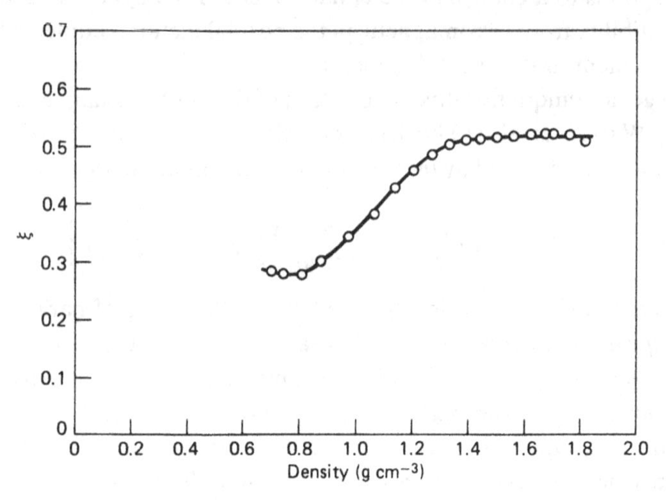

The density dependence of ξ shown in Figure 11.11 is contrary to naive expectations. It is obvious from its definition that ξ must approach the value 1 in the dilute (atomic) limit. However, the unexpected decrease of ξ in the range 1.4 to 0.8 g/cm^3 shows that this does not occur monotonically. The band calculations fail to account for this effect, since they indicate a simple monotonic increase of ξ from the value in the high-density metal to the atomic value (Kelly and Glötzel, 1986; Warren and Mattheiss, 1984). The reduction in ξ cannot be due to spin pairing in dimers or other species, since ξ is defined to be the charge density of only those electrons contributing to the susceptibility. Rather, the data show that there is a substantial movement of conduction electron charge away from the ions in the intermediate density range. The minimum in ξ at about 0.8 g/cm^3 indicates that this trend is reversed at very low density.

11.11. Evolution of the electronic structure of expanded liquid Cs

Liquid Cs near the melting point may be considered a normal liquid metal with properties typical of the condensed state. The small changes of most electronic properties on melting show that the electronic structure of the liquid is quite similar to that of *bcc* crystalline Cs close to the melting point. This remains basically true during expansion for densities in the range 1.4 to 1.6 g/cm^3 except for gradual changes in temperature and pressure coefficients from their solid-state values. These changes provide an indication of the fundamentally different character of thermal expansion in the solid and liquid states. Throughout this range of initial expansion, the Korringa enhancement parameter remains constant, and the susceptibility enhancement is consistent with the conventional Stoner theory. The distribution of electronic charge is affected little in this range and maintains a roughly constant value $\xi = 0.5$ for $\rho > 1.4$ g/cm^3.

As the density is reduced below about 1.4 to 1.6 g/cm^3, qualitative changes develop, and the liquid takes on unusual electronic characteristics. The susceptibility enhancement increases sharply, the Korringa relation breaks down, indicating a shift from ferromagnetic to antiferromagnetic enhancement, and the wave functions spread out so as to cause a substantial reduction of charge density at the nucleus. The latter effect is reversed only at the lowest densities, below 0.8 g/cm^3, roughly twice the critical density. Despite these changes, Cs retains metallic conductivity values down to 0.7 g/cm^3, and the Hall coefficient is within 10% of the free electron value to at least 1.1 g/cm^3. The electronic spin susceptibility at the lowest densities follows the Curie law expected for 1 electron/atom obeying classical Boltzmann statistics. Although the experimental value for the spin susceptibility is subject to some uncertainty because of the necessary diamagnetic correction, there is little evidence of a substantial fraction of paired spins in this range of the subcritical liquid.

The magnetic properties at low density are not to be explained by extension of the conventional Stoner enhancement mechanism. Rather, Cs becomes a highly correlated metal with antiferromagnetic spin fluctuations. As such it is related to other highly correlated systems, including high temperature superconducting cuprates.

To conclude this section it is of some interest to describe the way in which the Knight shift in liquid Hg reflects the change in electronic structure as the density is varied. El-Hanany and Warren (1975) have measured the Knight shift in liquid Hg from the normal liquid density to less than 8 g/cm^3 at 1730 K and 1411 bar. An abrupt metal-nonmetal transition between 9 and 8 g/cm^3 is preceded by ranges of (roughly) linearly decreasing shift ($11 < \rho < 13.6$ g/cm^3) and constant shift ($9 < \rho < 11$ g/cm^3), as shown in Figure 11.12.

Figure 11.12. Knight shift in liquid Hg as measured by El-Hanany and Warren (1975).

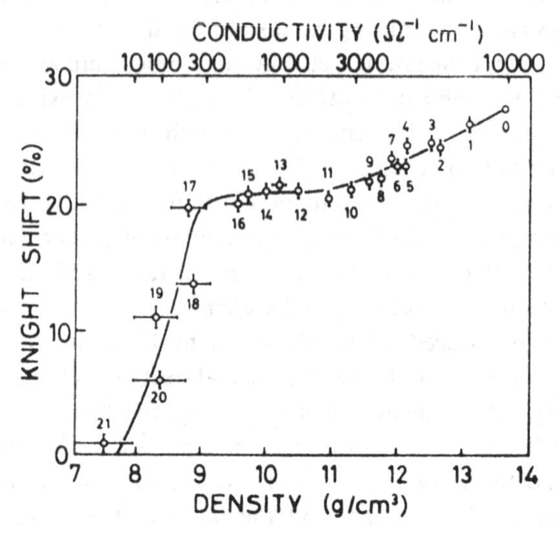

11.12. Phenomenology and heavy Fermion theory of magnetic susceptibility of expanded fluid alkali metals

As emphasized before, expanded alkali metals provide a testing ground of the understanding of the influence of electron-electron correlation on electrical and magnetic properties and the role of such interactions in the metal-insulator transition, which, in these materials, is believed to occur near to the liquid-vapour critical region (see, for example, Yonezawa and Ogawa, 1982).

The experiments of Freyland (1979, 1980; see also Bottyan et al., 1983) on the magnetic susceptibilities of alkali metals along the liquid-vapour coexistence curve have stimulated much interest, and the following treatment (see Chapman and March, 1988) focuses on these fluid metals. Warren (1984; 1987) extracted the paramagnetic contribution to the total susceptibility due to the conduction electrons and pointed out that the behaviour in the low-density liquid region is consistent with a correlation-enhanced Pauli paramagnetism, limited by the free-spin Curie value. This has already been illustrated in Figure 11.8, which incorporates the data of Jüngst et al. (1985) on the liquid-vapour coexistence curve of Cs. In what follows, the behaviour of the magnetic susceptibility in the vicinity of the metal-insulator transition will be considered, initially by means of a phenomenological description and then by utilizing a microscopic theory developed for heavy-electron systems.

11.12.1. Phenomenology

The starting point of the following description is the phenomenological treatment of a correlation-induced metal insulator transition at absolute zero $T = 0$ due to March, Suzuki and Parrinello (1979). Following these authors, let us write the ground-state energy per atom of a half-filled band near the metal-insulator transition (in the metallic phase) as an expansion in the magnetization per atom m and the quasi-particle renormalization factor q (see Chapman and March, 1988). The latter is the discontinuity in the single-particle occupation number at the Fermi surface, which decreases continuously to zero at the (assumed second-order) transition. Thus

$$E(m, q) = E_0 + am^2 + \cdots + bq + cq^2 + \cdots + eqm^2 + \cdots. \quad (11.30)$$

The volume magnetic susceptibility χ is then

$$\chi = \frac{n_0 \mu_0 \mu_B^2}{2(a + eq)}, \quad (11.31)$$

where n_0 is the number density. Hence if a is zero or vanishes at least as fast as q on approaching the transition, the susceptibility is enhanced as in the model of Brinkman and Rice (1970) for such a system. For this particular situation, one can identify readily the coefficients in the expansion (11.30) with microscopic parameters to find

$$a = 0, \qquad b = \frac{(U_c - U)}{8}, \qquad c = \frac{U}{32}; \qquad e = \frac{[1 + \frac{3}{2}\bar{\varepsilon}N(\varepsilon_f)]}{2N(\varepsilon_f)}, \quad (11.32)$$

giving $\chi \sim q^{-1}$, as noted by Brinkman and Rice (1970). Here as earlier, U is the Hubbard on-site interaction, $N(\varepsilon_f)$ is the electronic density of states per atom at the Fermi energy, U_c is the value of the Hubbard U at the metal-insulator transition, and $\bar{\varepsilon}$ is the band energy in the absence of correlation.

Let us next turn to consider the situation at elevated temperatures. A natural extension of (11.30) leads to the free energy per atom at temperature T as having the form

$$F(m, q, T) = E_0 + a(T)m^2 + \cdots + b(T)q + c(T)q^2 + \cdots + e(T)qm^2 + \cdots . \tag{11.33}$$

Here, of course, the interpretation of q in terms of the average number of doubly occupied sites (given explicitly in the Brinkman-Rice model referred to above) is more appropriate, since the discontinuity in the single-particle occupation number will not be such a well-defined quantity. Furthermore, as $T \to \infty$, one must expect a reversion to Curie-like behaviour in the susceptibility as the degeneracy temperature of the Fermi fluid is exceeded. This is achieved if the coefficient a in (11.33) is proportional to T, with the coefficient e much less strongly dependent on T and remaining finite as $T \to 0$. Thus, if $a = \alpha T$, then as T tends to infinity, $\chi \to n_0 \mu_0 \mu_B^2 / 2\alpha T$ and for lower temperatures is always less than this limiting value for all densities. If $\alpha = \frac{1}{2} k_B$, this is simply the Curie law for the electrons. As $T \to 0$, one has $\chi \to n_0 \mu_0 \mu_B^2 / 2eq$, and one has to require $1/2e \to N(\varepsilon_f)$ in the limit of high density ($q \to 1$) if one is to regain the Pauli susceptibility (compare Section 11.1). Thus one expects a gradual transition between these limiting forms when $a \approx eq$, i.e., $2eq \approx k_B T$.

11.12.2. Microscopic theory

To proceed further, Chapman and March (1988) draw on the work of Rice et al. (1985), who extended the original Brinkman-Rice model to elevated temperatures, in order to describe the normal-state properties of heavy-electron systems. For a half-filled band, these workers wrote the free energy per atom as (with σ labeling spin):

$$F = \sum_{\mathbf{k}\sigma} q_\sigma \varepsilon_{\mathbf{k}} n_{\mathbf{k}\sigma} + Ud + k_B T \sum_{\mathbf{k}\sigma} w_{\mathbf{k}} [n_{\mathbf{k}\sigma} \ln n_{\mathbf{k}\sigma} + (1 - n_{\mathbf{k}\sigma}) \ln(1 - n_{\mathbf{k}\sigma})], \tag{11.34}$$

where

$$n_{\mathbf{k}\sigma} = \left\{ 1 + \exp\left[\frac{q(\varepsilon_{\mathbf{k}} - \mu)}{w_{\mathbf{k}} k_B T} \right] \right\}^{-1}. \tag{11.35}$$

Here $w_{\mathbf{k}}$ is a renormalization factor in $\mathbf{k}$-space introduced to account for the nonorthogonality of the quasi-particle states $n_{\mathbf{k}\sigma}$ and d is the average number of doubly occupied sites. Although clearly the details of $w_{\mathbf{k}}$ are not fully known, the constraints

$$\bar{w} = \sum_{\mathbf{k}} w_{\mathbf{k}} = \frac{(\frac{1}{2} - d)\ln(\frac{1}{2} - d) + d\ln 2}{\ln 2}, \tag{11.36}$$

$$\bar{w}_{-1} = \sum_{\mathbf{k}} w_{\mathbf{k}}^{-1} = 2, \tag{11.37}$$

and

$$w \to 1 \quad \text{as} \quad k \to k_{\mathrm{f}} \tag{11.38}$$

hold for the situation considered here. With

$$q = \frac{\{[d(n_{\uparrow} - d)]^{1/2} + [d(n_{\downarrow} - d)]^{1/2}\}^2}{n_{\uparrow} n_{\downarrow}}, \tag{11.39}$$

the magnetic susceptibility may be obtained by expanding the free energy to $O(m^2)$:

$$\frac{n_0 \mu_0 \mu_B^2}{\chi} = 2q\left(1 - \frac{1}{4(1-2d)^2}\right)\sum_{\mathbf{k}\sigma} \varepsilon_{\mathbf{k}} n_{\mathbf{k}\sigma} + k_B T\left(\sum_{\mathbf{k}\sigma} w_{\mathbf{k}}^{-1} n_{\mathbf{k}\sigma}(1 - n_{\mathbf{k}\sigma})\right)^{-1}. \tag{11.40}$$

Now as T tends to zero, the second term becomes $q/N(\varepsilon_{\mathrm{f}})$, and writing $\bar{\varepsilon} = \sum_{\mathbf{k}\sigma} \varepsilon_{\mathbf{k}} n_{\mathbf{k}\sigma}$ in this limit, one then finds

$$\chi = \frac{n_0 \mu_0 \mu_B^2 N(\varepsilon_{\mathrm{f}})}{q}\left[1 + 2\bar{\varepsilon} N(\varepsilon_{\mathrm{f}})\left(1 - \frac{1}{4(1-2d)^2}\right)\right]^{-1}, \tag{11.41}$$

regaining the result of Brinkman and Rice (1970). For $k_B T \gg q\varepsilon_{\mathrm{f}}$, then $n_{k\sigma} \approx \frac{1}{2}$, so that

$$\chi = \frac{n_0 \mu_0 \mu_B^2}{k_{\mathrm{B}} T} \tag{11.42}$$

using the value for $\bar{w}_{-1}$ from (11.37). Hence there is a crossover between enhanced Pauli paramagnetism and Curie behaviour for $k_{\mathrm{B}} T \approx q\varepsilon_{\mathrm{f}}$, and the structure of the theory is the same as from the phenomenological description given before. One may also note that for the low-temperature, high-density region of Figure 11.8, (11.41) provides a mechanism for the observed enhancement of the susceptibility over the Pauli value and thus the upturn in the curve. Here, it may be assumed that the reduction of d from its uncorrelated value of $\frac{1}{4}$ is small. Writing $d = \frac{1}{4} - \delta$, where δ is small and positive, one finds from (11.39) that $q = 1$ to order δ; on expanding the bracket in (11.41), similarly:

$$\chi = \frac{n_0 \mu_0 \mu_B^2 N(\varepsilon_f)}{1 + 16\bar{\varepsilon} N(\varepsilon_f)\delta}. \tag{11.43}$$

Since $\bar{\varepsilon} < 0$, χ shows a Stonerlike enhancement (see (11.18)), the effect of which will tend to diminish as the density falls, whereas at the same time the $1/q$ factor becomes more important.

Referring to Figure 11.8, then at the maximum in the susceptibility curve, as Warren (1984) pointed out, the enhanced Pauli susceptibility appears to be restricted by the Curie limit. At this point $\rho \approx 0.8$ g cm^{-3} and $T \approx 1780$ K, giving $q \approx 0.18$ (i.e., $d \approx 0.02$). Thus the renormalization of the electron degeneracy temperature due to the effects of correlation is strongly marked. More generally, one will expect a crossover region in the behaviour of the susceptibility to the high-density side of the metal-insulator transition, as indicated schematically in Figure 11.13. The experimental data for Rb and Na (Freyland, 1980; Bottyan et al., 1983), which do not show a maximum such as in caesium (Freyland, 1979), do not extend to low enough densities for this to be seen, although it is to be expected from the preceding treatments that such maxima should occur before the metal-insulator transition is reached. It is to be noted, of course, that near to the liquid-vapour critical region, the susceptibility is likely to be subject to diamagnetic corrections, arising from the presence of aggregate species (see Section 12.6).

Figure 11.13. Schematic form of coexistence curve and metal-insulator transition (Chapman and March, 1988).

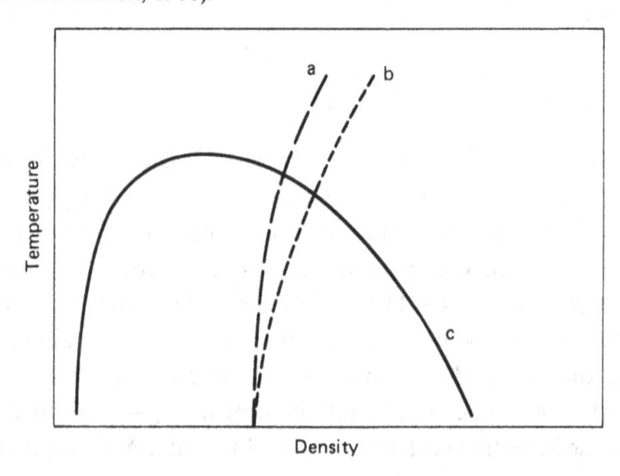

To summarize the preceding, it seems clear that the magnetic susceptibility data of expanded fluid Cs can be interpreted by means of a phenomenological model, which generalizes that given by March, Suzuki, and Parrinello (1979) for $T > 0$. This is supported, at a microscopic level, by adapting the work of Rice et al. (1985) on heavy-electron systems. If the peak in the susceptibility-density plot for Cs along the coexistence curve is interpreted as the limitation of enhanced Pauli behaviour and consequent crossover to a Curie-like regime, then q at this point is about 0.2. This figure constitutes then an upper bound for the value of q at the metal-insulator transition, and it is of interest (see Chapman and March, 1988) to note the marked difference from the prediction of the jellium model (see Chapter 14). There $q \approx 0.5$ at the corresponding density (Lantto, 1980) and the metal-insulator transition occurs at a much lower density (Ceperley and Alder, 1980). It is evident that the influence of electron-ion interaction in the highly expanded fluid metal is substantial.

12
Liquid-vapour surface

This chapter begins with a relatively brief discussion of the thermodynamics of liquid surfaces. Then the statistical mechanics of inhomogeneous systems, already developed for treating freezing in Chapter 6, are used to obtain some formally exact results for the surface tension of a liquid. These formulae will then, essentially, be developed by gradient expansion methods to yield an interesting relation between bulk and surface properties, related by the "width" of the liquid-vapour interface.

12.1. Thermodynamics of liquid surfaces

The atomic density profile, denoted by $\rho(z)$, must vary continuously across the interface from the value ρ_l of the bulk liquid to the value ρ_v of the bulk vapour. This variation can be expected to take place over a few atomic distances, at least when one is far from the critical point.

The anisotropy of the profile implies a net attraction to the liquid phase of an atom in the transition region: One must perform work to bring an atom from the bulk of the liquid to the surface; i.e., an excess of free energy is associated with the creation of the interface, namely, the surface free energy. It also implies that the tangential pressure, defined as the force per unit area transmitted perpendicularly across an area element in the yz or xz plane is a function $p_t(z)$ of position in the transition region. The difference between the components $p_t(z)$ and $p_n = p$ of the stress tensor in the transition region is negative, i.e., it has the nature of a tension, namely, surface tension.

The surface free energy, as discussed, for example, in the reviews of Brown and March (1976), or that of Tosi (1985), is defined by comparing the free energy F of the actual system with the sum of the free energies of suitably chosen amounts of homogeneous liquid and vapour phases. If f_l and f_v are the free energies per unit volume of the homogeneous phases, one can write the surface free energy σ per unit area as

$$\sigma = \frac{1}{A}(F - f_l V_l - f_v V_v) \tag{12.1}$$

where A is the surface area and V_l and V_v are the volumes of the homogeneous phases. These are fixed by

$$V_l + V_v = V, \qquad \rho_l V_l + \rho_v V_v = N, \tag{12.2}$$

which therefore attributes no excess matter to the interface. These conditions can be reexpressed as

$$\int_{-\infty}^{z_G} dz(\rho_l - \rho(z)) + \int_{z_G}^{\infty} dz[\rho_v - \rho(z)] = 0, \tag{12.3}$$

which fixes the location z_G of the Gibbs surface dividing the two hypothetical homogeneous fluids.

The thermodynamic definition (12.1) has some consequences:

1. $\sigma\, dA$ is the amount of work required to increase the surface area A by dA in any isothermal reversible process.
2. The excess surface entropy is given by $s = -d\sigma/dT$, and thus the excess surface energy is $u = \sigma - T\, d\sigma/dT$.

With regard to (1), it should be stressed that the work done against the surface tension in expanding the surface area by stretching is equal to the surface free energy of the same area of the new surface; i.e., the surface tension and the surface free energy, which are customarily expressed in dyn cm^{-1} and in erg cm^{-2}, respectively, are numerically the same for an interface between two fluids. This line of argument can be employed to express σ in terms of the integrated deficit of tangential pressure as

$$\sigma = \int dz[p - p_l(z)], \tag{12.4}$$

a derivation of this result being given, for instance, in the book by March and Tosi (1976). The argument assumes that alternative processes lead to the same equilibrium surface structure, as will be the case when diffusion is allowed.

Before turning to statistical mechanical theory of the atomic density profile through the liquid-vapour surface, it is worth summarizing here a few facts and some phenomenology. First, as treated by Faber (1972), the values of σ for liquid metals correlate with the latent heat of vaporization. This can be reconciled intuitively by the notion that the work done in bringing an atom from the bulk liquid up to the surface involves breaking a fraction of its bonds. The corresponding surface entropy, which can be obtained from the measured temperature dependence of σ, is about k_B per surface atom. Such an amount can be estimated by considering the effect of replacing an appropriate number of bulk sound waves by capillary waves, as discussed, for example, in the book by Faber (1972). However, some notable exceptions occur, such as Zn, where $d\sigma/dT$ is positive over a limited range of temperatures above the triple point. One anticipates that $\sigma(T)$ should decrease with increasing temperature and vanish at the critical point (see Section 9.2), where the distinction between liquid and vapour no longer obtains.

It is also of interest to record that a "law of corresponding states" is observed, as summarized by Buff and Lovett (1968), for the surface tension of simple atomic and molecular liquids. In particular, in terms of critical volume V_c and temperature T_c, the quantity $\sigma(T)V_c^{2/3}/T_c$ is practically a universal function of T/T_c. The data turn out to be fitted quite usefully by

$$\sigma(T) \propto \left(l - \frac{T}{T_c}\right)^{1.27 \pm 0.02}. \tag{12.5}$$

Following this general survey, the discussion will focus primarily on the way the liquid-vapour interface can be characterized, along, of course, with the surface tension, by the properties of the density profile (for earlier work, see Croxton and Ferrier, 1971).

12.2. Model using theory of inhomogeneous electron gas

The simplest theories of the surface tension of liquid metals focus all attention on the behaviour of the conduction electrons. At the most elementary level, one constructs, following Brown and March (1973; see also 1976), the total energy E of the inhomogeneous system as

$$E = \int d\mathbf{r} \left[\varepsilon(\rho) + \frac{\lambda \hbar^2}{8m} \frac{(\nabla \rho)^2}{\rho} \right] \tag{12.6}$$

where λ is a numerical factor. The slowly varying electron gas theory, in fact, gives $\lambda = \frac{1}{9}$, whereas von Weizsäcker's original work would correspond to the choice $\lambda = 1$ (see Lundqvist and March, 1983). In this equation (12.6) the local energy term $\varepsilon(\rho)$ is the kinetic, exchange and correlation energy of a homogeneous electron gas at density ρ, whereas the inhomogeneity term in $(\nabla\rho)^2$ is associated with kinetic energy. The variational principle for the energy E reads

$$\frac{\delta(E - \mu N)}{\delta\rho} = 0 \tag{12.7}$$

where the chemical potential μ of the electronic cloud is playing the role of a Lagrange multiplier, taking care of normalization. This yields, from (12.6), the Euler equation

$$-\frac{\lambda\hbar^2}{4m}\left[\frac{\rho''}{\rho} - \frac{1}{2}\left(\frac{\rho'}{\rho}\right)^2\right] + \frac{d\varepsilon}{d\rho} = \mu, \tag{12.8}$$

from which the conduction electron density profile $\rho(z)$ through the planar liquid metal surface in the xy plane can be determined. Setting $\rho = \psi^2$ this is formally equivalent to a Schrödinger equation for ψ, namely,

$$-\frac{\lambda\hbar^2}{2m}\frac{d^2\psi}{dz^2} + \left(\frac{d\varepsilon}{d\rho} - \mu\right)\psi = 0. \tag{12.9}$$

The surface tension, defined as the energy difference between the inhomogeneous electron gas and a homogeneous system per unit surface area (see the analogy with (12.38)) is given by

$$\sigma = \frac{\lambda\hbar^2}{m}\int_{-\infty}^{\infty} dz[\psi'(z)]^2. \tag{12.10}$$

This elementary approach has the merit that it can be completely solved analytically for certain simple forms of $\varepsilon(\rho)$ and, in particular, for the case

$$\varepsilon(\rho) = \rho\varepsilon_0\left[\left(\frac{\rho}{\rho_0}\right)^{2/3} - 2\left(\frac{\rho}{\rho_0}\right)^{1/3}\right], \tag{12.11}$$

which corresponds to the sum of a Thomas-Fermi-like kinetic term proportional to $\rho^{5/3}$ and a Dirac-Slater exchange term proportional to $\rho^{4/3}$ (see Lundqvist and March, 1983). Clearly ρ_0 is the equilibrium density at zero pressure and the corresponding electron gas compressibility is given by $K^{-1} = (2/9)\rho_0\varepsilon_0$. The solution of (12.9) is readily verified to be

$$\frac{\rho(z)}{\rho_0} = \left(1 + B\exp\left(\frac{z}{l}\right)\right)^{-3} \tag{12.12}$$

where B is a constant, whereas the length l, which clearly measures the surface thickness, is given by

$$l = \left(\frac{9\lambda\hbar^2}{8m\varepsilon_0}\right)^{1/2}. \tag{12.13}$$

From (12.10) one finally obtains

$$\sigma K = \tfrac{3}{4}l, \tag{12.14}$$

which connects a bulk property, the compressibility K, with a surface property σ via the surface thickness l. A formula of this kind will be brought into contact with experiment shortly, after a more basic discussion of the liquid-vapour surface, using statistical mechanics. One still invokes a low-order density gradient expansion, but now for the ionic profile, rather than for the electron density.

In the present context, let us refer to the fact that, in lowest order, the preceding treatment has assumed, by neglecting electrostatic energy, that the ionic and electron density profiles through the surface are the same. In fact, of course, for a given ionic density profile, the electrons will always spill out a little further than the ions in order to minimize the electronic kinetic energy.

If one takes a more refined approach, then one could calculate the surface electron density profile $\rho(z)$, avoiding a density gradient expansion of the electronic kinetic energy. This, for example, has been done in a semi-infinite jellium model (compare Chapter 14) by Lang and Kohn (1970, 1971; see also Lang, 1983 and Krotscheck, Qian, and Kohn, 1985) and, subsequently, by a computer experiment, by Ceperley (1989). These lead to accurate profiles for that model. In such refinements, as already noted above, a Hartree-like term for the electron-ion interaction will appear when the ionic and electronic density profiles are allowed to differ somewhat, as will indeed be the case in liquid metals. For specific work in this area, including some account of the transition from metallic to localized states in the transition region from liquid to vapour, the work of Rice et al. (1974) should be consulted. This is the point, however, at which attention will be focussed on the statistical mechanics of the liquid-vapour interface, following the discussion of the thermodynamics of surfaces in Section 12.1.

12.3. Formally exact pair potential theory

The burden of this section is to outline the results of the formally exact theory of Kirkwood and Buff (1949) within a framework of pairwise interatomic forces. This approach unavoidably rests on the knowledge of the pair function $g(\mathbf{r},\mathbf{r}')$ of atoms in the now inhomogeneous system. This is the generalization of the bulk liquid pair function, which is the much simpler limit where g defined before becomes $g(|\mathbf{r}-\mathbf{r}'|)$. In the case of a planar surface, which is the geometry dealt with below, one can conveniently view the full pair function as dependent on this vector difference $\mathbf{R}=\mathbf{r}-\mathbf{r}'$ and on the coordinate z of the first atom measured perpendicular to the planar surface. An interatomic potential of the form $\phi(R)$, depending, therefore, only on the relative separation R of a pair of atoms, is assumed to underlie the following treatment.

The tangential pressure $p_t(z)$ introduced in Section 12.1 can then be expressed as

$$p_t(z) = k_B T\rho(z) - \frac{1}{2}\int \mathrm{d}\mathbf{R}\frac{X^2}{R}\phi'(R)g(\mathbf{R},z) \tag{12.15}$$

where the components of $\mathbf{R}$ have been written as (X, Y, Z) and $\phi'(R) = \mathrm{d}\phi(R)/\mathrm{d}R$. When one is far from the interface, it is clear that the z dependence must disappear, and in this limit (12.15) reduces to the well-known formula for the bulk pressure, following almost immediately from the virial theorem:

$$p = \rho k_B T - \frac{1}{6}\int \mathrm{d}\mathbf{R}\, R\phi'(R)n_2(R); \qquad n_2 = \rho^2 g(r), \tag{12.16}$$

with $g(r)$ the usual pair function normalized so that it tends to unity at large r.

An expression similar to (12.15) holds for the normal pressure p_n. By imposing the condition for hydrostatic equilibrium $p_n(z) = p$, one derives an expression for the density profile $\rho(z)$:

$$k_B T\frac{\mathrm{d}\rho(z)}{\mathrm{d}z} = \int \mathrm{d}\mathbf{R}\frac{z}{R}\phi'(R)n_2(R,z). \tag{12.17}$$

Use of the result (12.4) discussed earlier in this chapter then yields the expression for the surface tension σ given by

$$\sigma = \frac{1}{2}\int_{-\infty}^{\infty}\mathrm{d}z\int \mathrm{d}\mathbf{R}\frac{X^2 - Z^2}{R}\phi'(R)n_2(R,z). \tag{12.18}$$

It remains a major source of difficulty to calculate the pair function n_2 in the presence of the surface; therefore, to date the analytical progress made in evaluating this formula, due to Kirkwood and Buff (1949), has related n_2 to the bulk pair function $g(r)$. The most drastic simplification is to take the liquid as homogeneous up to the Gibbs surface and to assume the vapour has negligible density. One thereby is led back from (12.18) to the much earlier result of Fowler (1937):

$$\sigma = \tfrac{1}{8}\pi\rho_1^2 \int_0^\infty dR\, \phi'(R) R^4 g(R). \tag{12.19}$$

In evaluating an integral such as in (12.19), it has emerged from a variety of directions that it is important to input a $g(r)$ wholly consistent with the assumed $\phi(R)$. This can be done by using computer-simulation data, an example being the study of Freeman and McDonald (1973). These workers obtain reasonable results from (12.19), and they exhibit an improvement when they allow a smooth density profile of finite thickness. For a detailed discussion of this approach, reference may be made to the work of Berry et al. (1972).

12.4. Triezenberg-Zwanzig formula for surface tension: direct correlation function

Let us consider, to be definite, a neutral monatomic fluid. In this case, a formally exact equation for the density profile $\rho(\mathbf{r})$ in an inhomogeneous fluid state can be written (see Appendix 13.3 for the derivation of the multicomponent generalization)

$$\nabla\rho(\mathbf{r}) = \rho(\mathbf{r}) \int d\mathbf{r}'\, c(\mathbf{r},\mathbf{r}')\nabla\rho(\mathbf{r}'), \tag{12.20}$$

where, however, paralleling the Kirkwood-Buff treatment, $c(\mathbf{r},\mathbf{r}')$ is the Ornstein-Zernike direct correlation function in the presence of the inhomogeneity.

For the example of a planar liquid-vapour interface (perpendicular to the z axis, say) the matrix $c(\mathbf{r},\mathbf{r}')$ depends only on the three quantities z, z', and $s = [(x-x')^2 + (y-y')^2 + (z-z')^2]^{1/2}$. Taking the two-dimensional Fourier transform

$$\hat{c}(k;z,z') = \int d^2s \exp(ik\cdot s) c(s;z,z'), \tag{12.21}$$

one then finds

$$\frac{d\rho(z)}{dz} = \rho(z) \int_{-\infty}^{\infty} dz' c_0(z,z') \frac{d\rho(z')}{dz'}, \tag{12.22}$$

where $c_0(z, z') \equiv \hat{c}(k = 0; z, z') = \int d^2s\, c(\mathbf{r}, \mathbf{r}')$. Equation (12.22) is the equilibrium condition in the fluctuation approach to surface tension of Triezenberg and Zwanzig (1972). Their derivation hinges on examining the effect of a long-wavelength fluctuation of the planar surface with wave-vector along the surface (see Appendix 13.3 for the generalization to the multicomponent case of what follows). To lowest order in the wave-number of the fluctuation, this represents a rigid translation of the system and (12.22) follows from the fact that neither a free-energy change nor a surface area change accompany such a translation.

These changes arise, instead, at the next order in the wave number, and their evaluation results in the desired expression for the surface tension. In particular the free-energy change ΔF for a small fluctuation $\Delta\rho(\mathbf{r})$ is given by

$$\Delta F = \tfrac{1}{2} k_B T \int\!\!\int d\mathbf{r}\, d\mathbf{r}'\, \Delta\rho(\mathbf{r}) K(\mathbf{r}, \mathbf{r}') \Delta\rho(\mathbf{r}'), \tag{12.23}$$

where

$$K(\mathbf{r}, \mathbf{r}') = \frac{\delta(\mathbf{r} - \mathbf{r}')}{\rho(\mathbf{r}')} - c(\mathbf{r}, \mathbf{r}') \tag{12.24}$$

for a classical fluid. A detailed calculation for the planar surface yields for the surface tension

$$\sigma = k_B T \int_{-\infty}^{\infty} \int dz\, dz' \frac{d\rho(z)}{dz} c_2(z, z') \frac{d\rho(z')}{dz'} \tag{12.25}$$

where

$$\begin{aligned} c_2(z, z') &= \frac{1}{4} \int d^2s\, s^2 c(\mathbf{r}, \mathbf{r}') \\ &= -\frac{1}{4} \left[\frac{d^2\hat{c}}{dk^2}(k; z, z') \right]_{k=0}. \end{aligned} \tag{12.26}$$

The reader who requires further details of the derivation of (12.25) may consult Appendix 13.3 on the generalization of the surface tension theory to multicomponent systems.

Equations (12.22) and (12.25) are formally exact and independent of the detailed nature of the interionic forces. Of course, $c(\mathbf{r}, \mathbf{r}')$ is not presently known in the inhomogeneous fluid, which means, in turn, that c_0 and c_2 are not currently available. The theory becomes practicable only when these quantities are related, inevitably approximately, to properties of the homogeneous fluid.

In concluding this section, it is of interest to record that the equivalence of the Kirkwood-Buff and Triezenberg-Zwanzig formulae for surface tension presented here has been demonstrated directly by Schofield (1979), for the case of pairwise additive forces already assumed by Kirkwood and Buff (1949).

12.5. Microscopic foundation of Cahn-Hilliard phenomenology

Though the discussion of Section 12.4 was exact, the difficulty of evaluation prompts a simpler approach. It will be helpful to consider the description of surface tension in terms of the atomic density profile by summarizing first the phenomenology associated with the names of Cahn and Hilliard (1958). These workers, as will be seen explicitly, essentially characterize the density profile by a single parameter, its width l, already referred to in Section 12.2. The precise meaning of this will become clearer when one turns from phenomenology to a microscopic theory based on the classical statistical mechanics of inhomogeneous fluids. The correlation established by Cahn and Hilliard—namely, that the product σK_T, with σ the surface tension and K_T the isothermal compressibility, is in essence the width l (compare (12.14))—was shown by Egelstaff and Widom (1970) to be amply confirmed experimentally for a whole class of liquids at or near the triple point. A compilation made subsequently by Alonso and March (1985) for many liquid metals near the melting point is reproduced in Table 12.1.

Because of this empirical correlation, it is of obvious interest to understand the problem more deeply from first-principles theory. However, since the motivation for the following argument, due to Bhatia and March (1978) lies in the Cahn-Hilliard approach, let us recall briefly that σ is written by these authors as the sum of two terms:

$$\begin{aligned}\sigma &= \sigma_1 + \sigma_2 \\ &= \left[\frac{l(\Delta\rho)^2}{2\rho^2 K_T}\right] + Bl\left[\frac{\Delta\rho}{l}\right]^2.\end{aligned} \tag{12.27}$$

In this expression, the first contribution arises from the treatment of the surface inhomogeneity as an "accidental fluctuation," whereas the second is evidently coming from the density gradient across the interface. In (12.27), in addition to the effective thickness of the interface, l, $\Delta\rho$ is the fluctuation in the number density ρ, whereas B is a constant.

Table 12.1. *Correlation between surface tension and bulk compressibility of metals near melting point [after Alonso and March (1985)].*

Metal	Surface tension σ (dyn cm^{-1})	Isothermal compressibility K_{T_m} (10^{-12} dyn^{-1} cm^2)	Product σK_{T_m} (Å)
Li	410	(11)	0.45
Na	200	18.6	0.37
K	110	38.2	0.42
Rb	85	49.3	0.42
Cs	70	68.8	0.48
Be	(1350)	(1.94)	0.26
Mg	570	5.06	0.29
Ca	350	11.0	0.38
Sr	295	13.1	0.39
Ba	255	17.8	0.45
Cu	1310	1.49	0.19
Ag	910	2.11	0.19
Zn	770	2.50	0.19
Cd	590	3.24	0.19
Hg	485	3.75	0.18
Al	865	2.42	0.21
Ga	715	2.19	0.16
In	560	2.96	0.17
Tl	465	3.83	0.18
Sn	570	2.71	0.15
Pb	460	3.49	0.16
Sb	390	4.90	0.19
Bi	380	4.21	0.16
Fe	1830	1.43	0.26

Minimizing σ with respect to l yields the alternative forms

$$\sigma_{\min} = 2Bl(\Delta\rho/l)^2 \tag{12.28}$$

or

$$\sigma_{\min} = \frac{l(\Delta\rho)^2}{\rho^2 K_{\mathrm{T}}}, \quad \text{or} \quad \sigma_{\min} K_{\mathrm{T}} \sim l. \tag{12.29}$$

The objective below is to expose the origin of this correlation (12.29) or (12.14) from first-principles theory. To do so, one seeks a parallel with the above phenomenology, in that

1. One wants to minimize the free energy of the inhomogeneous system with respect to the density profile $\rho(x)$; and
2. One again wishes to separate density gradient terms clearly from bulk or local density contributions.

Such an approach has been developed by Yang et al. (1976), who write the free energy of a nonuniform system in terms of a local free-energy density $\psi(\mathbf{r})$ as

$$F = \int d\mathbf{r}\, \psi(\mathbf{r}). \tag{12.30}$$

Gradient expansion of $\psi(\mathbf{r})$ then yields, for a flat interface of area $\tilde{a}$ in zero external potential:

$$F = \tilde{a} \int_a^b dx\, \psi(x) \tag{12.31}$$

where

$$\psi(x) = \psi[\rho(x)] + \tfrac{1}{2}A[\rho(x)]\rho'(x)^2, \tag{12.32}$$

with $A(\rho)$ given by

$$A(\rho) = \tfrac{1}{6}k_B T \int d\mathbf{r}\, r^2 c(\mathbf{r}, \rho). \tag{12.33}$$

Here a and b define the boundaries of the system of volume V. In (12.32), $\psi(\rho)$ is the free-energy density of a uniform system of density ρ, whereas $c(\mathbf{r}, \rho)$ in (12.33) is the Ornstein-Zernike direct correlation function of a uniform system of density ρ (see Section 2.3). Minimizing F, with the chemical potential μ introduced as the Lagrange multiplier taking care of the normalization of $\rho(x)$, Yang et al. (1976) obtain the Euler equation of the variational problem as

$$\mu = \mu[\rho(x)] - A[\rho(x)]\rho''(x) - \tfrac{1}{2}A'[\rho(x)]\rho'(x)^2. \tag{12.34}$$

In this equation, $\mu(\rho)$ is the chemical potential of a uniform system of density ρ, whereas

$$A'(\rho) = \frac{\partial A(\rho)}{\partial \rho}; \qquad \mu(\rho) = \frac{\partial \psi(\rho)}{\partial \rho}. \tag{12.35}$$

Equation (12.34) is equivalent to the constancy of the pressure p across the inhomogeneity, p being given by an integration of (12.34) as

$$p = \mu\rho(x) - \psi[\rho(x)] + \tfrac{1}{2}A[\rho(x)]\rho'(x)^2. \tag{12.36}$$

Integrating (12.36) over the total volume V, the free energy is obtained by Yang et al. (1976) as

$$F = \mu N - pV + \sigma\tilde{a}, \tag{12.37}$$

where

$$\sigma = \int_a^b \mathrm{d}x\, A[\rho(x)]\rho'(x)^2, \tag{12.38}$$

a result that goes back, in essence, to van der Waals.

It should be emphasized at this point that the form (12.38) is a more sophisticated version of the result (12.28), whereas the result analogous to (12.29) is what one seeks here. By analogy with the decomposition (12.27), one now divides the constant interfacial pressure p into two parts:

$$p = \mu\rho(x) - \psi[\rho(x)] + p_{\text{density gradient}}(x). \tag{12.39}$$

Using (12.36), (12.38), and (12.39) it follows first that

$$\sigma = 2\int_a^b \mathrm{d}x\, p_{\text{density gradient}}(x) \tag{12.40}$$

and hence one obtains the following alternative form for the surface tension:

$$\sigma = 2\int_a^b \mathrm{d}x\{p - \mu\rho(x) + \psi[\rho(x)]\}. \tag{12.41}$$

Equation (12.41) is the generalization of the form (12.29) one is seeking, as will now be demonstrated. First, since $p_{\text{density gradient}}(x)$ is nonzero only over the effective thickness of the interface, it follows from (12.40) that l is the extent of the range of integration in (12.41). Secondly, the pressure p is given by

$$p = \mu\rho_l - \psi(\rho_l) = \mu\rho_v - \psi(\rho_v), \tag{12.42}$$

ρ_l being the bulk liquid density and ρ_v the vapour density. One now makes a Taylor expansion of $\psi[\rho(x)]$ around the bulk liquid density, and correct to $O([\rho(x) - \rho_l]^2)$, one finds

$$\sigma \sim \frac{l}{K_T}\int_{\text{interface}} \frac{\mathrm{d}x[\rho(x) - \rho_l]^2}{\rho_l^2} + \text{higher-order terms}. \tag{12.43}$$

Here K_T^{-1} is $\rho_l^2(\partial^2\psi/\partial\rho^2)_{\rho_l}$. Thus, to lowest order, it has been established that (12.41) is the counterpart of the Cahn-Hilliard relation (12.29). One notes that (12.41) can be evaluated from a knowledge of the density profile $\rho(x)$ plus thermodynamic information as a function of density over the range from ρ_v to ρ_l. In contrast, (12.38) requires knowledge of $A(\rho)$ or the direct correlation function $c(r)$ over the same density range.

It is to be noted here that Stott and Young (1981) have made use of a similar approach in subsequent work on liquids with van der Waals interactions.

12.6. Nonequilibrium problems: condensation and evaporation

Kreuzer, Chapman, and March (1988) have considered the processes of condensation and evaporation (or sublimation) in a two-phase system, using the methods of nonequilibrium thermodynamics (Kreuzer, 1981). These workers set up the coupled balance equations for mass and energy flow across the interface between two phases of a one-component system. Their approach will be used below, and in particular, Onsager's reciprocity relations are imposed to relate the phenomenological coefficients that enter the theory to experimentally accessible quantities such as condensation coefficients and heats of evaporation and sublimation. The same is then done for a metal in contact with a vapour of atoms and dimers.

12.6.1. Balance equations

ONE-COMPONENT, TWO-PHASE SYSTEM

Consider a one-component system with N molecules enclosed in a volume V. In particular, one focuses on the regime of temperature and pressure where two phases, liquid and vapour (or solid-vapour), can coexist. Let us call N_1 and N_2 the number of particles in the condensed and dilute phases, respectively. One is, at first sight, tempted to isolate the system as well, i.e., to insist that the energy $U_1 + U_2$ in the two phases is conserved. Such a restriction would, however, prevent a proper treatment of isothermal processes in what follows. Therefore it is necessary to account for the coupling to a heat reservoir, the energy being U_R and the (constant) temperature being T.

Notationally X_i will be written for extensive variables, such that $X_1 = N_1$, $X_2 = N_2$, $X_3 = U_1$, $X_4 = U_2$, and $X_5 = U_\mathrm{R}$. Close to equilibrium, the thermodynamic fluxes across the interface may be written

$$\frac{\mathrm{d}X_i}{\mathrm{d}t} = \sum_{j=1}^{5} L_{ij} \left.\frac{\partial S}{\partial X_j}\right|_{U,V,N}, \tag{12.44}$$

where one can identify the thermodynamic forces as

$$\left.\frac{\partial S}{\partial N_j}\right|_{U,V,N} = -\frac{\mu_j}{T_j}, \tag{12.45}$$

$$\left.\frac{\partial S}{\partial U_j}\right|_{U,V,N} = \frac{1}{T_j}, \tag{12.46}$$

and

$$\left.\frac{\partial S}{\partial U_\mathrm{R}}\right|_{U,V,N} = \frac{1}{T}. \tag{12.47}$$

Here μ_j and T_j denote, respectively, the chemical potentials and temperatures in the two (homogeneous) phases, whereas T is the temperature of the reservoir. The phenomenological coefficients L_{ij} in (12.44) satisfy Onsager's reciprocity relations:

$$L_{ij} = L_{ji}. \tag{12.48}$$

At this stage one invokes the conservation laws:

$$\frac{\mathrm{d}N}{\mathrm{d}t} = \frac{\mathrm{d}}{\mathrm{d}t}(N_1 + N_2) = 0 \tag{12.49}$$

and

$$\frac{\mathrm{d}U}{\mathrm{d}t} = \frac{\mathrm{d}}{\mathrm{d}t}(U_1 + U_2 + U_\mathrm{R}) = 0. \tag{12.50}$$

Then one can eliminate all but 6 of the 25 coefficients in (12.44) to obtain the explicit equations:

$$\frac{\mathrm{d}N_1}{\mathrm{d}t} = -\frac{\mathrm{d}N_2}{\mathrm{d}t} = -L_{11}\left\{\frac{\mu_1}{T_1} - \frac{\mu_2}{T_2}\right\} + L_{13}\left\{\frac{1}{T_1} - \frac{1}{T}\right\} + L_{14}\left\{\frac{1}{T_2} - \frac{1}{T}\right\}, \tag{12.51}$$

$$\frac{\mathrm{d}U_1}{\mathrm{d}t} = -L_{13}\left\{\frac{\mu_1}{T_1} - \frac{\mu_2}{T_2}\right\} + L_{33}\left\{\frac{1}{T_1} - \frac{1}{T}\right\} + L_{34}\left\{\frac{1}{T_2} - \frac{1}{T}\right\}, \tag{12.52}$$

and

$$\frac{dU_2}{dt} = -L_{14}\left\{\frac{\mu_1}{T_1} - \frac{\mu_2}{T_2}\right\} + L_{34}\left\{\frac{1}{T_1} - \frac{1}{T}\right\} + L_{44}\left(\frac{1}{T_2} - \frac{1}{T}\right). \tag{12.53}$$

This is as far as the general methods of nonequilibrium thermodynamics can take us. It may be noted at this point that, as required, equilibrium conditions $T_1 = T_2 = T$ and $\mu_1 = \mu_2$ imply vanishing fluxes.

12.6.2. Balance equations for a metal

To describe evaporation (sublimation) and condensation in a metallic system, one must take explicit account of the facts that (1) the vapour phase typically consists of atoms and dimers and (2) the condensed phase contains ions and electrons (see also Chapter 14).

Below, ions and electrons in the vapour phase and atoms in the metal will be neglected. Let us denote by N_a and N_m the numbers of atoms and molecules, respectively, in the vapour phase and by N_+ and N_- the numbers of ions and electrons in the condensed phase. U_a, U_m, U_+, and U_- are the respective internal energies, where U_R again denotes the internal energy of a heat reservoir. The conservation laws (12.49) and (12.50) now read

$$\frac{d}{dt}(N_a + 2N_m + N_+) = 0 \tag{12.54}$$

and

$$\frac{d}{dt}(U_a + U_m + U_+ + U_- + U_R) = 0. \tag{12.55}$$

For the case of monovalent metals, to be considered here, $N_- = N_+$. Under isothermal conditions, the particle numbers obey the following rate equations:

$$\frac{dN_a}{dt} = \frac{L_{am}}{T}(\mu_m - 2\mu_a) + \frac{L_{a+}}{T}(\mu_+ + \mu_- - \mu_a) \tag{12.56}$$

$$\frac{dN_m}{dt} = \frac{L_{am}}{T}\left(\mu_a - \frac{1}{2}\mu_m\right) + \frac{L_{m+}}{T}\left(\mu_+ + \mu_- - \frac{1}{2}\mu_m\right) \tag{12.57}$$

and

$$\frac{dN_+}{dt} = \frac{dN_-}{dt} = \frac{L_{a+}}{T}(\mu_a - \mu_+ - \mu_-) + \frac{L_{m+}}{T}(\mu_m - 2\mu_+ - 2\mu_-). \quad (12.58)$$

To proceed beyond this stage one must evidently, in (12.51)–(12.53) and (12.56)–(12.58), (1) specify the chemical potentials and (2) relate the Onsager coefficients L_{ij} to measurable quantities; thus, this will be considered next.

12.6.3. Onsager coefficients

To return to develop further (12.51)–(12.53), one now specifies isothermal conditions $T_1 = T_2 = T$. Then the processes of condensation and evaporation are examined in turn.

Condensation

Let us assume that at temperature $T_1 = T_2 = T$ there are slightly more particles in the dilute gas phase than appropriate for the pertaining equilibrium vapour pressure P_0. One can then write

$$\mu_2 = \mu_2^0 + \left.\frac{\partial \mu_2}{\partial N_2}\right|_{T,V_2} \Delta N_2 \quad (12.59)$$

where $\mu_2^0 = \mu_1^0$ are the equilibrium values of the chemical potentials and V_2 is the volume available to the gas phase. For given values of V, N, T, and P_0, it can be obtained from the isotherm, e.g., via the Maxwell construction (see Chapter 9) in the case of the van der Waals isotherm for fluids.

Treating the gas phase as ideal, one can write

$$\mu_2 = k_B T \ln \frac{P\lambda^3}{k_B T} \quad (12.60)$$

with $\lambda = \mathrm{h}/(2\pi M k_B T)^{1/2}$ denoting the thermal de Broglie wavelength of an atom. Inserting $PV_2 = N_2 k_B T$ into (12.60), one calculates the derivative required in (12.59) as

$$\left.\frac{\partial \mu_2}{\partial N_2}\right|_{T,V_2} = \frac{k_B T}{N_2}. \quad (12.61)$$

Using (12.59) and (12.61) in (12.51), one finds

$$\frac{dN_1}{dt} = L_{11}\frac{k_B}{N_2}\Delta N_2, \quad (12.62)$$

implying that the excess particles ΔN_2 in the dilute gas phase will condense into the high-density phase. For this process of condensation, one usually writes empirically

$$\frac{dN_1}{dt} = \frac{S_c A \Delta P}{(2\pi M k_B T)^{1/2}}. \tag{12.63}$$

Here, it is implied that the excess flux of particles, the last term on the right-hand side involving the excess pressure $\Delta P = P - P_0$, hitting the interface area A between the two phases gets stuck (condensed) with a condensation probability satisfying $0 < S_c \leqslant 1$. To complete the connection between (12.62) and (12.63), one invokes the ideal gas law in the form

$$\Delta N_2 = \frac{V_2}{k_B T} \Delta P \tag{12.64}$$

to find

$$\frac{L_{11}}{T} = S_c \frac{N_2}{V_2} \left(\frac{1}{2\pi M k_B T} \right)^{1/2} A. \tag{12.65}$$

Equation (12.65) achieves the objective of relating the Onsager coefficient L_{11} to measurable quantities. It is to be stressed that of these quantities, the condensation coefficient S_c contains all the information about the dynamics of the sticking process. This quantity S_c must either be taken from experiment or calculated from a statistical theory.

EVAPORATION (AND SUBLIMATION)

In contrast to condensation, the reverse process of evaporation from a liquid (or sublimation from a solid) is always an activated process.

As a very simple model to illustrate the way one proceeds, let us treat the condensed matter (either liquid or solid) by the simplest thermodynamic model—namely, that of Einstein. Accordingly, one has for the free energy F the form

$$F = 3Nk_B T \ln\left(1 - \exp\left\{\frac{-h\nu}{k_B T}\right\}\right) - Nw \tag{12.66}$$

where $\nu = \nu_L(\nu_S)$ is the Einstein frequency characterizing the harmonic excitations in the liquid (solid). In (12.66), w is the heat of evaporation (sublimation). From (12.66) one finds for the chemical potential

$$\mu_1 = -w + 3k_B T \ln\left[1 - \exp\left(\frac{-h\nu}{k_B T}\right)\right]. \tag{12.67}$$

Thus the equilibrium vapour pressure is found to be

$$P_1 = \frac{k_B T}{\lambda^3} \exp\left(\frac{\mu_1}{k_B T}\right) \tag{12.68}$$

with μ_1 given by (12.67). One is then led to the rate equation (Kreuzer, Chapman, and March, 1988)

$$\frac{\mathrm{d}N_1}{\mathrm{d}t} = S_c A \left\{ \frac{k_B T}{(2\pi M k_B T)^{1/2}} \frac{N - N_1}{V - N_1 v_1} \right.$$
$$\left. - \frac{2\pi M}{h^3} (k_B T)^2 \left[1 - \exp\left(\frac{-h\nu}{k_B T}\right)\right]^3 \exp\left(\frac{-w}{k_B T}\right) \theta(N_1) \right\}. \tag{12.69}$$

Here one has again invoked the ideal gas law

$$P = \frac{N_2 k_B T}{V_2} = k_B T \frac{N - N_1}{V - N_1 v_1} \tag{12.70}$$

where v_1 is the constant volume per particle in the condensed phase. In (12.69) the Heaviside function $\theta(N_1)$ has been included in the evaporation (sublimation) term to account for the (trivial) fact that evaporation ceases once the condensate has disappeared. The evaporation term, as anticipated, has the form appropriate to an activated process, the activation energy in the Boltzmann factor being given by the latent heat w of evaporation (sublimation). The prefactor in the evaporation rate is still-T-dependent, albeit weakly so.

Kreuzer, Chapman, and March also consider a Debye rather than an Einstein model; the resulting rate law is of the same form as (12.69) but with a somewhat different prefactor in the evaporation term.

One notes that (12.69), or its analogue in the Debye model, can be written in the form

$$\frac{\mathrm{d}N_1}{\mathrm{d}t} = S_c A \frac{N_2}{V_2} \frac{k_B T}{(2\pi M k_B T)^{1/2}}$$
$$\times \left\{ 1 - a(T) \exp\left[\frac{(s_2 - s_1)}{k_B}\right] \exp\left(\frac{-w}{k_B T}\right) \theta(N_1) \right\}, \tag{12.71}$$

with s_1 and s_2 denoting the entropies per particle in the condensed and gas phase, respectively. The factor $a(T)$ is weakly temperature dependent and of order unity. This form (12.71) is, in fact, consistent with transition state theory (see, for example, Pechukas, 1981).

ENERGY BALANCE

To complete the discussion of isothermal condensation-evaporation processes, let us return to (12.51) and write the energy balances

$$\frac{dU_1}{dt} = -\frac{L_{13}}{T}(\mu_1 - \mu_2) = \frac{L_{13}}{L_{11}}\frac{dN_1}{dt} \tag{12.72}$$

and

$$\frac{dU_2}{dt} = -\frac{L_{14}}{T}(\mu_1 - \mu_2) = \frac{L_{14}}{L_{11}}\frac{dN_2}{dt}. \tag{12.73}$$

Treating the dilute phase again as an ideal gas, i.e., setting $U_1 = \frac{3}{2}N_1 k_B T$, allows the identification of the energy convection coefficient

$$L_{13} = \tfrac{3}{2}k_B T L_{11}. \tag{12.74}$$

If one adopts $U_2 \simeq -N_2 w$ for the condensed phase, then likewise one finds

$$L_{14} = -wL_{11}. \tag{12.75}$$

Thus transferring a particle isothermally from the condensed to the dilute phase requires a transfer of energy $w + \frac{3}{2}k_B T$ from the reservoir.

It may be seen at this stage why one had to include the coupling to a heat reservoir in setting up (12.51)–(12.53). Had one not done so but simply isolated the one-component system, energy conservation $U_1 + U_2 =$ constant would imply $L_{14} = L_{13}$. Evidently, this would lead to a contradiction between (12.74) and (12.75).

Turning briefly to nonisothermal situations, one first examines a case where $T_1 = T \neq T_2$. Equations (12.51)–(12.53) then yield

$$\frac{dU_1}{dt} = -L_{13}\left\{\frac{\mu_1}{T} - \frac{\mu_2}{T_2}\right\} + L_{34}\left\{\frac{1}{T_2} - \frac{1}{T}\right\} \tag{12.76}$$

and

$$\frac{dU_2}{dt} = -L_{14}\left\{\frac{\mu_1}{T} - \frac{\mu_2}{T_2}\right\} + L_{44}\left\{\frac{1}{T_2} - \frac{1}{T}\right\}. \tag{12.77}$$

Similarly, for $T_2 = T \neq T_1$ one finds

$$\frac{dU_1}{dt} = -L_{13}\left\{\frac{\mu_1}{T_1} - \frac{\mu_2}{T}\right\} + L_{33}\left\{\frac{1}{T_1} - \frac{1}{T}\right\} \tag{12.78}$$

and

$$\frac{dU_2}{dt} = -L_{14}\left\{\frac{\mu_1}{T_1} - \frac{\mu_2}{T}\right\} + L_{34}\left\{\frac{1}{T_1} - \frac{1}{T}\right\}. \tag{12.79}$$

This set of equations suggests an interpretation of L_{34} as the coefficient of heat conduction across the interface between the two phases. Likewise $L_{33} + L_{34}$ and $L_{44} + L_{34}$ are measures of the energy transfer to the reservoir.

12.6.4. Equilibrium properties

Let us next summarize the equilibrium properties implied by (12.69). For vanishing flux, one finds

$$N_1 = \frac{N - (V/\lambda^3)(1 - \exp(-h\nu/k_B T)^3 \exp(-w/k_B T)}{1 - (V_1/\lambda^3)(1 - \exp(-h\nu/k_B T)^3 \exp(-w/k_B T)} \tag{12.80}$$

for the number of particles in the condensed phase; they occupy a volume $V_1 = N_1 v_1$. Likewise, for the particle density in the gas phase,

$$\frac{N_2}{V_2} = \frac{1}{\lambda^3}\left(1 - \exp\left(\frac{-h\nu}{k_B T}\right)\right)^3 \exp\left(\frac{-w}{k_B T}\right). \tag{12.81}$$

These two equations permit the determination of the mass and volume fractions of the two phases. It should be noted that (12.70) gives the equilibrium pressure in the (ideal) gas phase.

12.6.5. Onsager coefficients in a metallic system

To discuss the balance equations (12.56)–(12.58) further, let us assume that the atoms and dimers in the vapour phase form ideal gases so that their respective chemical potentials read

$$\mu_a = k_B T \ln\left(\frac{N_a}{V_g}\lambda^3\right) \tag{12.82}$$

and

$$\mu_m = -\varepsilon_d + k_B T \ln\left\{\frac{1}{2^{1/2}}\frac{N_m}{V_g}\frac{\lambda^3}{Q_{int}}\right\} \tag{12.83}$$

where λ is the thermal de Broglie wavelength for atoms as given following (12.60). Here $\varepsilon_d > 0$ is the binding energy of the dimer and Q_{int} denotes its internal partition function accounting for vibrational and rotational degrees of freedom.

Dissociation and Recombination

To study dissociation in the gas phase, let us assume, following Kreuzer, Chapman, and March (1988), that there are slightly too many dimers, so that

$$\mu_m = \mu_m^0 + \frac{\partial \mu_m}{\partial N_m} \Delta N_m \tag{12.84}$$

with $\Delta N_m = N_m - N_m^0$. Inserting (12.84) into (12.57), one then expresses N_m^0 via (12.83) in terms of μ_m^0 and notes that $\mu_m^0 = 2\mu_a^0$, then introducing N_a via (12.82). To make contact with phenomenological rate theory, it may be noted that dissociation is customarily described by a rate equation

$$\frac{dN_m}{dt} = -k_{dis}(N_m - N_m^0), \tag{12.85}$$

thus introducing the rate constant k_{dis}. One thus finds for the dissociation-recombination terms in (12.57):

$$\frac{l_{am}}{T}\left(\mu_a - \frac{1}{2}\mu_m\right) = k_{dis}\left\{2^{1/2} Q_{int} \frac{\lambda^3}{V_g} \exp\left(\frac{\varepsilon_d}{k_B T}\right) N_a^2 - N_m\right\}. \tag{12.86}$$

Condensation and Evaporation

Let us first examine condensation from the atomic gas. With

$$\mu_a = \mu_a^0 + \frac{\partial \mu_a}{\partial N_a} \Delta N_a, \tag{12.87}$$

one finds from (12.58)

$$\begin{aligned} \left.\frac{dN_+}{dt}\right|_{cond} &= \frac{L_{a+}}{T} \frac{\partial \mu_a}{\partial N_a} \Delta N_a \\ &= S_a A \frac{\Delta P_a}{(2\pi M k_B T)^{1/2}} \end{aligned} \tag{12.88}$$

where

$$\Delta P_a = P_a - P_a^0 = P_a - \frac{k_B T}{\lambda^3} \exp\left(\frac{\mu_+ + \mu_-}{k_B T}\right). \tag{12.89}$$

In the last line of (12.88), the condensation coefficient S_a for atoms has been introduced. Arguments paralleling those just given eventually result in (12.58) taking the form

$$\frac{dN_+}{dt} = \frac{dN_-}{dt} = \frac{S_a A}{(2\pi M k_B T)^{1/2}}\left[P_a - \frac{k_B T}{\lambda^3}\exp\left(\frac{\mu_+ + \mu_-}{k_B T}\right)\right]$$
$$+ \frac{S_m A}{(2\pi M k_B T)^{1/2}}\left[2^{-1/2}P_m - \frac{k_B T}{\lambda^3}Q_{int}\exp\left(\frac{\varepsilon_d}{k_B T}\right)\exp 2\left(\frac{\mu_+ + \mu_-}{k_B T}\right)\right]. \tag{12.90}$$

It should be noted that as long as the (condensed) metal phase is present, the evaporation rates into atoms and molecules, r_a and r_d, remain constant. In particular, their ratio is given by

$$\frac{r_m}{r_a} = \frac{S_m}{S_a} Q_{int}\exp\left(\frac{\varepsilon_d}{k_B T}\right)\exp\left(\frac{\mu_+ + \mu_-}{k_B T}\right). \tag{12.91}$$

It may also be noted that in equilibrium $\mu_+^0 + \mu_-^0 = \mu_a^0 = \frac{1}{2}\mu_m^0$, so that (12.91) can also be written as

$$\frac{r_m}{r_a} = \frac{2^{-1/2}P_m S_m}{P_a S_a}. \tag{12.92}$$

This relation is a manifestation of detailed balance relating condensation and evaporation rates in equilibrium. The choice of the chemical potentials μ_+ and μ_- must, of course, reflect the details of the metal being studied. No matter how sophisticated the model, it is important to recognize that the chemical potentials μ_+ and μ_- reflect equilibrium properties of the condensed metal phase.

Kreuzer, Chapman, and March propose an experiment by means of which (12.92) might provide a means of extracting S_m/S_a, the ratio of the sticking coefficients for liquid metals, but it remains to be seen whether this can be accomplished in practice.

13
Binary liquid-metal alloys

In the previous chapters, the theory of pure liquid metals has been developed in some depth. The philosophy has been, in essence, to regard a pure liquid metal as a collection of suitably screened ions, interacting via effective ion-ion pair potentials. Recognizing that these pair potentials have features characteristic of liquid metals because the effective ion-ion interactions are mediated by the (almost totally degenerate) assembly of conduction electrons, nevertheless the treatment then is, essentially, that of a one-component liquid. In a more basic, fully first-principles treatment of a pure liquid metal, it is really to be viewed as a two-component system: ions, say in liquid Na, Na^+, and electrons, e^-.

In the present chapter, before turning to this approach to a pure liquid metal, some aspects of the theory of the previous chapters will be generalized to apply to binary liquid-metal alloys, such as Na-K or Na-Cs. Really, such alloys are three-component systems (compare Appendix 14.6), but in this chapter, following the philosophy of the earlier chapters, they will be treated as two-component systems. Thus, in liquid Na-K, the short-range order in the binary liquid metal alloy will be described by three partial pair correlation functions, namely, $g_{\text{Na-Na}}(r)$, $g_{\text{K-K}}(r)$ and $g_{\text{Na-K}}(r)$. The corresponding partial structure factors are $S_{\text{Na-Na}}(k)$, and so on. Various linear combinations of these structure factors are often very helpful, as emphasized especially in the work of Bhatia and Thornton (1970).

Of course, such a structural description has to be given for each concentration of the alloy. Furthermore, the effective ion-ion potentials $\phi_{\text{Na-Na}}(r)$, and so on, will be concentration-dependent, both directly so and, so to speak, indirectly because screening involves the Fermi energy; this energy is also expected to vary with concentration even in monovalent binary alloys mentioned above because of size differences between the components. Because of these very considerable complications, chemical solution theories of mixtures play an important role in the presentation of this subject.

13.1. Simple binary fluid mixtures

Following the presentation of Young (1987) closely, the same method of exposition will be adopted as for pure metals. Thus, as a starting point, let us begin by ignoring the specifically metallic features (an Ar-Kr mixture is then an obvious example). The metallic features will be incorporated at a later stage.

13.1.1. Hamiltonian

As for an Ar-Kr mixture, one is now dealing with a simple two-component system of N_1 atoms of type 1 and N_2 atoms of type 2 in a total volume V. The total particle number is $N = N_1 + N_2$, and the total particle density $\rho = N/V$. The partial number densities may be written $\rho_i = N_i/V = c_i\rho$, where $c_1 \equiv c$ and $c_2 \equiv 1 - c$ are the concentrations.

For the Hamiltonian, one writes

$$H = T_1 + T_2 + \tfrac{1}{2}\sum v_{11}(|\mathbf{r}_i^1 - \mathbf{r}_j^1|) + \sum v_{12}(|\mathbf{r}_i^1 - \mathbf{r}_j^2|) + \tfrac{1}{2}\sum v_{22}(|\mathbf{r}_i^2 - \mathbf{r}_j^2|) \tag{13.1}$$

where the instantaneous positions of the ions in the two subsystems are $\{\mathbf{r}_i^1\}$ and $\{\mathbf{r}_i^2\}$, the T_i are the kinetic energies, and the v_{ij} are the interaction energies. There is no volume term and the v_{ij} are independent of ρ. Evidently (13.1) is a generalization of (3.1); the following section generalizes Section 3.1.1.

Description of Atom-Atom Partial Structure Factors

Let $\rho_j g_{ij}(r)\,\mathbf{dr}$ be the probability of finding an atom of type j in volume element $\mathbf{dr}$ at position $\mathbf{r}$, given there is one of type i at the origin. Then (3.2) and (3.3) become

$$E = \tfrac{3}{2}k_B T + \tfrac{1}{2}\rho \sum c_i c_j \int v_{ij}(r) g_{ij}(r)\,\mathbf{dr} \tag{13.2}$$

and

$$P = \rho k_B T - \tfrac{1}{6}\rho \sum c_i c_j \int r \frac{\partial v_{ij}}{\partial r}(r) g_{ij}(r)\,\mathbf{dr} \tag{13.3}$$

and the equations can also be written in inverse space (see (3.5 and (3.6)).

The partial structure factors are defined by (see, for example, Ashcroft and Langreth, 1967)

$$S_{ij}(q) = \delta_{ij} + (\rho_i \rho_j)^{1/2} \int \{g_{ij}(r) - 1\} e^{i\mathbf{q}\cdot\mathbf{r}} \, d\mathbf{r} \tag{13.4}$$

which on Fourier inversion gives

$$g_{ij}(r) = 1 + \frac{1}{(2\pi^3)(\rho_i \rho_j)^{1/2}} \int \{S_{ij}(q) - \delta_{ij}\} e^{-i\mathbf{q}\cdot\mathbf{r}} \, d\mathbf{q}. \tag{13.5}$$

Equation (13.4) is evidently a generalization of (2.4). The $S_{ij}(q)$ are measurable in principle, but the practical difficulties are still considerable, and it is not feasible at the time of writing to use measured data in (13.2) and (13.3).

The informative low-q region is particularly inaccessible but, once again, thermodynamic measurements can help to determine the $S_{ij}(0)$. This is one consequence of the work of Bhatia and Thornton (1970) that will be detailed later; for the moment, let us record the analogue of (2.5), a result that has been known since the work of Kirkwood and Buff (1951) to be

$$\rho k_B T K_T = \frac{S_{11}(0) S_{22}(0) - S_{12}^2(0)}{c_2 S_{11}(0) + c_1 S_{22}(0) - 2(c_1 c_2)^{1/2} S_{12}(0)}. \tag{13.6}$$

As in Section 3.1.1 one now has virial (13.3) and compressibility routes to the equation of state; again it is to be noted that in an exact theory of the distribution functions, there would be compatibility, whereas, in practice, one must usually rely on one method or the other (see also Appendix 5.2).

The $S_{ij}(q)$ can also be interpreted as response functions. If the fluid is exposed to a small external potential $\delta\varphi_1(q)\cos\mathbf{q}\cdot\mathbf{r}$ acting on atoms of type 1 and another, $\delta\varphi_2(q)\cos\mathbf{q}\cdot\mathbf{r}$, acting on atoms of type 2 (i.e., a perturbation $\Sigma\delta\varphi_1(q)\cos(\mathbf{q}\cdot\mathbf{r}_i^1) + \Sigma\delta\varphi_2(q)\cos(\mathbf{q}\cdot\mathbf{r}_i^2)$ is added to (13.1)); then, to first order, the partial densities become $\rho_i + \delta\rho_i(q)\cos\mathbf{q}\cdot\mathbf{r}$, where (see (3.10))

$$\delta\rho_i(q) = -(k_B T)^{-1} \sum_j (\rho_i \rho_j)^{1/2} S_{ij}(q) \delta\varphi_j(q). \tag{13.7}$$

The often-used (Faber-Ziman) partial structure factors $a_{ij}(q)$, related to the $S_{ij}(q)$ by the simple linear formulae $S_{ij} = \delta_{ij} + (c_i c_j)^{1/2}(a_{ij} - 1)$, are not directly proportional to the $\delta\rho_i(q)$ and so are not response functions in the above sense.

Finally, direct correlation functions $c_{ij}(r)$ are again of considerable importance, the defining Ornstein-Zernike equations being (see Pearson and Rushbrooke, 1957)

$$h_{ij}(r) = c_{ij}(r) + \sum_k \rho_k \int h_{ik}(|\mathbf{r} - \mathbf{r}'|)c_{kj}(r')\,d\mathbf{r}' \tag{13.8}$$

where $h_{ij} = g_{ij} - 1$. Fourier transformation yields (Pearson and Rushbrooke, 1957)

$$\tilde{S}_{ij}(q) = (\rho_i\rho_j)^{1/2}\tilde{c}_{ij}(q) + \sum (\rho_k\rho_j)^{1/2}\tilde{S}_{ik}(q)\tilde{c}_{kj}(q) \tag{13.9}$$

where $\tilde{S}_{ij} = S_{ij} - \delta_{ij}$. This may be solved algebraically to obtain (see (2.8)) the S_{ij} in terms of the $\tilde{c}_{ij}$ or vice versa. Solving for the S_{ij} gives

$$\begin{aligned} S_{11}(q) &= \frac{1 - \rho_2\tilde{c}_{22}(q)}{D(q)} \\ S_{12}(q) &= \frac{(\rho_1\rho_2)^{1/2}\tilde{c}_{12}(q)}{D(q)} \\ S_{22}(q) &= \frac{1 - \rho_1\tilde{c}_{11}(q)}{D(q)} \end{aligned} \tag{13.10}$$

where

$$D(q) = [1 - \rho_1\tilde{c}_{11}(q)][1 - \rho_2\tilde{c}_{22}(q)] - \rho_1\rho_2\tilde{c}_{12}^2(q).$$

If, formally, one sets $S_{12}(q) = 0$ for all q, one returns to the simplicity of the one-component case (2.8). Clearly, this is not a physically realizable assumption, although within the transformed context of the description introduced by Bhatia and Thornton (see later) it will be found to have some importance.

As in the one-component case (3.10), there are asymptotic relationships to the potentials:

$$c_{ij}(r) \sim \frac{-v_{ij}(r)}{k_B T} \text{ (large } r) \tag{13.11}$$

and approximate theories often start from this point. Solving (13.9) for the $\tilde{c}_{ij}(q)$ in terms of the $S_{ij}(q)$ and then Fourier transforming could, in principle, be used to obtain experimental information on the interatomic potentials. Unfortunately, as was indicated before, complete sets of $S_{ij}(q)$ have not, at the time of writing, been measured with great accuracy (certainly not enough for that purpose).

Number-Concentration Description

A description in terms of the individual species of a mixture is not the only feasible approach. An alternative, providing different insights, is possible: let us turn then to this alternative point of view. As indicated in Section 13.1, Bhatia and his coworkers introduced total number (N) and concentration (c) variables.

One may begin by defining $g_{NN}(r)$, $g_{NC}(r)$, and $g_{CC}(r)$:

$$\begin{aligned} g_{NN}(r) &= c_1^2 g_{11}(r) + 2c_1c_2 g_{12}(r) + c_2^2 g_{22}(r) \\ g_{NC}(r) &= c_1c_2\{c_1 g_{11}(r) + (c_2 - c_1)g_{12}(r) - c_2 g_{22}(r)\} \\ g_{CC}(r) &= c_1^2c_2^2\{g_{11}(r) - 2g_{12}(r) + g_{22}(r)\}. \end{aligned} \tag{13.12}$$

These equations have direct physical interpretations (see Young, 1987). For example, $g_{NN}(r)$ describes the expected density of atoms (of either species) around a given atom (also of either species) centred at the origin. Also $g_{CC}(r)$, governing the concentration fluctuations, is species-sensitive and indicates (by size and sign) the likelihood of finding a like atom at $\mathbf{r}$ to that at the origin (+ for like, − for unlike). The orderings measured by g_{NN} and g_{CC} have been called the topological and chemical short-range orders, respectively, by Chieux and Ruppersberg (1980; see also Ruppersberg and Egger, 1975).

13.2. Thermodynamics in terms of number-concentration correlation functions

In terms of these new g's, (13.2) and (13.3) may be rewritten as

$$E = \tfrac{3}{2}k_BT + \tfrac{1}{2}\rho \int \{v_{NN}(r)g_{NN}(r) + 2v_{NC}(r)g_{NC}(r) + v_{CC}(r)g_{CC}(r)\}\,\mathbf{dr} \tag{13.13}$$

and

$$P = \rho k_BT - \tfrac{1}{6}\rho \int \left\{\frac{r\partial v_{NN}}{\partial r}g_{NN} + \frac{2r\partial v_{NC}}{\partial r}g_{NC} + \frac{r\partial v_{CC}}{\partial r}g_{CC}\right\}\mathbf{dr} \tag{13.14}$$

where

$$\begin{aligned} v_{NN}(r) &= c_1^2 v_{11}(r) + 2c_1c_2 v_{12}(r) + c_2^2 v_{22}(r) \\ v_{NC}(r) &= c_1 v_{11}(r) + (c_2 - c_1)v_{12}(r) - c_2 v_{22}(r) \\ v_{CC}(r) &= v_{11}(r) - 2v_{12}(r) + v_{22}(r). \end{aligned} \tag{13.15}$$

The number-concentration partial structure factors can now be defined by

(see Bhatia and Thornton, 1970)

$$S_{\mathrm{NN}}(q) = 1 + \rho \int \{g_{\mathrm{NN}}(r) - 1\} e^{i\mathbf{q}\cdot\mathbf{r}}\, d\mathbf{r}$$

$$S_{\mathrm{NC}}(q) = \rho \int g_{\mathrm{NC}}(r) e^{i\mathbf{q}\cdot\mathbf{r}}\, d\mathbf{r} \qquad (13.16)$$

$$S_{\mathrm{CC}}(q) = c_1 c_2 + \rho \int g_{\mathrm{CC}}(r) e^{i\mathbf{q}\cdot\mathbf{r}}\, d\mathbf{r}$$

and, using these, one can write (13.13) and (13.14) in inverse space. For example (see Ratti and Bhatia, 1977),

$$E = \tfrac{3}{2}k_B T + \tfrac{1}{2}\rho\tilde{v}_{\mathrm{NN}}(0) + \frac{1}{2(2\pi)^3}\int \{\tilde{v}_{\mathrm{NN}}(q)[S_{\mathrm{NN}}(q) - 1]$$
$$+ 2\tilde{v}_{\mathrm{NC}}(q)S_{\mathrm{NC}}(q) + \tilde{v}_{\mathrm{CC}}(q)[S_{\mathrm{CC}}(q) - c_1 c_2]\}\, dq. \qquad (13.17)$$

Equation (13.12) provides a bridge between the number-concentration structure factors, given by (13.16), and the "species" definition (13.4). Thus, one finds

$$S_{\mathrm{NN}} = c_1 S_{11} + 2(c_1 c_2)^{1/2} S_{12} + c_2 S_{22}$$

$$S_{\mathrm{NC}} = c_1 c_2 \left\{ S_{11} + \left[\left(\frac{c_2}{c_1}\right)^{1/2} - \left(\frac{c_1}{c_2}\right)^{1/2}\right] S_{12} - S_{22} \right\} \qquad (13.18)$$

$$S_{\mathrm{CC}} = c_1 c_2 \{c_2 S_{11} - 2(c_1 c_2)^{1/2} S_{12} + c_1 S_{22}\}$$

and, inversely,

$$S_{11} = c_1 S_{\mathrm{NN}} + 2S_{\mathrm{NC}} + \frac{S_{\mathrm{CC}}}{c_1}$$

$$S_{12} = (c_1 c_2)^{1/2} S_{\mathrm{NN}} + \left[\left(\frac{c_2}{c_1}\right)^{1/2} - \left(\frac{c_1}{c_2}\right)^{1/2}\right] S_{\mathrm{NC}} - \frac{S_{\mathrm{CC}}}{(c_1 c_2)^{1/2}} \qquad (13.19)$$

$$S_{22} = c_2 S_{\mathrm{NN}} - 2S_{\mathrm{NC}} + \frac{S_{\mathrm{CC}}}{c_2}.$$

In these two sets of equations, the q-argument of each S has been suppressed for notational convenience.

The practical difficulties of measuring partial structure factors by radiation spectroscopy, particularly at low q, have already been mentioned. At $q = 0$, however, Bhatia and Thornton (1970) were able to show that

$$S_{NN}(0) = \rho k_B T K_T + \delta^2 S_{CC}(0)$$
$$S_{NC}(0) = -\delta S_{CC}(0) \tag{13.20}$$
$$S_{CC}(0) = \frac{k_B T}{(\partial^2 G/\partial c^2)_{P,T,N}}$$

where $\delta = V^{-1}(\partial V/\partial c)_{P,T,N}$ and G is the Gibbs free energy per particle. This is the complete two-component generalization of (2.5). As in that case, the $q = 0$ limits of the partial structure factors are accessible to thermodynamic investigation, and these can provide checks on radiation spectroscopy data.

Equations (13.20) give, as a special case,

$$\rho k_B T K_T = \frac{S_{NN}(0) S_{CC}(0) - S_{NC}^2(0)}{S_{CC}(0)}. \tag{13.21}$$

This must be consistent with the Kirkwood-Buff result (13.6) and use of (13.18) or (13.19) provides the proof.

13.3. Number-concentration structure factors as response functions

The number-concentration structure factors act as response functions as follows (Young, 1987). Consider the perturbation relating to (13.7), but now focus on the responses

$$\delta\rho(q) = \delta\rho_1(q) + \delta\rho_2(q),$$
$$\delta c(q) = N^{-1}\{c_2 \delta\rho_1(q) - c_1 \delta\rho_2(q)\}, \tag{13.22}$$

which are, respectively, the total density and concentration fluctuations. (The first of these interpretations is obvious; the second follows by noting that since $N = N_1 + N_2$, $\delta(N_1/N) = N^{-1}\{(N_2/N)\delta N_1 - (N_1/N)\delta N_2\}$.) Then

$$-\left(\frac{k_B T}{\rho}\right)\delta\rho(q) = S_{NN}(q)\delta\bar{\varphi}(q) + S_{NC}(q)\delta\Delta\varphi(q)$$
$$-\left(\frac{k_B T}{\rho}\right)\delta c(q) = S_{NC}(q)\delta\bar{\varphi}(q) + S_{CC}(q)\delta\Delta\varphi(q) \tag{13.23}$$

where $\bar{\varphi} = c_1\varphi_1 + c_2\varphi_2$ and $\Delta\varphi = \varphi_1 - \varphi_2$.

It will be seen from (13.23) that large (small) $S_{CC}(q)$ is indicative of large (small) concentration fluctuations of the same wavelength (a precise result following when $S_{NC} = 0$, i.e., when the concentration and number fluctuations are uncorrelated). In fact, at $q = 0$ (infinite wavelength), general thermodynamic theory and the use of (13.20) leads to the conclusion that $S_{CC}(0) = \infty$ implies phase separation (PS), whereas $S_{CC}(0) = 0$ implies compound formation (CF). Thus, if $S_{CC}(0) \gg c_1 c_2$ (the ideal value), one follows Bhatia and coworkers by speaking of a PS tendency, whereas if $S_{CC}(0) \ll c_1 c_2$, one can likewise refer to a CF tendency. Ruppersberg and Egger (1975; Chieux and Ruppersberg, 1980) refer, equivalently, to self-coordinating and heterocoordinating tendencies for reasons that if not already obvious, will soon be so. This behaviour is analogous to that of $S(0)$ in relationship to K_T (see (2.5) and (3.9)) in a one-component system.

It is next appropriate to introduce the corresponding number-concentration direct correlation functions $c_{NN}(r)$, $c_{NC}(r)$, $c_{CC}(r)$. These are defined by

$$\begin{aligned} h_{NN} &= c_{NN} + \rho h_{NN}{}^{*}c_{NN} + \rho h_{NC}{}^{*}c_{CN} \\ h_{NC} &= c_{NC} + \rho h_{NN}{}^{*}c_{NC} + \rho h_{NC}{}^{*}c_{CC} \\ h_{CC} &= c_{CC} + \rho h_{CN}{}^{*}c_{NC} + \rho h_{CC}{}^{*}c_{CC} \end{aligned} \tag{13.24}$$

where

$$h_{\alpha\beta}{}^{*}c_{\beta\gamma} \equiv \int h_{\alpha\beta}(|\mathbf{r} - \mathbf{r}'|)c_{\beta\gamma}(\mathbf{r}')\,d\mathbf{r}'$$

and $h_{NN} = g_{NN} - 1$, $h_{NC} = g_{NC}$, $h_{CC} = g_{CC}$. The latter definitions ensure that each $h_{\alpha\beta} \to 0$ as $r \to \infty$. Equations (13.24) have the same structure as (13.8) and can be similarly analyzed.

The NC and "species" h's are linked (see (13.12)) by

$$\begin{aligned} h_{NN} &= c_1^2 h_{11} + 2c_1 c_2 h_{12} + c_2^2 h_{22} \\ h_{NC} &= c_1 c_2 [c_1 h_{11} + (c_2 - c_1) h_{12} - c_2 h_{22}] \\ h_{CC} &= c_1^2 c_2^2 (h_{11} - 2h_{12} + h_{22}). \end{aligned} \tag{13.25}$$

Thus, using (13.24), it follows that the direct correlation functions are related by

$$\begin{aligned} c_{NN} &= c_1^2 c_{11} + 2c_1 c_2 c_{12} + c_2^2 c_{22} \\ c_{NC} &= c_1 c_{11} + (c_2 - c_1) c_{12} - c_2 c_{22} \\ c_{CC} &= c_{11} - 2c_{12} + c_{22}. \end{aligned} \tag{13.26}$$

By noting (13.11), one can see that the new c's can be linked asymptotically to the potentials according to the relationships

$$c_{\alpha\beta}(r) \sim -\frac{v_{\alpha\beta}(r)}{k_B T} \qquad \text{(large } r\text{)}, \tag{13.27}$$

where the $v_{\alpha\beta}$ are given by the left-hand sides of (13.15). As was stated after (13.11), at the time of writing experiment is not yet developed enough to allow evaluation of the three partial potentials. However, in the exceptional cases of zero* alloys (Ratti and Bhatia, 1977), $S_{CC}(q)$ can be accurately measured and then an approximate $v_{cc}(r)$ can be deduced (see, e.g., Copestake et al., 1983) if (as is often the case) $S_{NC}(q) \approx 0$. For, by the remarks following (13.10) and (13.24), one has

$$S_{CC}(q) \approx \{(c_1 c_2)^{-1} - \rho\tilde{c}_{CC}(q)\}^{-1}, \tag{13.28}$$

and so, by (13.27),

$$v_{cc}(r) \approx \frac{k_B T}{(2\pi)^3 \rho} \int \left\{ \frac{1}{S_{CC}(q)} - \frac{1}{c_1 c_2} \right\} e^{-i\mathbf{q}\cdot\mathbf{r}} \, d\mathbf{q} \qquad \text{(large } r\text{)}. \tag{13.29}$$

The function $v_{cc}(r)$ is often called the ordering potential and has a long history. As one can see, it arises very naturally via the number-concentration formalism, with a rather direct link to $S_{CC}(q)$ and, therefore, as the commentary after (13.23) makes clear, to any PS or CF tendency that the system might have. Evidently, by (13.15), $v_{cc} < 0$ (around near-neighbour distances) corresponds to self-coordination and $v_{cc} > 0$ to heterocoordination (see Young, 1987).

13.4. Structure and forces in liquid-metal alloys

The potentials with which one is concerned in alloy theory are much less well understood than those of the pure metals of Section 3.3. However, shapes similar to those depicted in Figures 3.1 and 3.5 can be expected.

To develop a comprehensive statistical-mechanical theory of a binary *a-b* liquid mixture, let us first set up force equations that generalize the one-component equation (5.2). Consider specifically a binary liquid mixture with N atoms of type a at $\mathbf{R}_1 \ldots \mathbf{R}_N$ and n atoms of type b at $\mathbf{r}_1 \ldots \mathbf{r}_n$. For a configuration in which atom a is situated at R_1 and atom b at r_1, one now

* A zero alloy is defined by $c_1 b_1 + c_2 b_2 = 0$, where b_1 and b_2 are the neutron-nuclear scattering amplitudes

wishes to write down an equation for the mean force $-\partial U_{ab}(R_1, r_1)/\partial \mathbf{R}_1$ acting on atom a. First there is the direct interaction, say $-\partial \phi_{ab}(\mathbf{R}_1, \mathbf{r}_1)/\partial \mathbf{R}_1$, and second (see [5.2]) a contribution from the "rest" of the system, namely the indirect interaction.

13.4.1. Calculation of indirect interaction

Consider a second atom of type a at $\mathbf{R}_2$ acting with a force $-\partial \phi_{aa}(\mathbf{R}_1 - \mathbf{R}_2)/\partial \mathbf{R}_1$ and a second atom of type b at $\mathbf{r}_2$, with resulting force $-\partial \phi_{ab}(R_1, r_2)/\partial \mathbf{R}_1$. One must then sum these contributions, multiply by the probability of this four-atom configuration

$$\left[\frac{\rho^{(4)}_{abab}(\mathbf{R}_1 \mathbf{r}_1 \mathbf{R}_2 \mathbf{r}_2)}{\rho^{(2)}_{ab}(\mathbf{R}_1 \mathbf{r}_1)}\right],$$

and finally integrate over the positions of the second type of atom. The result is (Page, de Angelis, and March, 1982)

$$-\frac{\partial U_{ab}(\mathbf{R}_1 \mathbf{r}_1)}{\partial \mathbf{R}_1} = -\frac{\partial \phi_{ab}(R_1 r_1)}{\partial \mathbf{R}_1} - \int \frac{\rho^{(4)}_{abab}(\mathbf{R}_1 \mathbf{r}_1 \mathbf{R}_2 \mathbf{r}_2)}{\rho^{(2)}_{ab}(\mathbf{R}_1 \mathbf{r}_1)} \left[\frac{\partial \phi_{aa}(\mathbf{R}_1 - \mathbf{R}_2)}{\partial \mathbf{R}_1} + \frac{\partial \phi_{ab}(\mathbf{R}_1 \mathbf{r}_2)}{\partial \mathbf{R}_1}\right] d\mathbf{R}_2\, d\mathbf{r}_2. \tag{13.30}$$

Since three-atom correlations $\rho^{(3)}$ are related to $\rho^{(4)}$ by

$$\rho^{(3)}_{aba}(\mathbf{R}_1 \mathbf{r}_1 \mathbf{R}_2) = \int \rho^{(4)}_{abab}(\mathbf{R}_1 \mathbf{r}_1 \mathbf{R}_2 \mathbf{r}_2)\, d\mathbf{r}_2 \tag{13.31}$$

and

$$\rho^{(3)}_{abb}(\mathbf{R}_1 \mathbf{r}_1 \mathbf{r}_2) = \int \rho^{(4)}_{abab}(\mathbf{R}_1 \mathbf{r}_1 \mathbf{R}_2 \mathbf{r}_2)\, d\mathbf{R}_2 \tag{13.32}$$

one finally obtains (Page, de Angelis, and March, 1982)

$$-\frac{\partial U_{ab}(\mathbf{R}_1 \mathbf{r}_1)}{\partial \mathbf{R}_1} = -\frac{\partial \phi_{ab}(\mathbf{R}_1 \mathbf{r}_1)}{R_1} - \int \frac{\rho^{(3)}_{aab}(\mathbf{R}_1 \mathbf{r}_1 \mathbf{R}_2)}{\rho^{(2)}_{ab}(\mathbf{R}_1 \mathbf{r}_1)} \frac{\partial \phi_{aa}(\mathbf{R}_1 \mathbf{R}_2)}{\partial \mathbf{R}_1} d\mathbf{R}_2 - \int \frac{\rho^{(3)}_{abb}(\mathbf{R}_1 \mathbf{r}_1 \mathbf{r}_2)}{\rho^{(2)}_{ab}(\mathbf{R}_1 \mathbf{r}_1)} \frac{\partial \phi_{ab}(\mathbf{R}_1 \mathbf{r}_2)}{\partial \mathbf{R}_1} d\mathbf{r}_2. \tag{13.33}$$

A similar force equation can be written for two atoms of type a. The above approach is formally exact: an approximate treatment of (13.33), leading to a potential ϕ_{ab}, has been given by Page, de Angelis, and March (1982).

13.5. Chemical approaches

Although the pseudopotential approach of Section 3.4 has been generalized to binary alloys (Hafner, 1985: see also the review by Young, 1987), this is the point at which to turn to chemical approaches to binary liquid alloys. Since the work by Bhatia and Thornton (1970), the concentration fluctuations in the long wavelength limit $S_{CC}(0)$, discussed above as the limit $q \to 0$ of $S_{CC}(q)$, have been proved very useful in obtaining microscopic information for molten alloys. Attention has also been drawn to the fact that the knowledge of $S_{CC}(0)$ together with the local ordering (measured in terms of Cowley-Warren short-range order (SRO) parameter for the nearest-neighbour shell) may shed light on the phenomenon of easy glass formation in many binary molten alloys. Sections 13.5–13.10, therefore, aim to highlight the information that one can obtain for $S_{CC}(0)$ and SRO in binary molten alloys from the theoretical models available (see Singh, 1987).

From the viewpoint of such modeling, the binary molten alloys may be broadly grouped into two major categories: symmetric and asymmetric alloys. As the name implies, properties such as free energy of mixing, ΔG_M, heat of mixing ΔH_M, concentration fluctuations, $S_{CC}(0)$, etc., of symmetric alloys (for example CuZn, AlZn, BiPb, CdMg, and NaK) are symmetric or very close to it about the concentration $c = \frac{1}{2}$. These alloys are usually referred to as regular alloys. $S_{CC}(0)$ for these have been widely studied by using the conformal solution (Bhatia, Hargrove and March, 1973) of Longuet-Higgins (1951). Though this model has been successful to infer $S_{CC}(0)$ with the use of interchange energy w (which describes whether a pair of like atoms or a pair of unlike atoms is energetically preferred as nearest neighbour), it fails to provide any information on SRO because this conformal solution model (see Appendix 13.1) ignores the extent of the local ordering. Bhatia and Singh (1982) have subsequently shown that such difficulty might be overcome by working with a model that is essentially based on a method proposed by Bethe (1935) and Peierls (1936) to treat SRO in a solid.

The mixing properties of asymmetric alloys, on the other hand, are evidently not symmetric about $c = \frac{1}{2}$. A substantial effort has been made to understand the cause of asymmetry, which is well summarized by Bhatia

(1977). Much attention has been focused either on size effects as in NaCs (Bhatia and March, 1975) or on the existence of chemical complexes in the molten state (as in MgBi, CuSn, AgAl, and AlLa). The complex-forming alloys have been extensively studied both theoretically and experimentally and have been referred to in the literature under names such as complex-forming solution, compound-forming solution, or regular associated solution. (It is from this class that many systems are found to be suitable for forming metallic glasses.) The majority of theoretical models (see Singh, 1987) in use to date essentially assume the idea of the existence of chemical complexes or a privileged group of atoms, $A_\mu B_\nu$, in the molten state (μ and ν are small integers and denote the number of A atoms and B atoms, respectively, in the complex). It may be pointed out that these alloys have the characteristic that in the solid state they form compounds at one or more stoichiometric compositions, say $c_c = \mu/\mu + \nu$. This information is important in assessing the values of μ and ν, which may also be verified from the fact that $S_{CC}(0)$ has a dip and generally the equilibrium phase diagram (composition-temperature curve) exhibits a bump in the vicinity of c_c. Based on this complex-formation model, Bhatia, Hargrove, and Thornton (1974; see also Bhatia and Hargrove, 1974) were able to offer an explanation of the form of $S_{CC}(0)$ and other thermodynamic properties for varieties of compound-forming alloys, but the explanation of SRO was lacking. This prompted Bhatia and Singh (1982, 1984) to take a more microscopic approach with a view to obtaining simultaneous information on $S_{CC}(0)$ and SRO. This will be discussed in Section 13.10. This will be followed by a brief discussion of the importance of these physical quantities in the understanding of the phenomenon of glass formation (see especially Appendix 13.7).

13.6. Concentration fluctuations $S_{CC}(0)$

Bhatia and Thornton (1970) showed that the concentration fluctuations $S_{CC}(0)$ at $q = 0$ (long wavelength) in a binary mixture may be expressed as

$$S_{CC}(0) = N\langle(\Delta c)^2\rangle, \tag{13.34}$$

where $\langle(\Delta c)^2\rangle$ represent the mean square fluctuations in the concentration and can readily be derived from statistical mechanics* in terms of the Gibbs

* For a derivation, see Bhatia and Thornton (1970).

free energy G (see (13.20)):

$$S_{CC}(0) = \frac{Nk_B T}{(\partial^2 G/\partial c^2)_{T,P,N}}. \tag{13.35}$$

Furthermore, G for a binary mixture consisting of $N_A = Nc$ and $N_B = N(1-c)$ g moles of A and B atoms may be expressed as

$$G = N[cG_A^0 + (1-c)G_B^0] + G_M, \tag{13.36}$$

where G_A^0 and G_B^0 are the Gibbs free energies per atom of the species A and B, respectively. G_M is the free energy of mixing. In view of (13.36), $S_{CC}(0)$ then becomes

$$S_{CC}(0) = \frac{Nk_B T}{(\partial^2 G_M/\partial c^2)_{T,P,N}}. \tag{13.37}$$

By making use of standard thermodynamic relations, (13.37) can also be written as

$$S_{CC}(0) = \frac{(1-c)a_A}{(\partial a_A/\partial c)_{T,P,N}} = \frac{ca_B}{(\partial a_B/\partial(1-c))_{T,P,N}} \tag{13.38}$$

where a_A and a_B are the thermodynamic activities of the components A and B in the mixture. The quantity $(\partial^2 G/\partial c^2)$ is also known as the stability of the solution and was originally introduced by Darken (1967) following a different route. Thus the reciprocal quantity $Nk_B T/S_{CC}(0)$ is a measure of the stability of the mixture.

Equation (13.38) is customarily utilized to obtain experimental values of $S_{CC}(0)$ from measured activity data or from free energy of mixing data. Model expressions for G_M or activity, on the other hand, provide analytical forms for $S_{CC}(0)$.

In the case of ideal solutions, for example,

$$G_M = c\ln c + (1-c)\ln(1-c). \tag{13.39}$$

$S_{CC}(0)$, therefore, becomes

$$S_{CC}^{id}(0) = c(1-c). \tag{13.40}$$

Any deviation of $S_{CC}(0)$ from the ideal value $S_{CC}^{id}(0)$ is of interest in reflecting the extent of interactions in the mixture. As already noted in Section 13.3, if at a given composition $S_{CC}(0) \gg S_{CC}^{id}(0)$, there is a tendency toward segregation or phase separation, whereas $S_{CC}(0) \ll S_{CC}^{id}(0)$ is an indication of strong association or the existence of the chemical complexes in the mixture.

13.7. Short-range-order parameter

The short-range-order parameter α_1 (Warren, 1969; Cowley, 1950; see also March, Wilkins, and Tibballs, 1976) for the first neighbour shell is usually defined in terms of the conditional probability $[A/B]$, i.e.,

$$[A/B] = c(1 - \alpha_1), \tag{13.41}$$

where $[A/B]$ defines the probability that an A atom exists at site 2 as a nearest neighbour of a given B atom at site 1. Equation (13.41) thus provides immediate insight into the local arrangement of atoms in the mixture. For a random distribution of atoms, $[A/B]$ is simply c, and then $\alpha_1 = 0$. If $\alpha_1 < 0$, then A-B pairs of atoms are preferred over A-A or B-B pairs as nearest neighbours; for the converse case, $\alpha_1 > 0$. By adopting a probabilistic approach, the limiting values of α_1 can easily be shown to lie in the range

$$\begin{aligned} -\frac{c}{(1-c)} &\leqslant \alpha_1 \leqslant 1, \qquad c \leqslant \frac{1}{2}; \\ -\frac{(1-c)}{c} &\leqslant \alpha_1 \leqslant 1, \qquad c \geqslant \frac{1}{2}. \end{aligned} \tag{13.42}$$

For $c = \frac{1}{2}$, one has $-1 \leqslant \alpha_1 \leqslant 1$. The minimum possible value, $\alpha_1^{\min} = -1$, means complete ordering of A-B pairs in the melt, whereas the maximum value, $\alpha_1^{\max} = 1$, indicates that the A-A and B-B pairs in the melt are totally segregated.

Equation (13.41) can also be written in terms of the generalized probability, X_{AB}, which denotes that the one lattice site of a nearest-neighbour pair is occupied by an A atom and the other by a B atom. Since $X_{\mathrm{AB}} = (1 - c)[A/B]$, (13.41) therefore becomes

$$X_{\mathrm{AB}} = c(1 - c)(1 - \alpha_1), \tag{13.43}$$

where X_{AB} can be evaluated from knowledge of thermodynamic quantities, as will be discussed in Sections 13.8 and 13.9 of this chapter.

Experimentally, α_1 is determined from the concentration-concentration structure factors $S_{\mathrm{CC}}(q)$ and the number-number structure factors $S_{\mathrm{NN}}(q)$ given in (13.18). For melts in which the A and B atoms have nearly the same size, Ruppersberg and Egger (1975; for reviews see Chieux and Ruppersberg, 1980; Steeb, Falch, and Lamparter, 1984) showed that

$$\int_{r_1-\varepsilon}^{r_1+\varepsilon} 4\pi r^2 \rho_{\mathrm{CC}}(r)\,\mathrm{d}r = \alpha_1 z, \tag{13.44}$$

where r_1 is the mean distance of the first neighbour shell from a given atom. The actual distance, however, is taken between $r_1 - \varepsilon$ and $r_1 + \varepsilon$. z is the number of atoms in the first shell, usually known as the coordination number, and is determined from $S_{NN}(q)$. $\rho_{CC}(r)$ is referred to as the radial concentration correlation function, which is obtained from the Fourier transform (FT) of $S_{CC}(q)$, i.e. (see Appendix 13.1 for a model),

$$4\pi r^2 \rho_{CC}(r) = 4\pi r \, \mathrm{FT}\left[\frac{S_{CC}(q)}{c(1-c)} - 1\right]. \tag{13.45}$$

13.8. Regular and conformal solutions

13.8.1. *Model and basic formalism*

As mentioned earlier, regular (or conformal; see Appendix 13.1) alloys usually have symmetries in $S_{CC}(0)$ or G_M about $c = 1/2$, and the constituent atoms A and B have approximately the same size. If atoms A and B are assumed to be located at sites of a lattice, the grand partition function of the system is then written as

$$\Xi = \sum_E q_A^{N_A}(T) q_B^{N_B}(T) e^{(\mu_A N_A + \mu_B N_B - E)/k_B T}, \tag{13.46}$$

where the $q_i(T)$ $(i = A, B)$ denote the partition functions associated with inner and translational degrees of freedom of atoms i and the μ_i are the chemical potentials. N_A and N_B are the number of A and B atoms in the alloy with the configurational energy E.

In order to make (13.46) tractable, resort will be made to the following simplifications. The whole set of lattice sites is divided into two groups, a cluster of few lattice sites, say, in domain 1 and the remainder, say, in domain 2. Thus one expresses Ξ as the product of the grand partition functions for each domain; i.e., (13.46) may be written as

$$\Xi = \Xi_1' \cdot \Xi_2, \tag{13.47}$$

where the prime on Ξ_1 indicates that the interaction of the domain (cluster) 1 with domain 2 is incorporated in it. Ξ_1' is the quantity of interest and (see Singh, 1987) it can be expressed in the form

$$\Xi_1' = \sum_{E_1} \xi_A^{N_{1A}} \xi_B^{N_{1B}} \phi_A^{\nu_A} \phi_B^{\nu_B} e^{-E_1/k_B T} \tag{13.48}$$

where

$$\xi_A = q_A(T)e^{\mu_A/k_BT}; \qquad \xi_B = q_B(T)e^{\mu_B/k_BT}. \tag{13.49}$$

The energy E_1 represents the configurational energy of the cluster. ν_A is the number of lattice sites in domain 2 which are the nearest neighbours of the A atoms in the cluster; a similar statement can be made for ν_B. ϕ_A and ϕ_B are constants, which in fact may be eliminated from the final result. Ξ_1' may now be used to estimate the average values of A and B atoms in the cluster, i.e.,

$$\langle N_{1A}\rangle = k_BT\frac{\partial \ln \Xi_1'}{\partial \mu_A} = \xi_A\frac{\partial \ln \Xi_1'}{\partial \xi_A}, \tag{13.50}$$

$$\langle N_{1B}\rangle = k_BT\frac{\partial \ln \Xi_1'}{\partial \mu_B} = \xi_B\frac{\partial \ln \Xi_1'}{\partial \xi_B}. \tag{13.51}$$

By taking the ratio of the above expressions one finds

$$\frac{\langle N_{1B}\rangle}{\langle N_{1A}\rangle} = \frac{(1-c)}{c} = \frac{\dfrac{\xi_B}{\xi_A}\dfrac{\partial \ln \Xi_1'}{\partial \xi_B}}{\dfrac{\partial \ln \Xi_1'}{\partial \xi_A}}. \tag{13.52}$$

Obviously this is independent of the size of the cluster. Therefore clusters of two different sizes can be used to evaluate (13.52). First, consider just one lattice site in the cluster, for which (13.48) reduces to

$$\Xi_1'^{(1)} = \xi_A\phi_A^z + \xi_B\phi_B^z, \tag{13.53}$$

where z is the coordination number. For a cluster of two lattice sites, all possible arrangements of atoms on sites within the cluster are depicted in Table 13.1. Equation (13.48) for a cluster of two sites is then

Table 13.1. *Arrangement of A and B atoms in a cluster of two lattice sites [after Singh (1987)].*

Site 1	Site 2	ν_A	ν_B	E_1
A	A	$2(z-1)$	0	ε_{AA}[a]
B	B	0	$2(z-1)$	ε_{BB}
A	B	$(z-1)$	$(z-1)$	ε_{AB}
B	A	$(z-1)$	$(z-1)$	ε_{AB}

[a] ε_{ij} denotes the energy of the ij bond.
z is the coordination number.

$$\Xi_1'^{(2)} = \xi_A^2 \phi_A^{2(z-1)} e^{-\varepsilon_{AA}/k_B T} + \xi_B^2 \phi_B^{2(z-1)} e^{-\varepsilon_{BB}/k_B T} + 2\xi_A \xi_B (\phi_A \phi_B)^{z-1} e^{-\varepsilon_{AB}/k_B T}. \tag{13.54}$$

From (13.52) and (13.53) one has

$$\frac{(1-c)}{c} = \frac{\xi_B}{\xi_A}\left(\frac{\phi_B}{\phi_A}\right)^z. \tag{13.55}$$

Similarly (13.52) and (13.54) give

$$\frac{(1-c)}{c} = \frac{\xi_B}{\xi_A} \frac{\xi_B \phi_B^{2(z-1)} e^{-\varepsilon_{BB}/k_B T} + \xi_A (\phi_A \phi_B)^{z-1} e^{-\varepsilon_{AB}/k_B T}}{\xi_A \phi_A^{2(z-1)} e^{-\varepsilon_{AA}/k_B T} + \xi_B (\phi_A \phi_B)^{z-1} e^{-\varepsilon_{AB}/k_B T}}. \tag{13.56}$$

By setting

$$\sigma = \left(\frac{\phi_B}{\phi_A}\right) \exp\left[\frac{1}{2k_B T}(\varepsilon_{BB} - \varepsilon_{AA})\right], \tag{13.57}$$

(13.55) and (13.56) yield

$$\sigma^2 + \left(\frac{1-2c}{\eta c}\right)\sigma - \frac{(1-c)}{c} = 0, \tag{13.58}$$

where

$$\eta = \exp\left[\frac{1}{2k_B T}(2\varepsilon_{AB} - \varepsilon_{AA} - \varepsilon_{BB})\right] = \exp\left(\frac{w}{zk_B T}\right). \tag{13.59}$$

$w = z(\varepsilon_{AB} - \varepsilon_{AA}/2 - \varepsilon_{BB}/2)$ is usually known as the interchange energy, where ε_{AB}, ε_{AA}, and ε_{BB} are the energies for *AB*, *AA*, and *BB* pairs of atoms. Obviously if $w < 0$, there is a tendency to form pairs of unlike atoms, and if $w > 0$, then like atoms will tend to pair together. $w = 0$, however, indicates that atoms in the mixture are perfectly disordered.

Equation (13.58) is readily solved to yield

$$\sigma = \frac{\beta - 1 + 2c}{2\eta c} \tag{13.60}$$

with

$$\beta = [1 + 4c(1-c)(\eta^2 - 1)]^{1/2}. \tag{13.61}$$

From (13.60) and (13.61) one may also write

$$\sigma = \left[\frac{(1-c)}{c} \frac{\beta - 1 + 2c}{\beta + 1 - 2c}\right]^{1/2}. \tag{13.62}$$

In the following section it will be shown that σ is directly related to the ratio of the activity coefficients of the constituent species, and hence various thermodynamic quantities result.

13.9. Activity, free energy of mixing, and concentration fluctuations

To obtain these expressions, let us first note some standard thermodynamic relations, i.e.,

$$\mu_A - \mu_A^0 = k_B T \ln c\gamma_A \tag{13.63}$$

$$\mu_B - \mu_B^0 = k_B T \ln(1-c)\gamma_B \tag{13.64}$$

where the γ_i $(i = A, B)$ are activity coefficients of the species i in the alloy and μ_i^0 are the chemical potentials of the pure species i. Within the framework of the model, use of the partition function readily gives

$$\mu_A^0 = -k_B T \ln q_A(T) + \frac{z\varepsilon_{AA}}{2}, \tag{13.65}$$

$$\mu_B^0 = -k_B T \ln q_B(T) + \frac{z\varepsilon_{BB}}{2}. \tag{13.66}$$

Equations (13.63) to (13.66), together with (13.49), yield the ratio of the two activity coefficients $\gamma = \gamma_A/\gamma_B$ as

$$\gamma = \frac{(1-c)}{c}\frac{\xi_A}{\xi_B}\exp\left[\frac{z}{2k_B T}(\varepsilon_{BB} - \varepsilon_{AA})\right]. \tag{13.67}$$

By eliminating ξ_A/ξ_B with the help of (13.55) and (13.57), this reduces to

$$\gamma = \sigma^z \tag{13.68}$$

or

$$\gamma = \left[\frac{(1-c)}{c}\frac{\beta - 1 + 2c}{\beta + 1 - 2c}\right]^{z/2}. \tag{13.69}$$

Thus for equiatomic composition ($c = \frac{1}{2}$) of regular alloys $\gamma = 1.0$. It may be noted that although the above formalism seems somewhat different from the quasi-chemical approximation (QCA; see Guggenheim, 1952), the expression for γ is identical in the two cases.

Let us next obtain an expression for the excess free energy of mixing, which is defined as

$$G_{\mathrm{M}}^{xs} = G_{\mathrm{M}} - Nk_BT\{c\ln c + (1-c)\ln(1-c)\}. \qquad (13.70)$$

$G_{\mathrm{M}}^{xs}/Nk_BT$ can be found from γ by using the thermodynamic relation (see Singh, 1987)

$$\frac{\mathrm{d}}{\mathrm{d}c}\left\{\frac{G_{\mathrm{M}}^{xs}}{Nk_BT}\right\} = \ln\gamma \qquad (13.71)$$

or

$$\frac{G_{\mathrm{M}}^{xs}}{Nk_BT} = \int_0^c \ln\gamma\,\mathrm{d}c = z\int_0^c \ln\sigma\,\mathrm{d}c, \qquad (13.72)$$

where $\ln\sigma$ is expressed via (13.68) and (13.69). Since the interchange energy w is considered independent of c, (13.72) is readily integrated to yield

$$\frac{G_{\mathrm{M}}^{xs}}{Nk_BT} = c\ln\left[\frac{\beta-1+2c}{c(\beta+1)}\right]^{z/2} + (1-c)\ln\left[\frac{\beta+1-2c}{(1-c)(1+\beta)}\right]^{z/2}, \qquad (13.73)$$

which is the same as that obtained in QCA (Singh, 1987). The preceding expressions for G_{M}^{xs} and γ can be seen to depend entirely on the coordination number, z, and the interchange energy w. z is usually chosen from the knowledge of the structure of the alloy (for example, $z = 6$, 8, and 12 for simple cubic, *bcc*, and *fcc* structures, respectively) and w is to be determined from experimental data on activity at a given concentration. Independent theoretical evaluation of w is evidently required in order to gain deeper insight. It should be emphasized that in the framework of the theory, w is explicitly considered to be independent of concentration but may depend on temperature and pressure. There are, however, some indications that w should weakly depend on concentration (see Singh, 1987, and references there).

From the definition of interchange energy it is seen that if $w > 0$, there is a tendency toward segregation, which may lead to separation of phases dominated by A atoms and B atoms. The condition of critical mixing for equiatomic composition of the alloy is given by

$$\frac{w}{k_BT_c} = z\ln\frac{z}{z-2}. \qquad (13.74)$$

For $T > T_c$, the alloy remains in one phase, whereas at a temperature below the critical temperature, it separates out into two phases.

Finally $S_{CC}(0)$ is obtained by using equations (13.37) and (13.73) as

$$S_{CC}(0) = \frac{c(1-c)}{1+\frac{1}{2}z(1/\beta - 1)}. \tag{13.75}$$

Equation (13.75) is fully discussed by Bhatia and Singh (1982) for regular alloys, where β is defined in (13.61). For such an alloy, $S_{CC}(0)$ is symmetric about $c = \frac{1}{2}$ and reduces to its ideal value; i.e., $S_{CC}^{id}(0) = c(1-c)$ for $w = 0$. If $w > 0$, $S_{CC}(0) > S_{CC}^{id}(0)$ and for $w < 0$, $S_{CC}(0) < S_{CC}^{id}(0)$. The plot of $S_{CC}(0)$ for $z = 8$ for different values of w is shown in Figure 13.1. It may be noted that the $S_{CC}(0)$-c curve has a point of inflexion for a given $w/k_B T$. Above this, there is only one maximum in $S_{CC}(0)$, namely, at $c = \frac{1}{2}$, but below this value of $w/k_B T$, one gets a dip in $S_{CC}(0)$ at $c = \frac{1}{2}$ with a peak on either side.

Figure 13.1. Concentration fluctuations $S_{CC}(0)$ versus concentration from regular solution theory (see (13.75)) for different values of the (dimensionless) interchange energy $w/k_B T$ for the coordination number $z = 8$ (after Singh, 1987).

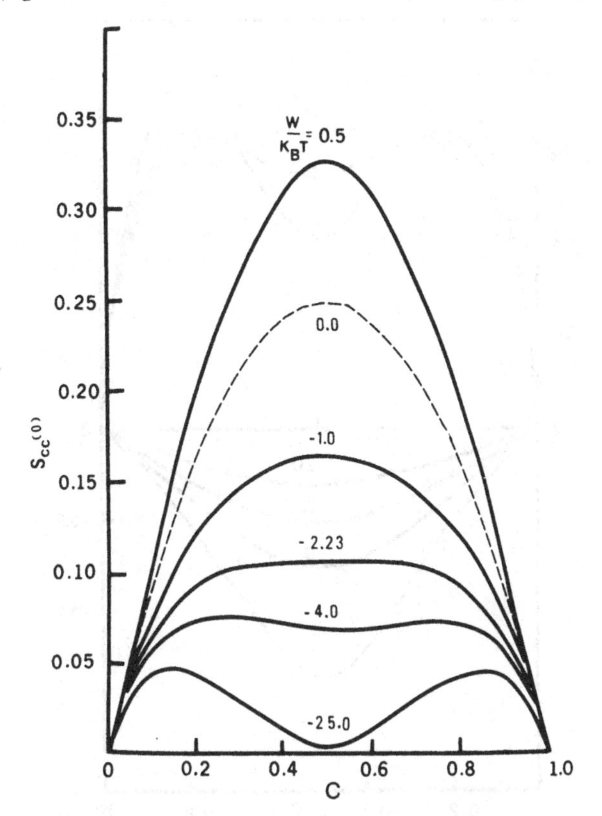

$S_{CC}(0) \to 0$ as $w/k_B T \to -\infty$. It is to be emphasized that these features of $S_{CC}(0)$ are in contrast to those obtained by using the zeroth or conformal solution expression (13.80), where $S_{CC}(0)$ always has its maximum at $c = \frac{1}{2}$ for all values of $w/k_B T$.

The point of inflexion in the $S_{CC}(0)$-c curve can be obtained from the condition that $\partial^2 S_{CC}(0)/\partial c^2 = 0$ at $c = \frac{1}{2}$, which yields

$$2(1 - \theta)\eta^3 + 3\theta\eta^2 - \theta = 0, \qquad \theta = \frac{z}{2}. \tag{13.76}$$

For $z = 8$, the point of inflexion in $S_{CC}(0)$ computed from (13.76) corresponds to $w/k_B T \simeq -2.6$ and for $z = 12$, $w/k_B T \simeq -3.3$.

Figure 13.2. Concentration fluctuations $S_{CC}(0)$ and short-range order parameter α_1 plotted versus concentration for a given (dimensionless) interchange energy $w/k_B T = -5.0$ but for different values of the coordination number z (after Singh, 1987).

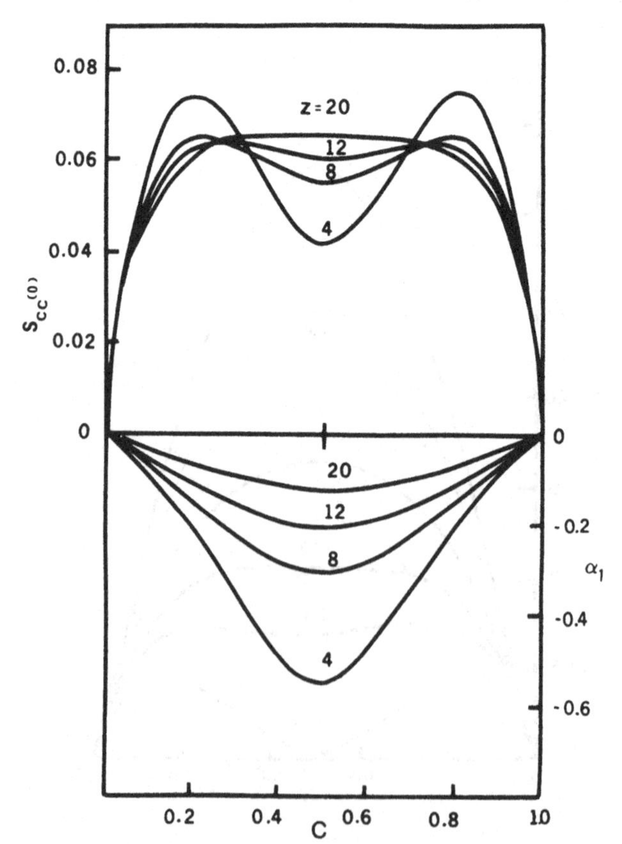

Equation (13.75) can also be utilized to understand the dependence of $S_{CC}(0)$ on z. The plot of $S_{CC}(0)$, for a given $w/k_BT = -5.0$, is shown for different values of z in the upper part of Figure 13.2. For low z, two maxima and a dip in $S_{CC}(0)$ are well pronounced. However, as z increases, this distinct feature becomes diffuse.

It is also to be emphasized that as $z \to \infty$, the various thermodynamic relations just given reduce to what one terms the regular solution model in the zeroth approximation (Guggenheim, 1952) or the conformal solution model (Longuet-Higgins, 1951; see also Appendix 13.1). For instance,

$$\left.\begin{aligned} \gamma &= \exp\left\{(1-2c)\frac{w}{k_BT}\right\} &\quad (13.77)\\ \frac{G_M^{xs}}{Nk_BT} &= c(1-c)\frac{w}{k_BT} &\quad (13.78)\\ \frac{w}{k_BT_c} &= 2.0 &\quad (13.79)\\ S_{CC}(0) &= \frac{c(1-c)}{1-c(1-c)(2w/k_BT)} &\quad (13.80) \end{aligned}\right\} \text{ for } z \to \infty.$$

Though these results of the conformal solution model have been widely used to discuss the thermodynamics of simple binary molten alloys (see Bhatia, Hargrove and March, 1973; Bhatia, 1977, and references given there), they cannot be used to interpret SRO because the model ignores the local ordering and always leads to $\alpha_1 \to 0$.

13.9.1. *Short-range-order parameter*

Let us first recall expression (13.43) where the SRO parameter, α_1, for the first-neighbour shell is expressed in terms of X_{AB}, the probability that two neighbouring lattice sites are occupied by A and B atoms. The latter can now be evaluated by employing the grand partition function as (Singh, 1987)

$$X_{AB} = X_{BA} = \frac{\xi_A \xi_B (\phi_A \phi_B)^{z-1} e^{-\varepsilon_{AB}/k_BT}}{\Xi_1'^{(2)}} \qquad (13.81)$$

and, similarly,

$$X_{AA} = \frac{\zeta_A^2 \phi_A^{2(z-1)} e^{-\varepsilon_{AA}/k_B T}}{\Xi_1'^{(2)}}, \tag{13.82}$$

$$X_{BB} = \frac{\zeta_B^2 \phi_B^{2(z-1)} e^{-\varepsilon_{BB}/k_B T}}{\Xi_1'^{(2)}}. \tag{13.83}$$

Making use of these relations one finds (Bhatia and Singh, 1982)

$$X_{AB} = \frac{2c(1-c)}{\beta + 1} \tag{13.84}$$

where β is defined in (13.61). Equation (13.84) and the relation $X_{AB} = c(1-c)(1-\alpha_1)$ (see, for example, Singh, 1987) immediately yield

Figure 13.3. Plot of short-range order parameter α_1 versus concentration for regular solutions for different values of $w/k_B T$ (after Singh, 1987).

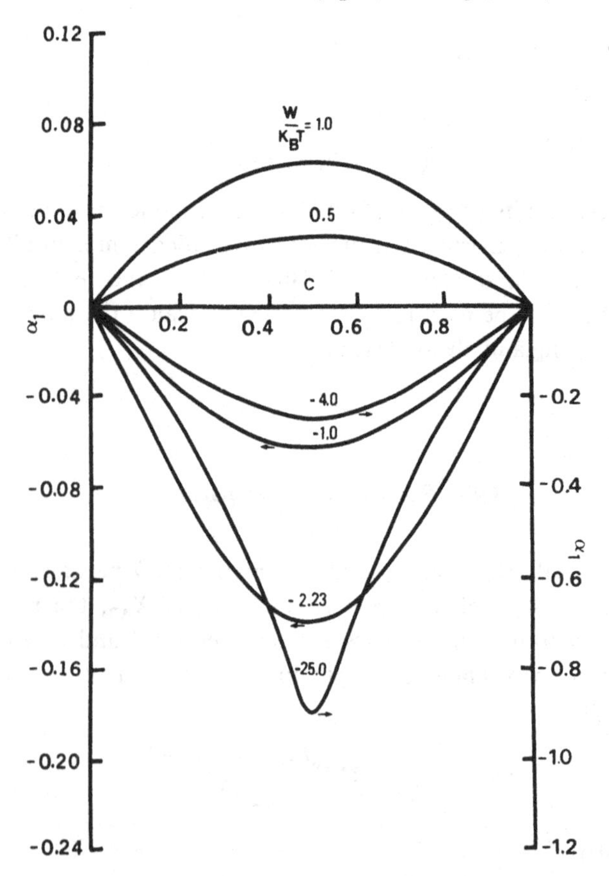

$$\alpha_1 = \frac{\beta - 1}{\beta + 1}. \tag{13.85}$$

Thus one is led to an expression for the short-range-order parameter α_1 that may be directly evaluated from knowledge of the coordination number z and the (dimensionless measure of) interchange energy $w/k_B T$.

This expression for α_1 as a function of concentration is plotted for different values of $w/k_B T$ in Figure 13.3. As for $S_{CC}(0)$ for regular solutions, α_1 is symmetrical about $c = \frac{1}{2}$. However, unlike $S_{CC}(0)$, the peak always lies at $c = \frac{1}{2}$. Singh (1987) gives the example of the CdMg system as illustrative of the use of the preceding treatment (see his Figures 4-6), but further detail is not pursued here. Rather, a brief summary will be given of an approach based on a model of complex formation, which can be useful when regular solution theory fails.

13.10. Complex formation as model for $S_{CC}(0)$

When such compound-forming molten alloys are considered, it can be assumed that they are strongly interacting systems. Then, in contrast to the regular (or conformal: see Appendix 13.1) solution models, G_M^{xs} and $S_{CC}(0)$ are often found to be asymmetric about $c = \frac{1}{2}$, even though the constituent species do not differ greatly in size. Singh (1987) cites a number of alloys systems including MgBi and AgAl, which are considered briefly next.

The fact that various alloys, including the preceding examples, form compounds in the solid state at one or more stoichiometric compositions led Bhatia and coworkers (Bhatia, Hargrove, and Thornton, 1974; see also Gerling, Pool, and Predel, 1983; Singh, 1987) to assume the existence of chemical complexes $A_\mu B_\nu$ in the liquid state. Here A and B denote the constituent atoms, while μ and ν are their respective numbers in the complex. The binary alloy therefore consists of a mixture of A atoms, B atoms, and a number of complexes $A_\mu B_\nu$. Bhatia and Singh (1982; see also Singh, 1987), while developing a quasi-chemical model for such systems, further assumed that the energy of an AB, AA, or BB bond depends on whether that bond is part of a complex or not. A brief account of the way $S_{CC}(0)$ is then calculated, together with the main result, is given in Appendix 13.2. The result for $S_{CC}(0)$ in (A13.2.12) has been applied to the MgBi alloy, where the chemical complexes are Mg_3Bi_2 ($\mu = 3, \nu = 2$). The results are shown in Figure 13.4, taken from Singh (1987). The model gives a good account of the data, though, of course, it contains adjustable parameters.

Figure 13.4. Concentration fluctuations $S_{CC}(0)$ and short-range order parameter α_1 for a MgBi melt at 975 K. Dashed curves are ideal solution theory. The dot is from Boos and Steeb (1977) (after Singh, 1987).

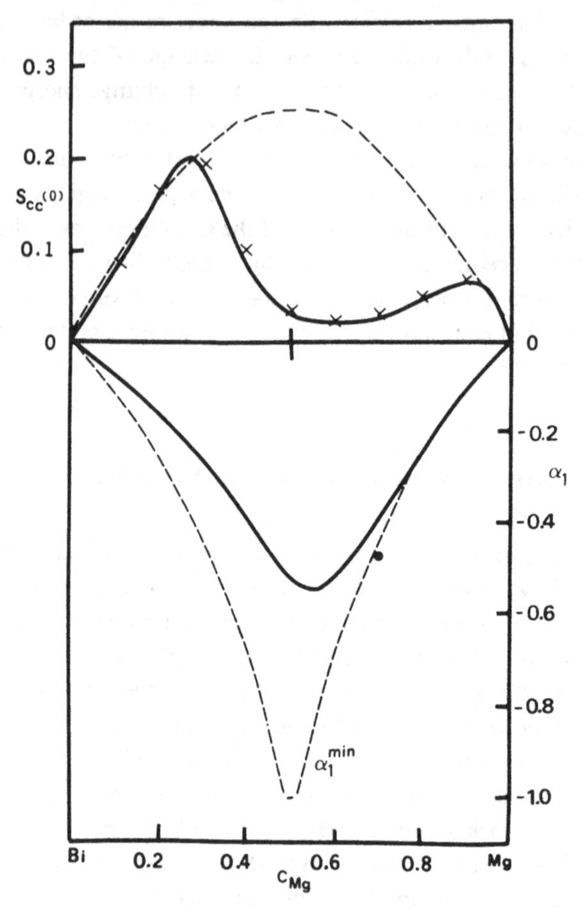

For weakly interacting compound-forming molten alloys, the quasi-chemical expressions can be simplified considerably (see Appendix 13.2). The expression $S_{CC}(0)$ thus obtained which is recorded in (A13.2.18), has been used for the AgAl alloy in Figure 13.5, taken from Singh (1987). Again, the model is clearly a useful one. Bhatia and Singh (1984) have also developed a quasi-lattice theory. This is of interest in that it can handle successfully strongly interacting compound-forming alloys (see Singh, 1987; Figure 9), but further detail will not be given.

Finally, toward the end of Appendix 13.7, a discussion is included about the role of the concentration fluctuations $S_{CC}(0)$ in metallic glass formation.

Figure 13.5. Shows concentration fluctuations $S_{CC}(0)$ and short-range order parameter α_1 as function of concentration for an AgAl melt at 1173 K. The dashed curve represents ideal solution theory (after Singh, 1987).

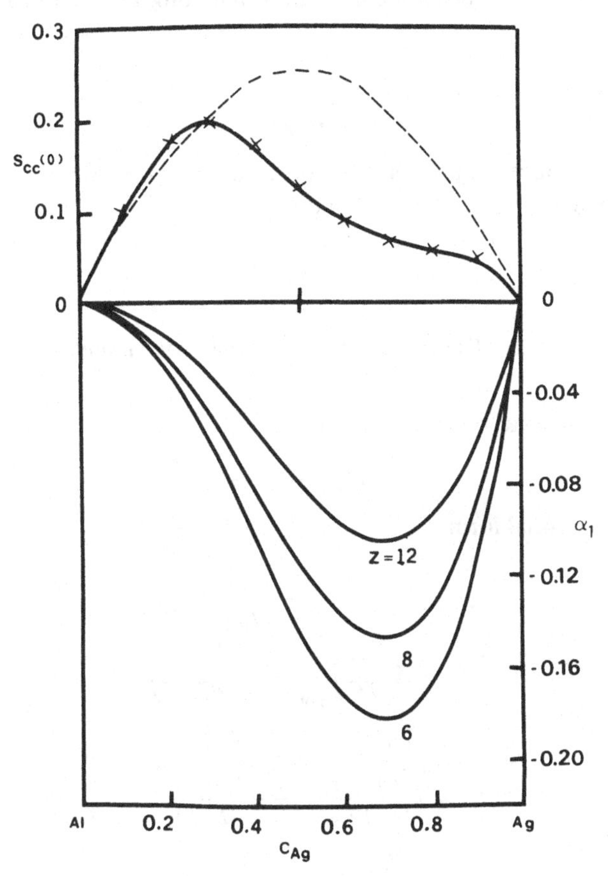

13.11. Phase diagrams

The use of the concentration fluctuations in the calculation of phase diagrams has been formulated by Bhatia and March (1972; 1975). The following discussion mainly follows the subsequent work of Geerstma (1985); however as an introduction let us consider the approach of Bhatia and March (1972) in relation to the eutectic mixture of liquid Na-K.

Returning to the discussion of Section 13.8, as an application of regular conformal-solution theory (compare Appendix 13.1) an outline will now be given of the calculation of the phase diagram of such liquid Na-K alloys (Bhatia and March, 1972).

The liquidus curve of an ideal solution has long been known to have the form

$$\ln(1 - c_2) = \left(\frac{L_{10}}{R}\right)(T_1^{-1} - T^{-1}) \tag{13.86}$$

c_2 being the concentration of element 2 and L_{10} the latent heat at the freezing temperature T_1 of pure liquid 1.

13.11.1. Thermodynamic equations along liquidus

In terms of chemical potentials

$$\mu_{10}^s(\mathrm{T}) = \mu_1(T, c_2) \tag{13.87}$$

with the differential form

$$\frac{\Delta T}{\Delta c_2} = -\frac{\left(\dfrac{\partial \mu_1}{\partial c_2}\right)_{\mathrm{p,T}}}{\left(\dfrac{\partial \mu_1}{\partial T}\right)_{c_2,\mathrm{p}} - \left(\dfrac{\partial \mu_{10}^s(T)}{\partial T}\right)_{\mathrm{p}}}$$

$$= \frac{\left(\dfrac{\partial \mu_1}{\partial c_2}\right)_{\mathrm{p,T}}}{(L/T)} = -\frac{c_2\left(\dfrac{\partial^2 G}{\partial c_2^2}\right)_{\mathrm{p,T}}}{(L/T)}$$

$$= -\frac{RT^2 c_2}{S_{\mathrm{cc}}(0)L} \tag{13.88}$$

where the subscript zero refers to a pure substance and superscript s denotes solid. Hence it can be seen that the liquidus curve depends crucially on the concentration fluctuations $(\Delta c)^2$. In the preceding equation for the slope of the liquidus curve, L is a generalized concentration-dependent latent heat defined by

$$\frac{L}{T} = \frac{L_{10}}{T_1} + \int_{T_1}^{T} \frac{\Delta c_{\mathrm{p10}}}{T}\,\mathrm{d}T - \left(\frac{\partial}{\partial T}[RT\ln(\gamma_1\{1 - c_2\})]\right)_{c_1,c_2,\mathrm{p}}, \tag{13.89}$$

γ_1 being the activity, whereas $c_{\mathrm{p10}}^L - c_{\mathrm{p10}}^S = \Delta c_{\mathrm{p10}}$.

If one now expands Δc_{p10} around T_1 and neglects higher terms than the first, calculating the activity γ_1 from conformal-solution theory ($RT \ln \gamma_1 \simeq wc^2$), then the equation of the liquidus curve is obtained in the form

$$c(t) - 1 = (c_{\text{ideal}} - 1)\exp\left(-\frac{Wc^2}{t}\right) \tag{13.90}$$

where $t = T/T_1$ and $W = w/RT_1$. If the value $w/RT_1 = 1.1$ is used (Bhatia, Hargrove, and March, 1973), the liquidus curves of the eutectic Na-K mixture are found to be in good agreement with experiment, as shown in Figure 13.6.

It is of interest to add a comment here on the Na-Cs system. The liquidus curves of this alloy are described usefully by Flory's (1942) model (see Bhatia and March, 1975). The slope of the liquidus curve is given by (13.88), as seen earlier, so it is evident that a rather flat portion of the liquidus curve will be found where there is a huge peak in $S_{CC}(0)$ as in Figure 13.7 (see also Neale and Cusack, 1982, and Singh and Bhatia, 1984); this is observed near $c = 0.75$–0.8.

Figure 13.6. Liquidus curves of the eutectic mixture, as calculated from conformal solution theory by Bhatia and March (1972), compared with experimental curves.

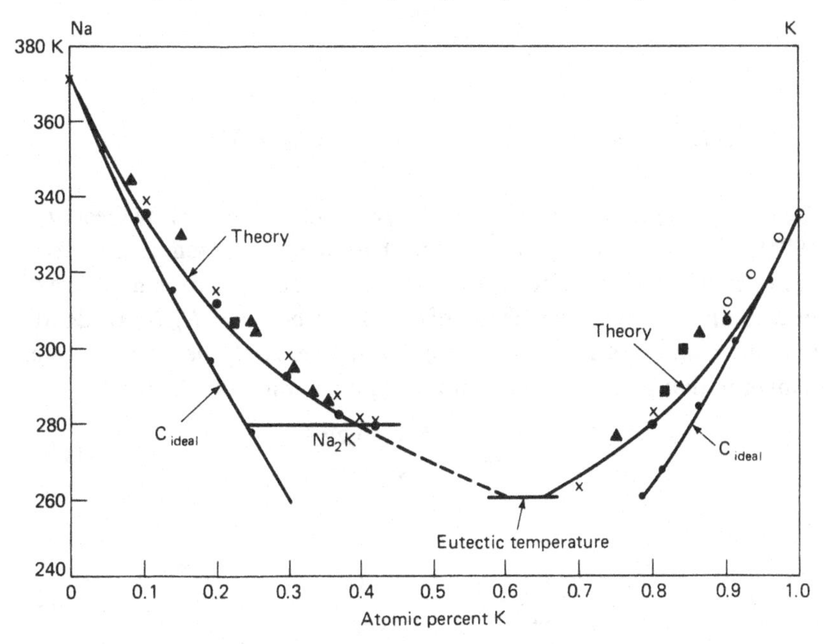

Figure 13.7. Concentration fluctuations $S_{CC}(0)$ versus concentration in a liquid Na-Cs alloy. Experimental results are from Ichikawa, Granstaff and Thompson (1974) whereas the theoretical curve is due to Bhatia and March (1975).

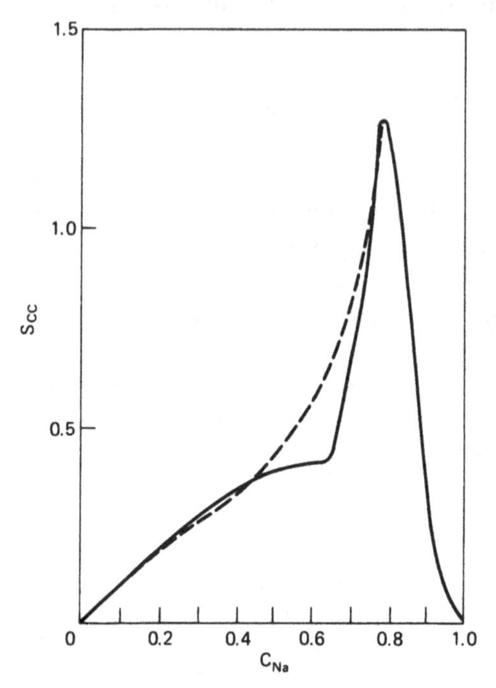

13.11.2. *More general properties, including melting extrema*

The equations given by Bhatia and March (1975) were used by Geertsma (1985) to describe some properties of phase diagrams. Denoting the two phases considered by I and II, the A-B alloy is taken to have fractions of the B component in the two phases of c^{I} and c^{II}, where $c^{\mathrm{I}} = N_{\mathrm{B}}^{\mathrm{I}}/N^{\mathrm{I}}$. Evidently $N_{\mathrm{B}}^{\mathrm{I}}$ is the number of B particles in phase I, whereas N^{I} denotes the total number in that phase. Then the preceding equations take the form

$$\frac{\mathrm{d}T}{\mathrm{d}c^{\mathrm{I}}} = -\frac{(c^{\mathrm{I}} - c^{\mathrm{II}})RT^2}{S_{\mathrm{CC}}^{\mathrm{I}} L^{\mathrm{II}}} \tag{13.91}$$

and

$$\frac{\mathrm{d}T}{\mathrm{d}c^{\mathrm{II}}} = -\frac{(c^{\mathrm{I}} - c^{\mathrm{II}})RT^2}{S_{\mathrm{CC}}^{\mathrm{II}} L^{\mathrm{I}}}, \tag{13.92}$$

these serving to determine the phase boundaries in the temperature-concentration (T-c) plane. In these equations

$$L^{\mathrm{I}} = (1 - c^{\mathrm{I}})L_{\mathrm{A}} + c^{\mathrm{I}}L_{\mathrm{B}} \tag{13.93}$$

and

$$L^{\mathrm{II}} = (1 - c^{\mathrm{II}})L_{\mathrm{A}} + c^{\mathrm{II}}L_{\mathrm{B}}, \tag{13.94}$$

whereas

$$\frac{L_{\mathrm{A}}}{T} = \left[\frac{\partial \mu_{\mathrm{A}}^{\mathrm{II}}(T, c^{\mathrm{II}})}{\partial T}\right]_{c^{\mathrm{II}}, p} - \left[\frac{\partial \mu_{\mathrm{A}}^{\mathrm{I}}(T, c^{\mathrm{I}})}{\partial T}\right]_{c^{\mathrm{I}}, p}. \tag{13.95}$$

Further, μ is the chemical potential as usual: $\mu_{\mathrm{A}}(T, c) = \mu_{0\mathrm{A}}(T) + RT \ln a_{\mathrm{A}}(T, c)$, where $\mu_{0\mathrm{A}}(T)$ is the chemical potential of pure component A. a_{A} above has now been written for the activity of component A. L_{A} may be interpreted as a generalized latent heat. Finally, the concentration fluctuations S_{CC} can be written as $(1\text{-}c)/[(\partial \ln a_{\mathrm{B}})/\partial c]_{p,T}$.

Geertsma discusses, on the basis of the preceding equations, some general properties of phase diagrams with melting extrema (see Figure 13.8a); Equation (13.91) describes the liquidus, and (13.92) the solidus ($I \to l$, $II \to s$). A melting extremum is characterized by $\mathrm{d}T/\mathrm{d}c^l = 0$; as $c^l = c^s$, so $\mathrm{d}T/\mathrm{d}c^s = 0$; i.e., the solidus and liquidus touch at the melting extremum denoted by T_0, c_0.

For such a melting extremum, one can derive, from (13.91) and (13.92):

$$\frac{\mathrm{d}^2 T}{\mathrm{d}c^{l2}} = -RT^2 \frac{(1 - S_{\mathrm{CC}}^s L^l / S_{\mathrm{CC}}^l L^s)}{(S_{\mathrm{CC}}^l L^s)} \tag{13.96}$$

and

$$\frac{\mathrm{d}^2 T}{\mathrm{d}c^{s2}} = RT^2 \frac{(1 - S_{\mathrm{CC}}^l L^s / S_{\mathrm{CC}}^s L^l)}{(S_{\mathrm{CC}}^s L^l)}. \tag{13.97}$$

These yield

$$\frac{\mathrm{d}^2 T/\mathrm{d}c^{l2}}{\mathrm{d}^2 T/\mathrm{d}c^{s2}} = \left(\frac{S_{\mathrm{CC}}^s}{S_{\mathrm{CC}}^l}\right)^2 \tag{13.98}$$

because at such an extremum $L_0 = L^l = L^s$.

A maximum in the phase diagram evidently occurs for $\mathrm{d}^2 T/\mathrm{d}c^{l2} < 0$ and a minimum for $\mathrm{d}^2 T/\mathrm{d}c^{l2} > 0$, when $L_0 > 0$, as is usually the case. This implies ((13.96) and (13.97)) that a maximum (minimum) occurs for

Figure 13.8. Schematic forms of various types of phase diagrams, following Geertsma (1985). (a) Melting minimum. (b) Eutectic point. (c) Phase separation. (d) A congruently melting component c_c. (e) An uncongruently melting compound c_{c_2} (after Geertsma, 1985).

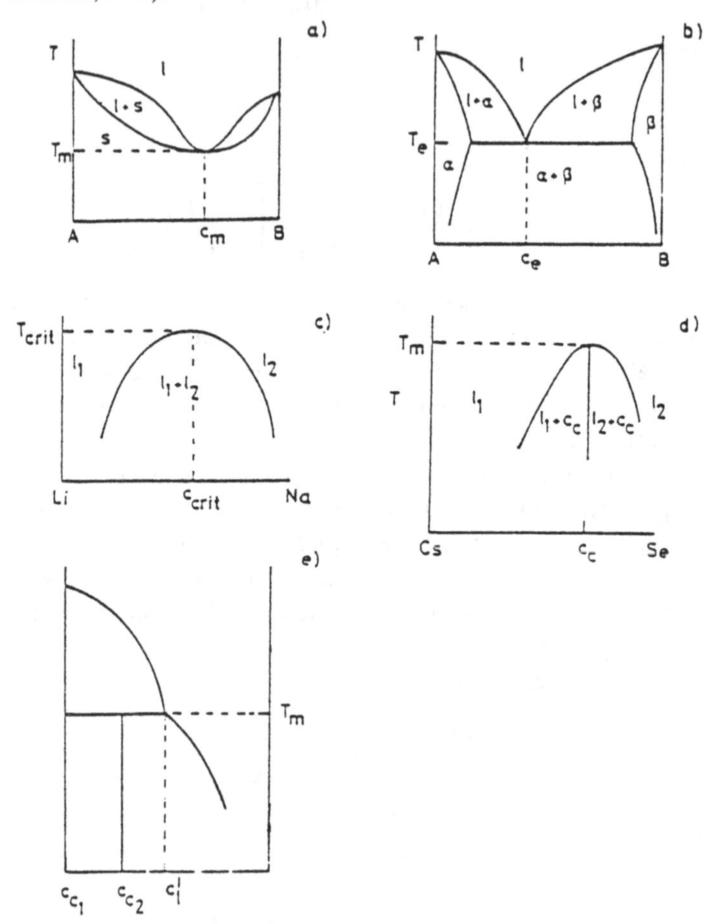

$$S_{CC}^{s} < S_{CC}^{l} (S_{CC}^{s} > S_{CC}^{l}) \quad \text{or} \quad \frac{\partial^2 G}{\partial c^{s2}} > \frac{\partial^2 G}{\partial c^{l2}} \left(\frac{\partial^2 G}{\partial c^{s2}} < \frac{\partial^2 G}{\partial c^{l2}} \right)$$

with G, as usual, the Gibbs free energy.

The last inequalities are usually derived from graphical considerations using G-c curves; the phase diagram has a melting minimum (maximum) when the curvature of the solid G curve is smaller (larger) than the curvature of the liquid G curve. In terms of S_{CC}: when the concentration fluctuations in the solid are larger (smaller) than in the liquid, then there will be a melting point minimum (maximum). From (13.98) it can be seen that the curvatures of the liquidus and the solidus have the same sign around the extremum.

One can derive (Geertsma, 1985) from (13.98), for small deviations around c_0:

$$\frac{c_0 - c^s}{c_0 - c^l} = \pm \frac{S^s_{CC}}{S^l_{CC}}. \tag{13.99}$$

This equation is rather accurate near a melting extremum. Equation (13.99) can also be derived by Taylor or series expansion of the solidus $T^s(c^s)$ and the liquidus $T^l(c_l)$ around the point T_0, c_0.

Geertsma (1985) gives a somewhat more comprehensive discussion of the eutectic diagram Figure 13.8b than that just presented, as well as a treatment of the other classes of diagram in Figure 13.8 (see also Bhatia and March, 1975), but the reader should refer to the original papers for further details.

13.12. Vacancies and melting curve in binary alloys

The argument of Section 6.6, in which the vacancy formation energy of a solid was related to the melting temperature, has been generalized by March and Tosi (1986) to metallic alloys and mixtures. Their argument and the main consequences of it will now be outlined.

Their starting point is that the total potential energy of the ionic array in a metal alloy can be written (see Section 13.2) as

$$E = Nu(\rho) + \tfrac{1}{2} \sum_{i_\alpha \neq j_\beta} \phi_{\alpha\beta}(|\mathbf{r}_{i_\alpha} - \mathbf{r}_{j_\beta}|) \tag{13.100}$$

where N is the total number of atoms in the system of volume V, $\rho = N/V$ is the average number density, and $\phi_{\alpha\beta}(|\mathbf{r}_{i_\alpha} - \mathbf{r}_{j_\beta}|)$ is the pair potential between the ith particle of type α and the jth particle of type β.

The liquid pair-distribution functions $g_{\alpha\beta}$ are now introduced: In other words, one will be interested in the hot binary systems near melting. Of course, in this step, it is implied that the theory is restricted to those systems where there is no major change in local coordination on melting, as is the case for close-packed metals. For the expectation value of the total potential energy per particle, this gives the result

$$\frac{\bar{E}}{N} = u(\rho) + \tfrac{1}{2} \sum_{\alpha\beta} \rho_\alpha \rho_\beta \int g_{\alpha\beta}(r)\phi_{\alpha\beta}(r)\, d\mathbf{r} \tag{13.101}$$

where, in metallic alloys, not only u but also the pair potentials $\phi_{\alpha\beta}$ are density-dependent.

As for pure metal, one considers (see Section 6.8.1) the creation of the vacancy in two stages. In step 1, with associated energy $E_1^{(1)}$, say, the volume of the metal is expanded by the partial volume Ω_1 to yield

$$E_1^{(1)} = -\Omega_1\left[\rho^2\frac{\partial u}{\partial\rho} + \tfrac{1}{2}\sum_{\alpha\beta}\rho_\alpha\rho_\beta\int g_{\alpha\beta}(r)\rho\frac{\partial\phi_{\alpha\beta}}{\partial\rho}(r)\,\mathrm{d}\mathbf{r}\right]. \qquad (13.102)$$

No term in $\partial g/\partial\rho$ appears in (13.102) because in this first step the ions are specifically held fixed.

Turning to the second step, i.e., the removal of an ion from the vacancy site to its position on the surface, the corresponding energy may be written

$$E_1^{(2)} = -\tfrac{1}{2}\sum_\alpha\rho_\alpha\int g_{1\alpha}(r)\phi_{1\alpha}(r)\,\mathrm{d}\mathbf{r}, \qquad (13.103)$$

the vacancy formation energy $E_{\mathrm{V}1}$ for an atom of type 1 then being

$$E_{\mathrm{v}1} = E_1^{(1)} + E_1^{(2)}. \qquad (13.104)$$

Equations (13.102)–(13.104) constitute a solution of the problem. However, following March and Tosi (1986), it is again useful to invoke the thermodynamic expression for the pressure, as in the pure metal case (see Minchin et al. (1974) for that limit.)

13.12.1. Introduction of pressure

In a binary alloy, and for the model expressed through (13.100), the pressure p can be written in the form (compare Section 13.5) in terms of $g_{\alpha\beta}$ and $\phi_{\alpha\beta}$ as

$$p = \rho^2\frac{\mathrm{d}u}{\mathrm{d}\rho} - \tfrac{1}{2}\sum_{\alpha\beta}\rho_\alpha\rho_\beta\int g_{\alpha\beta}(r)\times\left[\frac{1}{3}r\frac{\partial\phi_{\alpha\beta}(r)}{\partial r} - \rho\frac{\partial\phi_{\alpha\beta}}{\partial\rho}\right]. \qquad (13.105)$$

This expression then allows the calculation of the formation enthalpy of the vacancy, namely, $E_{\mathrm{v}1} + p\Omega_1$, as

$$E_{\mathrm{v}1} + p\Omega_1 = -\tfrac{1}{2}\sum_\alpha\rho_\alpha\int g_{1\alpha}(r)\phi_{1\alpha}(r)\,\mathrm{d}\mathbf{r} - \tfrac{1}{6}\Omega_1\sum_{\alpha\beta}\rho_\alpha\rho_\beta\int g_{\alpha\beta}(r)r\frac{\partial\phi_{\alpha\beta}(r)}{\partial r}\,\mathrm{d}\mathbf{r}. \qquad (13.106)$$

This expression (13.106) is applicable to both metallic binary alloys and insulating mixtures.

13.12.2. Use of a model obeying the mean spherical approximation

To get the gist of the essential new features that come in for the binary system, the drastic mean spherical approximation (MSA) will be made for the liquid structure (see Appendix 6.1 for a monatomic crystalline model of vacancy in terms of liquid structure using MSA).

13.12.3. Model using MSA

The MSA means that one is to insert in the first term on the right-hand side of (13.106) the results

$$\phi_{\alpha\beta}(r) = -k_B T c_{\alpha\beta}(r) \tag{13.107}$$

outside the cores, this approximation being valid, since $g_{\alpha\beta}(r)$ enters the integrand. The $c_{\alpha\beta}$'s are the partial Ornstein-Zernike direct correlation functions related to $h_{\alpha\beta}(r) = g_{\alpha\beta}(r) - 1$ by the Pearson-Rushbrooke (1957) relations

$$h_{\alpha\beta}(r) = c_{\alpha\beta}(r) + \sum_{\gamma} \rho_{\gamma} \int \mathrm{d}\mathbf{r}'\, h_{\alpha\gamma}(|\mathbf{r} - \mathbf{r}'|) c_{\gamma\beta}(r')\, \mathrm{d}\mathbf{r}'. \tag{13.108}$$

Then one readily obtains

$$\begin{aligned} &-\tfrac{1}{2} \sum_{\alpha} \rho_{\alpha} \int g_{1\alpha}(r)\phi_{1\alpha}(r)\, \mathrm{d}\mathbf{r} \\ &= \tfrac{1}{2} k_B T \left[\sum_{\alpha} \rho_{\alpha} \int \mathrm{d}\mathbf{r}\, h_{1\alpha}(r) c_{1\alpha}(r) + \sum_{\alpha} \rho_{\alpha} \int \mathrm{d}\mathbf{r}\, c_{1\alpha}(\mathbf{r}) \right], \end{aligned} \tag{13.109}$$

where T is the temperature of the system, which is conveniently taken at the melting of the mixture. Using the definition of the Fourier transform,

$$\tilde{c}_{\alpha\beta}(k) = (\rho_{\alpha}\rho_{\beta})^{1/2} \int \mathrm{d}\mathbf{r}\, c_{\alpha\beta}(r), \tag{13.110}$$

and utilizing (13.108) at $r = 0$, where $h_{\alpha\beta}(r = 0) = -1$, one can rewrite (13.109) as

$$\begin{aligned} -\tfrac{1}{2} \sum_{\alpha} \rho_{\alpha} \int g_{1\alpha}(r)\phi_{1\alpha}(r)\, \mathrm{d}\mathbf{r} &= -\tfrac{1}{2} k_B T [1 + c_{11}(r = 0)] \\ &\quad + \tfrac{1}{2} k_B T \sum_{\alpha} \left(\frac{\rho_{\alpha}}{\rho_1} \right)^{1/2} \tilde{c}_{1\alpha}(k = 0). \end{aligned} \tag{13.111}$$

This has achieved the purpose of replacing the potentials in this term by quantities that are accessible from diffraction and thermodynamic data.

The other term on the right-hand side of (13.106) can be handled by using the virial expression for the pressure, namely,

$$p = \rho k_B T - \tfrac{1}{6} \sum_{\alpha\beta} \rho_\alpha \rho_\beta \int g_{\alpha\beta}(r) r \frac{\partial \phi_{\alpha\beta}(r)}{\partial r} \, d\mathbf{r}, \tag{13.112}$$

which then from (13.111), (13.112), and (13.106) yields

$$E_{v1} = -\Omega_1 \rho k_B T - \tfrac{1}{2} k_B T [1 + c_{11}(r=0)] + \tfrac{1}{2} k_B T \sum_\alpha \left(\frac{\rho_\alpha}{\rho_1} \right)^{1/2} \tilde{c}_{1\alpha}(k=0). \tag{13.113}$$

for the vacancy formation energy. Although this formula can, as mentioned before, be evaluated using diffraction and thermodynamic data, it will be developed next in terms of the difference in partial volumes of the two components. From experience with the one-component case (Bhatia and March, 1984a; Bernasconi and March, 1986), the appropriate model to use to evaluate $c_{11}(r = 0)$ if one wishes to avoid the use of diffraction data is the Percus-Yevick solution for a binary mixture of neutral hard spheres (see Appendix 3.1 for the one-component Percus-Yevick theory).

13.12.4. Introduction of thermodynamic quantities

First, the last term in (13.113) will be rewritten in terms of the quantity (Bhatia and Thornton, 1970)

$$\delta = \frac{1}{V} \left(\frac{\partial V}{\partial c_2} \right)_{p,T,N} \tag{13.114}$$

where c_2 is the concentration of component 2.

The way to relate $\tilde{c}_{1\alpha}(k = 0)$ first to the partial structure factors of the mixture and then to δ and the isothermal compressibility K_T of the mixture is described in the book by March and Tosi (1976). The result thereby obtained is

$$\sum_\alpha \left(\frac{\rho_\alpha}{\rho_1} \right)^{1/2} \tilde{c}_{1\alpha}(k=0) = 1 - \frac{1 + c_2 \delta}{\rho k_B T K_T}. \tag{13.115}$$

Hence (13.113) becomes

$$\frac{E_{v1}}{k_B T} = -\rho\Omega_1 - \tfrac{1}{2}c_{11}(r=0) - \frac{1+c_2\delta}{2\rho k_B T}. \tag{13.116}$$

This reduces to the result of Bhatia and March (1984a; see also Appendix 6.1) in the limit $c_2 \to 0$.

13.12.5. *Use of Percus-Yevick binary hard sphere model*

Although (13.116) is calculable from experimental data, it is worthwhile using a hard sphere model for evaluation of $c_{11}(r=0)$. This can be found from the work of Lebowitz (1964) as

$$c_{11}(r=0) = -\frac{\Omega_1}{k_B T K_T^{hs}} \tag{13.117}$$

where it must be emphasized that K_T^{hs}, the Percus-Yevick hard sphere compressibility, must not be equated with the mixture compressibility K_T, which crucially involves attractive forces. Rather, K_T^{hs} must be evaluated in terms of hard sphere diameters σ_1 and σ_2 from the equation of state of the mixture given by Lebowitz (1964).

Hence, substituting (13.117) in (13.116), one finds

$$\frac{E_{v1}}{k_B T} = -\rho\Omega_1\left[1 - \frac{1}{2\rho k_B T K_T^{hs}}\right] - \frac{1+c_2\delta}{2\rho k_B T K_T}. \tag{13.118}$$

It is to be stressed that (13.118) expresses the concentration dependence of the vacancy formation energy E_{v1} in units of $k_B T$ in terms of the compressibility K_T of the mixture and the size factors involved both through δ and through the hard sphere compressibility. Although the work of Bernasconi, March, and Tosi (1986) cautions against using the MSA in metals, some evidence in favour of the size effect displayed in (13.118) has been emphasized by the writer (March, 1987e).

13.13. Theory of freezing

In this section, based on the work of March and Tosi ((1981a, b, c); see also March, 1985) the freezing of two-component systems will be discussed. This can be viewed as a generalization of the theory of freezing of pure liquid metals set up in Chapter 6. Their starting point is the set of equations relating the singlet densities ρ_i of a two-component system to the three partial direct correlation functions c_{ij}. These equations are derived in Appendix 13.3 for ν components. Following the work of Lovett (1977), one now integrates (A13.3.12) under the assumption that $c_{ij}(\mathbf{r}_1, \mathbf{r})$ depends only on $|\mathbf{r}_1 - \mathbf{r}|$, as in a bulk liquid. The result is then (March and Tosi, 1981a)

$$\ln \rho_i(\mathbf{r}_1) = \sum_{j=1}^{\nu} \int \mathrm{d}\mathbf{r}\, c_{ij}(|\mathbf{r}_1 - \mathbf{r}|)\rho_j(\mathbf{r}) + A_i \tag{13.119}$$

where A_i is a constant of integration.

The question one now poses is whether, given the liquid partial direct correlation functions c_{ij}, (13.119) can exhibit a periodic solution, such as $\rho_{ip}(\mathbf{r})$, in coexistence with the obvious liquid homogeneous solution ρ_{il}, when the constant A_i is the same in the two phases. To answer this question, it is helpful to construct the difference between (13.119) applied to periodic and to homogeneous one-particle (singlet) densities in turn. Regarding the result

$$\frac{\ln \rho_{ip}(\mathbf{r}_1)}{\rho_{il}} = \sum_{j=1}^{\nu} \int \mathrm{d}\mathbf{r}\, c_{ij}(|\mathbf{r}_1 - \mathbf{r}|)[\rho_{jp}(\mathbf{r}) - \rho_{jl}] \tag{13.120}$$

as the Euler equation of a variational problem again, one constructs the difference $\Delta\Omega$ in thermodynamic potential, Ω being now specifically $F - \sum_i N_i\mu_i$, with F the Helmholtz free energy, N_i the number of particles, and μ_i the chemical potential, both of species i. Using the separation into free-particle terms and those arising from the interparticle interactions reflected in c_{ij}, one obtains (see (6.6))

$$\frac{\Delta\Omega}{k_B T} = \sum_{i=1}^{\nu} \int \mathrm{d}\tau \left\{\rho_{ip}(\mathbf{r}) \ln \left|\frac{\rho_{ip}(\mathbf{r})}{\rho_{il}}\right| - [\rho_{ip}(\mathbf{r}) - \rho_{il}]\right\}$$
$$- \tfrac{1}{2} \sum_{i,j=1}^{\nu} \iint \mathrm{d}\tau_1\, \mathrm{d}\tau_2 [\rho_{ip}(\mathbf{r}_1) - \rho_{il}] c_{ij}(|\mathbf{r}_1 - \mathbf{r}_2|)[\rho_{jp}(\mathbf{r}_2) - \rho_{jl}]. \tag{13.121}$$

Varying this expression for $\Delta\Omega$ with respect to the periodic singlet densities, the correct Euler equation is regained.

One now uses (13.120) to remove $\ln[\rho_{ip}(\mathbf{r})/\rho_{il}]$ from (13.121), and one finds

$$\frac{\Delta\Omega_{\min}}{k_B T} = -\sum_{i=1}^{\nu} \int d\tau [\rho_{ip}(\mathbf{r}) - \rho_{il}] + \frac{1}{2} \sum_{i,j=1}^{\nu} \iint d\tau_1 \, d\tau_2$$
$$\times [\rho_{ip}(\mathbf{r}_1) + \rho_{il}] c_{ij}(|\mathbf{r}_1 - \mathbf{r}_2|) [\rho_{jp}(\mathbf{r}_2) - \rho_{jl}]. \quad (13.122)$$

This quantity should vanish at the coexistence point at which the periodic singlet densities are to be found from (13.120).

Though this theory has been applied to alkali halides as well as to the freezing of materials such as $BaCl_2$ into a fast-ion conducting state (see March, 1985, for a review), it has not, to the writer's knowledge, been applied numerically to binary metallic alloys to date.

13.14. Surface segregation

The same basic approach, using the equations of the statistical mechanics of inhomogeneous systems given in Appendix 13.3, can be applied, at least in principle, to the problem of surface segregation. In fact, the following approach, due to Bhatia and March (1978), is to generalize the phenomenology of Cahn and Hilliard, set out in Chapter 12, to treat binary alloys. Then, in Appendix 13.4, the way in which this phenomenology can be related to the equations for the statistical mechanics of inhomogeneous systems will be explained.

It should be noted here that Gibbs (see Adamson, 1960, p.73) long ago pointed out using thermodynamics that the excess surface concentration is related to the dependence of surface tension σ on concentration c, so that the problem of surface segregation is that of finding $d\sigma/dc$.

Again following the Cahn-Hilliard approach (see Section 12.5), one considers a fluctuation Δn in the number density n, occurring in a volume V, say. Then from fluctuation theory one associates with it a free-energy contribution F_1 given by (Section 12.5)

$$F_1 = \left(\frac{AV}{K_T}\right)\left(\frac{\Delta n}{n}\right)^2$$

where A is equal to $\frac{1}{2}$ for "accidental" fluctuations. For the generalization to binary systems affected below, it is useful at this point to write F_1 in an alternative, but equivalent, form:

$$F_1 = A\left(\frac{\partial^2 F}{\partial N^2}\right)_{T,V} (\Delta N^2).$$

This expression would lead to a contribution σ_1 to the surface energy: $\sigma_1 = Al(\Delta n)^2/n^2 K_T$. Minimization of σ_1 with respect to the thickness l would yield the manifestly erroneous result that $l = 0$. Therefore, in the Cahn-Hilliard phenomenology, one adds a term to the free energy which is proportional to the square of the density gradient (see, for example, Bhatia and March, 1979)

$$F_2 = BV(\Delta n/l)^2$$

with a corresponding contribution $\sigma_2 = Bl/(\Delta n/l)^2$ to the surface energy.

13.14.1. Generalization to two components

For two components, the above expression for F_1 can be generalized to read

$$F_1 = A\left[\left(\frac{\partial^2 F}{\partial N_1^2}\right)_{T,V,N_2}(\Delta N_1)^2 + \left(\frac{\partial^2 F}{\partial N_2^2}\right)(\Delta N_2)^2 + 2\left(\frac{\partial^2 F}{\partial N_1 \partial N_2}\right)\Delta N_1 \Delta N_2\right], \tag{13.123}$$

N_1 and N_2 being the number of particles of types 1 and 2, respectively. ΔN_1 and ΔN_2 measure fluctuations in them. Next, F_2 above is similarly generalized to read

$$F_2 = \left(\frac{1}{l^2}\right)[b_{11}(\Delta N_1)^2 + 2b_{12}\Delta N_1 \Delta N_2 + b_{22}(\Delta N_2)^2]. \tag{13.124}$$

In (13.123), by analogy with the one-component case, $A = \frac{1}{2}$ for fluctuation theory. Hence

$$\sigma = \left(\frac{1}{V}\right)\left[AXl + \left(\frac{1}{l}\right)Y\right] \tag{13.125}$$

where X and Y denote the contents of the brackets in the preceding equations for F_1 and F_2, respectively. As before, the minimum $d\sigma/dl = 0$ yields

$$l^2 = \frac{Y}{XA}, \tag{13.126}$$

or

$$\sigma = 2Al\left(\frac{X}{V}\right). \tag{13.127}$$

Following Bhatia and March (1978), one now obtains

$$\frac{X}{V} = \frac{1}{K_T N^2}[\Delta N + N(\Delta c)\delta]^2 + \frac{1}{V}\left(\frac{\partial^2 G}{\partial c^2}\right)_{T,p,N}(\Delta c)^2 \tag{13.128}$$

where G is the Gibbs free energy for volume V, K_T is the alloy isothermal compressibility, i.e.,

$$K_T = -\frac{1}{V}\left(\frac{\partial V}{\partial p}\right)_{T,N,c}, \tag{13.129}$$

while δ is the size factor (see (13.115)):

$$\delta = \frac{1}{V}\left(\frac{\partial V}{\partial c}\right)_{T,p,N} = \frac{v_1 - v_2}{cv_1 + (1-c)v_2} \tag{13.130}$$

where v_1 and v_2 are now written for the partial molar volumes of the two species. Furthermore, in (13.128)

$$\Delta N = \Delta N_1 + \Delta N_2 \tag{13.131}$$

while

$$\Delta c = N^{-1}[(1-c)\Delta N_1 - c\Delta N_2] \tag{13.132}$$

with $N = N_1 + N_2$ and c denoting the concentration of species 1.

Hence, from (13.127) and (13.128), one finds

$$\sigma \sim \frac{2Al}{K_T N^2}\left(\{\Delta N + N(\Delta c)\delta\}^2 + \frac{K_T N^2}{V}\left(\frac{\partial^2 G}{\partial c^2}\right)_{T,p,N}(\Delta c)^2\right). \tag{13.133}$$

As a result of treating a liquid alloy, it can be seen from (13.133) that a specifically new feature arises, namely, the determination of Δc. Clearly, however, in accord with the previous minimization of σ, Δc should be allowed to adjust in order again to minimize the surface free energy. Assuming l constant, one determines Δc from

$$\frac{\partial \sigma}{\partial \Delta c} = 0. \tag{13.134}$$

Using (13.133) in (13.134), it is found that

$$\Delta c = \left(\frac{-\delta/K_T}{(1/V)(\partial^2 G/\partial c^2) + (\delta^2/K_T)}\right)\frac{\Delta N}{N}. \tag{13.135}$$

This result for Δc is not to be confused with the excess surface concentration, which is defined differently. The corresponding minimum value of σ is then found by substituting this value for Δc into (13.133). With A taken again as $\sim\frac{1}{2}$ and $\Delta N \sim N$ following precisely the Cahn-Hilliard work, plus $\Delta N_i \sim N_i$, $i = 1, 2$ ($|\Delta c| \ll 1$), the desired result

$$\sigma \sim \frac{l}{K_T}\left[1 + \frac{\delta^2 S_{CC}(0)}{\rho k_B T K_T}\right]^{-1} \tag{13.136}$$

is obtained. In this formula, $S_{CC}(0)$ has been introduced in place of $\partial^2 G/\partial c^2$ through (see (13.35))

$$S_{CC}(0) = nk_B T\left(\frac{1}{V}\frac{\partial^2 G}{\partial c^2}\right)^{-1}, \tag{13.137}$$

which represents the zero wavenumber limit of the liquid structure factor describing concentration correlations or fluctuations (Bhatia and Thornton, 1970).

Via (13.136), one is provided with an almost quantitative route to σ as a function of concentration from a knowledge of:

1. The isothermal compressibility as a function of concentration c,
2. The concentration fluctuations $S_{CC}(0)$, and
3. The size factor δ.

Choosing as an illustrative example the nonconducting system acetone–chloroform, shown in Figure 11.8 from Adamson (1960), there is no difficulty in explaining these results by means of (13.136); here $\delta^2/\theta \sim 0.25$ and $(\delta^2/\theta)S_{CC}(0)$, with $\theta = nk_B TK_T$, is expected to be negligible at all concentrations.

In view of this, it will be useful of emphasize some of the salient qualitative features predicted by (13.136), which one should be able to bring into contact with experiment.

Qualitative Predictions of Surface-Tension Formula

Let us construct from (13.136) the slope $d\sigma/dc$, when one finds

$$\frac{1}{\sigma}\frac{d\sigma}{dc} \sim \frac{1}{l}\frac{dl}{dc} - \frac{1}{K_T}\frac{dK_T}{dc} - \frac{S_{CC}(0)\dfrac{d}{dc}\left(\dfrac{\delta^2}{\theta}\right) + \dfrac{\delta^2}{\theta}\dfrac{dS_{CC}(0)}{dc}}{\left[1 + \left(\dfrac{\delta^2}{\theta}\right)S_{CC}(0)\right]}. \tag{13.138}$$

It is useful first to examine the limiting cases $c \to 0$ and $c \to 1$ from (13.138). Because the concentration fluctuations $S_{CC}(0) \to c$ as $c \to 0$ and to $(1 - c)$ as

$c \to 1$, one finds

$$\frac{1}{\sigma}\frac{\mathrm{d}\sigma}{\mathrm{d}c} \sim \frac{1}{l}\frac{\mathrm{d}l}{\mathrm{d}c} - \frac{1}{K_T}\frac{\mathrm{d}K_T}{\mathrm{d}c} - \frac{\delta^2}{\theta} \quad \text{at } c = 0 \tag{13.139}$$

and

$$\frac{1}{\sigma}\frac{\mathrm{d}\sigma}{\mathrm{d}c} \sim \frac{1}{l}\frac{\mathrm{d}l}{\mathrm{d}c} - \frac{1}{K_T}\frac{\mathrm{d}K_T}{\mathrm{d}c} + \frac{\delta^2}{\theta} \quad \text{at } c = 1. \tag{13.140}$$

Dilute alkali solutions in mercury. As an example, consider the rather remarkable variation of surface tension in amalgams of the alkali metals. Thus, in Figure 13.9, the results of Pugachevich and Timofeevicheva (1951, 1954) for dilute Na, K and Cs in Hg are reproduced. From the size factor term δ^2/θ, one estimates that for K, $-(1/\sigma)\,\mathrm{d}\sigma/\mathrm{d}c|_{c=0}$ is of the order of 100, while for Cs a value three or four times as large results. It seems evident that the behaviour shown in Figure 13.9 is therefore accounted for in a natural physical manner as due to the size factor appearing in (13.139).

Mg-Sn and Mg-Pb systems. Though somewhat less striking than the preceding example of alkalis in Hg, the systems Mg-Sn and Mg-Pb will next be considered, since there are measurements of surface tension available across the entire concentration range. According to (13.139) and (13.140), even if $\mathrm{d}l/\mathrm{d}c$ and $\mathrm{d}K_T/\mathrm{d}c$ retain the same sign over the whole concentration

Figure 13.9. Variation of surface tension with concentration for amalgams of alkali metals (after Bhatia and March, 1978). This behaviour can be accounted for in a natural way in terms of the size factor δ in the phenomenological formula (13.136).

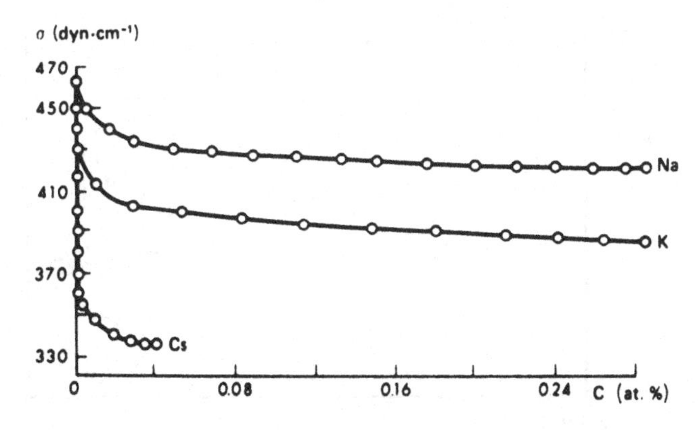

range, which is to be expected physically in many, though not necessarily all, liquid binary alloys, one sees that the size-factor term δ^2/θ could lead to a change in sign from negative to positive as one goes across from $c = 0$ to $c = 1$, and such behaviour is indeed observed in Mg-Sn and Mg-Pb.

Data on surface segregation and their interpretation. Although, as already stressed, the problem of surface segregation in dilute alloys is the same as the problem of the sign of $\mathrm{d}\sigma/\mathrm{d}c|_{c=0}$, where c is the solute concentration, the problem is of sufficient importance in materials science and for aspects of catalysis (see March, 1986) to warrant dwelling a little further on the preceding phenomenological theory in this area.

Equation (13.136) is evidently a suitable basis for the discussion of surface segregation in dilute alloys. It shows immediately that the condition favouring surface segregation is

$$\left.\frac{\mathrm{d}K_\mathrm{T}}{\mathrm{d}c}\right|_{c=0} > 0. \tag{13.141}$$

Figure 13.10. The behaviour of compressibility K_T and surface tension for a liquid binary metallic alloy near pure metal limits $c = 0$ and $c = 1$. Solid lines depict possible behaviour of the slope of K_T^{-1} near the endpoints. Dashed lines indicate predictions as to possible slopes of surface tension as a function of concentration (schematic only).

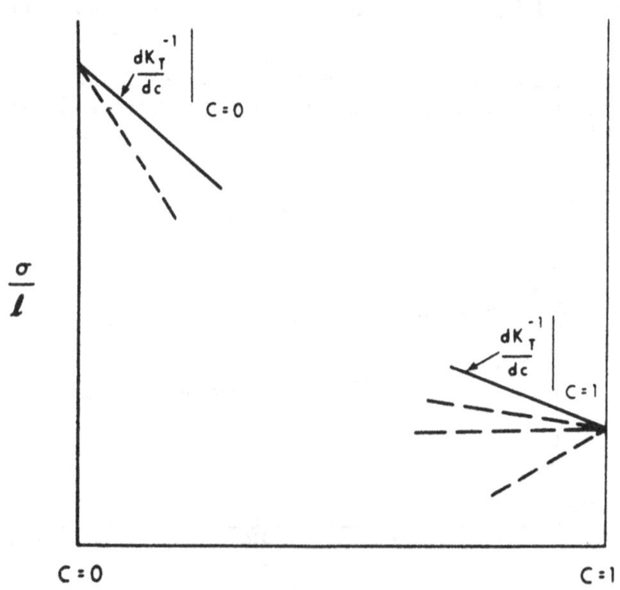

It should be noted that, according to phenomenological theory, this inequality (13.141) is a sufficient condition for surface segregation provided $(\mathrm{d}l/\mathrm{d}c)_{c=0}$ can be taken as zero. Bhatia and March (1978) give arguments that Friedel screening requires this condition for liquid-metal binary alloys with atoms of different valence. The size factor always assists and, hence, as shown schematically in Figure 13.10, makes the slope of $\mathrm{d}\sigma/\mathrm{d}c$ more negative.

Of course, the size factor $-\delta^2/\theta$ can be sufficiently large and negative in (13.139) to allow $\mathrm{d}\sigma/\mathrm{d}c$ to be negative as well. Thus, the inequality (13.141) is not a necessary condition for surface segregation.

Table 13.2 contains a collection of data (Wilson, 1965, Bhatia and March, 1978). It shows the effect of alloying on the surface tension in dilute solutions of two metals, as given in the first two columns. The third column records the prediction of the phenomenology, obtained in the simplest possible way by examining if $K_1 > K_2$, where $K = K_T$; the 2 referring to the solvent metal and 1 to the solute in its own pure metal. Since accurate data on the liquids are still not always available, Table 13.2 has been constructed using the data in Mott and Jones (1936) at 20°C for K_1 and K_2.

As can be seen from detailed inspection of the table, the correlation with the criterion (13.141) is quite encouraging. The most decisive cases are those that are surface inactive (reference may be made to the notes below Table II in the paper by Bhatia and March (1978) for further details of these).

Binary, liquid transition-metal alloys containing Mn and Cr. March (1984) has drawn attention anew to the fact that the bulk moduli of the 3*d*, 4*d*, and 5*d* transition series in the solid state exhibit a pronounced dependence on filling of the *d* band, as is clear, for example, from the review by Gschneider (1964). In contrast to the 4*d* and 5*d* series, the 3*d* series exhibits a pronounced dip at Mn. Formula (13.136) in the pure-metal limit predicts a close correlation between σ and B; it therefore comes as no surprise that σ plotted against *d*-shell occupancy also shows a dip at Mn for the 3*d* series, as seen, for instance, in Figure 5.6 of the review by Brown and March (1976).

Rewriting the criterion (13.141) as

$$\left(\frac{\mathrm{d}B}{\mathrm{d}c}\right)_{c=0} < 0, \tag{13.142}$$

in the absence of evidence to the contrary the assumptions will now be made that (1) for pure transition metals, the bulk-modulus variation with the filling of the *d*-shell retains the same character just above the melting point as in the solid phase and (2) based on known specific examples, the sign of

Table 13.2. *Effect of alloying on surface tension in dilute solutions of two metals (solvent is 2, solute is 1) [after Bhatia and March (1978)].*

Solvent (2)	Type	Compressibility properties $K_T = K$ (10^{-12} cm$^2\cdot$dyn^{-1})
Hg	Surface-active: Cd, Ag, Sn, Pb	$K_{Hg} = K_2 = 3.8$
	Mg, Tl, Sr, Ba, Na, Li, K, Rb, Cs	$K_1 > K_2$ for Sr, Ba, and alkalis. For others, $K_1 < K_2$ and size effect must dominate.
	Surface-inactive: Co, Bi, Zn, Cu	All have $K_1 < K_2$.
Sn	Active: Bi, Na, Pb, Sb, Tl	$K_2 = 1.9$. All have $K_1 > K_2$.
	Inactive: Cd, Zn, Al, Mn, Cu	Cd has $K_1 > K_2$: rest have $K_1 < K_2$.
Bi	Active: Na, K	$K_2 = 3.0$ $K_1 > K_2$.
	Inactive: Pb, Zn	$K_1 < K_2$.
Fe	Active: Cu, Sn	$K_2 = 0.6$, $K_1 > K_2$
	Inactive: —	—
Al	Active: Zn, Li, Bi, Pb	$K_2 = 1.4$, $K_1 > K_2$.
	Inactive: Mg, Sb, Sn	All have $K_1 > K_2$.
Cu	Active: Sb, Sn, Ag, Au	$K_2 = 0.75$, $K_1 > K_2$ except for Au for which size difference is important.
	Inactive: —	—
Ag	Active: —	—
	Inactive: Cu	$K_2 = 1.0$, $K_1 < K_2$.
Sb	Active: —	—
	Inactive: Cd, Zn, Pb	$K_2 = 2.7$, $K_1 < K_2$.
Zn	Active: Sb, Sn, Bi, Pb, Li	$K_2 = 1.7$, $K_1 > K_2$.
	Inactive: —	—
Pb	Active: Bi, K, Na, Ca	$K_2 = 2.3$, $K_1 > K_2$.
	Inactive: Sn	$K_1 < K_2$.

Note: Out of the 19 surface-inactive cases, four alloys disagree, on this basis, with the predictions of this theory. These are Cd in Sn and Mg, Sb, and Sn in Al. It would be of interest to have measurements of K_T vs. c in the liquid at the temperatures at which the surface inactivity is recorded experimentally, for these four cases.

dB/dc can be predicted from the bulk moduli of the two pure metals involved. Then given assumptions (1) and (2), the criterion (13.142) predicts that in binary liquid transition-metal alloys with Mn as a solute in dilute concentration, Mn will segregate to the surface in 21 out of the theoretically possible 26 alloys. The remaining 5 are specifically Sc, Y, La, Tc, and Re and because condition (13.142) is sufficient but not necessary, surface segregation of Mn may or may not occur in these cases.

Though less interesting than Mn, it is also worth adding that according to the criterion (13.142), nine dilute Cr-liquid transition-metal alloys should show Cr segregation to the surface and the remaining 17 possibilities, listed in March (1984), may or may not.

13.15. Metal-insulator transitions

In this section, a brief review will be given of the topic of metal-insulator transitions in the liquid state involving binary systems. Already, some considerable discussion has been given concerning the question of such transitions in pure liquid metals, such as caesium, as they are taken along the coexistence curve, for example, toward the critical point.

Endo, Tamura, and Yao (1987) have given an extensive review of this area; the following brief account draws on this article. Inevitably though, in such a large area, examples will have to be selected for illustrative purposes and complete coverage is not feasible. Thus, liquid Se-Te mixtures will be considered first. This will be followed by a discussion, related to that of Chapters 9 through 11 on expanded liquid Hg, of electronic and thermodynamic properties of amalgams near the liquid-gas critical point. In such two-component systems, a metal-nonmetal transition can occur by changing temperature, pressure, density, and composition.

Following Endo, Tamura, and Yao (1987), let us consider then the example of Te-Se liquid mixtures.

13.15.1. Liquid chalcogenides

The crystal structure of trigonal Se consists of helical chains. Each atom within a chain is bonded covalently to two neighbours, and the bonding between the chains is weak compared with the intrachain bonding. The twofold coordination in the chains of crystalline Se is largely preserved in the liquid state. Liquid Se is a typical liquid semiconductor.

It is known that Se and Te form a continuous series of solid solutions consisting of mixed chains. The ratio of intrachain to interchain distance is larger in Te than in Se, which implies that the bonding character of Te is less anisotropic. The structure of liquid Te is believed (Endo, Tamura, and Yao, 1987) to contain a large fraction of threefold coordinated sites.

Figure 13.11 shows the equal-conductivity curve for liquid Se-Te mixtures on the concentration (x)-temperature plane (Yao et al., 1980). In Figure 13.12, the semiconductor to metal transition region in the same plane is shown by the cross-hatching. It is to be noted from Figure 13.11 that the value of the conductivity σ is nearly $10^2 \Omega^{-1}$ cm^{-1} near this cross-hatched region in Figure 13.12, confirming that the hatched region

Figure 13.11. Equal-conductivity curves on the concentration- (x) temperature plane for liquid Se-Te mixtures. The numbers on the lines denote the conductivity (after Endo, Tamura, and Yao, 1987).

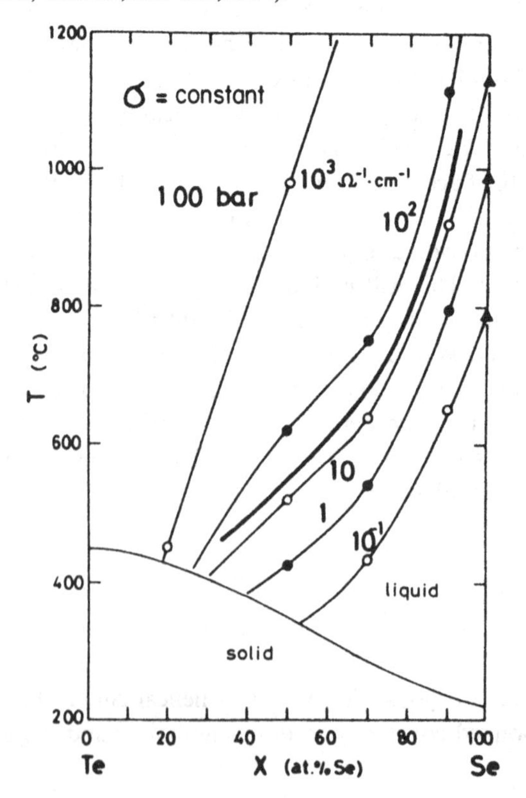

Figure 13.12. The semiconductor-to-metal transition region (shown by cross-hatched lines) in the concentration- (x) temperature plane for liquid Se-Te mixtures (After Endo, Tamura, and Yao, 1987).

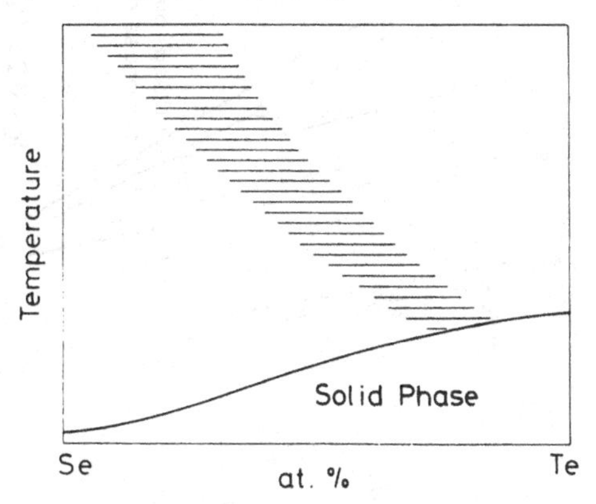

corresponds to the boundary between the semiconducting and metallic regions. The way this semiconductor-metal transition region changes with pressure has been discussed by Endo, Tamura, and Yao (1987).

13.15.2. Bi amalgam

As a second example, let us consider—again following Endo, Tamura, and Yao (1987)—the isothermal conductivities at 1.6 kbar of liquid Bi amalgam, as shown in Figure 13.13. A substantial increase in conductivity is brought about by the addition of Bi above 1450°C, the temperature at which the metal-nonmetal transition occurs. The increase in conductivity is considered (see Endo, Tamura, and Yao, 1987) to be associated with the increase in the density of states at the Fermi level, $N(E_f)$.

These two examples are intended only to indicate that there is a rich area here, where progress in both theory and experiment can be expected.

13.16. Magnetic properties of alloys

To conclude this chapter on binary liquid metal alloys, the topic of magnetism in liquid metals covered in Chapter 11 will be taken up again for alloys (see also March and Sayers, 1979). There is, of course, a substantial body

Figure 13.13. Concentration variation of the isothermal conductivities at 1.6 kbar for liquid Bi amalgam (LHS). Isochore lines for liquid 0.2 atomic % Bi amalgam with those for Hg denoted by dashed lines. Circle indicates critical point of Hg (RHS). (After Endo, Tamura, and Yao, 1987.)

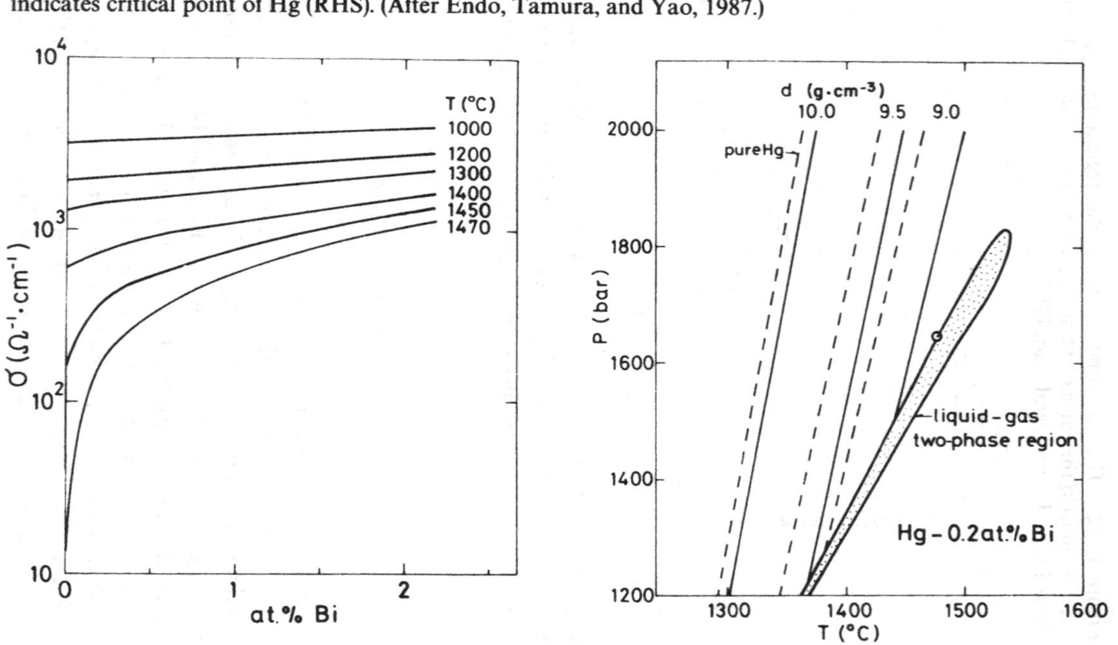

of literature in this area of the magnetic properties of liquid metal alloys. In this section a few topics will be selected as examples of magnetic behaviour that can occur. An outline of some general theoretical background that is useful will be given together with a brief discussion of differences between liquid and solid alloys.

One rather general point to note is that there is, at the time of writing, quite limited knowledge, at a quantitative level, of the structure of liquid binary alloys. Although in principle the technique of neutron diffraction, combined with isotopic enrichment, can lead to the extraction of three partial structure factors $S_{ij}(k)$ required to define the pair correlations in a liquid binary alloy of a given composition, such data are appearing only slowly because of the difficulties involved in such a programme, one being the high cost of suitable isotopes. An example in which the partial structure factors are known is a specific Cu-Sn alloy studied by Enderby, North, and Egelstaff (1966).

Because of such lack of detail, the message in this area is to seek out gross correlations rather than to stress finer details. However, as one example of the way in which liquid structure can enter the theory of magnetic susceptibility, one can cite the work of Dupree and Sholl (1975), who carried out detailed calculations on noble metal-tin alloys. Because this work is reviewed by March and Sayers (1979) and extends the pseudopotential treatment referred to in Section 11.2 for simple, pure liquid metals, details are not given here.

Rather, attention will be directed to the problems of liquid transition metal alloys A and B, with both A and B taken as transition elements as well as with B chosen as a specific nontransition element (e.g., Cu and Al). Some admittedly primitive models and classifications will then be summarized.

13.16.1. Liquid noble-metal-based alloys: 3d impurities in liquid Cu

Gardner and Flynn (1967) reported measurements of the impurity susceptibility of $3d$ impurities in liquid Cu and of the electronic polarization in the solvent near the localized moments associated with the impurity. Information on the latter is not so readily obtained in solid alloys because of the enormous Knight shift changes caused by the polarization when not time-averaged by rapid diffusion. Owen et al. (1957) examined the spin resonance of Mn impurities in solid Cu and found a g value close to 2, indicating that

Table 13.3. *The number of unpaired electrons per impurity atom in* Cu.

Impurity	n (liquid)	n (solid)
Cr	3.55 ± 0.4	—
Mn	4.40 ± 0.15	4.26 ± 0.15
Fe	3.64 ± 0.20	3.51 ± 0.20
Co	2.9 ± 0.4	2.8 ± 0.4

the orbital angular momentum is quenched. In addition, the resonance linewidth indicates a lifetime of 10^{-10} s for Zeeman levels in a magnetic field, which sets a lower limit on the length of time for which the configuration of the electrons forming the moment exists. Since the configuration is thus stable for the duration of many Larmor periods in an applied field of 10 kG, one regards the z component of spin as a good quantum number for Mn in Cu. Gardner and Flynn (1967) found Cr, Mn, Fe, and Co impurities in liquid Cu to be magnetic and to obey the Curie-Weiss law with impurity susceptibility given by $\chi_d = cN\mu_{\text{eff}}^2/3k_B(T - \theta)$. The number of unpaired electrons per impurity atom obtained using this equation are recorded in Table 13.3 for both the solid and liquid phase assuming that $\theta \simeq 0$, which corresponds to neglecting the interaction between impurities.

The behaviour of these impurities in both solid and liquid Cu is seen to be similar, but there is a discontinuity in χ_d^{-1} of about 5% at the melting transition, χ_d^{-1} being lower in the liquid phase at the transition than in the solid phase.

The Cu Knight shift changes at 1100°C due to $3d$ transition impurities were found to be linear in the impurity concentration. Very large changes were observed and Mn concentrations greater than 5% were sufficient to reverse the sign of the ^{63}Cu resonance Knight shift. Similar measurements were also made as a function of temperature from 1160°C down to 950°C in the supercooled state for some alloys. Sc, Ti, and Ni alloys were found to have temperature-independent shifts, but each of the magnetic solutes gave rise to considerable temperature variation, a plot of the reciprocal Knight shift change against temperature revealing Curie-Weiss behaviour.

13.16.2. Al-based liquid alloys

Howe et al. (1972) have reported magnetic susceptibility and Knight shift measurements of Al-based liquid host metals with Cr and Mn as dilute impurities. The solvent alloys contained additions of up to 30% of Cu, Zn, Ga, Ge, Ag, or Si, thus extending the work of Flynn, Rigney, and Gardner (1967) on Al-based alloys. At 960°C the impurity susceptibilities of Cr and Mn in liquid Al were found to be 10.1×10^{-4} and 7.0×10^{-4} mol^{-1} for Mn and Cr, respectively, and, when corrected for the observed T-dependence of χ, agree well with the earlier work of Flynn, Rigney, and Gardner (1967). The results for Cr and Mn in liquid alloys of Al with up to 30 at% of Cu, Zn, Ga, Ge, Ag, or Si of the impurity susceptibility are presented by Howe et al. (1972) as a plot of β_{Cr} against β_{Mn}, where β is the solution of

$$\chi(\beta, T) = \frac{g^2\mu_B^2}{3\beta} - \frac{g^2\mu_B^2 S_m(S_m + 1)}{3k_B T\{\exp[\beta S_m(S_m + 1)/k_B T] - 1\}} \qquad (13.143)$$

with $S_m = \frac{5}{2}$ for Mn and 2 for Cr. The reason for choosing such a representation of the susceptibility measurements is that Flynn, Peters, and Wert (1971) and Peters and Flynn (1972) obtained such an expression for the susceptibility in a model which will be discussed later. Howe et al. (1972) claim that the plot thereby obtained agrees with the plot of β_{Cr} against β_{Mn} for liquid Ge, Cu, Ga-Ge, Zn-Ga, Cu-Zn, Zn, and Au hosts reported by Peters and Flynn (1972), except for alloys with a host consisting of Zn in Al, even though Zn itself falls on the plot. To explain this, Howe et al. (1972) suggest that the Zn 3*d* bands may mix significantly with the Al conduction band even though the Zn 3*d* level lies below the 5*p* band in pure Zn. If this explanation is correct, it is difficult to understand why Al-Cu-based alloys lie on the plot.

Relevant to the anomalous temperature dependence of the susceptibility of liquid alloys are measurements of the electrical resistivity of the solid alloys, a topic briefly referred to in Appendix 13.8.

13.16.3. Liquid 3d-3d transition metal alloys

The behaviour of the susceptibility of 3*d*-3*d* transition metal alloys, as well as that of the pure elements, was investigated by Nakagawa (1956).

Busch et al. (1973) reported measurements of the susceptibility of the pure transition metals just above the melting points, the results being shown in Figure 13.14.

The susceptibility of alloys of transition metals next to one another in the 3*d* series as measured by Nakagawa (1956) appear to follow the trend exhibited by the pure metals (Busch et al., 1973), increasing from Mn to Fe, reaching a maximum at Co, and decreasing from Ni to Cu. Alloys of 3*d*

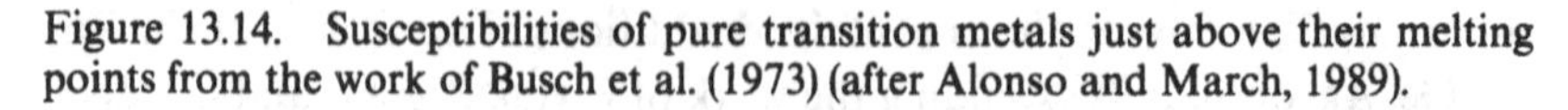

Figure 13.14. Susceptibilities of pure transition metals just above their melting points from the work of Busch et al. (1973) (after Alonso and March, 1989).

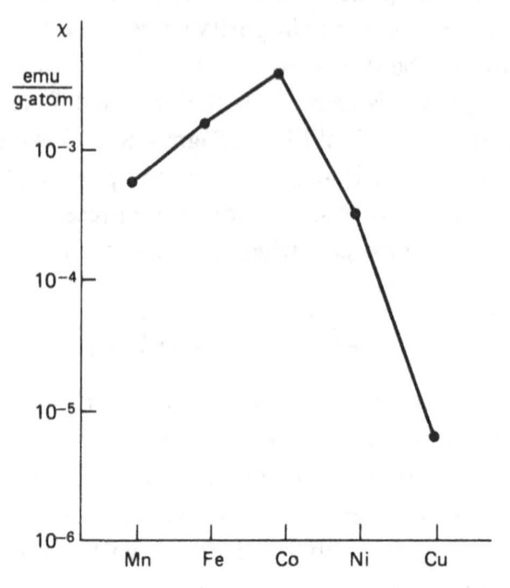

elements in the liquid state differing in atomic number by more than unity, however, show strong departures from this behaviour. Fe-Ni alloys show only a weak maximum in the region of Fe_3Ni compared with that seen on passing from Fe to Co to Ni, and the susceptibility of Cr alloys decreases in the order $CrFe_3$, $CrCo_3$, $CrNi_3$.

A striking result is that only a small change in χ is observed on melting, the results of Nakagawa being given in Table 13.4. The main points to be noted are as follows:

1. The chief exception to the observation that the change in χ on melting is small is in Fe, for which χ is markedly reduced from its value in the high-temperature *bcc* δ phase on melting. In the *bcc* phase of Fe, there is a large peak in the density of states $N(E)$ (see, for example, Pettifor, 1970), which is not present in the *hcp* or *fcc* phases. Furthermore, the difference in the density-of-states curves of *hcp* and *fcc* is much smaller than between *fcc* and *bcc*. Since the local symmetry in the liquid resembles that in the close-packed phases, it would appear, therefore, that the decrease in χ of Fe on melting is due to the decrease in $N(E_f)$ expected from the change in local symmetry. Corresponding to this, Fe-Cr and Fe-V alloys in the ratio 3 : 1, which are also *bcc* below T_m, exhibit a large decrease in χ on melting, similar to that in δ Fe.
2. The susceptibility of Mn was found to be almost independent of T in each phase at high temperatures, i.e., in the β phase (600–1050°C), γ phase (1050–1150°C), δ phase (1150–1250°C), and the liquid phase. A discontinuous increase in χ

Table 13.4. *Changes in χ and $dT/d\,(1/\chi)$ at melting temperature T_m. $\Delta\chi = \chi_l - \chi_s$, χ_l and χ_s are the values of χ at T_m in the liquid and solid, respectively, and c_l and c_s are the values of $dT/d(1/\chi)$ at T_m in the liquid and solid, respectively.*

	fcc structure just below T_m	T_m	$\frac{\Delta\chi}{\chi_s}$ (%)	$\frac{c_l}{c_s}$
Ni		1455	0	1.0
Co		1500	−3	1.0
Ni-Co	(3 : 1)	1460	−3	1.0
	(1 : 1)	1480	−8	1.0
	(1 : 3)	1485	−6	1.0
Ni-Fe	(3 : 1)	1445	−3	1.0
	(1 : 1)	1440	+5	1.0
	(1 : 3)	1460	+8	0.8
Ni-Mn	(3 : 1)	1220	+10	1.0
Ni-Cr	(3 : 1)	1410	+18	1.0
Co-Fe	(3 : 1)	1470	+4	0.9
	(1 : 1)	1470	+10	0.8
	(1 : 3)	1510	+8	0.6
Co-Mn	(3 : 1)	1360	+14	0.8
Co-Cr	(3 : 1)	1440	+10	0.6
Fe-Mn	(3 : 1)	1440	+9	0.6
	bcc structure just below T_m			
Mn		1250	+2	1.0
Fe		1540	−20	1.5
Fe-Cr	(3 : 1)	1510	−14	2.3
Fe-V	(3 : 1)	1475	−15	3.4

occurred during heating at each transition temperature. The increase in χ on melting Mn may arise from a narrowing of the 3*d* bands and a consequent increase in $N(E_f)$ due to the volume expansion on melting (El-Hanany and Warren, 1975).

3. The susceptibility of Ni shows no change on melting and behaves as

$$\chi = \alpha + \frac{C}{T - \theta} \tag{13.144}$$

where α is weakly temperature-dependent. Co and Ni-Co alloys are *fcc* and obey the Curie law (i.e., with α set equal to zero in (13.144)) in both solid and liquid with a small decrease in χ on melting.

4. At high temperatures Ni-Fe alloys are *fcc*, those that contain 25 and 50% Ni showing an increase in χ on melting, whereas those with 75% Ni show a decrease in χ. Ni-Mn and Ni-Cr alloys with 75% Ni and high-temperature Fe-Mn and Co-Fe alloys are *fcc*, and all exhibit an increase in χ on melting. A similar variation is also shown by Co-Mn and Co-Cr alloys, which are also *fcc* in the ratio 3 : 1 at high temperatures.

Briane (1973a, b) studied the magnetic susceptibility of Fe-Ni alloys through the melting temperature. The results for the change in χ are in agreement with the earlier measurements by Nakagawa (1956) except in the region of pure Ni, where a 2% increase in χ was found on melting, the slope of the $1/\chi$ versus T curve changing from 1.17×10^2 g/cm^{-3}K^{-1} in the solid to 0.60×10^2 g/cm^{-3}K^{-1} in the liquid, in contrast to the zero change in the slope found by Nakagawa.

Having summarized some salient features of experimental data of a number of liquid alloy systems, it will be useful next to refer to some models and to Friedel's classification for crystalline hosts.

13.16.4. Some useful models and classifications

Friedel's concept of a virtual bound state around an impurity in a metal is valuable in interpreting some of the specific data on liquid metal alloys. While Friedel (1956, 1958, 1962) specifically had in mind crystalline hosts, we note that he suggested three possible effects that may act to split the (assumed) virtual bound state.

CRYSTAL FIELD

In a field of cubic symmetry, this would act to split the d state into $d(x^2 - y^2, x^2 - z^2)$ and $d(xy, yz, zx)$. This effect is of order 1 eV and could play a significant role in alloys with matrices of small valence.

d-d COULOMB CORRELATIONS

If d-d correlations are strong, the effect will be to stabilize states with integral numbers of d electrons. This would act to split the d^{10} level into ten d^n states ($n = 1, \ldots, 10$), well separated in energy. If such a splitting is not observed, it may be due to the screening of the d-d interactions by the valence electrons of the host. Such screening is known to reduce considerably the energy difference between the $3d^9$ and $3d^{10}$ configurations in pure Ni, as reviewed by Herring (1966). It is not expected to reduce the considerable energy cost of forming a $3d^8$, $3d^9$ or $3d^{10}$ configuration in pure Mn or the $3d^9$ and $3d^{10}$ configuration in pure Fe, but when such atoms are dissolved in a host with a large number of valence electrons, such screening may be more effective.

d-d Exchange Interactions

The *d-d* exchange interactions could lead to a splitting of the d^{10} state into two d^5 states with opposite spin, which would be successively filled in a transitional series. This is what is observed in free atoms and ions (Hund's rule). In the Hartree-Fock approximation and according to Blandin and Friedel (1959), the condition for the virtual bound state to develop a moment is

$$(U + 4J)N_d(E_F) \geqslant 10 \tag{13.145}$$

where U is the average electron interaction between two electrons of opposite spin on the atom and J is the energy gained by aligning the spins. With $N_d(E)$ given in terms of the position of the peak in the local density of states, say $E_{0\sigma}$, as

$$N_{l\sigma}(E) = \frac{2l(l+1)}{\pi} \frac{\Delta}{\Delta^2 + (E_{0\sigma} - E)^2}, \tag{13.146}$$

where Δ is the width of the local density of states at the impurity site, with $l = 2$, the condition (13.145) is nearly equivalent to

$$p(U + 4J) \geqslant W \tag{13.147}$$

where $p = Z_d$ if $Z_d \leqslant 5$ and $(10 - Z_d)$ if $Z_d > 5$, W being the effective width of the virtual bound state, which was obtained by Friedel by comparing the width of the state with the maximum exchange energy obtained for total splitting. The approximate equivalence of the two criteria suggests that, when stable, the splitting is nearly always complete.

Anderson (1961) developed this further, introducing the impurity as a localized extra orbital representing the *d*-state of the impurity in the conduction band of the host metal. The one-electron part of the Hamiltonian includes the localized and band states together with a mixing interaction $V_{\mathbf{k}d}$ coupling the two types of state and leading to a broadening of the impurity level, the Hamiltonian being taken as

$$H = \sum_{\sigma} \varepsilon_{\mathbf{k}} a^+_{\mathbf{k}\sigma} a_{\mathbf{k}\sigma} + \sum_{\sigma} \varepsilon_d a^+_{d\sigma} a_{d\sigma} + \sum_{\mathbf{k}\sigma} (V_{\mathbf{k}d} a^+_{\mathbf{k}\sigma} a_{d\sigma} + c \cdot c) + U n_{d\sigma} n_{d-\sigma}, \tag{13.148}$$

which was treated by Anderson in the Hartree-Fock approximation.

Extending this model to the case of a fivefold degenerate impurity state, Klein and Heeger (1966) calculated the additional susceptibility due to the impurity to be

$$\chi_d = \frac{2\mu_B^2 N_d(E_f)}{1 - (U + 4J)N_d(E_f)/10}. \tag{13.149}$$

March and Sayers (1979) discuss the observed behaviour of transition metal impurities in noble metal and nontransition metal hosts in the light of these results. Interesting differences arise between the behaviour of solid alloys on the one hand and liquid alloys on the other; these are briefly reviewed next.

13.16.5. *Difference between solid and liquid alloys*

Two rather different types of behaviour of transition-element impurities in noble and simple liquid-metal hosts emerge from experimental studies, some of which have been summarized earlier.

1. Impurities such as Mn in liquid Bi, In, Sn, and Sb or Cr, Mn, Fe, and Co in liquid Cu are magnetic, the magnetic susceptibility exhibiting a Curie-Weiss dependence on T, with only a small change in the Curie constant on melting, χ^{-1} being about 5% lower in the liquid phase at T_m than in the solid phase. This discontinuity seems to correspond to a slight increase in the local density of states upon melting, arising from a decrease in V_{kd} in the Anderson model (see (13.148)) due to the increase in the nearest-neighbour distance on melting.
2. Fe, Co, and Ni in liquid Sn and Sb, for example, are nonmagnetic in the solid, but in the liquid χ is quite strongly dependent on T, $d\chi/dT$ being negative for Fe and positive for Co and Ni. Tamaki (1967, 1968) offered an explanation from a T^2 term obtained by expanding the Fermi distribution function. However, there is the difficulty that Gardner and Ardary (1976) found a linear increase of χ with T for Co impurities in liquid Sn. Also temperature-independent susceptibility χ has been observed in the solid phase at low temperature.

Another example is liquid Al with $3d$ impurities, the solid resistivity having a single maximum at low T corresponding to an unsplit bound state but a double-peaked structure at high T corresponding to an exchange splitting for elements in the middle of the series. Flynn, Rigney, and Gardner (1967) propose that the observed increase of χ with T in the liquid may be due to an increase of local moment on the impurity with increasing T to a value determined by charge neutrality. Whereas theoretical work is still limited to date on such magnetic behaviour in the liquid state, the impurity susceptibility in the solid phase has been interpreted in one of the two limiting cases of the Anderson model, namely, the *s-d* model and spin fluctuation theory (see March and Sayers (1979)).

14

Two-component theory of pure liquid metals

In this chapter we shall report on developments in liquid-metal theory in which, instead of dealing with ions coupled by effective interactions mediated by the conduction electrons, the liquid metal is treated as a two-component system, conduction electrons and positive ions (Cowan and Kirkwood, 1958; Watabe and Hasegawa, 1973; Chihara, 1973; March and Tosi, 1973). Earlier, evidence has been presented for well-defined collective modes of the positive ions in liquid Rb near its melting point. There is, of course, no doubt that, in any dense assembly of conduction electrons, there are well-defined plasma modes. Therefore, a workable model to keep in mind for such a two-component system begins with the assumption that there are two types of well-defined collective excitations; density fluctuations (analogous to phonons in a crystal) and plasmons (Tosi and March, 1973a; for details see Appendix 14.1).

Following the approach of March and Tosi (1973), the two-component theory will be developed in terms of three partial structure factors, $S_{\mathrm{ii}} = S$, the nuclear-nuclear structure factor as observed by neutron scattering, the electron-electron structure factor S_{ee}, and the "cross" correlations represented by $S_{\mathrm{ie}}(k)$.

14.1. Electron-ion Hamiltonian and density fluctuation operators

To implement the programme just outlined, let us set up a description of the liquid metal in terms of the local number densities $\rho_{\mathrm{e}}(\mathbf{r}, t)$ and $\rho_{\mathrm{i}}(\mathbf{r}, t)$ of electrons and ions, respectively, at position $\mathbf{r}$ and time t. Then the density fluctuation operators $\rho_{\mathbf{k}}(t)$ for electrons and ions are introduced through

$$\rho_{\mathrm{e}}(\mathbf{r}, t) = \sum_i \delta(\mathbf{r} - \mathbf{r}_i(t)) = \sum_{\mathbf{k}} \rho_{\mathrm{e}\mathbf{k}}(t) \exp(i\mathbf{k}\cdot\mathbf{r}) \tag{14.1}$$

and

$$\rho_{\mathrm{i}}(\mathbf{r}, t) = \sum_j \delta(\mathbf{r} - \mathbf{R}_j(t)) = \sum_{\mathbf{k}} \rho_{\mathrm{i}\mathbf{k}}(t) \exp(i\mathbf{k}\cdot\mathbf{r}). \tag{14.2}$$

Inverting these relations yields

$$\rho_{e\mathbf{k}}(t) = \mathscr{V}^{-1} \sum_i \exp[-i\mathbf{k}\cdot\mathbf{r}_i(t)] \tag{14.3}$$

and

$$\rho_{i\mathbf{k}}(t) = \mathscr{V}^{-1} \sum_j \exp[-i\mathbf{k}\cdot\mathbf{R}_j(t)], \tag{14.4}$$

$\mathbf{r}_i$ denoting the position of the ith electron, $\mathbf{R}_j$ denoting that of the jth ion, and $\mathscr{V}$ being the total volume of the liquid metal.

Denoting electron-electron, ion-ion, and electron-ion interactions by v, V, and V_{ie}, respectively, and with ions of mass M, the total Hamiltonian reads

$$\begin{aligned} H &= \sum_i \frac{p_i^2}{2m} + \sum_j \frac{p_j^2}{2M} + \frac{1}{2}\sum_{i\neq i'} v(|\mathbf{r}_i - \mathbf{r}_{i'}|) \\ &\quad + \frac{1}{2}\sum_{j\neq j'} V(\mathbf{R}_j - \mathbf{R}_{j'}) + \sum_{ij} V_{ie}(\mathbf{r}_i - \mathbf{R}_j) \\ &= \sum_i \frac{p_i^2}{2m} + \sum_j \frac{p_j^2}{2M} + \frac{1}{2}\sum_{\mathbf{k}} v_{\mathbf{k}}(\rho_{e\mathbf{k}}^{+}\rho_{e\mathbf{k}} - n_e) \\ &\quad + \frac{1}{2}\sum_{\mathbf{k}} V_{\mathbf{k}}(\rho_{i\mathbf{k}}^{+}\rho_{i\mathbf{k}} - n_i) + \sum_{\mathbf{k}} V_{ie}(\mathbf{k})\rho_{e\mathbf{k}}^{+}\rho_{i\mathbf{k}}, \end{aligned} \tag{14.5}$$

the density of electrons n_e and the ionic density n_i being related through the valence Z by $n_e = Zn_i$. It will often be convenient in what follows to work with unit volume, i.e., $\mathscr{V} = 1$.

14.1.1. Response functions

For use later, let us define the response function

$$\begin{aligned} \chi_{ee}(\mathbf{q},\omega) &= \int_0^\infty dt \exp(-i\omega t)\left(\frac{1}{i\hbar}\right)\langle[\rho_{e\mathbf{q}}^{+}(0), \rho_{e\mathbf{q}}(t)]\rangle \\ &= \beta\int_0^\infty dt \exp(-i\omega t)\langle\dot{\rho}_{e\mathbf{q}}^{+}(0); \rho_{e\mathbf{q}}(t)\rangle \end{aligned} \tag{14.6}$$

with appropriate generalizations for χ_{ie} and χ_{ii} (March and Tosi, 1973). Here, as usual, $\beta = (k_B T)^{-1}$, whereas $\langle X; Y\rangle$ is the canonical correlation function of Kubo, namely,

$$\langle X; Y\rangle = \beta^{-1}\int_0^\beta d\lambda\langle \exp(\lambda H)X\exp(-\lambda H)Y\rangle \tag{14.7}$$

with

$$\langle [X(0), Y(t)]\rangle = i\hbar\beta\langle \mathring{X}(0); Y(t)\rangle. \tag{14.8}$$

It is sometimes useful to define response functions for mass densities. This is then related to the discussion of correlation functions in a classical binary alloy given by Bhatia, Thornton and March (1974; see also Appendix 13.6). These quantities, along with a few of their properties, will be dealt with in Appendix 14.2.

14.1.2. Equations of motion for density matrices

Next it will be useful to introduce field operators $\psi(\mathbf{r}, t)$ and $\Psi(\mathbf{r}, t)$ for electrons and ions, respectively, and then to form the equations of motion for the density matrices, defined as $\langle\psi^+(\mathbf{x}, t)\psi(\mathbf{x}', t)\rangle$ and $\langle\Psi^+(\mathbf{x}, t)\Psi(\mathbf{x}', t)\rangle$. These equations are

$$\begin{aligned}\left\{i\hbar\left(\frac{\partial}{\partial t}\right) - \left(\frac{\hbar^2}{2m}\right)(\nabla_{\mathbf{x}}^2 - \nabla_{\mathbf{x}'}^2)\right\}&\langle\psi^+(\mathbf{x}, t)\psi(\mathbf{x}', t)\rangle \\ = -\int d\mathbf{x}''[v(\mathbf{x} - \mathbf{x}'') - v(\mathbf{x}' - \mathbf{x}'')]&\langle\psi^+(\mathbf{x}, t)\psi(\mathbf{x}', t)\rho_e(\mathbf{x}'', t)\rangle \\ -\int d\mathbf{x}''[V_{ie}(\mathbf{x} - \mathbf{x}'') - V_{ie}(\mathbf{x}' - \mathbf{x}'')]&\langle\psi^+(\mathbf{x}, t)\psi(\mathbf{x}', t)\rho_i(\mathbf{x}'', t)\rangle\end{aligned} \tag{14.9}$$

and

$$\begin{aligned}\{i\hbar(\partial/\partial t) - (\hbar^2/2M)(\nabla_{\mathbf{x}}^2 - \nabla_{\mathbf{x}'}^2)\}&\langle\Psi^+(\mathbf{x}, t)\Psi(\mathbf{x}', t')\rangle \\ = -\int d\mathbf{x}''[V(\mathbf{x} - \mathbf{x}'') - V(\mathbf{x}' - \mathbf{x}'')]&\langle\Psi^+(\mathbf{x}, t)\Psi(\mathbf{x}', t)\rho_i(\mathbf{x}'', t)\rangle \\ -\int d\mathbf{x}''[V_{ie}(\mathbf{x} - \mathbf{x}'') - V_{ie}(\mathbf{x}' - \mathbf{x}'')]&\langle\Psi^+(\mathbf{x}, t)\Psi(\mathbf{x}', t)\rho_e(\mathbf{x}'', t)\rangle.\end{aligned} \tag{14.10}$$

Since these two equations have basically the same form, the calculation need be carried out only for the electronic equation of motion (14.9), and the final result will then be written for (14.10).

14.2. Wigner distribution functions

With the aim of setting up the theory in terms of current and energy densities, let us next introduce the mixed density matrix, or Wigner distribution function, $f(\mathbf{p}, \mathbf{R}, t)$, defined by

$$f(\mathbf{p}, \mathbf{R}, t) = \int d\mathbf{r} \exp\left(\frac{i\mathbf{p}\cdot\mathbf{r}}{\hbar}\right)\left\langle \psi^{+}\left(\mathbf{R} + \frac{1}{2}\mathbf{r}, t\right)\psi\left(\mathbf{R} - \frac{1}{2}\mathbf{r}, t\right)\right\rangle \quad (14.11)$$

with a similar function $F(\mathbf{p}, \mathbf{R}, t)$ for the ions.

From this Wigner distribution function, one can now construct the density $n(\mathbf{R}, t)$, the current density $j(\mathbf{R}, t)$, and the "kinetic energy tensor" $\pi(\mathbf{R}, t)$ as

$$n(\mathbf{R}, t) = \sum_{\mathbf{p}} f(\mathbf{p}, \mathbf{R}, t) \quad (14.12)$$

$$\mathbf{j}(\mathbf{R}, t) = \left(\frac{1}{m}\right)\sum_{\mathbf{p}} \mathbf{p} f(\mathbf{p}, \mathbf{R}, t) \quad (14.13)$$

and

$$\pi_{\alpha\beta}(\mathbf{R}, t) = \left(\frac{1}{m}\right)\sum_{\mathbf{p}} p_\alpha p_\beta f(\mathbf{p}, \mathbf{R}, t) \quad (14.14)$$

with analogous definitions of N, $\mathbf{J}$, and Π for the ions in terms of F and the ion mass M.

The equation of motion (14.9) can then be expanded about the diagonal to yield

$$\left\{i\hbar\left(\frac{\partial}{\partial t}\right) - \left(\frac{\hbar^2}{m}\right)\nabla_{\mathbf{R}}\cdot\nabla_{\mathbf{r}}\right\}\sum_{\mathbf{p}} \exp\left(\frac{-i\mathbf{p}\cdot\mathbf{r}}{\hbar}\right) f(\mathbf{p}, \mathbf{R}, t)$$

$$= -\int d\mathbf{x}''[\mathbf{r}\cdot\nabla_{\mathbf{R}} v(\mathbf{R} - \mathbf{x}'') + O(r^3)]$$

$$\times\left[\langle\rho_e(\mathbf{R}, t)\rho_e(\mathbf{x}'', t)\rangle + \left(\frac{m}{i\hbar}\right)\mathbf{r}\cdot\langle\mathbf{j}(\mathbf{R}, t)\rho_e(\mathbf{x}'', t)\rangle + O(r^2)\right]$$

$$-\int d\mathbf{x}''[\mathbf{r}\cdot\nabla_{\mathbf{R}} V_{ie}(\mathbf{R} - \mathbf{x}'') + O(r^3)]$$

$$\times\left[\langle\rho_e(\mathbf{R}, t)\rho_i(\mathbf{x}'', t)\rangle + \left(\frac{m}{i\hbar}\right)\mathbf{r}\cdot\langle\mathbf{j}(\mathbf{R}, t)\rho_i(\mathbf{x}'', t)\rangle + O(r^2)\right], \quad (14.15)$$

the current density operator $\mathbf{j}(\mathbf{R}, t)$ being given by

$$\mathbf{j}(\mathbf{R},t) = \left(\frac{i\hbar}{2m}\right)\{[\nabla_{\mathbf{R}}\psi^+(\mathbf{R},t)]\psi(\mathbf{R},t) - \psi^+(\mathbf{R},t)[\nabla_{\mathbf{R}}\psi(\mathbf{R},t)]\}. \quad (14.16)$$

Equation (14.15) is the central equation from which continuity and conservation equations can now be obtained, following March and Tosi (1973).

14.2.1. Continuity equation and current conservation

Equating terms independent of $\mathbf{r}$ on both sides of (14.15), one finds almost immediately that

$$\frac{\partial n(\mathbf{R},t)}{\partial t} + \nabla_{\mathbf{R}}\cdot\mathbf{j}(\mathbf{R},t) = 0, \quad (14.17)$$

which is the usual continuity equation.

Terms of $O(r)$ similarly yield

$$\frac{m\partial\mathbf{j}(\mathbf{R},t)}{\partial t} + \nabla_{\mathbf{R}}\cdot\pi(\mathbf{R},t) = -\int d\mathbf{x}''\nabla_{\mathbf{R}}v(\mathbf{R}-\mathbf{x}'')\langle\rho_e(\mathbf{R},t)\rho_e(\mathbf{x}'',t)\rangle$$
$$-\int d\mathbf{x}''\nabla_{\mathbf{R}}V_{ie}(\mathbf{R}-\mathbf{x}'')\langle\rho_e(\mathbf{R},t)\rho_i(\mathbf{x}'',t)\rangle. \quad (14.18)$$

When one passes over to the jellium model, which corresponds to the granular ions in a real liquid metal being smeared out into a uniform neutralizing background in which interacting electrons move, then (14.17) and (14.18) provide the basis for the electron gas theory of Singwi et al. (1970; see also Singwi and Tosi, 1981).

If one introduces a velocity field $\mathbf{v}(\mathbf{R},t)$ through

$$\mathbf{j}(\mathbf{R},t) = n(\mathbf{R},t)\mathbf{v}(\mathbf{R},t) \quad (14.19)$$

and separates the momentum flux tensor into two parts,

$$\pi_{\alpha\beta}(\mathbf{R},t) = mj_\alpha(\mathbf{R},t)v_\beta(\mathbf{R},t) + \pi^0_{\alpha\beta}(\mathbf{R},t), \quad (14.20)$$

then (14.18) can be written as

$$mn(\mathbf{R},t)\left(\frac{D\mathbf{v}(\mathbf{R},t)}{Dt}\right) = -\nabla_{\mathbf{R}}\cdot\pi^0(\mathbf{R},t) - \int d\mathbf{x}''\,\nabla_{\mathbf{R}}v(\mathbf{R}-\mathbf{x}'')\langle\rho_e(\mathbf{R},t)\rho_e(\mathbf{x}'',t)\rangle$$
$$-\int d\mathbf{x}''\,\nabla_{\mathbf{R}}V_{ie}(\mathbf{R}-\mathbf{x}'')\langle\rho_e(\mathbf{R},t)\rho_i(\mathbf{x}'',t)\rangle, \quad (14.21)$$

where

$$\frac{D\mathbf{v}}{Dt} = \frac{\partial \mathbf{v}}{\partial t} + (\mathbf{v} \cdot \nabla_{\mathbf{R}})\mathbf{v}. \tag{14.22}$$

It will be seen later that the Navier-Stokes equation for the liquid metal derives from a combination of (14.21) with the corresponding equation for the ions.

14.2.2. Energy transport equation

The final equation can be derived from (14.15) by working to $O(r^2)$, which then yields

$$\begin{aligned} &\frac{i\hbar}{2}\frac{\partial}{\partial t}\left(-\frac{i}{\hbar}\right)^2 \sum_{\mathbf{p}} p_\alpha^2 f(\mathbf{p}, \mathbf{R}, t) - \frac{\hbar^2}{2m}\left(-\frac{i}{\hbar}\right)^3 \sum_{\mathbf{p}} \mathbf{p} \cdot \nabla_{\mathbf{R}} p_\alpha^2 f(\mathbf{p}, \mathbf{R}, t) \\ &= -\frac{m}{i\hbar}\int d\mathbf{x}'' \, \nabla_{\mathbf{R}_\alpha} v(\mathbf{R} - \mathbf{x}'') \langle j_\alpha(\mathbf{R}, t)\rho_e(\mathbf{x}'', t)\rangle \\ &\quad - \frac{m}{i\hbar}\int d\mathbf{x}'' \, \nabla_{\mathbf{R}_\alpha} V_{ie}(\mathbf{R} - \mathbf{x}'') \langle j_\alpha(\mathbf{R}, t)\rho_i(\mathbf{x}'', t)\rangle. \end{aligned} \tag{14.23}$$

Taking the trace then leads to

$$\begin{aligned} \left(\frac{\partial}{\partial t}\right) k(\mathbf{R}, t) + \nabla_{\mathbf{R}} \cdot \mathbf{j}_k(\mathbf{R}, t) &= -\int d\mathbf{x}'' \, \nabla_{\mathbf{R}} v(\mathbf{R} - \mathbf{x}'') \cdot \langle \mathbf{j}(\mathbf{R}, t)\rho_e(\mathbf{x}'', t)\rangle \\ &\quad - \int d\mathbf{x}'' \, \nabla_{\mathbf{R}} V_{ie}(\mathbf{R} - \mathbf{x}'') \cdot \langle \mathbf{j}(\mathbf{R}, t)\rho_i(\mathbf{x}'', t)\rangle, \end{aligned} \tag{14.24}$$

where the kinetic energy density $k(\mathbf{R}, t)$ is given by

$$k(\mathbf{R}, t) = \sum_{\mathbf{p}} \left(\frac{p^2}{2m}\right) f(\mathbf{p}, \mathbf{R}, t) = \frac{1}{2}\operatorname{Tr} \pi(\mathbf{R}, t) \tag{14.25}$$

and the kinetic energy flux $\mathbf{j}_k$ is simply

$$\mathbf{j}_k(\mathbf{R}, t) = \sum_{\mathbf{p}} \left(\frac{\mathbf{p}}{m}\right)\left(\frac{p^2}{2m}\right) f(\mathbf{p}, \mathbf{R}, t). \tag{14.26}$$

To see that (14.24) is, in fact, an energy-transport equation, consider the following identity for the first term on the right-hand side of this equation:

$$
\begin{aligned}
-\int d\mathbf{x}''\, \nabla_{\mathbf{R}} v(\mathbf{R}-\mathbf{x}'')\cdot\langle \mathbf{j}(\mathbf{R},t)\rho_e(\mathbf{x}'',t)\rangle
&= -\nabla_{\mathbf{R}}\cdot\int d\mathbf{x}''\, v(\mathbf{R}-\mathbf{x}'')\langle \mathbf{j}(\mathbf{R},t)\rho_e(\mathbf{x}'',t)\rangle \\
&\quad + \int d\mathbf{x}''\, v(\mathbf{R}-\mathbf{x}'')\langle \nabla_{\mathbf{R}}\cdot\mathbf{j}(\mathbf{R},t)\rho_e(\mathbf{x}'',t)\rangle. \qquad (14.27)
\end{aligned}
$$

The continuity equation (14.17), written in operator form, gives

$$
\langle \nabla_{\mathbf{R}}\cdot\mathbf{j}(\mathbf{R},t)\rho_e(\mathbf{x}'',t)\rangle = -\left[\left(\frac{\partial}{\partial t}\right)\langle \rho_e(\mathbf{R},t)\rho_e(\mathbf{x}'',t')\rangle\right]_{t'=t} \qquad (14.28)
$$

and hence the last term on the right-hand side of (14.27) can be evaluated as

$$
\begin{aligned}
&\int d\mathbf{x}''\, v(\mathbf{R}-\mathbf{x}'')\langle \nabla_{\mathbf{R}}\cdot\mathbf{j}(\mathbf{R},t)\rho_e(\mathbf{x}'',t)\rangle \\
&\qquad = -\frac{1}{2}\int d\mathbf{x}''\, v(\mathbf{R}-\mathbf{x}'')\left(\frac{\partial}{\partial t}\right)\langle \rho_e(\mathbf{R},t)\rho_e(\mathbf{x}'',t)\rangle, \qquad (14.29)
\end{aligned}
$$

this following from (14.28) because of the symmetry of the correlation function $\langle \rho_e(\mathbf{R},t)\rho_e(\mathbf{x}'',t)\rangle$. But the electron-electron potential energy density $v_{ee}(\mathbf{R},t)$ is simply

$$
v_{ee}(\mathbf{R},t) = \frac{1}{2}\int d\mathbf{x}''\, v(\mathbf{R}-\mathbf{x}'')\langle \rho_e(\mathbf{R},t)\rho_e(\mathbf{x}'',t)\rangle \qquad (14.30)
$$

and so the right-hand side of (14.29) becomes $-(\partial/\partial t)v_{ee}(\mathbf{R},t)$.

Similarly, the second term on the right-hand side of (14.24) can be evaluated, yielding

$$
\begin{aligned}
&-\int d\mathbf{x}''\, \nabla_{\mathbf{R}} V_{ie}(\mathbf{R}-\mathbf{x}'')\cdot\langle \mathbf{j}(\mathbf{R},t)\rho_i(\mathbf{x}'',t)\rangle \\
&\quad = -\nabla_{\mathbf{R}}\cdot\int d\mathbf{x}''\, V_{ie}(\mathbf{R}-\mathbf{x}'')\langle \mathbf{j}(\mathbf{R},t)\rho_i(\mathbf{x}'',t)\rangle - \left(\frac{\partial}{\partial t}\right)V_{ie}(\mathbf{R},t) \qquad (14.31)
\end{aligned}
$$

where

$$
V_{ie}(\mathbf{R},t) = \int d\mathbf{x}''\, V_{ie}(\mathbf{R}-\mathbf{x}'')\langle \rho_e(\mathbf{R},t)\rho_i(\mathbf{x}'',t^-)\rangle. \qquad (14.32)
$$

Thus one can rewrite (14.24) as an energy transport equation for the electrons

$$\frac{\partial}{\partial t}\varepsilon(\mathbf{R},t) + \nabla_{\mathbf{R}}\cdot\mathbf{j}_\varepsilon(\mathbf{R},t) = 0, \tag{14.33}$$

whereas for the ions one has

$$\frac{\partial}{\partial t}\mathscr{E}(\mathbf{R},t) + \nabla_{\mathbf{R}}\cdot\mathbf{J}_{\mathscr{E}}(\mathbf{R},t) = 0. \tag{14.34}$$

Here the explicit forms for the energy densities for electrons and ions are

$$\varepsilon(\mathbf{R},t) = \sum_p \left(\frac{p^2}{2m}\right) f(\mathbf{p},\mathbf{R},t) + \int \mathrm{d}\mathbf{x}\, v(\mathbf{R}-\mathbf{x})\langle\rho_e(\mathbf{R},t)\rho_e(\mathbf{x},t^-)\rangle$$
$$+ \int \mathrm{d}\mathbf{x}\, V_{ie}(\mathbf{R}-\mathbf{x})\langle\rho_e(\mathbf{R},t)\rho_i(\mathbf{x},t^-) \tag{14.35}$$

and

$$\mathscr{E}(\mathbf{R},t) = \sum_{\mathbf{p}} \left(\frac{p^2}{2M}\right) F(\mathbf{p},\mathbf{R},t) + \int \mathrm{d}\mathbf{x}\, V(\mathbf{R}-\mathbf{x})\langle\rho_i(\mathbf{R},t)\rho_i(\mathbf{x},t^-)\rangle.$$
$$+ \int \mathrm{d}\mathbf{x}\, V_{ie}(\mathbf{R}-\mathbf{x})\langle\rho_i(\mathbf{R},t)\rho_e(\mathbf{x},t^-)\rangle. \tag{14.36}$$

Similarly, for the energy fluxes one has

$$j_\varepsilon(\mathbf{R},t) = \sum_{\mathbf{p}} \left(\frac{\mathbf{p}}{m}\right)\left(\frac{p^2}{2m}\right) f(\mathbf{p},\mathbf{R},t) + \int \mathrm{d}\mathbf{x}\, v(\mathbf{R}-\mathbf{x})\langle\mathbf{j}(\mathbf{R},t)\rho_e(\mathbf{x},t)\rangle$$
$$+ \int \mathrm{d}\mathbf{x}\, V_{ie}(\mathbf{R}-\mathbf{x})\langle\mathbf{j}(\mathbf{R},t)\rho_i(\mathbf{x},t)\rangle \tag{14.37}$$

and

$$J_\varepsilon(\mathbf{R},t) = \sum_{\mathbf{p}} \left(\frac{\mathbf{p}}{M}\right)\left(\frac{p^2}{2M}\right) F(\mathbf{p},\mathbf{R},t) + \int \mathrm{d}\mathbf{x}\, V(\mathbf{R}-\mathbf{x})\langle J(\mathbf{R},t)\rho_i(\mathbf{x},t)\rangle$$
$$+ \int \mathrm{d}\mathbf{x}\, V_{ie}(\mathbf{R}-\mathbf{x})\langle\mathbf{J}(\mathbf{R},t)\rho_e(\mathbf{x},t)\rangle. \tag{14.38}$$

This completes the set of continuity and conservation equations that was the objective of the above work. These equations are, of course, fully microscopic, and the hydrodynamic equations can be obtained from them by gradient expansions (March and Tosi, 1973; see also Appendix 14.3).

Next, the theory developed thus far will be applied to a variety of properties of liquid metals.

14.3. Electronic effects in dynamical structure

Here, the two-component theory just developed will be used to derive the partial dynamical structure factors for a pure liquid metal. The usual procedure assumes the hydrodynamic form for the ion-ion dynamical structure factor $S_{ii}(\mathbf{q}, \omega)$ to be the same as for a classical one-component liquid. In contrast, in the present treatment, which follows Tosi, Parrinello, and March (1974), the contribution to the sound-wave attenuation from single-particle excitations in the conduction electron system is derived explicitly.

Then the relation to neutron elastic scattering (see Chapter 8) will also be established and some contact with measurement on liquid gallium will prove possible. By studying the charge-charge correlation function for the two-component liquid metal, a theory of electrical resistivity is afforded. This is shown to reduce to Baym's (1964) treatment of the scattering of electrons by the density fluctuations in the limit of weak electron-ion interaction. The relation between the high-frequency conductivity, obtained for arbitrary strength of the electron-ion coupling, and the classical Drude-Zener theory is then demonstrated, and the sum rules for conductivity are seen to be related to those for $S_{ii}(\mathbf{q}, \omega)$ as the limit q tends to zero. This is followed by some discussion of the thermodynamics (see Chapter 3) of the two-component system, and this is again related to results for weak electron-ion coupling.

The outline of what follows consists first of the definition immediately below of longitudinal response functions, and formal solutions of the equations of motion, which are the analogues of classical continuity equations (see Section 14.1), are obtained. Naturally, to derive explicit results, decoupling procedures must subsequently be invoked.

Following this, the random phase approximation (RPA) and its generalizations are discussed. Although useful for some purposes, this approach is not adequate for treating transport phenomena. After the discussion of thermodynamics mentioned already, the response functions are cast into a form resembling hydrodynamics, and the properties of the generalized transport coefficients thereby introduced are developed. The account concludes with the application of this approach to single-particle motion and diffusion.

14.4. Longitudinal response of a two-component pure liquid metal

Let us consider the two-component liquid under the influence of weak, space- and time-dependent external potentials $V_j^{\text{ext}}(\mathbf{q}, \omega)$ acting on the density fluctuations of the two components ($j = 1, 2$). The density response functions $\chi_{jl}(\mathbf{q}, \omega)$ relate the variations in density of the two components to the external potentials through

$$n_j(\mathbf{q}, \omega) = \sum_l \chi_{jl}(\mathbf{q}, \omega) V_l^{\text{ext}}(\mathbf{q}, \omega). \tag{14.39}$$

Later, expressions for the response functions will be derived in terms of three-body dynamical correlation functions by a formal solution of the equations of motion for the density fluctuations, under the assumption that density fluctuations are not coupled to energy fluctuations. This is true when the ratio of the specific heats γ is near to unity. As seen in Chapter 8, this is not a bad approximation for simple s-p liquid metals near the melting point. The partial dynamical structure factors $S_{jl}(\mathbf{q}, \omega)$ are directly related to the response functions by the fluctuation-dissipation theorem

$$S_{jl}(\mathbf{q}, \omega) = -\frac{1}{\pi}(1 - \exp(-\beta\omega))^{-1} \operatorname{Im} \chi_{jl}(\mathbf{q}, \omega). \tag{14.40}$$

For the liquid metal, which is, of course, a two-component liquid of charged particles, it is also useful to introduce the Hartree potentials

$$V_j^H(\mathbf{q}, \omega) = V_j^{\text{ext}}(\mathbf{q}, \omega) + \sum_l v_{jl}(\mathbf{q}) n_l(\mathbf{q}, \omega) \tag{14.41}$$

where $v_{jl}(\mathbf{q})$ is the Fourier transform of the bare interaction potential between particles of types j and l, assumed to depend only on the relative distance between the two particles. One can then introduce, following Watabe and Hasegawa (1973), "proper" density response functions $\tilde{\chi}_{jl}(\mathbf{q}, \omega)$, which relate the density variations to the Hartree potentials:

$$n_j(\mathbf{q}, \omega) = \sum_l \tilde{\chi}_{jl}(\mathbf{q}, \omega) V_l^H(\mathbf{q}, \omega). \tag{14.42}$$

From (14.39)–(14.42), one then finds

$$\chi_{jj}(\mathbf{q},\omega) = \tilde{\chi}_{jj}(\mathbf{q},\omega)[1 + v_{jj}(\mathbf{q})\chi_{jj}(\mathbf{q},\omega) + v_{j\bar{j}}(\mathbf{q})\chi_{jj}(\mathbf{q},\omega)]$$
$$+ \tilde{\chi}_{j\bar{j}}(\mathbf{q},\omega)[v_{\bar{j}j}(\mathbf{q})\chi_{jj}(\mathbf{q},\omega) + v_{\bar{j}\bar{j}}(\mathbf{q})\chi_{\bar{j}j}(\mathbf{q},\omega)]; \quad (14.43)$$

and for the unlike components, $\bar{j}$ denoting a component of a type different from j:

$$\chi_{j\bar{j}}(\mathbf{q},\omega) = \tilde{\chi}_{jj}(\mathbf{q},\omega)[v_{jj}(\mathbf{q})\chi_{j\bar{j}}(\mathbf{q},\omega) + v_{j\bar{j}}(\mathbf{q})\chi_{\bar{j}\bar{j}}(\mathbf{q},\omega)]$$
$$+ \tilde{\chi}_{j\bar{j}}(\mathbf{q},\omega)[1 + v_{\bar{j}j}(\mathbf{q})\chi_{j\bar{j}}(\mathbf{q},\omega) + v_{\bar{j}\bar{j}}(\mathbf{q})\chi_{\bar{j}\bar{j}}(\mathbf{q},\omega)]. \quad (14.44)$$

The solution of (14.43) and (14.44) then yields the like and unlike χ's in terms of the proper response functions, the particle interactions v, and the quantity $\Delta(\mathbf{q},\omega)$ defined as

$$\Delta(\mathbf{q},\omega) = [1 - v_{11}(\mathbf{q})\tilde{\chi}_{11}(\mathbf{q},\omega) - v_{12}(\mathbf{q})\tilde{\chi}_{21}(\mathbf{q},\omega)]$$
$$\cdot[1 - v_{22}(\mathbf{q})\tilde{\chi}_{22}(\mathbf{q},\omega) - v_{21}(\mathbf{q})\tilde{\chi}_{12}(\mathbf{q},\omega)]$$
$$- [v_{11}(\mathbf{q})\tilde{\chi}_{12}(\mathbf{q},\omega) + v_{12}(\mathbf{q})\tilde{\chi}_{22}(\mathbf{q},\omega)]$$
$$\cdot[v_{21}(\mathbf{q})\tilde{\chi}_{11}(\mathbf{q},\omega) + v_{22}(\mathbf{q})\tilde{\chi}_{21}(\mathbf{q},\omega)]. \quad (14.45)$$

For a two-component plasma (one has $v_{jj}(\mathbf{q}) = -v_{j\bar{j}}(\mathbf{q})$) under an external electric field, $V_1^{\mathrm{ext}}(\mathbf{q},\omega) = -V_2^{\mathrm{ext}}(\mathbf{q},\omega)$, and the Hartree potentials are then given by

$$V_j^H(\mathbf{q},\omega) = \Delta^{-1}(\mathbf{q},\omega)V_j^{\mathrm{ext}}(\mathbf{q},\omega), \quad (14.46)$$

so that in this simple case the function $\Delta(\mathbf{q},\omega)$ reduces to the dielectric function of the plasma (see Section 10.6).

In a liquid metal like Na^+, e^-, it will be convenient to include in the Hartree potentials only the long-range Coulomb interactions and to account for the short-range parts of the bare ion-ion and electron-ion potentials through appropriate modification of the proper responses; this is illustrated later. Nonlocality of the short-range potentials can also be accounted for at the expense of some mathematical complication.

14.4.1. Formal solution of equations of motion

By combining the continuity equations for the particle densities of the two components with the equations of motion for the current densities derived in Section 14.2, one finds the equations of motion for the particle densities, following Singwi et al. (1970), as

$$\frac{\partial^2 n_j(\mathbf{R},t)}{\partial t^2} = \frac{1}{m_j}\nabla_\alpha\nabla_\beta\pi^j_{\alpha\beta}(\mathbf{R},t) + \frac{n_j}{m_j}\nabla^2 V_j^{\text{ext}}(\mathbf{R},t)$$

$$+ \frac{1}{m_j}\sum_l \int d\mathbf{x}\, \nabla_\alpha[\nabla_\alpha v_{je}(\mathbf{R}-\mathbf{x})\langle\rho_j(\mathbf{R},t)\rho_l(\mathbf{x},t)\rangle] \qquad (14.47)$$

where $\pi^j_{\alpha\beta}(\mathbf{R},t)$ is the kinetic-energy tensor for the jth component defined in terms of the Wigner single-particle distribution function $f_j(\mathbf{p},\mathbf{R},t)$ by

$$\pi^j_{\alpha\beta}(\mathbf{R},t) = \sum_{\mathbf{p}} \frac{p^j_\alpha p^j_\beta}{m_j} f_j(\mathbf{p},\mathbf{R},t) \qquad (14.48)$$

and $\langle\rho_j(\mathbf{R},t)\rho_l(\mathbf{x},t)\rangle$ are the two-body correlation functions in the perturbed liquid.

The kinetic-energy tensor and the two-body correlation functions in (14.47) are to be evaluated in the presence of the external potentials. For weak potentials, one may expand these quantities to linear terms in the density variations and formally solve (14.47) in terms of functional derivatives evaluated on the unperturbed liquid. One can thus write for the variations produced by the external fields

$$\delta\pi^j_{\alpha\beta}(\mathbf{R},t) = \iint d\mathbf{y}\, d\tau \sum_l \left.\frac{\delta\pi^j_{\alpha\beta}(\mathbf{R},t)}{\delta n_l(\mathbf{y},\tau)}\right|_{\text{eq}} \delta n_l(\mathbf{y},\tau)$$

$$\equiv \iint d\mathbf{y}\, d\tau \sum_l \lambda^{jl}_{\alpha\beta}(\mathbf{R}-\mathbf{y}, t-\tau)\delta n_l(\mathbf{y},\tau) \qquad (14.49)$$

$$\delta\langle\rho_j(\mathbf{R},t)\rho_l(\mathbf{x}t)\rangle_c = \iint d\mathbf{y}\, d\tau \sum_s \left.\frac{\delta\langle\rho_j(\mathbf{R},t)\rho_e(\mathbf{x},t)\rangle_c}{\delta n_s(\mathbf{y},\tau)}\right|_{\text{eq}} \delta n_s(\mathbf{y},\tau)$$

$$\equiv \iint d\mathbf{y}\, d\tau \sum_s F_{jls}(\mathbf{R}-\mathbf{x}, \mathbf{x}-\mathbf{y}, t-\tau)\delta n_s(\mathbf{y},\tau) \qquad (14.50)$$

where the suffix c denotes the cluster part of the two-body correlation function (Tosi, Parrinello, and March, 1974).

One can then solve the equations of motion (14.47) in Fourier transform to find

$$\chi_{jj}(\mathbf{q},\omega) = \frac{\dfrac{n_j q^2}{m_j}\left[\omega^2 - \dfrac{1}{m_{\bar{j}}}\Gamma_{\bar{j}\bar{j}}(\mathbf{q},\omega)\right]}{\Gamma(\mathbf{q},\omega)} \qquad (14.51)$$

$$\chi_{j\bar{j}}(\mathbf{q},\omega) = \frac{\dfrac{n_{\bar{j}}q^2}{m_{\bar{j}}m_j}\Gamma_{j\bar{j}}(\mathbf{q},\omega)}{\Gamma(\mathbf{q},\omega)} \tag{14.52}$$

and

$$\tilde{\chi}_{jj}(\mathbf{q},\omega) = \frac{\dfrac{n_j q^2}{m_j}\left[\omega^2 - \dfrac{1}{m_{\bar{j}}}\tilde{\Gamma}_{\bar{j}\bar{j}}(\mathbf{q},\omega)\right]}{\tilde{\Gamma}(\mathbf{q},\omega)} \tag{14.53}$$

$$\tilde{\chi}_{j\bar{j}}(\mathbf{q},\omega) = \frac{\dfrac{n_{\bar{j}}q^2}{m_{\bar{j}}m_j}\tilde{\Gamma}_{j\bar{j}}(\mathbf{q},\omega)}{\tilde{\Gamma}(\mathbf{q},\omega)}. \tag{14.54}$$

In these equations the functions $\tilde{\Gamma}_{jl}(\mathbf{q},\omega)$ are defined in terms of the Fourier transforms of the functions given in (14.49) and (14.50) by

$$\tilde{\Gamma}_{jl}(\mathbf{q},\omega) = q_\alpha q_\beta \lambda^{jl}_{\alpha\beta}(\mathbf{q},\omega) + \sum_s \sum_{\mathbf{k}} \mathbf{q}\cdot\mathbf{k}\, v_{js}(\mathbf{k}) F_{jsl}(\mathbf{q}-\mathbf{k},\mathbf{q},\omega). \tag{14.55}$$

Furthermore,

$$\Gamma_{jl}(\mathbf{q},\omega) = n_j q^2 v_{jl}(\mathbf{q}) + \tilde{\Gamma}_{jl}(\mathbf{q},\omega), \tag{14.56}$$

$$\tilde{\Gamma}(\mathbf{q},\omega) = \left[\omega^2 - \frac{1}{m_1}\tilde{\Gamma}_{11}(\mathbf{q},\omega)\right] \times \left[\omega^2 - \frac{1}{m_2}\tilde{\Gamma}_{22}(\mathbf{q},\omega)\right] - \frac{1}{m_1 m_2}\tilde{\Gamma}_{12}(\mathbf{q},\omega)\tilde{\Gamma}_{21}(\mathbf{q},\omega) \tag{14.57}$$

and

$$\Gamma(\mathbf{q},\omega) = \left[\omega^2 - \frac{1}{m_1}\Gamma_{11}(\mathbf{q},\omega)\right]\left[\omega^2 - \frac{1}{m_2}\Gamma_{22}(\mathbf{q},\omega)\right] - \frac{1}{m_1 m_2}\Gamma_{12}(\mathbf{q},\omega)\Gamma_{21}(\mathbf{q},\omega). \tag{14.58}$$

The generalized dielectric function is given by the ratio

$$\Delta(\mathbf{q},\omega) = \frac{\Gamma(\mathbf{q},\omega)}{\tilde{\Gamma}(\mathbf{q},\omega)}. \tag{14.59}$$

Equations (14.51) and (14.52), or alternatively, (14.53) and (14.54), embody the generalized hydrodynamics for longitudinal phenomena in the two-component liquid, in which the functions $\tilde{\Gamma}_{jl}(\mathbf{q},\omega)$ play the role of generalized transport coefficients. In the long-wavelength limit, their imaginary part will give the usual transport coefficients (sound attenuation, electrical resistivity, and plasmon damping), whereas their real part will contribute to determine the energies of the collective excitations in the liquid (sound velocity and plasmon frequency shift and dispersion). Exact expressions for these functions are provided through (14.55) in terms of functional derivatives, that is, in terms of higher-order correlation functions (see Parrinello and Tosi, 1972). One can, in fact, write

$$\begin{aligned}\frac{\delta\langle\rho_j(\mathbf{R},t)\rho_s(\mathbf{x},t)\rangle_c}{\delta n_l(\mathbf{y},\tau)} &= \iint d\mathbf{y}'\,d\tau' \sum_m \frac{\delta\langle\rho_j(\mathbf{R},t)\rho_s(\mathbf{x},t)\rangle_c}{\delta V_m^{\text{ext}}(\mathbf{y}',\tau')}\,\frac{\delta V_m^{\text{ext}}(\mathbf{y}',\tau')}{\delta n_l(\mathbf{y},\tau)} \\ &= -\frac{i}{\hbar}\iint d\mathbf{y}'\,d\tau' \sum_m \chi_{ml}^{-1}(\mathbf{y}'-\mathbf{y},\tau'-\tau) \\ &\quad\times \langle T[\rho_m(\mathbf{y}',\tau')\rho_j(\mathbf{R},t)\rho_s(\mathbf{x},t)]\rangle_c \end{aligned} \tag{14.60}$$

or

$$F_{jsl}(\mathbf{q}-\mathbf{k},\mathbf{q},\omega) = -\frac{i}{\hbar}\sum_m \chi_{ml}^{-1}(\mathbf{q},\omega)\,T_{jsm}(\mathbf{q}-\mathbf{k},\mathbf{q},\omega), \tag{14.61}$$

where $\chi_{ml}^{-1}(\mathbf{y}'-\mathbf{y},\tau'-\tau)$ is the inverse response matrix and $T_{jsm}(\mathbf{q}-\mathbf{k},\mathbf{q},\omega)$ is the Fourier transform of the dynamic triplet correlation function introduced in (14.60). Similarly one has

$$\begin{aligned}\lambda_{\alpha\beta}^{jl}(\mathbf{R}-\mathbf{y},t-\tau) &= \frac{i\hbar}{m_j}\iint d\mathbf{y}'\,d\tau' \sum_s \chi_{sl}^{-1}(\mathbf{y}'-\mathbf{y},\tau',\tau) \\ &\quad\cdot\{\nabla_\alpha^{(\mathbf{r})}\nabla_\beta^{(\mathbf{r})}\langle T[\psi_j^+(\mathbf{R}+\tfrac{1}{2}\mathbf{r},t)\psi_j(\mathbf{R}-\tfrac{1}{2}\mathbf{r},t)\rho_s(\mathbf{y}',\tau')]\rangle_c\}_{r=0}\end{aligned} \tag{14.62}$$

The appearance of the inverse response function in (14.60) indicates that the functional derivative is to be evaluated as a triplet function in which the vertex at $\rho_l(\mathbf{y},\tau)$ is to be taken as a proper vertex—namely, as a triplet function in which the densities $\rho_j(\mathbf{R},t)$ and $\rho_s(\mathbf{x},t)$ correlate with the screened density $\rho_l^{sc}(\mathbf{y},\tau)$. Similarly, (14.62) involves correlations between the kinetic-tensor operator and the screened density.

In actual calculations, appropriate decouplings of the functional derivatives will have to be devised, depending on the physical situation. A simple example of decoupling of the expression (14.60) for weak electron-ion and electron-electron interactions, which is equivalent to Baym's (1964) treatment by the Boltzmann equation, will be given in the following discussion of electrical resistivity. Decoupling schemes for strong electron-electron interactions in the homogeneous electron liquid have been developed and analyzed by a number of workers (see also Vashishta and Singwi, 1972;

Goodman and Sjölander, 1975). For the kinetic terms (14.62), experience on the homogeneous electron liquid suggests that a single-particle approximation may be sufficient for the electrons, provided that the dynamical effects are preserved in the expression (14.60). On the other hand, the treatment of the kinetic terms for the ions may appeal to approximations developed for such contributions in liquid argon (Hubbard and Beeby, 1969; also see Chapter 8: Singwi, Sköld and Tosi 1970; Pathak and Singwi, 1970).

It is of interest here to note that the above formal solution of the equations of motion (Tosi, Parrinello, and March, 1974) is akin of Mori's formulation of transport (Mori, 1965a) in terms of a generalized Langevin equation (see also Kubo, 1966). The developments just given show that the relevant quantities in the theory are the perturbations of the kinetic tensors and of the internal forces in the liquid.

14.4.2. Effective potentials

The physical significance of (14.51) and (14.52) becomes apparent if one rewrites them in terms of effective single-component response functions and effective potentials. One defines

$$\tilde{\chi}_j(\mathbf{q},\omega) = \frac{n_j q^2}{m_j}\left[\omega^2 - \frac{1}{m_j}\tilde{\Gamma}_{jj}(\mathbf{q},\omega)\right]^{-1} \tag{14.63}$$

and

$$\chi_j(\mathbf{q},\omega) = \tilde{\chi}_j(\mathbf{q},\omega)\left[1 - v_{jj}(\mathbf{q})\tilde{\chi}_j(\mathbf{q},\omega)\right]^{-1} \equiv \frac{\tilde{\chi}_j(\mathbf{q},\omega)}{\varepsilon_j(\mathbf{q},\omega)} \tag{14.64}$$

as effective response functions for the jth component. One also defines the effective potentials

$$\tilde{v}_{jj}(\mathbf{q},\omega) = v_{jj}(\mathbf{q}) + \frac{|\tilde{v}_{j\bar{j}}(\mathbf{q},\omega)|^2}{v_{\bar{j}\bar{j}}(\mathbf{q})}\left[\frac{1}{\varepsilon_{\bar{j}}(\mathbf{q},\omega)} - 1\right] \tag{14.65}$$

and

$$\tilde{v}_{j\bar{j}}(\mathbf{q},\omega) = v_{j\bar{j}}(\mathbf{q}) + \frac{1}{n_j q^2}\tilde{\Gamma}_{j\bar{j}}(\mathbf{q},\omega). \tag{14.66}$$

Equations (14.51) and (14.52) can then be written in the forms

$$\chi_{jj}(\mathbf{q},\omega) = \frac{\tilde{\chi}_j(\mathbf{q},\omega)}{1 - v_{jj}(\mathbf{q},\omega)\tilde{\chi}_j(\mathbf{q},\omega)} \tag{14.67}$$

and

$$\chi_{j\bar{j}}(\mathbf{q},\omega) = \chi_j(\mathbf{q},\omega)\tilde{v}_{j\bar{j}}(\mathbf{q},\omega)\chi_{\bar{j}\bar{j}}(\mathbf{q},\omega). \tag{14.68}$$

According to (14.67) and (14.65), each component responds to a direct perturbation as if it were effectively a single component with an interparticle potential screened by the other component. Similarly, according to (14.68), each component responds to an indirect perturbation as if it were effectively a single component responding to a direct external field produced by the polarization of the other component, to which it is coupled via a modified interaction potential given by (14.66). It is to be noted that the same effective intercomponent potential $\tilde{v}_{j\bar{j}}(\mathbf{q},\omega)$ enters both (14.68) and (14.65).

Of course, the proper response (14.63), although formally defined for a single component, still contains all the effects of the electron-ion interactions entering in $\tilde{\Gamma}_{jj}(\mathbf{q},\omega)$. Such effects of the electron-ion interaction on the electronic dielectric function have been discussed in detail for the solid state (see Pethick, 1970), in connection with phonon and plasmon properties. However, according to (14.66), liquid (or solid-) state effects also enter to modify the electron-ion potential, and such effects are usually not explicitly accounted for in pseudopotential models of the effective ion-ion interaction. As is clear from (14.66) and (14.55), the modifications to the electron-ion potential depend on the electron-ion correlations in the liquid metal and furthermore are frequency-dependent.

In particular, anticipating a result that will be derived shortly, at long wavelengths and low frequencies one finds

$$\lim_{\omega\to 0}\lim_{\mathbf{q}\to 0}\tilde{v}_{ie}(\mathbf{q},\omega) = -\frac{4\pi Ze^2}{q^2}\left(1 - \frac{i\omega}{4\pi\sigma}\right) \sim \frac{-4\pi Ze^2}{q^2}\left(1 - \frac{1}{\varepsilon(0,\omega)}\right), \tag{14.69}$$

where σ is the electrical conductivity and $\varepsilon(0,\omega)$ the dynamic dielectric function of the liquid metal.

14.5. Random phase approximation (RPA) for response functions

The random phase approximation (RPA) replaces the proper response functions by the response functions of an ideal gas; i.e., it sets $F_{jsl} = 0$ in

(14.55) and replaces $\lambda_{\alpha\beta}^{jl}(\mathbf{q}, \omega)$ by its free-particle value. Then one finds

$$\tilde{\chi}_{jl}^{\mathrm{RPA}}(\mathbf{q}, \omega) = \delta_{jl}\chi_j^0(\mathbf{q}, \omega) = \delta_{jl}\frac{n_j q^2}{m_j}\left[\omega^2 - \frac{l}{m_j} q_\alpha q_\beta \lambda_{\alpha\beta}^{jj}(\mathbf{q}, \omega)\right]_{\mathrm{free}}^{-1} \tag{14.70}$$

where $\chi_j^0(\mathbf{q}, \omega)$ is the free-particle polarization function. Equations (14.63) and (14.64) correspondingly give (see Watabe and Hasegawa, 1973; Chihara, 1973)

$$\chi_{jj}^{\mathrm{RPA}}(\mathbf{q}, \omega) = \chi_j^0(\mathbf{q}, \omega)\frac{l - v_{\bar{j}\bar{j}}(\mathbf{q})\chi_{\bar{j}}^0(\mathbf{q}, \omega)}{\Delta^{\mathrm{RPA}}(\mathbf{q}, \omega)} \tag{14.71}$$

and

$$\chi_{j\bar{j}}^{\mathrm{RPA}}(\mathbf{q}, \omega) = \frac{\chi_j^0(\mathbf{q}, \omega) v_{j\bar{j}}(\mathbf{q})\chi_{\bar{j}}^0(\mathbf{q}, \omega)}{\Delta^{\mathrm{RPA}}(\mathbf{q}, \omega)} \tag{14.72}$$

where

$$\Delta^{\mathrm{RPA}}(\mathbf{q}, \omega) = [1 - v_{\mathrm{ii}}(\mathbf{q})\chi_{\mathrm{i}}^0(\mathbf{q}, \omega)][1 - v_{\mathrm{ee}}(\mathbf{q})\chi_{\mathrm{e}}^0(\mathbf{q}, \omega)] - |v_{\mathrm{ie}}(\mathbf{q})|^2\chi_{\mathrm{i}}^0(\mathbf{q}, \omega)\chi_e^0(\mathbf{q}, \omega). \tag{14.73}$$

If one assumes that the ions behave classically and that the electron liquid is fully degenerate, plus the assumption that the electronic mean free path is sufficiently long compared with q^{-1} (which is not the case in liquid metals, as discussed further below), then one has in the hydrodynamic limit

$$\chi_i^0(\mathbf{q}, \omega) \to \frac{n_{\mathrm{i}} q^2}{m_{\mathrm{i}}\omega^2} - i n_{\mathrm{i}}\beta\left(\frac{\pi m_{\mathrm{i}}\beta\omega^2}{2q^2}\right)^{1/2} \exp\left(\frac{-m_{\mathrm{i}}\beta\omega^2}{2q^2}\right) \tag{14.74}$$

for the ions and

$$\chi_{\mathrm{e}}^0(\mathbf{q}, \omega) \to \begin{cases} -\dfrac{3n_{\mathrm{e}}}{m_{\mathrm{e}} v_{\mathrm{f}}^2}\left(1 + i\dfrac{\pi}{2}\dfrac{\omega}{q v_{\mathrm{f}}}\right), & \omega < q v_{\mathrm{f}} \\ \dfrac{n_e q^2}{m_{\mathrm{e}}\omega^2}, & \omega > q v_{\mathrm{f}} \end{cases} \tag{14.75}$$

for the electrons. For a two-component plasma one then finds the well-known result

$$\chi_{\mathrm{ii}}^{\mathrm{RPA}}(\mathbf{q}, \omega) \to \frac{n_{\mathrm{i}} q^2/m_{\mathrm{i}}}{\omega^2 - s_0^2 q^2(1 - i(\pi/2)(\omega/q v_{\mathrm{f}})}, \tag{14.76}$$

where $s_0 = v_{\mathrm{f}}(Zm_{\mathrm{e}}/3m_{\mathrm{i}})^{1/2}$ is the Bohm-Staver sound velocity. Thus the RPA accounts for sound attenuation through the excitation of single electron-hole pairs. It is worth remarking that the damping in space of a sound wave, written conventionally as $\exp(-\frac{1}{2}\alpha z)$, is then characterized by $\alpha = (\pi/2)(\omega/v_f)$, which is the same result as given by Pippard (1960) for electron mean free path $\gg q^{-1}$. On the other hand, it is easily seen from (14.75) that the usual viscosity, as well as plasmon damping and electrical resistivity, are missing from the theory.

It is to be noted here that the generalized RPA theories that have been developed for the homogeneous electron liquid (Singwi et al., 1970, see also Singwi and Tosi, 1981) and for electron-hole drops in semiconductors (see, for example, Vashishta et al., 1973) still adopt the RPA expression for the proper polarization functions (14.70) but take into account the short-range correlations between the particles by replacing the bare potentials by static effective potentials. The latter involve the static structure factors, according to expressions that can be obtained by appropriate approximations (see Parrinello and Tosi, 1972) on the functional derivatives introduced in (14.49) and (14.50). Such theories seem to lead to rather reliable results for the static structure factors but are equivalent to RPA in the hydrodynamic limit, albeit with a modified sound velocity. The frequency dependence of the effective potentials must be included for a description of transport phenomena. Applications of such theories to liquid metals (see Watabe and Hasegawa, 1973; Chihara, 1973) are nevertheless useful from the point of view of improving the description of the arrangement of the electrons around the ions beyond linear screening theory; such arrangement being of primary concern in the determination of the effective ion-ion interaction.

Before turning to the discussion of hydrodynamic behaviour, it will be useful next to deal with the thermodynamics resulting from this two-component theory—that is, the zero-frequency and long-wavelength regime of the response functions.

14.6. Thermodynamics

To introduce the thermodynamics, it is useful to recall briefly that there are a variety of arguments, both static (see (1) and (2)) and dynamic (3) by which one is led to the Bohm-Staver formula (see Appendix 6.2) for the velocity of sound, referred to earlier in connection with (14.76). Examples of these are:

1. Derivation from the equation of state $P \propto n_e^{5/3}$ of a free Fermi gas, by calculating $\partial P/\partial d$, d being the mass density,

2. Use of the compressibility formula, via the direct or Ornstein-Zernike function $C_{ii}(r)$, and
3. Starting out from the ionic plasma frequency, which, by pileup of electrons around the ions, is then converted from an optic to an acoustic mode.

All these apparently dissimilar methods lead, in linear response, to the same Bohm-Staver result. For example, for point ions of charge Ze, method (2) worked out semiclassically leads to an effective ion-ion interaction

$$\tilde{v}_{ii}(r) = \frac{Z^2 e^2 \exp(-q_{TF} r)}{r} \tag{14.77}$$

q_{TF}^{-1} being the Thomas-Fermi screening radius derived in Chapter 4. The Ornstein-Zernike function is then constructed in RPA as

$$C_{ii}(q) = -n_i \beta \tilde{v}_{ii}(q). \tag{14.78}$$

Hence $C_{ii}(0)$ is found and using $S_{ii}(0) \doteqdot 1/C_{ii}(0)$, the usual result is recovered.

Method (1) uses only the electronic properties in the first step leading to the equation of state, whereas method (2) focuses on the ionic assembly alone. These approaches are, in fact, special cases of statistical thermodynamic results discussed, for instance, by Gray (1973). In these methods one can obtain the isothermal compressibility by starting from the chemical potential of either electrons or ions, say μ_e and μ_i, respectively. The electronic route is then to differentiate μ_e with respect to n_e at constant μ_i. If one denotes the kinetic energy density of the conduction electrons in the presence of the ions by $t_e[n_e]$, one can write the following for the electron chemical potential, in terms of the pair functions $g_{ee}(r)$ and $g_{ie}(r)$ for electron-electron and electron-ion correlations, respectively:

$$\mu_e = \frac{\delta t_e[n_e]}{\delta n_e} + \frac{\partial}{\partial n_e}\left\{\frac{1}{2} n_e \int d\mathbf{r}[g_{ee}(r) - 1] v_{ee}(r) + n_i \int d\mathbf{r}[g_{ie}(r) - 1] v_{ie}^{C}(r) + n_i \int d\mathbf{r}\, g_{ie}(r) v_{ie}^{s}(r)\right\} \tag{14.79}$$

where $v_{ie}^{C}(r)$ and $v_{ie}^{s}(r)$ are the Coulomb and short-range parts of the electron-ion potential.

The Bohm-Staver formula includes simply the first term, the kinetic energy density t_e being taken as the free Fermi gas form. On the other hand, the chemical potential of the ions is given by

$$\mu_i = k_B T + \frac{\partial}{\partial n_i}\left\{n_e \int d\mathbf{r}\, \bar{g}_{ie}(r) v_{ie}(r) + \frac{1}{2} n_i \int d\mathbf{r}\, \bar{g}_{ii}(r) v_{ii}(r)\right\} \tag{14.80}$$

where the notation $\bar{g}(r)$ has been used to indicate that $g(r)$ should be replaced by $g(r) - 1$ in the Coulomb part of the integrals. Again, in principle, it is possible to calculate the compressibility from (14.80). However, to the writer's knowledge, no practical application of (14.80) has been made at the time of writing.

This is the point at which to summarize the main thermodynamic results of the two-component theory. First, in the thermodynamic limit (zero frequency and long wavelength), one can write

$$\lim_{\mathbf{q}\to 0} \Gamma_{jl}(\mathbf{q}, 0) = 4\pi n_j e_j e_l + \alpha_{jl} q^2 \tag{14.81}$$

and then one finds from (14.51) and (14.52) that

$$\lim_{\mathbf{q}\to 0} \left[\chi_{\text{ii}}(\mathbf{q}, 0) = \frac{1}{Z} \chi_{\text{ie}}(\mathbf{q}, 0) = \frac{1}{Z^2} \chi_{\text{ee}}(\mathbf{q}, 0) \right]$$
$$= -n_{\text{i}}(\alpha_{\text{ii}} + \alpha_{\text{ei}} + Z\alpha_{\text{ie}} + Z\alpha_{\text{ee}})^{-1}. \tag{14.82}$$

On the other hand, the requirement that density fluctuations of either component be completely neutralized in this limit leads to the relations (see Watabe and Hasegawa, 1973; Chihara, 1973)

$$\lim_{\mathbf{q}\to 0} \left[S_{\text{ii}}(\mathbf{q}) = \frac{1}{Z^{1/2}} S_{\text{ie}}(\mathbf{q}) = \frac{1}{Z} S_{\text{ee}}(\mathbf{q}) \right] = n_{\text{i}} k_{\text{B}} T K_{\text{T}} \tag{14.83}$$

between the static structure factors, where K_{T} is the isothermal compressibility of the liquid. If one takes the ions to behave classically so that

$$\lim_{\mathbf{q}\to 0} [\chi_{\text{ii}}(\mathbf{q}, 0) = -n_{\text{i}} \beta S_{\text{ii}}(\mathbf{q})], \tag{14.84}$$

then one finds that the electrons also behave classically in this limit, namely,

$$\lim_{\mathbf{q}\to 0} [\chi_{jl}(\mathbf{q}, 0) = -(n_j n_l)^{1/2} \beta S_{jl}(\mathbf{q})] = -n_j n_l K_{\text{T}}. \tag{14.85}$$

This equation represents the compressibility sum rule for the liquid metal.

It is straightforward to show from the definitions of the functions $\Gamma_{jl}(\mathbf{q}, \omega)$ in terms of functional derivatives that the quantity $n_{\text{i}}(\alpha_{\text{ii}} + \alpha_{\text{ei}} + Z\alpha_{\text{ie}} + Z\alpha_{\text{ee}})$ gives the isothermal bulk modulus of the liquid. In the thermodynamic limit the kinetic tensor and the two-body correlations depend only on the local density and its derivatives, and taking due account of the symmetry properties of the pair correlations, one can write (Fröhlich, 1967)

$$\delta\pi^j_{\alpha\beta}(\mathbf{R}, t) = \sum_l \frac{\partial \pi^j_{\alpha\beta}}{\partial n_l} \delta n_l(\mathbf{R}, t) + \cdots \tag{14.86}$$

and

$$\delta\langle\rho_j(\mathbf{R},t)\rho_l(\mathbf{x},t)\rangle_c = \tfrac{1}{2}[\delta n_j(\mathbf{R},t)\gamma_{jl}(r) + \delta n_l(\mathbf{x},t)\gamma_{lj}(r)] + \cdots, \quad (14.87)$$

where $\mathbf{r} = \mathbf{x} - \mathbf{R}$ and

$$\gamma_{jl}(r) = \frac{\partial\langle\rho_j(\mathbf{R})\rho_l(\mathbf{x})\rangle_c}{\partial n_j}. \quad (14.88)$$

Still taking the ions to behave classically, one finds

$$\lim_{\mathbf{q}\to 0} q_\alpha q_\beta \lambda^{\mathrm{ii}}_{\alpha\beta}(\mathbf{q},0) = \frac{\partial}{\partial n_{\mathrm{i}}}\sum_{\mathbf{p}}\frac{(\mathbf{q}\cdot\mathbf{p})^2}{m_{\mathrm{i}}}f_{\mathrm{i}}(\mathbf{p}) = \frac{2}{3}q^2\frac{\partial t_{\mathrm{i}}}{\partial n_{\mathrm{i}}} = q^2 k_{\mathrm{B}}T; \quad (14.89)$$

similarly,

$$\lim_{\mathbf{q}\to 0} q_\alpha q_\beta \lambda^{\mathrm{ie}}_{\alpha\beta}(\mathbf{q},0) = 0 \quad (14.90)$$

$$\lim_{\mathbf{q}\to 0} q_\alpha q_\beta \lambda^{\mathrm{ee}}_{\alpha\beta}(\mathbf{q},0) = \frac{2}{3}q^2\frac{\partial t_{\mathrm{e}}}{\partial n_{\mathrm{e}}}. \quad (14.91)$$

From (14.87) one also finds

$$\lim_{\mathbf{q}\to 0}\sum_s\sum_{\mathbf{k}}\mathbf{q}\cdot\mathbf{k}v_{js}(\mathbf{k})F_{jsl}(\mathbf{q}-\mathbf{k},\mathbf{q},0)$$

$$= \lim_{\mathbf{q}\to 0}\frac{1}{2}\sum_s\sum_{\mathbf{k}}\mathbf{q}\cdot\mathbf{k}v_{js}(\mathbf{k})\int d\mathbf{r}[\delta_{jl}\gamma_{js}(r)\exp(i\mathbf{k}\cdot\mathbf{r}) + \delta_{sl}\gamma_{sj}(r)\exp(i\{\mathbf{k}-\mathbf{q}\}\cdot\mathbf{r})]$$

$$= \lim_{\mathbf{q}\to 0}\frac{1}{2}\sum_{\mathbf{k}}\mathbf{q}\cdot\mathbf{k}v_{jl}(\mathbf{k})\int d\mathbf{r}\,\gamma_{lj}(r)\exp(i\{\mathbf{k}-\mathbf{q}\}\cdot\mathbf{r})$$

$$= -\frac{i}{2}\sum_{\mathbf{k}}\mathbf{q}\cdot\mathbf{k}v_{jl}(\mathbf{k})\int d\mathbf{r}\,\gamma_{lj}(r)\mathbf{q}\cdot\mathbf{r}$$

$$= -\frac{1}{6}q^2\int d\mathbf{r}\,\gamma_{lj}(r)\mathbf{r}\cdot\nabla v_{jl}(r). \quad (14.92)$$

The short-range deviations of the bare ion-ion and electron-ion potentials from a Coulomb potential, which contribute to the α_{jl}'s according to (14.81), can be accounted for by replacing the cluster correlation $g_{jl}(r) - 1$ by the full correlation function $g_{jl}(r)$ in the evaluation of $\gamma_{jl}(r)$ for the short-range contributions to (14.92). The final expression is

$$n_{\mathrm{i}}(\alpha_{\mathrm{ii}} + \alpha_{\mathrm{ei}} + Z\alpha_{\mathrm{ie}} + Z\alpha_{\mathrm{ee}})$$

$$= nk_{\mathrm{B}}T + n\frac{\partial}{\partial n}\left[\frac{2}{3}t_{\mathrm{e}} - \frac{1}{6}\sum_{jl}n_j n_l\int d\mathbf{r}\,\bar{g}_{jl}(r)\mathbf{r}\cdot\nabla v_{jl}(r)\right], \quad (14.93)$$

where the right-hand side is clearly the value of the bulk modulus according to the virial theorem.

As mentioned before, K_T can also be evaluated as the compressibility of the electron liquid (or, alternatively, of the ion liquid) in the presence of the electron-ion interactions by evaluating the density derivative of the electron chemical potential. However, it is then necessary to account for the shift of the electronic chemical potential due to the electron-ion interactions, as has been emphasized for solid metals by Pethick (1970). This alternative route to the evaluation of K_T requires, in contrast to (14.93), knowledge of the second density derivative of the pair correlations.

Finally, it is of interest to comment on the relation of (14.93) to the result for weak electron-ion interaction. First, one may note that, for a classical one-component system with pair potential $\phi(r)$, the familiar expression for the pressure P (see Chapter 3) is

$$P = nk_B T - \tfrac{1}{6}n^2 \int d\mathbf{r}\, g(r)\mathbf{r} \cdot \nabla\phi(r). \tag{14.94}$$

In the customary treatment of a metal, the effective potential energy of the ionic system is written in the form of a volume-dependent term, which is structure independent, and a pair potential term, the pair potential being density dependent, as stressed in Chapter 3 on dielectric screening. Thus, if one makes the Born-Oppenheimer approximation and then writes

$$U = Nu_0(n) + \tfrac{1}{2}\sum_{i \neq i'} \phi(r_{ii'}, n) + \tfrac{3}{2}Nk_B T, \tag{14.95}$$

the expression (14.94) is generalized to read (Ascarelli and Harrison, 1969; see also Chapter 3)

$$P = nk_B T + n^2 \frac{du_0}{dn} - \frac{1}{2}n^2 \int d\mathbf{r}\, g(r)\left[\frac{1}{3}\mathbf{r} \cdot \nabla\phi(r, n) - n\frac{\partial}{\partial n}\phi(r, n)\right]. \tag{14.96}$$

The result (14.93) represents the generalization of this formula when (a) the Born-Oppenheimer approximation is not invoked and (b) the electron-ion interaction is not weak.

To demonstrate the way in which (14.95) follows from the general result, let us write the total internal energy density of the liquid metal as

$$u = \tfrac{3}{2}n_i k_B T + t_e + \tfrac{1}{2}\int d\mathbf{r}\{n_e^2[g_{ee}(r) - 1]v_{ee}(r) + n_i^2 \bar{g}_{ii}(r)v_{ii}(r)$$
$$+ 2n_i n_e \bar{g}_{ie}(r)v_{ie}(r)\}, \tag{14.97}$$

where t_e and $g_{ee}(r)$ are the kinetic-energy density and the pair correlation function for the electrons in the presence of electron-ion interactions. Under the assumptions (a) and (b) specified above, one can consider the switching on of the electron-ion interaction as a process of polarization of the electron liquid by the ions in their fixed configuration. By a standard result of linear response theory, one-half of the energy gained by the interaction of the polarization with the polarizing field cancels against the "quasi-elastic" energy spent in creating the polarization. The former energy in the present case is

$$\frac{1}{2}\int d\mathbf{r}\, 2n_i n_e[g_{ie}(r) - 1]v_{ie}(r),$$

whereas the latter is

$$t_e + \frac{1}{2}\int d\mathbf{r}\, n_e^2[g_{ee}(r) - 1]v_{ee}(r) - u_h,$$

where u_h is the energy density of the homogeneous electron liquid at the electronic density obtaining in the liquid metal. One can therefore write

$$u = \frac{3}{2}n_i k_B T + u_h + \frac{1}{2}\int d\mathbf{r}\{n_i^2\bar{g}_{ii}(r)v_{ii}(r) + n_i n_e \bar{g}_{ie}(r)v_{ie}(r) + n_i n_e v_{ie}^s(r)\}. \tag{14.98}$$

Still taking the electron-ion interaction as weak, the electron-ion structure factor is related to the ion-ion structure factor by (Tosi and March, 1973a)

$$S_{ie}(\mathbf{k}) = \left(\frac{n_i}{n_e}\right)^{1/2} S_{ii}(\mathbf{k})\frac{v_{ie}(\mathbf{k})}{v_{ee}(\mathbf{k})}\left[\frac{1}{\varepsilon(\mathbf{k})} - 1\right] \tag{14.99}$$

where $\varepsilon(k)$, as in Chapter 4, is the static dielectric function of the electron liquid. Equation (14.98) becomes

$$\begin{aligned} u = {} & \frac{3}{2}n_i k_B T + u_h + \frac{1}{2}n_i[\phi_{ii}(r) - v_{ii}(r)]_{r=0} \\ & + \frac{1}{2}\int d\mathbf{r}[n_i^2 v_{ii}^s(r) + 2n_i n_e v_{ie}^s(r) - n_i^2\phi_{ii}(r)] \\ & + \frac{1}{2}n_i^2\int d\mathbf{r}\, g_{ii}(r)\phi_{ii}(r), \end{aligned} \tag{14.100}$$

where $\phi_{ii}(r)$ is the screened ion-ion potential, given as usual in Fourier transform by

$$\phi_{ii}(\mathbf{k}) = v_{ii}(\mathbf{k}) + \frac{|v_{ie}(\mathbf{k})|^2}{v_{ee}(\mathbf{k})}\left[\frac{1}{\varepsilon(\mathbf{k})} - 1\right]. \tag{14.101}$$

This completes the identification of the structure-independent part of the energy in (14.95). Equation (14.100) reduces for a solid metal to the expression for the internal energy density given by Ashcroft and Langreth (1967).

Needless to say, a full treatment based on (14.97) would require a solution for the electron states in the presence of strong electron-ion interaction (see Chapter 10)

14.7. Hydrodynamics in two-component theory

As previously noted, the expressions for the response functions derived in Section 14.5 describe the generalized hydrodynamics of the liquid metal for longitudinal phenomena under the specific assumption that the ratio γ of the specific heats can be put equal to unity. In this section, the structure of these response functions will be recast, following Tosi, Parrinello, and March (1974), into a form that is closer to hydrodynamics, and then their behaviour will be analyzed in the hydrodynamic limit of long wavelength and low frequency in order to derive the usual transport coefficients.

Let us first build linear combinations of the response functions defined in (14.39) by considering the response of the mass density to an external potential coupled to mass density fluctuations:

$$\chi_{\mathrm{MM}}(\mathbf{q}, \omega) = \sum_{jl} m_j m_l \chi_{jl}(\mathbf{q}, \omega), \tag{14.102}$$

the response of the charge density to an electric potential

$$\chi_{\mathrm{QQ}}(\mathbf{q}, \omega) = \sum_{jl} e_j e_l \chi_{jl}(\mathbf{q}, \omega), \tag{14.102a}$$

and the nondiagonal response describing coupling between mass and charge density fluctuations

$$\chi_{\mathrm{mQ}}(\mathbf{q}, \omega) = \chi_{\mathrm{Qm}}(\mathbf{q}, \omega) = \tfrac{1}{2} \sum_{jl} (e_j m_l + e_l m_j) \chi_{jl}(\mathbf{q}, \omega). \tag{14.103}$$

From (14.51) and (14.52), one finds

$$\chi_{\mathrm{mm}}(\mathbf{q}, \omega) = \frac{n_{\mathrm{i}} M q^2 [\omega^2 - \omega_p^2 - \alpha_p(\mathbf{q}, \omega)]}{[\omega^2 - \alpha_{\mathrm{s}}(\mathbf{q}, \omega)][\omega^2 - \omega_{\mathrm{p}}^2 - \alpha_{\mathrm{p}}(\mathbf{q}, \omega)] - [\alpha_{\mathrm{r}}(\mathbf{q}, \omega)]^2} \tag{14.104}$$

$$\chi_{\mathrm{QQ}}(\mathbf{q}, \omega) = \frac{q^2 \omega_{\mathrm{p}}^2 [\omega^2 - \alpha_{\mathrm{s}}(\mathbf{q}, \omega)]/4\pi}{[\omega^2 - \alpha_{\mathrm{s}}(\mathbf{q}, \omega)][\omega^2 - \omega_{\mathrm{p}}^2 - \alpha_{\mathrm{p}}(\mathbf{q}, \omega)] - [\alpha_{\mathrm{r}}(\mathbf{q}, \omega)]^2} \tag{14.105}$$

and

$$\chi_{\mathrm{mQ}}(\mathbf{q}, \omega) = \frac{e M q^2 (n_{\mathrm{i}} n_{\mathrm{e}}/m_{\mathrm{i}} m_{\mathrm{e}})^{1/2} \alpha_{\mathrm{r}}(\mathbf{q}, \omega)}{[\omega^2 - \alpha_{\mathrm{s}}(\mathbf{q}, \omega)][\omega^2 - \omega_{\mathrm{p}}^2 - \alpha_{\mathrm{p}}(\mathbf{q}, \omega)] - [\alpha_{\mathrm{r}}(\mathbf{q}, \omega)]^2}. \tag{14.106}$$

In these equations, $M = m_i + Zm_e$, $\omega_p^2 = 4\pi e^2(Z^2 n_i/m_i + n_e/m_e)$, and

$$\alpha_s(\mathbf{q},\omega) = \frac{1}{M}[\tilde{\Gamma}_{ii}(\mathbf{q},\omega) + \tilde{\Gamma}_{ei}(\mathbf{q},\omega) + Z\tilde{\Gamma}_{ie}(\mathbf{q},\omega) + Z\tilde{\Gamma}_{ee}(\mathbf{q},\omega)] \quad (14.107)$$

$$\alpha_p(\mathbf{q},\omega) = \frac{1}{M}\left[\frac{Zm_e}{m_i}\tilde{\Gamma}_{ii}(\mathbf{q},\omega) + \frac{m_i}{m_e}\tilde{\Gamma}_{ee}(\mathbf{q},\omega) - Z\tilde{\Gamma}_{ie}(\mathbf{q},\omega) - \tilde{\Gamma}_{ei}(\mathbf{q},\omega)\right] \quad (14.108)$$

$$\alpha_r(\mathbf{q},\omega) = \frac{1}{M}(Zm_i m_e)^{-1/2}\left\{Zm_e\tilde{\Gamma}_{ii}(\mathbf{q},\omega) - m_i\tilde{\Gamma}_{ee}(\mathbf{q},\omega) + \frac{1}{2}(Zm_e - m_i)[Z\tilde{\Gamma}_{ie}(\mathbf{q},\omega) + \Gamma_{ei}(\mathbf{q},\omega)]\right\}. \quad (14.109)$$

Using the procedure of Section 14.5, the $\tilde{\Gamma}_{jl}$'s now contain the non-Coulomb parts of the Hartree potential; namely, the cluster function in (14.50) is to be replaced by the full pair correlation function in evaluating the contributions of the short-range parts of the electron-ion and ion-ion potentials.

The significance of the functions $\alpha(\mathbf{q},\omega)$ is evident from (14.104) to (14.106). The function $\alpha_r(\mathbf{q},\omega)$ arises because of coupling between mass and charge density fluctuations, which can, however, be neglected in the long-wavelength limit, since $\alpha_r(\mathbf{q},\omega)$ behaves as q^2 (see below). The function $\alpha_s(\mathbf{q},\omega)$ then gives in this limit the frequency and damping of sound waves (compare (14.107) with (14.82)), whereas $\alpha_p(\mathbf{q},\omega)$ gives the frequency shift and damping, as well as the dispersion, of the plasmon waves. As will be seen, the same function gives the longitudinal conductivity $\sigma(0,\omega)$ and thus the electrical resistance of the liquid metal when calculated in the low-frequency limit. Equations (14.107)–(14.109) express these properties through functional derivatives of the kinetic tensors and of the pair correlation functions. In the hydrodynamic limit, locality in the density may be assumed in the evaluation of the real part of the α's, when one recovers the results of Section 14.5 (see (14.90)–(14.93)), and, in particular, the correct value of the sound-wave frequency. (Here, it is to be noted that the hydrodynamic limit does not describe correctly the properties of long wavelength plasmons, since the plasmon excitation is a high-frequency phenomenon.) Nonlocality, it may be added, is instead crucial for the imaginary parts.

Nonlocality implies an expansion in $\dot{n}_j$ and their space derivatives or in the currents and their space derivatives. A special situation arises in an electron-ion liquid as compared with normal liquids because of the special characteristics of the electronic component, which is, of course, a quantum liquid with a continuum of single-particle excited states without a gap, and with Fermi velocity much higher than the sound velocity. If one first takes the limit $q \to 0$ in all terms in the theory, as is usually done for normal liquids (see Chapter 7; in particular Kubo-Green formulae there), one would be missing the damping of sound waves by electron-hole creation processes, which are certainly operative since $s \ll v_f$. Following Tosi, Parrinello, and March (1974), this behaviour will be called the quasi-homogeneous limit for the liquid metal and will be treated explicitly, since the results will be used in the subsequent discussion and do indeed correctly apply to the hydrodynamic limit of other two-component charged liquids, such as the (electronically insulating) molten salts.

In order to describe correctly the hydrodynamic behaviour of the ions in a liquid metal, one must instead retain terms in the electronic contributions at long wavelength for which $qv_f > \omega$. There are now two possible limiting behaviours for the electrons. If $ql_e \gg 1$, where l_e is the electronic mean free path, then the electrons behave quasi-statically rather than hydrodynamically. This is the situation that one tries to realize, for example, for studies of the Fermi surface in solid metals by sound attenuation. In the opposite regime, $ql_e \ll 1$, the electrons behave hydrodynamically, just as the ions do. This situation appears to be realized for all liquid metals at the wavelengths of interest in ultrasonic experiments.

14.7.1. Quasi-homogeneous limit

Nonlocality in time is easily handled in a normal liquid. For the evaluation of the imaginary part of the α's, it is convenient to use the continuity equation in operator form and write

$$\tilde{\Gamma}_{jl}(\mathbf{q}, \omega) = \frac{\omega}{q} \text{F.T.} \left\{ \nabla_\alpha \nabla_\beta \frac{\delta \pi^j_{\alpha\beta}(\mathbf{R}, t)}{\delta J_l(\mathbf{y}, \tau)} \bigg|_{\text{eq}} \right.$$
$$\left. + \sum_s \int d\mathbf{x} \, \nabla_\alpha \left[\nabla_\alpha v_{js}(\mathbf{R} - \mathbf{x}) \frac{\delta \langle \rho_j(\mathbf{R}, t) \rho_s(\mathbf{x}, t) \rangle_c}{\delta J_l(\mathbf{y}, \tau)} \right]_{\text{eq}} \right\}_{\mathbf{q}, \omega}, \tag{14.110}$$

where $J_l(\mathbf{y}, \tau)$ is the longitudinal current density fluctuation for the lth component. One can now evaluate the functional derivatives in this equation by treating them as local in time—namely, by expanding the kinetic tensor and the two-body correlations in the current densities and their space derivatives. Indeed, the usual conditions for the hydrodynamic regime can be written at sound wavelength as

$$\omega\tau \ll 1 \quad \text{and} \quad \omega\tau \ll \frac{s}{v}$$

where v is the root-mean-square velocity of the particles. This shows that both conditions are simultaneously satisfied at low frequencies ($\omega\tau \ll 1$) if $v \lesssim s$, as is the case for normal liquids. This is also the situation for the ions in a liquid metal, since $(k_B T/m_i s^2)^{1/2} = [\lim_{\mathbf{q}\to 0} S_{ii}(\mathbf{q})]^{1/2} \sim 0.2$, but certainly not for the electrons, since $v_f/s \sim 10^3$.

The actual evaluation of the functional derivatives in (14.110) for a normal two-component liquid or for the electron-ion liquid in the quasi-homogeneous limit, where one neglects the consequences of the inequality $s \ll v_f$, follows the lines of the analogous calculation by Fröhlich (1967) for a one-component liquid. The kinetic part vanishes at equilibrium, being second order in the current. Thus the evaluation of the transport coefficients in a normal liquid requires only the evaluation of static triplet correlation functions between particle densities and screened current densities or their spatial derivatives.

Explicitly, one may write

$$\begin{aligned} \delta\langle\rho_j(\mathbf{R},t)\rho_s(\mathbf{x},t)\rangle_c &= [v_s(\mathbf{x},t) - v_j(\mathbf{R},t)]\sigma_{sj}(\mathbf{r}) + \cdots \\ &= [v_s(\mathbf{R},t) - v_j(\mathbf{R},t)]\sigma_{js}(r) + \cdots \end{aligned} \tag{14.111}$$

where $v_j(\mathbf{R},t) = (l/n_j)J_j(\mathbf{R},t)$ and

$$\sigma_{sj}(r) = -\sigma_{js}(r) = \frac{\partial\langle\rho_j(\mathbf{R})\rho_s(\mathbf{x})\rangle_c}{\partial v_s}. \tag{14.112}$$

The terms omitted in (14.111) can be handled in the same way as Fröhlich's and lead to viscositylike terms in the $\tilde{\Gamma}_{jl}$'s behaving as ωq^2; whereas the term that has been displayed in (14.111) is characteristic of the two-component liquid and leads to a dissipation term behaving as ω. The calculation is detailed in Appendix 14.4. One then finds

$$\begin{aligned} \text{F.T.}&\left\{\int d\mathbf{x}\, \nabla_\alpha[\nabla_\alpha v_{js}(\mathbf{R}-\mathbf{x})\delta\langle\rho_j(\mathbf{R},t)\rho_s(\mathbf{x},t)\rangle_c]_{eq}\right\}_{\mathbf{q},\omega} \\ &= iq\gamma(1-\delta_{js})[v_s(\mathbf{q},\omega) - v_j(\mathbf{q},\omega)] \\ &\quad - iq^3\left\{\sum_m \eta_{jm}v_m(\mathbf{q},\omega) + (1-\delta_{js})\eta_1[v_s(\mathbf{q},\omega) - v_j(\mathbf{q},\omega)]\right\}, \end{aligned} \tag{14.113}$$

where

$$\gamma = \int \mathrm{d}\mathbf{r}\, \sigma_{\mathrm{ei}}(\mathbf{r})(\hat{q}\cdot\nabla)v_{\mathrm{ei}}(r), \tag{14.114}$$

and the expressions for the η's are given in Appendix 14.4. The final results can be expressed in the form:

$$\operatorname{Im}\tilde{\Gamma}_{\mathrm{jj}}(\mathbf{q},\omega) \to -\frac{\omega}{n_{\mathrm{j}}}[\gamma + q^2(\eta_{\mathrm{jj}} - \eta_1)] \tag{14.115}$$

and

$$\operatorname{Im}\tilde{\Gamma}_{\mathrm{j}\bar{\mathrm{j}}}(\mathbf{q},\omega) \to \frac{\omega}{n_{\bar{\mathrm{j}}}}[\gamma - q^2(\eta_{\mathrm{j}\bar{\mathrm{j}}} + \eta_1)], \tag{14.116}$$

from which one finds

$$\operatorname{Im}\alpha_s(\mathbf{q},\omega) \to -\frac{\omega q^2}{n_{\mathrm{i}}M}(\eta_{\mathrm{ii}} + 2\eta_{\mathrm{ie}} + \eta_{\mathrm{ee}}) \tag{14.117}$$

$$\operatorname{Im}\alpha_{\mathrm{p}}(\mathbf{q},\omega) \to -\frac{M\gamma\omega}{n_{\mathrm{e}}m_{\mathrm{i}}m_{\mathrm{e}}} - \frac{\omega q^2}{n_{\mathrm{e}}m_{\mathrm{i}}m_{\mathrm{e}}M} \cdot[(Zm_{\mathrm{e}})^2\eta_{\mathrm{ii}} + m_{\mathrm{i}}^2\eta_{\mathrm{ee}} - 2Zm_{\mathrm{i}}m_{\mathrm{e}}\eta_{\mathrm{ie}} - M^2\eta_1] \tag{14.118}$$

and

$$\operatorname{Im}\alpha_{\mathrm{r}}(\mathbf{q},\omega) \to -\frac{\omega q^2}{n_{\mathrm{i}}M}(Zm_{\mathrm{i}}m_{\mathrm{e}})^{-1/2}[Zm_{\mathrm{e}}(\eta_{\mathrm{ii}} + \eta_{\mathrm{ie}}) - m_{\mathrm{i}}(\eta_{\mathrm{ee}} + \eta_{\mathrm{ie}})]. \tag{14.119}$$

It should be stressed that, while by the calculation of $\operatorname{Im}\alpha_s$ one has identified the contribution of viscosity to sound wave damping, it seems experimentally (Smith et al., 1967) that such damping in liquid metals is dominantly related to the coefficient of thermal conductivity. In the method employed above, although $c_{\mathrm{p}}/c_{\mathrm{v}}$ has been put equal to unity, the damping due to the electrical conductivity enters the theory in the short-mean free-path limit that is treated next. It should also be stressed again that the expression (14.118) applies at low frequencies and therefore cannot be used to evaluate plasmon properties. The frequency shift from ω_{p} and the damping of the plasmon due to the electron-ion interaction are determined by the value of the function

$$\gamma(\omega) = n_{\mathrm{e}} \int \mathrm{d}\mathbf{r}(\hat{\mathbf{q}}\cdot\nabla)v_{\mathrm{ie}}(r) \int \mathrm{d}(t-\tau)\exp(i\omega\{t-\tau\}) \frac{\delta\langle\rho_{\mathrm{i}}(\mathbf{R},t)\rho_{\mathrm{e}}(\mathbf{x},t)\rangle_{\mathrm{c}}}{\delta J_{\mathrm{e}}(\tau)} \tag{14.120}$$

evaluated at the plasmon frequency; this function at finite frequency has both real and imaginary parts. This is the extension then to the disordered case of the earlier study by March and Tosi (1972) of plasmons in metallic lattices.

HYDRODYNAMIC BEHAVIOUR WITH QUASI-STATIC ELECTRONS

Terms proportional to ω/qv_{f} appear in the hydrodynamic description of ions in liquid metals, which are associated with the physical processes of electron-hole pair excitation and thus, in essence, with deviations from the Born-Oppenheimer approximation. The main new term arises from the kinetic tensor for electrons in (14.110), and this will first be evaluated using the random-phase approximation (RPA). This assumes that $ql_{\mathrm{e}} \gg 1$ and furthermore neglects quasi-particle effects, namely, the quantum influence of the interactions on the kinetic tensor. Although $ql_{\mathrm{e}} \ll 1$ under normal ultrasonic attenuation conditions in liquid metals, it will be convenient to introduce that case by first studying the long free-path regime.

In RPA the change of the Wigner distribution due to a longitudinal electric field $E(\mathbf{q}, \omega)$, for $\omega \ll qv_{\mathrm{f}}$ and taking a sharp Fermi surface, is given by (Pines and Nozières, 1966)

$$\delta f(\mathbf{p}, \mathbf{q}, \omega) = -\frac{ie}{q}\delta(\varepsilon_{\mathbf{p}} - \mu)E(\mathbf{q}, \omega)\left[1 + \frac{\omega}{\mathbf{q}\cdot\mathbf{v}_p - i\varepsilon} + \left(\frac{\omega}{\mathbf{q}\cdot\mathbf{v}_p - i\varepsilon}\right)^2 + \cdots\right] \tag{14.121}$$

where $\mathbf{v}_{\mathbf{p}} = \mathbf{p}/m_{\mathrm{e}}$ and ε is a positive infinitesimal. The current density is, correspondingly,

$$\begin{aligned}
\delta J(\mathbf{q}, \omega) &= \sum_{\mathbf{p}} \frac{\mathbf{p}\cdot\hat{\mathbf{q}}}{m_{\mathrm{e}}}\delta f(\mathbf{p}, \mathbf{q}, \omega) \\
&= -\frac{ie\omega}{qm_e}E(\mathbf{q}, \omega)\sum_{\mathbf{p}}\delta(\varepsilon_{\mathbf{p}} - \mu)\frac{\mathbf{p}\cdot\hat{\mathbf{q}}}{\mathbf{q}\cdot\mathbf{v}_p - i\varepsilon}\left[1 + \frac{\omega}{\mathbf{q}\cdot\mathbf{v}_p - i\varepsilon}\right] \\
&= -\frac{ie\omega}{q^2}E(\mathbf{q}, \omega)\left[S_{\mathrm{f}} + i\pi\frac{\omega}{q}\sum_{\mathbf{p}}\delta(\varepsilon_{\mathbf{p}} - \mu)\delta(\mathbf{v}_p\cdot\hat{\mathbf{q}})\right] \\
&= -\frac{ie\omega}{q^2}E(\mathbf{q}, \omega)\left(S_{\mathrm{f}} + i\pi m_{\mathrm{e}}\frac{\omega}{q}d_{\mathrm{f}}\right),
\end{aligned} \tag{14.122}$$

where S_{f} and d_{f} are the area and diameter of the Fermi surface. The second term in this expression represents a dissipative current, contributed only by the electrons on the maximal cross section of the Fermi surface orthogonal to $\mathbf{q}$, as is required for momentum and energy conservation in the limit of vanishing $\mathbf{q}$ and ω. The evaluation of the functional derivative in (14.110) then gives

$$q_\alpha q_\beta \lambda^{\text{ee}}_{\alpha\beta}(\mathbf{q},\omega) = \frac{\omega}{q}\sum_{\mathbf{p}}\frac{(\mathbf{q}\cdot\mathbf{p})^2}{m_{\text{e}}}\frac{\delta f(\mathbf{p},\mathbf{q},\omega)}{\delta J(\mathbf{q},\omega)} \to -\frac{1}{m_{\text{e}}}\sum_{\mathbf{p}}\frac{(\mathbf{q}\cdot\mathbf{p})^2\delta(\varepsilon_{\mathbf{p}}-\mu)}{S_{\text{f}}+i\pi m_{\text{e}}(\omega/q)d_{\text{f}}}$$

$$\to \frac{q^2p_{\text{f}}^2}{3m_{\text{e}}}\left(1-\frac{i\pi}{2}\frac{\omega}{qv_{\text{f}}}\right), \qquad (14.123)$$

in agreement with (14.70) and (14.75). The leading term of this equation was, of course, already included in the evaluation of the real part of the functions $\alpha(\mathbf{q},\omega)$ and corresponds in fact to the constant α_{ee} in (14.81).

It is difficult to assess the corrections of order ω/qv_{f} to the $\tilde{\Gamma}_{jl}$'s in the interacting electron-ion system. It may be argued, however, that such deviations from the Born-Oppenheimer approximation should be relevant only for $\tilde{\Gamma}_{\text{ee}}(\mathbf{q},\omega)$. The evaluation of $\tilde{\Gamma}_{\text{ee}}(\mathbf{q},\omega)$ for the fully interacting homogeneous electron liquid in the quasi-static limit has been carried out by means of Landau's theory (Pines and Nozières, 1966), with the result

$$\operatorname{Im}\tilde{\Gamma}_{\text{ee}}(\mathbf{q},\omega) = -\frac{\pi}{6}q^2p_{\text{f}}^2\frac{\omega}{qp_{\text{f}}}+\cdots. \qquad (14.124)$$

Because of its simplicity, which reflects the geometric-resonance character of the ultrasonic attenuation by conduction electrons, it seems probable that this result would remain valid in a liquid metal under the conditions specified above ($ql_{\text{e}} \gg 1$ and sharp Fermi surface).

As is apparent by comparison with (14.117) and (14.119), the deviations from the Born-Oppenheimer approximation are then dominant in the quasi-static limit over the viscosity terms in the sound attenuation and in the mass-charge coupling. The leading term in $\operatorname{Im}\alpha_{\text{p}}(\mathbf{q},\omega)$ still remains, instead, the leading term in (14.118).

Fully Hydrodynamic Behaviour

The behaviour of the electrons in the liquid metal will now be considered along the simple lines followed above, but under the opposite condition—that strong scattering of the electrons is prevalent, namely, $ql_{\text{e}} \ll 1$, as one obtains under normal circumstances encountered in sound-wave experiments. Still neglecting the blurring of the Fermi surface due to disorder scattering (see Section 7.4, however) the result for a free-electron gas is well known from the theory of sound attenuation in solids (Pippard, 1960) and is

$$\operatorname{Im} \tilde{\Gamma}_{ee}(\mathbf{q}, \omega) = -\frac{4}{15} q^2 p_f^2 \frac{\omega l_e}{p_f} + \cdots. \tag{14.125}$$

The structure of the expression (14.117) for sound-wave attenuation is thereby recovered, the only change being that the electronic contribution η_{ee} to the viscosity-like coefficient is modified by the addition of a term equal to $\frac{4}{15} n_e p_f l_e$.

Therefore, one expects that in the limit $q l_e \ll 1$, the hydrodynamic behaviour of a liquid metal will have the same frequency and wave-vector dependence as in other two-component fluids. However, one may anticipate that the electronic terms in sound attenuation, and in particular the nonadiabatic term (14.125), will have a different ionic mass dependence from the normal viscosity term of (14.117). This point will be discussed a little further later.

Summary of Results for Dynamic Structure Factor of Ions

It will be useful at this point to pull together the main results for the dynamic structure factor of the ions in a liquid metal. The general structure of this quantity, in the classical limit, is

$$S_{ii}(\mathbf{q}, \omega) = -\frac{k_B T}{\pi \omega M^2} \operatorname{Im} \left[\chi_{MM}(\mathbf{q}, \omega) + \frac{2m_e}{e} \chi_{MQ}(\mathbf{q}, \omega) + \left(\frac{m_e}{e} \right)^2 \chi_{QQ}(\mathbf{q}, \omega) \right]. \tag{14.126}$$

In particular:

1. The elastic-scattering function $S_{ii}(q, 0)$ is obviously dominated by the sound-wave behaviour of $\chi_{MM}(\mathbf{q}, \omega)$, since ω_p^2 is very large. Then

$$S_{ii}(\mathbf{q}, 0) \simeq \frac{k_B T}{\pi M^2} \frac{\eta(\mathbf{q})}{[s(\mathbf{q})]^4}, \tag{14.127}$$

where

$$\eta(\mathbf{q}) = -\frac{n_i M}{q^2} \lim_{\omega \to 0} \left[\frac{1}{\omega} \operatorname{Im} \alpha_s(q, \omega) \right] \tag{14.128}$$

plays the role of a wave-vector-dependent viscosity (see Section 8.6) and

$$s(\mathbf{q}) = \left[\frac{1}{q^2} \lim_{\omega \to 0} \operatorname{Re} \alpha_s(\mathbf{q}, \omega) \right]^{1/2}, \tag{14.129}$$

that of a wave-vector-dependent sound velocity.

2. The static structure factor is given by

$$S_{\text{ii}}(\mathbf{q}) = -k_B T\chi_{\text{ii}}(\mathbf{q},0)$$
$$= \frac{n_i k_B T q^2}{M}\frac{\omega_p^2 + \alpha_p(\mathbf{q},0) - 2(Zm_e/m_i)^{1/2}\alpha_r(\mathbf{q},0) + (Zm_e/m_i)\alpha_s(\mathbf{q},0)}{\alpha_s(\mathbf{q},0)[\omega_p^2 + \alpha_p(\mathbf{q},0)] - [\alpha_r(\mathbf{q},0)]^2}$$
$$\simeq \frac{n_i k_B T}{Ms^2(\mathbf{q})}, \qquad (14.130)$$

where the inequality $Zm_e \ll m_i$ has been used.

3. In the hydrodynamic limit the dynamic structure factor becomes

$$S_{\text{ii}}(q,\omega) \to \frac{k_B T q^2}{\pi M^2}\left\{\frac{q^2\eta}{(\omega^2 - s^2q^2)^2 + (\omega q^2\eta/n_i M)^2} + \frac{M^2\gamma/m_i^2}{(\omega^2 - \omega_p^2)^2 + (M\omega\gamma/n_e m_i m_e)^2}\right\}, \qquad (14.131)$$

where $\eta = \lim_{q\to 0}\eta(\mathbf{q})$ is the sound-wave viscosity, conventionally written in terms of the shear viscosity η_s and of the bulk viscosity ζ as $\frac{4}{3}\eta_s + \zeta$. The last term of (14.131), which is the low-frequency remnant of the contribution of the ionic motions to the plasmon peak, is in fact of order m_e/m_i, since γ is related to the resistivity relaxation time through a factor $n_e m_e$ (see below).

14.8. Electrical resistivity

This is the point at which to return to the theory of electrical resistivity of liquid metals introduced in Chapter 7. The approach here will employ the charge-charge response function $\chi_{QQ}(q,\omega)$ defined by (14.103) to discuss the electrical conductivity of a liquid metal. The dielectric function and the longitudinal conductivity of the liquid metal are defined through the response of all the charges to a screened electric potential. Thus one can write

$$\varepsilon(\mathbf{q},\omega) = 1 + \frac{4\pi i}{\omega}\sigma(\mathbf{q},\omega) \qquad (14.132)$$

and in terms of $\chi_{QQ}(\mathbf{q},\omega)$, this is equal to

$$\varepsilon(q,\omega) = 1 - \frac{4\pi}{q^2}\chi_{QQ}(\mathbf{q},\omega)\left[1 + \frac{4\pi}{q^2}\chi_{QQ}(q,\omega)\right]^{-1}. \qquad (14.133)$$

If one uses the form of $\chi_{QQ}(\mathbf{q},\omega)$ in (14.105), it follows that

$$\varepsilon(\mathbf{q},\omega) = 1 - \frac{\omega_p^2[\omega^2 - \alpha_s(\mathbf{q},\omega)]}{[\omega^2 - \alpha_s(\mathbf{q},\omega)][\omega^2 - \alpha_p(\mathbf{q},\omega)] - [\alpha_r(\mathbf{q},\omega)]^2}. \qquad (14.134)$$

In particular, the longitudinal conductivity in the long-wavelength limit is given by

$$\sigma(0,\omega) = \frac{n_e e^2}{m_e} \frac{M/m_i}{-i\omega + M\gamma(\omega)/n_e m_i m_e}, \tag{14.135}$$

the function $\gamma(\omega)$ being as defined in (14.120). In this limit the "random currents" of Kubo (1966) coincide with the "screened currents" of the preceding presentation (Tosi, Parrinello and March, 1974), so that the function $\gamma(\omega)$ is proportional to the canonical correlation function between the "random forces" in the liquid. Finally, the relaxation time for electrical resistivity, defined through $\sigma = n_e e^2 \tau / m_e$, is given by

$$\tau^{-1} = \gamma / n_e m_e. \tag{14.136}$$

The Drude-Zener theory, which is remarkably successful for liquid metals (see, for example, Faber, 1972), follows if $\gamma(\omega)$ is assumed to vary only slowly with frequency over the range of ω of interest (often $0.1 < \omega\tau \lesssim 3$).

14.8.1. Sum rules on frequency-dependent conductivity and relation to dynamic structure

Before returning to discuss the way in which (14.135) can be used for weak electron-ion interaction, when it will be seen that the formula of Baym (1964) follows, let us consider briefly attempts that have been made to relate the electrical conductivity or resistivity of a liquid metal to appropriate limits of the dynamical structure factor $S_{ii}(q,\omega)$ for the ions and thus to relate electronic transport coefficients directly to inelastic neutron scattering, without the explicit appearance of the electron-ion interaction.

One such relation is implied in (14.126) together with the expressions for the three response functions χ_{MM}, χ_{MQ}, and χ_{QQ}. This leads to the result

$$\mathrm{Re}\,\sigma_e(0,\omega) = \frac{\pi e^2 M^2 \omega}{m_e^2}[1 - \exp(-\beta\omega)] \lim_{q\to 0} \frac{1}{q^2} S_{ii}(\mathbf{q},\omega) - \frac{\pi n_i M e^2}{m_e^2}\delta(\omega). \tag{14.137}$$

Here it must be emphasized that $\sigma_e(0,\omega) = \sigma(0,\omega)/\varepsilon(0,\omega)$ is the external conductivity in the sense of Kubo (1966). The relation (14.137) has been given by Hinkelmann (1970). Equation (14.131) also directly gives a Kubo formula for the measured resistivity

$$\sigma^{-1} = \pi\beta\left(\frac{4\pi e M}{m_e}\right)^2 \lim_{\omega\to 0}\lim_{q\to 0}\left[\frac{1}{q^2}S_{ii}(\mathbf{q},\omega)\right]. \tag{14.138}$$

These equations relate the conductivity to the plasmon part of the ionic structure factor. It turns out that they will not be appropriate, therefore, to derive electronic transport properties from neutron-scattering data. This comment, it should be added here, would not apply to the proposal of Jones (1973), who sets up what is essentially an approximate way of calculating $\varepsilon(\mathbf{q},\omega)$ in a liquid metal. Returning, however, to (14.137), this has some implications in connection with the second-moment rule for the ionic structure factor, which—it has sometimes been suggested (see Randolph, 1964)—is violated in some experiments performed on liquid lead. Indeed, with the conductivity sum rule for an assembly of charged carriers:

$$\int_{-\infty}^{\infty}\frac{d\omega}{2\pi}\operatorname{Re}\sigma_e(\mathbf{q},\omega) = \frac{\omega_p^2}{8\pi}; \tag{14.139}$$

(14.137) gives

$$\lim_{q\to 0}\int_{-\infty}^{\infty} d\omega\frac{\omega}{q^2}[1-\exp(-\beta\omega)]S_{ii}(\mathbf{q},\omega) = \frac{n_i}{m_i}, \tag{14.140}$$

showing that the usual second-moment sum rule is valid in liquid metals at long wavelengths. Though this proof is only for the limit $q \to 0$, it is to be noted that the usual classical sum rule gives the second moment proportional to q^2 for all q.

14.8.2 Weak scattering limit of electrical resistivity

The relation between electronic transport and ionic scattering becomes explicit, instead, in calculations of the electronic transport for weak electron-ion interactions, as demonstrated by Baym (1964; see also March, 1968). It will now be shown that the above treatment contains Baym's formula when the appropriate assumptions of weak electron-ion interaction plus the Born-Oppenheimer approximation are made.

Equations (14.135) and (14.114) yield for the relaxation time τ the result

$$\tau^{-1} = \frac{1}{n_e m_e}\int d\mathbf{r}(\hat{\mathbf{q}}\cdot\nabla)v_{ie}(r)\frac{\partial\langle\rho_i(\mathbf{R})\rho_e(\mathbf{x})\rangle_c}{\partial v_e}. \tag{14.141}$$

It is shown in Appendix 14.5 that under the above assumptions one has

$$\text{Im}[\text{F.T.}\{\langle\rho_i(\mathbf{R},t)\rho_e(\mathbf{x},t)\rangle_c\}_{\mathbf{k}}]$$
$$= \frac{1}{2}v_{ie}(\mathbf{k})\int_{-\infty}^{\infty}\frac{d\omega}{2\pi}S_{ii}(\mathbf{k},\omega)S'_{ee}(-\mathbf{k},-\omega) - S_{ii}(-\mathbf{k},-\omega)S'_{ee}(\mathbf{k},\omega) \tag{14.142}$$

where $S'_{ee}(\mathbf{k},\omega)$ is the nonequilibrium structure factor for the electrons. This equation clearly reflects the nature of the scattering process as an exchange of energy and momentum between the drifting electrons and the ionic-density fluctuations. In order to make contact with the Boltzmann-equation approach of Baym (1964), it is to be further assumed that the electron-electron interaction is weak, when one can adopt an RPA form for $S'_{ee}(\mathbf{k},\omega)$:

$$S'_{ee}(\mathbf{k},\omega) = \frac{[S'_{ee}(\mathbf{k},\omega)]_{\text{free}}}{|\varepsilon(\mathbf{k},\omega)|^2} \tag{14.143}$$

where $[S_{ee}(\mathbf{k},\omega)]_{\text{free}}$ is the nonequilibrium structure factor for free electrons. In terms of the distorted Fermi distribution $f(q)$, this is given by

$$[S'_{ee}(\mathbf{k},\omega)]_{\text{free}} = 2\pi\sum_{\mathbf{p}} f(\mathbf{p})[1 - f(\mathbf{p}+\mathbf{k})]\delta(\omega - \varepsilon_{\mathbf{p}+\mathbf{k}} + \varepsilon_{\mathbf{p}}). \tag{14.144}$$

Equation (14.141) can finally be written as

$$\tau^{-1} = \frac{1}{n_e m_e}\sum_{\mathbf{k}}\hat{\mathbf{q}}\cdot\mathbf{k}\,|v_{ie}(\mathbf{k})|^2\int_{-\infty}^{\infty}\frac{d\omega}{2\pi}S_{ii}(\mathbf{k},\omega)\frac{\partial}{\partial v_e}S'_{ee}(-\mathbf{k},\omega)$$
$$= \frac{1}{n_e m_e}\sum_{\mathbf{kp}}\hat{\mathbf{q}}\cdot\mathbf{k}\int_{-\infty}^{\infty}d\omega\,|v_{ie}^{sc}(\mathbf{k},\omega)|^2 S_{ii}(\mathbf{k},\omega)\delta(\omega - \varepsilon_{\mathbf{p}+\mathbf{k}} + \varepsilon_{\mathbf{p}})$$
$$\times\frac{\partial}{\partial v_e}\{f(\mathbf{p}+\mathbf{k})[1 - f(\mathbf{p})]\}. \tag{14.145}$$

It is readily demonstrated, by taking the usual form for the distorted Fermi distribution, namely,

$$f(\mathbf{q}) = f_{\text{equil}}(\varepsilon_{\mathbf{p}} - \mathbf{q}\cdot\mathbf{v}_e) \tag{14.146}$$

that (14.145) is equivalent to the formula of Baym (1964) for the electrical resistivity of a liquid metal. It is stressed again that the formula rests on (1) weak electron-ion and electron-electron interactions and (2) the Born-Oppenheimer approximation.

14.9. Elastic scattering of neutrons by liquid metals

At this point, it is of interest to consider a little further the information that can be derived from measurements of neutron elastic scattering from liquid metals. Some data are available (Barker et al., 1973) at the time of writing for liquid Rb and Ga, and Figure 14.1(a) shows the Ga results along with the static structure factor $S_{ii}(q)$. The main characteristic that can be seen

Figure 14.1. Shows neutron elastic scattering from liquid Ga (from Barker et al., 1973). (a) Static structure factor (solid curve) shown for comparison. (b) Plot of the ratio $S_{ii}(q, 0)/S_{ii}^2(q)$.

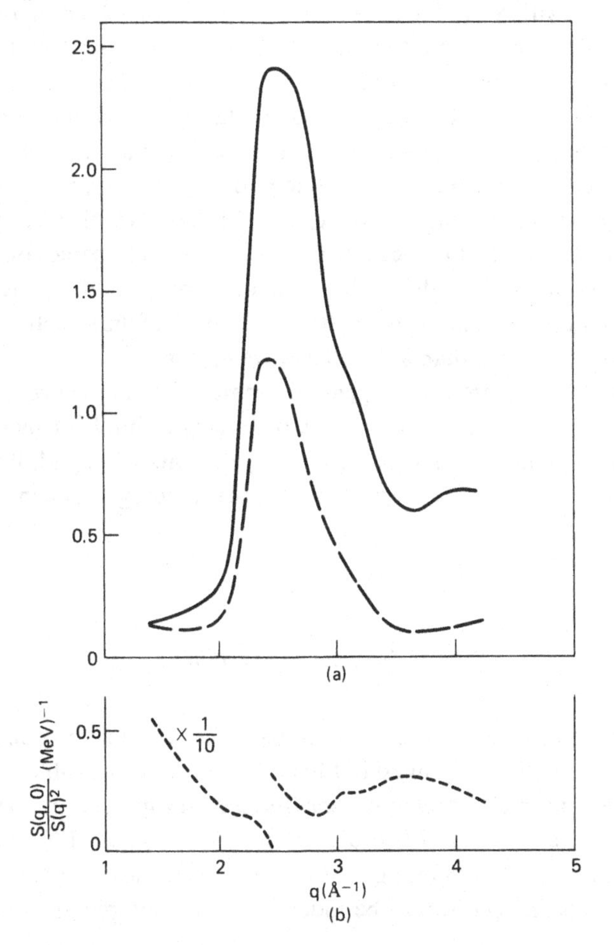

from this figure is the way the peaks and valleys in the static structure factor $S_{ii}(q)$ are reflected in the elastic scattering data.

The elastic scattering is given, in fact, by the result (14.127) in terms of the q-dependent viscosity $\eta(\mathbf{q})$ and sound velocity $s(\mathbf{q})$, and if one uses this equation in conjunction with (14.129), one finds

$$\pi n_i^2 k_B T \frac{S_{ii}(q,0)}{[S_{ii}(q)]^2} = \eta(\mathbf{q}). \tag{14.147}$$

Thus, if one assumes $\eta(\mathbf{q})$ to be slowly varying with q, then $S_{ii}(\mathbf{q},0)$ would be proportional to $S_{ii}^2(\mathbf{q})$; the q-dependence of the ratio $S_{ii}(\mathbf{q},0)/S_{ii}^2(\mathbf{q})$ is plotted in Figure 14.1(b) from the results of Figure 14.1(a). It can be seen there that appreciable variation with q occurs, and now the function $\eta(\mathbf{q})$ is out of phase with $S_{ii}(\mathbf{q})$, the minimum in $\eta(\mathbf{q})$ being at the peak of $S_{ii}(\mathbf{q})$.

As Tosi, Parrinello, and March (1974) point out, the fact that the collective-wave damping ought to vary qualitatively in such a manner is rather clear by analogy with (1) crystals and (2) liquid helium four. Thus, the first peak in the structure factor corresponds roughly to $2\pi/a$, where a is the near-neighbour distance. The zone boundary in the crystal being at π/a, one expects that the phonon damping will be appreciable there but will decrease by the time the first peak in $S_{ii}(q)$ is reached. The same conclusion can be arrived at by considering liquid helium four, the damping of the collective excitations again being smaller at the peak of the structure factor, i.e., in the roton region, than at smaller wave vectors.

These conclusions are also in general accord with the neutron elastic-scattering data on liquid Ga and on Rb. In passing it should be mentioned that the same behaviour is observed in experiments on liquid argon, to which one would also have anticipated that the preceding arguments would apply.

14.9.1. Isotopic mass scaling

In connection with the elastic scattering and its relation to sound-wave attenuation, it should be pointed out here that it may be possible to throw light on the observed anomalous mass-scaling properties of the shear viscosities of the isotopes of liquid lithium (see Ban et al., 1962; see also Ginoza and March, 1985; Omini, 1986). Thus in a classical one-component liquid, one must expect $S(q)$ to be independent of isotopic mass m when measured on pure isotopes of the same element. Since also (Brown and March, 1968) in this case

$$S(q, \omega) = m^{1/2} f(q, \omega m^{1/2}), \tag{14.148}$$

one sees that $S(q, 0)$ should scale simply as $m^{1/2}$. In the Green-Kubo formula (7.20) for $\frac{4}{3}\eta_s + \zeta$, it follows, of course, from (14.148) that $\frac{4}{3}\eta_s + \zeta \alpha m^{1/2}$. However, in the liquid metal, as has been discussed at some length, the electrons contribute to the sound-wave attenuation in a manner that depends on the electronic mean-free-path unless $q l_e \ll 1$. In this situation, the ratio m_e/m_i of the electron-to-ion mass must influence the simple result for the isotopic-mass dependence of the viscosity.

In summary, it seems possible that single-particle excitations in the conduction electron system may play a role in the departures from root-mass scaling in a liquid metal (see, however, Omini, 1986).

14.10. Single-particle motion from two-component theory

While the theory has so far not included thermal conduction, it is clear that this is because of the simplification that one chose to make at the outset of neglecting coupling between particle and energy-density fluctuations. There is no difficulty, at least in principle, in dealing with thermal conduction within the two-component theory developed in this chapter (see Tosi, Parrinello, and March, 1974).

However, the theory is posed in collective language, and it might seem that it is not therefore suitable for discussing the self-diffusion coefficient D (see Chapter 7), which is a single-particle property.

Therefore, this chapter will conclude by using the two-component theory developed before for a neutral one-component fluid, to illustrate the route by which one can tackle this problem of the meanderings of a labelled ion in time in a liquid metal. The procedure is to label an ion or atom by giving it an isotopic mass, denoted by m_0, different at first from the bulk liquid atoms. Such a situation in liquid argon would of course simulate the tracer method of measuring self-diffusion and so has direct physical significance. Also, in a metal, this introduction of a labeled isotope will lead into the theory of the Haeffner effect (see also Appendix 13.8) and evidently to a three-component system. It is because of this additional complication that the following treatment is restricted to diffusion in a neutral liquid.

14.10.1. Formulation of diffusion in a neutral liquid

As is well known (Kubo, 1966), a microscopic form of the Langevin equation describing diffusion in a neutral liquid can be derived from the Liouville equation. One is able then to write

$$\dot{\mathbf{v}}(t) + \frac{\beta}{3m_0}\int_0^t \mathrm{d}t' \langle \mathbf{k}(t)\mathbf{k}(t')\rangle \cdot \mathbf{v}(t') = \frac{1}{m_0}\mathbf{F}(t). \tag{14.149}$$

Here, as above, m_0 is the mass of the labelled atom, $\mathbf{F}(t)$ includes a stochastic component as well as the external force, and $\langle \mathbf{k}(t)\mathbf{k}(t')\rangle$ is the correlation function for the actual forces on the atom. In the macroscopic limit, (14.149) takes the form of the customary Langevin equation (see also Chapter 7):

$$\dot{\mathbf{v}}(t) + \zeta \mathbf{v}(t) = \frac{1}{m_0}\mathbf{F}(t) \tag{14.150}$$

with a friction constant ζ given by (see (7.22))

$$\zeta = \frac{\beta}{3m_0}\int_0^\infty \mathrm{d}t \langle \mathbf{k}(0)\mathbf{k}(t)\rangle. \tag{14.151}$$

From (14.150) one can write

$$\langle \mathbf{v}(0)\cdot \mathbf{v}(t)\rangle = \frac{3}{\beta m_0}\exp(-\zeta|t|). \tag{14.152}$$

Evidently, therefore, one has that the macroscopic self-diffusion constant D is given by (see (7.20))

$$D = \tfrac{1}{3}\int \mathrm{d}t \langle \mathbf{v}(0)\mathbf{v}(t)\rangle = (m_0\beta\zeta)^{-1}. \tag{14.153}$$

This equation already connects current-current and force-force correlation functions through an inverse relation.

Shortly, one will thereby identify the force-force correlation function for diffusion and also deal with the frequency spectrum (or spectral function) $g(\omega)$ of a liquid, already introduced and discussed in Section 7.6. If one defines this in the customary way through the frequency-dependent mobility $\mu(\omega)$ by

$$g(\omega) = \int_{-\infty}^{\infty} \mathrm{d}t \exp(-i\omega t)\langle \mathbf{v}(0)\mathbf{v}(t)\rangle = 2\beta^{-1}\,\mathrm{Re}\,\mu(\omega) \tag{14.154}$$

then knowledge of $g(\omega)$ for all ω is sufficient to determine the force-force correlation function completely, when use is made of the Kramers-Krönig relation (see March and Tosi, 1984).

Let us then consider, at first, a labeled atom of mass m_0, which may be either an isotope or an atom of the pure liquid. The currents associated with the isotope and the solvent, j_0 and j_s, for instance, in the presence of an external force F_0^{ext} acting on the isotope may then be written in terms of response functions by using (14.6). In particular, one has

$$j_0(\mathbf{q},\omega) = -\frac{i\omega}{q^2}\chi_{00}(\mathbf{q},\omega)F_0^{\text{ext}}(\mathbf{q},\omega) = n_0\mu_0(\mathbf{q},\omega)F_0^{\text{ext}}(q,\omega) \quad (14.155)$$

defining the generalized mobility of the tagged atom. Using (14.20) in the limit $q \to 0$, one finds the mobility in the form of a Mori expression (see Mori, 1965a, b)

$$\beta^{-1}\mu_0(\omega) = \frac{(\beta m_0)^{-1}}{i\omega + \gamma_D(\omega)}, \quad (14.156)$$

where the force-force correlation function for diffusion $\gamma_D(\omega)$ is given by

$$\gamma_D(\omega) = \gamma_{00}(\omega) - \frac{\gamma_{0s}(\omega)\gamma_{s0}(\omega)}{i\omega + \gamma_{ss}(\omega)}. \quad (14.157)$$

Here one has

$$\gamma_{jl}(\omega) = -\frac{i}{m_j\omega}\lim_{\mathbf{q}\to 0}\Gamma_{jl}(\mathbf{q},\omega) \quad (14.158)$$

and by using (14.20) and (14.15), these functions can be expressed in terms of functional derivatives of pair-correlation functions with respect to currents:

$$\gamma_{jl}(\omega) = \frac{1}{m_j}\sum_m \int d\mathbf{r}\,\hat{\mathbf{q}}\cdot\nabla v_{jm}(r)\int d(t-\tau)\exp(i\omega(t-\tau))\frac{\delta\langle\rho_j(\mathbf{R},t)\rho_m(\mathbf{x},t)\rangle}{\delta J_l(\tau)}. \quad (14.159)$$

It appears possible, at least in principle, to discuss properly the mass scaling of radioactive tracer diffusion within this formulation.

It should be emphasized that $g(\omega)$ at small ω determines not only the self-diffusion constant D (see 7.20) but also the shear viscosity η_s through (7.26). Clearly the small-ω form of the γ's must determine the shear viscosity η_0 as well as D when $q \to 0$. However, the low-frequency behaviour of the γ's, which will lead to the $\omega^{1/2}$ dependence in $g(\omega)$, is (as far as the writer is aware at the time of writing) not fully understood in terms of the preceding microscopic theory, although it must, from the discussion of long-time tails in Chapter 7, (see also Appendices 8.1 and 8.2), clearly be connected to the hydrodynamic limit of the theory developed in this chapter.

14.10.2 Force-force correlation functions for sodium and argon

To supplement the introductory comments in Chapter 7, let us consider briefly the results of molecular dynamical calculations of $g(\omega)$ by Paskin (1967), who treated both Na and Ar as one-component systems.

First the qualitative difference due to the softer core and longer-range effective potential in Na is that the frequency spectrum has a fairly well defined Debye edge ω_D, whereas in Ar there is a long high-frequency tail. To simulate this feature for Na, as well as to have a model capable of producing a dip at $\omega = 0$, which Paskin also found for Na, Tosi, Parrinello, and March (1974) used a model form:

$$g(\omega) = \frac{2}{\pi\omega_D}\frac{1 - a\omega^2}{1 - a\omega_D^2/5}\left(1 - \frac{\omega^2}{\omega_D^2}\right) \qquad (|\omega| < \omega_D)$$
$$= 0 \qquad (|\omega| \geqslant \omega_D) \qquad (14.160)$$

with cutoff at ω_D. This is shown schematically in Figure 14.2(a) for $a = 1$,

Figure 14.2. (a) Model of frequency spectrum $g(\omega)$ as given by (14.160) for values of parameter $a = 1, 0$ and -1. (b) Real part of force-force correlation function given in (14.162).

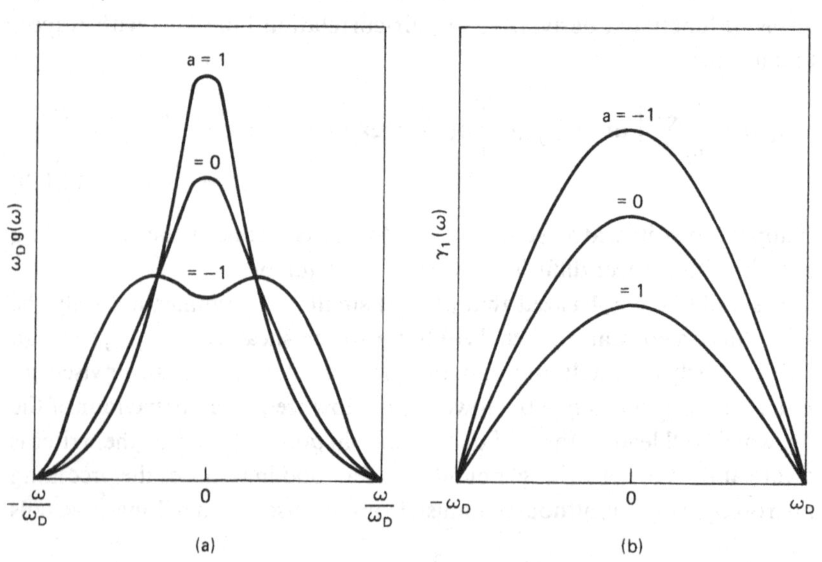

0, and -1. The case $a = -1$ has the main features of Paskin's spectral function for Na. The imaginary part of the mobility can be constructed from the Hilbert transform

$$2\beta^{-1} \operatorname{Im} \mu(\omega) = \frac{1}{\pi} \int_{-\infty}^{\infty} d\omega' \frac{g(\omega')}{\omega' - \omega}. \tag{14.161}$$

Tosi, Parrinello, and March (1974) then construct the real part $\gamma_1(\omega) = \operatorname{Re} \gamma_D(\omega)$ of the force-force correlation function, given by

$$\gamma_1(\omega) = m_0^{-1} \frac{\operatorname{Re} \mu(\omega)}{|\mu(\omega)|^2} \tag{14.162}$$

and this has the schematic form shown in Figure 14.2(b). Obviously, when $g(\omega)$ has finite range in ω, $\gamma_1(\omega)$ has the same range. It is worth noting that the dip in $g(\omega)$ at the origin found in Na results from the more peaked $\gamma_1(\omega)$ in Figure 14.2(b).

In contrast, when there is a relatively long high-frequency tail, as in Ar, the ranges of $g(\omega)$ and $\gamma_1(\omega)$ can be quite different, as with the form

$$g(\omega) = \frac{D\alpha^2}{\pi(\omega^2 + \alpha^2)^2} \tag{14.163}$$

which leads to $\gamma_1(\omega)$ constant for all ω.

The preceding models do not incorporate the $\omega^{1/2}$ singularity in $g(\omega)$ near $\omega = 0$ established in Chapter 7 (see also Appendices 8.1 and 8.2), and it is of interest to record here, following Tosi, Parrinello, and March (1974), the nature of the force-force correlation function that gives rise to such a "spike" at the origin in $g(\omega)$. From the work of Lighthill (1958), it can be shown that, as a consequence of the form (7.20), the imaginary part of the mobility has the low-frequency form

$$2\beta^{-1} \operatorname{Im} \mu(\omega) \simeq d_1 |\omega|^{1/2} \operatorname{sgn} \omega. \tag{14.164}$$

The real and imaginary parts of $\gamma_D(\omega)$ at small ω are then

$$\gamma_1(\omega) = \frac{\pi k_B T}{D} \left[1 - \frac{\pi}{D} d_1 \omega^{1/2} + \cdots \right] \tag{14.165}$$

and

$$\gamma_2(\omega) \simeq -k_B T \left(\frac{\pi}{D} \right)^2 d_1 |\omega|^{1/2} \operatorname{sgn} \omega. \tag{14.166}$$

Thus the addition of a spike in $g(\omega)$ for small ω must arise from a force-force correlation function $\gamma_1(\omega)$ rising cusplike at first from its value $\pi k_B T/D$ at zero frequency.

It seems that such behaviour should be characteristic of all liquids, as one of the more subtle consequences of the existence of the hydrodynamic regime.

To summarize this extended discussion of the two-component theory of liquid metals, the following four points are worth reiterating:

1. Expressions can be obtained for the partial structure factors of a pure liquid metal, and one can thereby demonstrate the way electronic effects enter the effective interionic potential treated approximately in Chapters 3 through 5 as well as the ionic dynamics (see Chapter 8).
2. Single-particle electronic excitations enter the sound-wave attenuation in an important way. The interest in studying further the dependence on isotopic mass of ultrasonic attenuation and neutron elastic scattering when feasible from pure isotopes of the same liquid metal clearly emerges (see Appendix 13.8).
3. The way that electrical conductivity enters the ionic dynamical structure factor is displayed. Baym's (1964) result follows as a limiting case.
4. The thermodynamics, discussed at some length in Chapter 3, can be expressed in terms of pair functions and bare interactions. The relation to the usual weak electron-ion interaction limit is indicated. The precise meaning of the volume-dependent, but structure-independent, term in the effective ionic potential energy is thereby clarified.

14.11. Diffraction evidence on pair correlation functions in two-component theory

To conclude this chapter, let us note, quite briefly, that it is possible, at least in principle, to extract the three pair-correlation functions $S_{ii}(k) = S(k)$, $S_{ie}(k)$, and $S_{ee}(k)$ from diffraction experiments. Indeed neutrons measure $S(k)$ for the ions directly, so that one requires two further diffraction studies to determine the electron-ion and electron-electron correlation functions. Egelstaff, March, and McGill (1974) proposed a way, using X ray and electron diffraction data, to get these two functions. It has to be noted, though, that for a long time, experimentalists sought to get precise agreement between $S(k)$'s derived from X rays on the one hand and neutrons on the other. This means that one is dealing with extracting information from small differences. Nevertheless, by suitably differencing available data at that time for X rays and neutrons, Egelstaff, March, and McGill were able to conclude that there was substantial ordering of the electrons in liquid metals. Subsequent work of Dobson (1978) on Na and Johnson (1990) on K have confirmed this conclusion, and this looks like a promising area for further work (see March and Tosi, 1980), both from experiment and theory.

15
Shock-wave studies

In this chapter, the focus is the study of the properties of liquid metals at high pressure and temperature, with special emphasis on shock-wave studies. Naturally, theory can be invoked to bring the experimental observations into contact with what has been learned about liquid metals in the earlier chapters. Then, in the concluding chapter, some emphasis is placed on the relevance of such properties of the lightest and most abundant elements, H and He, in considering the giant planets Jupiter and Saturn.

In a shock-wave experiment, one measures the shock and particle velocities; from these, the pressure and density of the final state can be obtained directly. For some materials, it is also possible to measure temperature and optical properties. But, in the main, detailed information about atomic and molecular processes must come from theoretical studies.

In Section 15.2 the fundamental relations of shock physics are introduced. This is followed by a discussion of some specific results on hot expanded metals: data necessarily limited by large binding energies and high values of the critical constants.

15.1. Shock compression

Here some of the essential physics of shock compression are summarized, following the account of Ross (1985). A shock wave is a disturbance propagating at supersonic speed in a material, preceded by an extremely rapid rise in pressure, density, and temperature. It seems natural, at first sight, to associate shock waves with explosions and other uncontrolled and irreversible processes. Although shock waves are irreversible, the process is well understood and can be controlled to produce a desired response. Shock-compression studies obtain high pressures by introducing a rapid impulse through the detonation of a high explosive, the impact of a high-speed projectile, or the absorption of an intense pulse of radiation. A shock wave not sustained loses energy through viscous dissipation and reduces to a sound wave (e.g., thunder). High-speed optical and electronic methods are necessary to measure certain dynamical variables that determine pressure, density, and energy.

It is relevant here to note that in shock-wave experiments the passage time of the shock is short compared with the disassembly or "fly-away" time of the sample. As a result, the attainable pressures for a given material are limited only by the energy density supplied by the driver. Chemical explosives have been used to obtain pressures up to 1 Mbar in liquids and up to 13 Mbar in metals, with corresponding temperatures of tens of thousands of degrees Kelvin. Final pressures ranging from 20–158 Mbar have been reported using underground nuclear explosions. In inertial confinement studies, pellets of liquid deuterium are subjected briefly to laser-driven dynamic pressures of about 1000 Mbar and temperatures in excess of 10^7 K [see also the two-volume work by Zeldovitch and Raizer (1966)].

15.2. Dynamics of shock waves

Again following the account of Ross (1985), let us consider a fluid (or solid) at rest with constant density and pressure ρ_0 and P_0, respectively, bounded on the left by a plane piston in a cylinder of area A. Assume that at t_0, the piston is set in motion with a constant velocity u_p. This motion compresses the material before it and a disturbance propagates. The first infinitesimal compression at the piston face results in the propagation of a sound wave with velocity C. However, subsequent compressions at the piston face take place with the material at higher densities and result in higher local sound speeds. This produces a train of waves in which the first is at the speed of sound in the undisturbed material and the last, closest to the piston face, is supersonic. Because the last wave can catch but not pass the first, all the waves eventually coalesce into a single steady wave front, across which exists a sharp discontinuity in pressure, density, and temperature. The width of the discontinuity is generally a few molecular mean-free-paths.

Figure 15.1. Coordinate system used, in which observer moves with shock front.

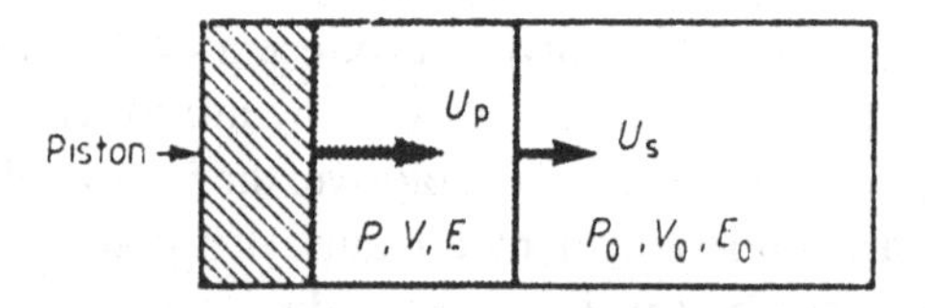

One can now apply the laws of conservation of mass, momentum, and energy in order to determine the pressure, density, and velocity of the disturbance. The coordinate system to be used, in which the observer is moving with the shock front, is shown in Figure 15.1. During a period δt, a mass of $\rho_0 u_s A\,\delta t$ passes through the shock front. Conservation of mass then requires that

$$\rho_0 u_s A t = \rho_l (u_s - u_p) A t, \tag{15.1}$$

leading to the expression for the density change

$$\frac{\rho_0}{\rho_l} = 1 - \frac{u_p}{u_s}. \tag{15.2}$$

The pressure jump across the shock front may be determined from the conservation of momentum. To the observer the momentum flow from the unshocked fluid into the shock front is

$$(\rho_0 u_s A\,\delta t) u_s.$$

The momentum flow away from the shock front is given by

$$\rho_l (u_s - u_p) A t (u_s - u_p) = \rho_0 u_s A\,\delta t (u_s - u_p) \tag{15.3}$$

where the right-hand side of (15.3) results from using (15.1). The change in momentum must equal the difference in the forces across the front $(P_l - P_0) A\,\delta t$, and hence conservation of momentum leads to the following expression for the pressure change:

$$P_l - P_0 = \rho_0 u_s u_p. \tag{15.4}$$

Similarly, the law of conservation of energy

$$\rho_l (u_s - u_p) E_l + \frac{P_l}{\rho_l} + \frac{(u_s - u_p)^2}{2} = \rho_0 u_s E_0 + \frac{P_0}{\rho_0} + \frac{u_s^2}{2} \tag{15.5}$$

leads to the energy equation

$$E_l - E_0 = \tfrac{1}{2}(P_l + P_0)(V_0 - V) \tag{15.6}$$

in which the specific volumes $V_0 = 1/\rho_0$ and $V_l = 1/\rho_l$ have been introduced. Equations (15.2), (15.4), and (15.6) are referred to as the Hugoniot equations.

It is to be emphasized here that the derivation of these Rankine-Hugoniot equations, from conservation of mass, momentum and energy, considers the shock front as a discontinuity but says nothing about its width.

Only the equilibrium properties on either side are considered in the flow properties.

The preceding Hugoniot equations represent the locus of all final states that can be reached by shock-compressing a material from a given initial state: The resultant curve of pressure against volume is termed a Hugoniot. If the initial state is known, then the final state properties may be determined from a measurement of any two of these five properties (u_s, u_p, P, V, and E). Of these, the shock and particle velocities are the most commonly measured [see also Nellis and Mitchell (1980)].

Thus, a measurement of u_s and u_p, with a knowledge of the initial conditions, is sufficient to determine the pressure P, volume V and energy E of the compressed state. The resulting $P-V$ curve, or equivalently a Hugoniot, represents the locus of points that can be reached from an initial condition. (Figure 3 of Ross (1985) shows experimental Hugoniots for the metals Li, Na, Al, Ti, Fe, Cu, Mo, Ag, Pb, Ta, and W, as illustrations of the extensive data now available from this technique.)

In the $u_s u_p$ plane, experimental shock data may be approximated by:

$$u_s = a_0 + a_1 u_p + a_2 u_p^2 + \cdots . \tag{15.7}$$

These are often linear over the entire experimental region, or sometimes the data can be represented using several linear segments. One expects a_0 in (15.7) to correlate with the low-pressure speed of sound.

The temperature, however, cannot be obtained from the Rankine-Hugoniot equations but must be derived from an equation of state. In a few cases, it has been determined from a direct measurement of the emitted radiation [see Ross (1985)].

Returning briefly to the Hugoniot equations, it is worth emphasizing two assumptions on which these rest. One is that of one-dimensional motion, and this can be met in the experiments. A second assumption is that of thermodynamic equilibrium immediately ahead of and behind the shock front. This will hold if the electronic and atomic relaxation times are much less than the experimental time scale of about 0.2 μs.

15.3. Some results on hot expanded metals

Experimental studies of metals are still restricted by the large binding energies and the high values of critical constants. As discussed in Chapter 9, at the time of writing critical data has been measured only for Hg, Cs, Rb, and K [compare Hensel (1981)]. Critical values for Mo, Pt, Ir, and V have been estimated from the results of isobaric experiments [see Ross (1985)] and those for Ni, Al, Cu, and Pb from the expansion isentropes of shock-wave studies on the porous metals (Bushman and Fortov, 1983). Static methods have been employed to study thermophysical, optical, and transport properties of Hg, Cs, Rb, and K over the liquid-vapour range up to 20 kbar and 2000 K (Hensel, 1981). Whereas dynamic measurements are made in a sufficiently short time to avoid containment problems, static methods are limited by sample reactivity at high temperatures and by vessel strength at high pressures.

The isobaric expansion technique (see, for example, Gathers, 1983) has led to expansion data for Al (see Figure 15.2), Ta, Mo, U, Pb, W, V, Ir and Cu. Also shown in Figure 15.2(b) is electrical conductivity, again for expanded liquid Al.

Ross (1985) points out that experiments that start by first shocking a porous metal sample achieve higher temperatures and lower densities than those with solidlike density and consequently undergo much larger expansions. Measurements have been made for porous Ni, Al, Cu, and Pb. (See Figure 15.3 for the phase diagram of Pb thereby determined.)

Figure 15.2. Thermal expansion (a) and electrical conductivity (b) of isobarically (~3 kbar) expanded liquid Al. The normal solid volume is 10 cm^3 mol^{-1}. Full curves are theoretical calculations made from nearly free electron theory (Gathers and Ross, 1984; after Ross, 1985).

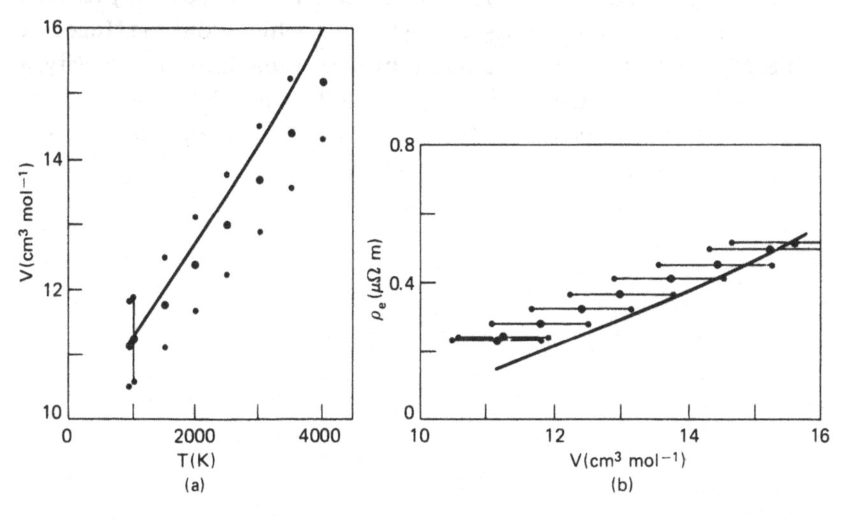

Figure 15.3. Shows phase diagram of Pb. Porous shock adiabat, degree of porosity $m = 1.25$. S is the expansion isentrope. Point CR is estimate of critical point (see Fortov et al., 1974; after Ross, 1985).

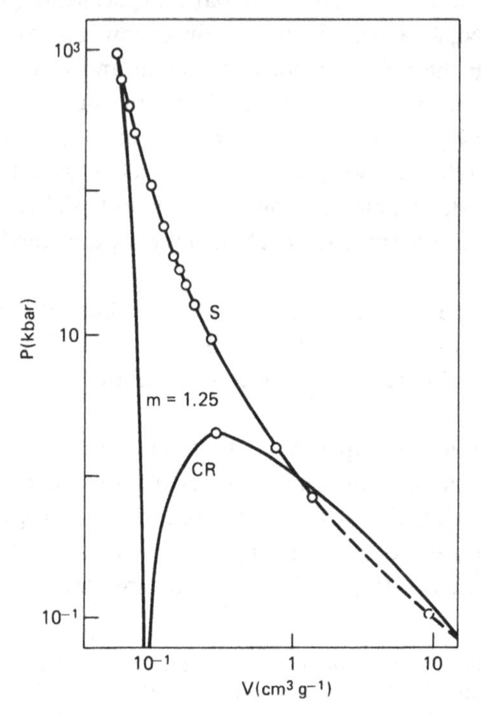

Some other results of such shock-wave techniques will be briefly referred to in the final chapter. As the last example here, melting along the Hugoniot of Fe, Al, and Ta has been determined by measuring the sound velocity; a sensitive indicator of the loss in shear strength from solid to liquid (Asay and Hayes, 1975; Brown and McQueen, 1980, 1982; Shaner, Brown, and McQueen (1984)).

16

Liquid hydrogen plasmas and constitution of Jupiter

In this chapter, a discussion will first be given of the theory of liquid hydrogen plasmas. Following early work on liquid metals as two-component systems, in which the semiclassical Thomas-Fermi approximation was used (Cowan and Kirkwood, 1958) to find pair-correlation functions for ions and electrons, March and Tosi (1973) described a fully quantal approach to this area which has been described in Chapter 14. This has subsequently been developed in a number of groups, especially through the work of Chihara (1984) and of Dharma-wardana and Perrot (1982). This two-component approach to a pure liquid metal and its correlation functions is, of course, to be contrasted with an approach to structure and forces as in Chapter 5. There the ions were treated as a one-component assembly in which, however, the effective interactions were mediated by the conduction electrons. In the two-component theory, the ions may be assumed, above the melting point of a liquid metal, to be a classical fluid. The electrons, as in Chapters 4 and 7, are usually to be treated as a degenerate quantal fluid.

Following a discussion of integral equation and density functional treatments of metallic liquid hydrogen, some discussion will be given of hydrogen-helium mixtures, their phase diagram, and the relevance to the constitution of Jupiter. Indeed, the properties of these two, the most abundant elements, are important for modeling both the giant planets Jupiter and Saturn. These elements are the major constituents of these planets and are subjected to pressures up to 45 million atmospheres and temperatures up to 20,000 K in Jupiter and about 10 million atmospheres and 14,000 K in Saturn.

Shock compression experiments on helium (see also Chapter 15) at Livermore (see Ross, 1985) have achieved higher helium pressures than ever before in the laboratory in an attempt to understand more about the behaviour of helium under such extreme conditions. The high densities and temperatures achieved in the shock-wave experiments led to the setting up of an effective helium pair potential in a previously unexplored region, which can then be utilized to generate a helium equation of state for the study of planetary interiors. This, it turns out, could be of particular interest in the case of Saturn. It has been proposed that the planet has an internal energy source associated with the unmixing and gravitational separation of the hydrogen-helium fluid at pressures below 10 million atmospheres. The existence of this phase transition is very sensitive to the hydrogen and helium equations of state.

16.1. Bijl-Jastrow variational theory of liquid metal hydrogen

In this section, an account will be given of a treatment of liquid-metallic hydrogen, following the work of Campbell and Zabolitzky (1984; see also Chakraborty et al., 1983). These workers use as framework the variational theory of Fermi liquids and thereby construct a relationship between properties of jellium (see Appendix 6.1: also 11.1) and liquid metal hydrogen. In particular the ground-state energy, and excitation spectra, have been calculated with a Bijl-Jastrow type of wave function without recourse to effective screened potentials but directly from Coulomb interactions. These results have the important features that they preserve the proper long-wavelength and high-density limits, which are of prime importance when dealing with Coulomb liquids.

16.1.1. Hamiltonian and trial wave function

One starts out from the basic Hamiltonian H given by

$$H = \sum_{i=1}^{N} \left(\frac{\hbar^2}{2m_e} \nabla_{\mathbf{r}_i}^2 + \frac{\hbar^2}{2m_p} \nabla_{\mathbf{R}_i}^2 \right) + \sum_{\substack{i,j=1 \\ i<j}}^{N} \left(\frac{e^2}{r_{ij}} + \frac{e^2}{R_{ij}} \right) - \sum_{i,j=1}^{N} \frac{e^2}{|\mathbf{r}_i - \mathbf{R}_j|}. \tag{16.1}$$

Here the electron mass is written as m_e and the proton mass as m_p. The vectors $\mathbf{r}_i$ refer to the electrons, and $\mathbf{R}_j$ refers to the protons.

Then one notes next that a (largely) analytic many-body theory that has proved to be successful in treating highly correlated quantum fluids is the variational procedure based on a Jastrow "product of two-body functions" type of total wave function [see, for example, Feenberg (1969) and Zabolitzky (1981)]. In the present case of a two-component mixture of N particles, the Jastrow variational wave function takes the explicit form

$$\psi(\mathbf{r}_1 \cdots \mathbf{r}_N, \mathbf{R}_1 \cdots \mathbf{R}_N) = F\phi = \left(\prod_{\substack{i,j=1 \\ i<j}}^{N} f_{\mathrm{ee}}(r_{ij}) f_{\mathrm{pp}}(R_{ij}) \right) \left(\prod_{ij=1}^{N} f_{\mathrm{ep}}(|\mathbf{r}_i - \mathbf{R}_j|) \phi \right). \tag{16.2}$$

In (16.2), the function ϕ has the form

$$\phi = \phi_0(\mathbf{r}_1 \cdots \mathbf{r}_N)\phi_0(\mathbf{R}_1 \cdots \mathbf{R}_N) \tag{16.3}$$

where ϕ_0 is the Slater determinant of plane waves appropriate to the ground-state of N noninteracting electrons or protons. The absence of momentum dependence in the two-body function $f(r)$ produces a small error in the high-density limit, but Campbell and Zabolitzky (1984) argue that it is not important in the density range of metallic hydrogen considered in their work. In addition, these workers assert that because the bare Coulomb interaction is long-range and, furthermore, because there is no hard core in liquid metallic hydrogen, then explicit three-body correlations (other than those that are representable as a product of two-body correlations) are unimportant and are not included in (16.2).

Given the space of Jastrow trial functions defined by (16.3), the objective is to formulate and to solve the set of Euler-Lagrange equations that follows from the variational principle:

$$0 = \frac{\delta}{\delta f_{\alpha\beta}(r)} \frac{\langle \psi | H | \psi \rangle}{\langle \psi | \psi \rangle}, \tag{16.4}$$

with H given by (16.1).

POTENTIAL ENERGY

Paralleling arguments used frequently in the early chapters of this book, the potential energy in the state ψ can be expressed in terms of the component radial distribution functions

$$V = \frac{\langle V \rangle}{N} = \frac{\rho}{2} \int \mathrm{d}^3 r \sum_{\alpha,\beta} v_{\alpha\beta}(r) g_{\beta\alpha}(r) \tag{16.5}$$

where ρ is the number density of electrons or protons, $v_{\alpha\beta}(r)$ is the Coulomb potential between type-α and type-β particles, and the radial distributions are to be calculated from the total wave function according to

$$g_{\alpha\beta}(\mathbf{r}_1^\alpha - \mathbf{r}_2^\beta) = \frac{N(N - \delta_{\alpha\beta})}{I\rho^2} \int \left(\frac{\mathrm{d}\tau^{2N}}{\mathrm{d}^3 r_1^\alpha \, \mathrm{d}^3 r_2^\beta} \right) |\psi|^2. \tag{16.6}$$

Here I is the normalization integral for ψ and $\mathrm{d}\tau$ includes summation over the spin indices. It is to be noted that one has divided all macroscopic energies by the number of electrons (or protons), not the total number of particles.

KINETIC ENERGY

Following earlier workers, Campbell and Zabolitzky (1984) write the kinetic energy T as a sum of three parts:

$$T = \frac{\langle T \rangle}{N} = T_\phi + T_2 + T_3 \tag{16.7}$$

where

$$T_\phi = \frac{3}{5}\left(\frac{l}{m_e} + \frac{1}{m_p}\right)\frac{\hbar^2 k_f^2}{2}, \tag{16.8}$$

$$T_2 = \frac{\rho}{2}\int d^3r \sum_{\alpha,\beta} g_{\alpha\beta}(r)\frac{\hbar^2}{4m_\alpha}|\nabla u_{\alpha\beta}(r)|^2 \tag{16.9}$$

and

$$T_3 = \frac{\rho^2}{2}\int d^3r_{12}d^3r_{13} \sum_{\alpha,\beta,\gamma} g^{(3)}_{\alpha\beta\gamma}(\mathbf{r}_1,\mathbf{r}_2,\mathbf{r}_3)\frac{\hbar^2}{4m_\alpha}\nabla_1 u_{\alpha\beta}(r_{12})\nabla_1 u_{\alpha\gamma}(r_{13}), \tag{16.10}$$

where the trial form $f_{\alpha\beta}(r)$ has been expressed as

$$f_{\alpha\beta}(r) = \exp\left[\frac{u_{\alpha\beta}(r)}{2}\right]. \tag{16.11}$$

In (16.10), the component three-body distribution functions are defined by

$$g^{(3)}_{\alpha\beta\gamma}(\mathbf{r}_1^\alpha, \mathbf{r}_2^\beta, \mathbf{r}_3^\gamma) = \rho^{-3}N(N - \delta_{\alpha\beta})(N - \delta_{\alpha\gamma} - \delta_{\beta\gamma})\int\left(\frac{d\tau^{2N}}{d^3r_1^\alpha\, d^3r_2^\beta\, d^3r_3^\gamma}\right)|\psi|^2. \tag{16.12}$$

This shows that it suffices to know the two- and three-body distribution functions as functionals of the Jastrow "potentials" $u_{\alpha\beta}(r)$ in order to calculate the energy and derive the Euler-Lagrange equations [see Campbell (1972)]. The hypernetted chain method, referred to in Chapter 5, and Born-Green-Yvon equations have been generalized to give the relation between the Jastrow potentials u and the distribution functions in Fermion quantum fluids [see, for example, Campbell et al. (1982)] and in multi-component systems [see Campbell (1972) and Campbell and Zabolitzky (1984)]. These latter workers make use of the fact that $|\phi_0|^2$ is a positive function and can be expressed following Feenberg as

$$|\phi_0|^2 = \left(\prod_{\substack{i,j=1\\i<j}}^{N} \exp(W_2(r_{ij}))\right)\left(\prod_{\substack{i,j,k=1\\i<j<k}}^{N} \exp(W_3(\mathbf{r}_i,\mathbf{r}_j,\mathbf{r}_k))\right)\cdots. \quad (16.13)$$

In principle the functions W_n are defined by the requirement that they produce the correct free-particle distribution functions $g_m^F(\mathbf{r}_1 \cdots \mathbf{r}_m)$ (defined as in (16.6) for $m = 2$ and in (16.12) for $m = 3, \ldots$, but with $|\psi|^2 = |\phi_0|^2$). However, the essential features of the idea are retained by replacing (16.13) by its first factor so that it has the Jastrow form and then making the choice of $W_2(r)$, which gives the correct free-Fermion radial distribution function $g_2^F(r)$. The calculational advantage of this approach (Campbell and Zabolitzky, 1984) is that it reduces the Fermion Jastrow wave function to an effective Boson Jastrow wave function squared, which is considerably simpler to deal with. Moreover, it preserves the correct long-wavelength structure of the two-body distribution functions, which is an essential feature when dealing with Coulomb interactions.

It is also necessary, of course, to evaluate the three-body distribution functions to obtain the component T_3 in the kinetic energy. Numerous approximational schemes have been proposed to do so. Campbell and Zabolitzky (1984) appeal to the convolution approximation, which works well in the present context [see Zabolitzky (1980)].

Results at Long Wavelength or High Density

Following these workers, in these limits one need retain only terms that are linear in $g - 1$:

$$g_{\alpha\beta\gamma}(\mathbf{r}_1, \mathbf{r}_2, \mathbf{r}_3) = 1 + h_{\alpha\beta}(r_{12}) + h_{\beta\gamma}(r_{23}) + h_{\gamma\alpha}(r_{31}) \quad (16.14)$$

where

$$h_{\alpha\beta}(r) = g_{\alpha\beta}(r) - 1. \quad (16.15)$$

It is then convenient to work entirely in wave-number space and to use matrix notation. Defining a structure-function matrix $\mathscr{S}$, which conveniently incorporates the masses by

$$\mathscr{S}_{\alpha\beta}(k) = (m_\alpha m_\beta)^{1/2} S_{\alpha\beta}(k) \quad (16.16)$$

and bare potential and Jastrow potential matrices by

$$\mathscr{V}_{\alpha\beta}(k) = \frac{\rho \tilde{v}_{\alpha\beta}(k)}{(m_\alpha m_\beta)^{1/2}} \quad (16.17)$$

and

$$\mathscr{U}_{\alpha\beta}(k) = \frac{\rho \tilde{u}_{\alpha\beta}(k)}{(m_\alpha m_\beta)^{1/2}} \tag{16.18}$$

the correlation potential energy becomes

$$V_c = \frac{1}{2\rho}\frac{1}{2\pi^2}\int_0^\infty k^2 \,\mathrm{Tr}[(\mathscr{S} - \mathscr{S}_F)\mathscr{V}]\,dk. \tag{16.19}$$

In the approximation of (16.13) the correlation kinetic energy is

$$T - T_\phi = \frac{1}{2\rho}\frac{1}{2\pi^2}\int_0^\infty k^2 \frac{\hbar^2 k^2}{4}\mathrm{Tr}(\mathscr{S}\mathscr{U}^2)\,dk. \tag{16.20}$$

Finally one needs to specify the relation between the Jastrow potentials and the radial distribution functions. An examination of the multicomponent hypernetted chain (HNC) theory reveals that in the long-wavelength limit or in the degenerate (i.e., high-density) limit for all wave numbers, this relationship may be usefully approximated by

$$\mathscr{U}(k) = \mathscr{S}_F(k)^{-1} - \mathscr{S}(k)^{-1}. \tag{16.21}$$

Minimizing the energy with respect to $\mathscr{U}$ (or, equivalently, $\mathscr{S}$) yields the result

$$\mathscr{S}^2 = \frac{\hbar^2 k^2}{4}\left[\mathscr{V} + \frac{\hbar^2 k^2}{4}\mathscr{S}_F^{-2}\right]^{-1}. \tag{16.22}$$

Although it is a straightforward matter to solve this equation for $\mathscr{S}$ and examine the resultant energy, it is well worth first analyzing this equation in terms of the collective modes of the system.

16.1.2. Collective behaviour and long-wavelength limit

It has been established that the long-wavelength behaviour of the Jastrow potential is a manifestation of the zero-point motion of density fluctuations and that the analytic behaviour of the pseudopotential is determined by the dispersion relation of the collective mode [see Enderby, Gaskell, and March, (1965), March (1968), Reatto and Chester (1970); see also Campbell and Zabolitzky for more recent references]. For single-component systems this relationship is conveniently expressed in terms of the Bijl-Feynman expression for the dispersion relation:

$$\lim_{k\to 0} \varepsilon(k) = \frac{\hbar^2 k^2}{2mS(k)}. \tag{16.23}$$

Thus for short-ranged interactions, $\varepsilon \sim \hbar kc$ and $S \sim \hbar k/2mc$, with c the velocity of sound, while for a Coulomb assembly, where ε is given in terms of the plasma frequency, $S(k)$ goes quadratically to zero as $\hbar k^2/2m\omega_p$. This behaviour, as discussed in detail, for example, by March and Tosi (1984), is caused by dynamical screening in the electron liquid.

In the two-component system (see Chapters 13 and 14), one can have two independent density fluctuation modes, one corresponding to the total density fluctuations and the other to concentration fluctuations. If the interactions are short-range, both modes will have linear dispersion relations. In the case of the two-component plasma of interest here, however, the concentration mode will have a finite frequency at long wavelengths given by the plasma frequency for the reduced mass of the constituent charges. The total density fluctuation mode will continue to exhibit linear behaviour.

This analytic behaviour is contained in (16.22), which can be confirmed by noting first that the Bijl-Feynman relationship for the collective modes of a multicomponent system is just the matrix generalization of (16.23), namely,

$$\lim_{k\to 0} \varepsilon(k) = \frac{\hbar^2 k^2}{2} \mathscr{S}^{-1}(k). \tag{16.24}$$

Of course, this matrix must be diagonalized in order to discover the dispersion relations. First, following Campbell and Zabolitzky (1984), one makes use of the solution for $\mathscr{S}$ given by (16.22) to rewrite this equation as

$$\varepsilon(k)^2 = \varepsilon_F(k)^2 + \hbar^2 k^2 \mathscr{V}(k) \tag{16.25}$$

where the diagonal matrix involved is defined by

$$\varepsilon_F(k) = \frac{\hbar^2 k^2}{2} \mathscr{S}_F^{-1}(k). \tag{16.26}$$

With Coulomb interactions the k^2 in the second term of (16.25) cancels the Coulomb divergence, producing the result

$$\varepsilon(k)^2 = \varepsilon_F(k)^2 + \hbar^2 \begin{pmatrix} \omega_{pe}^2 & -\omega_{pe}\omega_{pp} \\ -\omega_{pe}\omega_{pp} & \omega_{pp}^2 \end{pmatrix} \tag{16.27}$$

where the individual plasma frequencies are given by

$$\omega_{\mathrm{p}\alpha}^2 = \frac{4\pi e^2 \rho}{m_\alpha}. \tag{16.28}$$

Since $\mathscr{S}_\mathrm{F}$ has the form

$$\mathscr{S}_{\mathrm{F}\alpha\beta} = \begin{cases} m_\alpha \delta_{\alpha\beta}\left[\frac{3}{4}k\left(k_\mathrm{F} - \frac{1}{16}\left(\frac{k}{k_\mathrm{F}}\right)^3\right)\right], & k \leqslant 2k_\mathrm{F} \\ m_\alpha \delta_{\alpha\beta}, & k \geqslant 2k_\mathrm{F} \end{cases} \tag{16.29}$$

the first term in (16.27) is quadratic in k and thus is dominated by the second term at small k. However, since the determinant of this second term vanishes, the first term must be retained in diagonalizing (16.27). Consequently, there is but one plasma mode,

$$\lim_{k\to 0} \varepsilon_\mathrm{p}(k) = \hbar\omega_\mathrm{p} + \frac{\hbar^2 k^2}{2m^*}. \tag{16.30}$$

Here ω_p is the plasma frequency with reduced mass $\mu = m_\mathrm{e} m_\mathrm{p}/(m_\mathrm{e} + m_\mathrm{p})$ and the curvature is given by

$$\frac{1}{m^*} = \frac{4}{9}\frac{\hbar k_\mathrm{F}^2}{\omega_\mathrm{p}}\left(\frac{\mu}{m_\mathrm{e}^3} + \frac{\mu}{m_\mathrm{p}^3}\right). \tag{16.31}$$

The second collective mode has a linear dispersion relation in the long wavelength limit

$$\lim_{k\to 0} \varepsilon_\mathrm{s}(k) = \hbar c k, \tag{16.32}$$

with velocity

$$c = \frac{2}{3}\frac{\hbar k_\mathrm{F}}{(m_\mathrm{e} m_\mathrm{p})^{1/2}}. \tag{16.33}$$

The solutions at several values of mean particle spacing r_s are shown in Figures 16.1 and 16.2, r_s being measured in units of the Bohr radius $a_0 = \hbar^2/me^2$. The comparison in these figures with the jellium model for electrons and protons makes it quite clear that the lower mode is primarily associated with the (screened) motion of the protons, whereas the higher-energy mode is almost identical with the electron jellium plasma mode. In the former case the electron screening removes the proton plasmon in favour of the photon for $k < 2k_\mathrm{F}$ (compare the derivation of the Bohm-Staver formula for the velocity of sound in a simple liquid metal in Appendix 6.2). In the latter, the electron mass is replaced by the (almost equal) reduced mass.

The appearance of the correct plasma frequency in (16.24) demonstrates that the dynamical screening is properly included in the present approximate formulation. Moreover, it is to be noted that if the electron-proton interaction is either reduced in strength or made short-range, the deter-

Figure 16.1. Plasmon dispersion relation of solid metallic hydrogen. Jellium model is identical within line thickness.

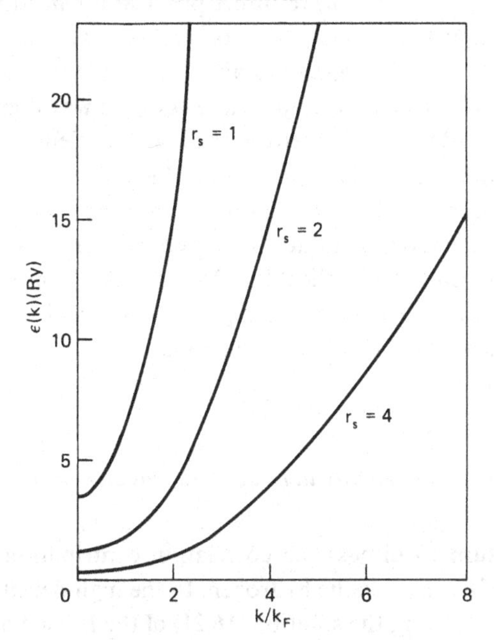

Figure 16.2. Further results for dispersion, but now for sound mode.

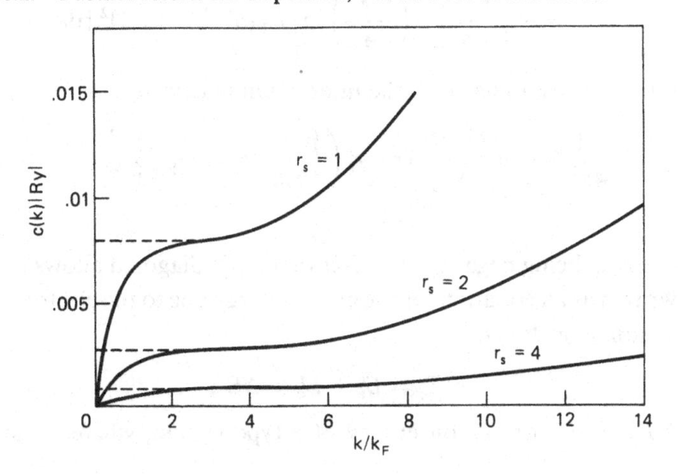

minant of $\mathscr{S}(k)$ no longer vanishes, and the collective modes both have finite frequency in the long wavelength limit.

RESULTS AT LOWER DENSITY

For the present purposes, it is important to discuss here which of the results presented here are preserved as one moves away from the extreme high-density limit into values of r_s that have direct physical application. It is readily demonstrated that the linearity of $S_{\alpha\beta}(k)$ and of $\varepsilon_s(k)$ is correct for all values of r_s for which the liquid phase is stable, although the coefficients of k in these quantities will depend on r_s in a more complex way than given, for example, in (16.31) and (16.33). Moreover, the long-wavelength limit of $\varepsilon_p(k)$ is the plasma frequency, as calculated above, which depends on r_s only through the density factor in ω_p^2. Furthermore the mean-spherical approximation (MSA) used here gives, qualitatively, the correct long-wavelength behaviour of the more generally applicable HNC method. Campbell and Zabolitzky give quantitative arguments to support these conclusions; further details will not be pursued here on these points.

16.1.3. Ground-state energy in relation to jellium model

This is the point to turn to discuss the correlation contribution to the ground-state energy of liquid metallic hydrogen. In the high-density limit, this is obtained by substituting the solution (16.21) of the Euler-Lagrange equation into the sum of (16.19) and (16.20). Hence one is led to the result

$$E_c = -\frac{1}{4\pi^2\rho}\int_0^\infty \left(\frac{\hbar^2k^4}{4}\right)\mathrm{Tr}[\mathscr{S}_F(\mathscr{S}^{-1} - \mathscr{S}_F^{-1})^2]\,dk \tag{16.34}$$

which is rewritten in terms of the interaction matrix as

$$E_c = -\frac{1}{4\pi^2\rho}\int_0^\infty \left(\frac{\hbar^2k^4}{4}\right)\mathrm{Tr}\left\{\mathscr{S}_F\left(\left\{\frac{4}{\hbar^2k^2}\mathscr{V} + \mathscr{S}_F^{-2}\right\}^{1/2} - \mathscr{S}_F^{-1}\right)^2 dk, \tag{16.35}$$

E_c as defined being negative. The fact that $\mathscr{S}_F$ is diagonal allows (16.34) to be rewritten as an equation for the excess energy due to the electron-proton interaction, namely, as

$$E_c = E_c^e + E_c^p + \Delta E_c, \tag{16.36}$$

where E_c^α is the correlation energy of α-type jellium, whereas the excess energy ΔE_c is given by

$$\Delta E_c = \frac{1}{4\pi^2 \rho} \int_0^\infty \left(\frac{\hbar^2 k^4}{2}\right)\left(\operatorname{Tr} \mathscr{S}^{-1} - \frac{1}{m_e S_e} - \frac{1}{m_p S_p}\right) dk$$

$$= \frac{1}{4\pi^2 \rho} \int_0^\infty k^2 [\varepsilon_p(k) + \varepsilon_s(k) - \varepsilon_{p0}(k) - \varepsilon_{e0}(k)]\, dk. \qquad (16.37)$$

In this equation $\varepsilon_p(k)$ and $\varepsilon_s(k)$ are the eigenvalues of the Bijl-Feynman excitation matrix defined in (16.24), $S_\alpha(k)$ is the liquid-structure function of α jellium and $\varepsilon_{\alpha 0}(k)$ is the corresponding single-component Bijl-Feynman spectrum of (16.23).

The second line in (16.37) indicates that ΔE_c arises from the shift in the zero-point energy of the collective modes due to the electron-proton interaction.

In the limit r_s tends to zero, the dominant contribution of each term in (16.34) is of order $\ln r_s$.

Although the Euler-Lagrange equations solved here are strictly valid only at small r_s, Campbell and Zabolitzky demonstrate that they still have well-defined solutions when used to extrapolate into the range of r_s of more direct physical interest. The energy versus r_s curve these workers obtain in this way is shown in Figure 16.3, where the interpolation formula of Vosko et al. (1980) has been used for E_c^e and E_c^p. The agreement with other calculations of E_c is good considering the large values of r_s shown in the figure.

It is relevant here, in concluding this section, to note that the preceding analysis on liquid metallic hydrogen could also be adapted to treat fully

Figure 16.3. Energy per electron versus mean interelectronic spacing r_s Solid line is work of Campbell and Zabolitzky; dashed line is from Mon et al. (1980); see also Chakravarty and Ashcroft (1978; dot-dashed line is from Chakraborty et al. (1983).

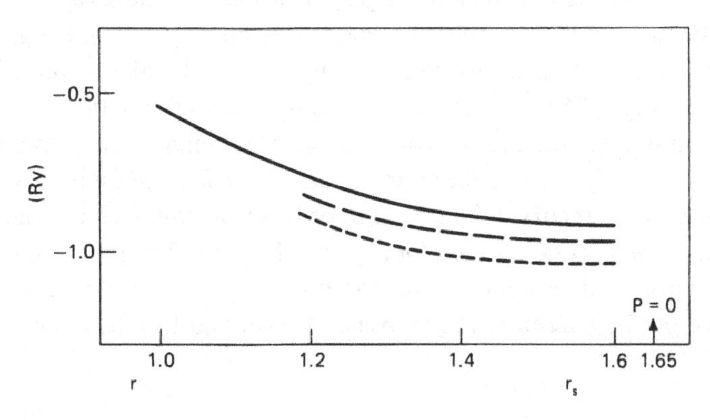

ionized multielectron liquids. For a nucleus of charge Ze, the potential energy $v_{\alpha\beta}(r)$ must include the appropriate Z factors, and in (16.17) and (16.18) the factor ρ must be replaced by $(\rho_\alpha \rho_\beta)^{1/2}$, where, of course, the electron density is Z times the density of nuclei. Finally, the Fermi wave number that appears in the expression for the free-particle structure matrix must be replaced by k_F^α. The result of all these changes is that the liquid metals with Fermion nuclei are formally equivalent to liquid metallic hydrogen, whereas those with Boson nuclei are formally equivalent to liquid metallic deuterium, with the only changes being in the values of the slopes of $S_{\alpha\beta}(k)$ and $\varepsilon_s(k)$ at long wavelength and the mass dependence of the coefficients of the asymptotic terms in the correlation energy.

Following these largely analytical developments on liquid metallic hydrogen, a brief summary is given in Appendix 16.2 of the quantum Monte Carlo simulation carried out by Ceperley and Alder (1987) on solid hydrogen, whose ground state, according to the work of Ashcroft et al. (1974), must be very close in energy to the liquid ground state treated above. A brief discussion of hydrogen-helium mixtures, of interest in the largest planets, is given in Section 16.3.

16.2. Inverse problem: proton-proton interaction in hydrogen plasmas

The concept of effective interionic potentials has been widely used and was given at least partial justification in Chapters 3, 4, and 14. As in Chapters 3 and 4, one approach to the construction of such interactions is via second-order perturbation theory using pseudopotentials, which describe the electron-ion interaction as a weak coupling. Unfortunately, in the problem of interest in this chapter, which in its simplest form is that of protons embedded in an electron gas, the usual procedures are not applicable because of the strongly nonlinear electron response of the gas to the point charge of the proton (see, for example, the early calculations of March and Murray (1960) on the electron density actually at the proton, as a function of electron gas density). This highly nonlinear response is of the essence not only for metallic hydrogen, the main focal point in this chapter, but also in the theory of hydrogen in ordinary metals and in some surface physics problems as well [see, for example, March (1986)]. At zero temperature, work to determine the proton-proton effective potential has been carried out by a number of groups (see Perrot and Rasolt, 1983; Norskov, 1979).

The present section is based on the elevated temperatures study of Dharma-wardana, Perrot, and Aers (DPA) (1983). These works use the idea of the inversion procedure applied to ion-ion interactions in Chapter 5. There, the most fruitful approach was based on using as input the measured liquid-structure factor, the resulting inversion of data for Na then leading to the effective ion-ion interaction shown in Figure 5.2. In the work of DPA, what is inverted is the proton-proton pair distribution of an electron-proton two-component system obtained by a full nonlinear finite temperature density functional calculation (also see the review by Callaway and March, 1984).

16.2.1. Density functional equations for the ion-electron system

The elevated temperature density functional theory (see Mermin, 1965) has at its heart the fact that the equilibrium grand-canonical potential, $\Omega[n, \rho]$ say, is a unique functional of the density distributions n and ρ and is a minimum for the exact distributions of electrons and of ions.

Then, as shown by Dharma-wardana and Perrot (1982), the grand-canonical potential can be usefully separated into four pieces:

$$\Omega[n, \rho] = T[n, \rho] + \Omega_e + \Omega_{ei} + \Omega_i. \tag{16.38}$$

Here $T[n, \rho]$ is the kinetic energy functional of the noninteracting system, with the (formally) exact interacting densities n and ρ.

$T[n, \rho]$ can be written in the form

$$T[n, \rho] = \int F^{\circ}[n, \rho]\, d\mathbf{r}. \tag{16.39}$$

The other contributions to Ω as obtained by DPA may be expressed in the forms

$$\Omega_e = -\int \frac{Z}{r} n(\mathbf{r})\, d\mathbf{r} + \frac{1}{2}\int \frac{n(\mathbf{r})n(\mathbf{r}')}{|\mathbf{r} - \mathbf{r}'|}\, d\mathbf{r}\, d\mathbf{r}' + \int F^{e}_{xc}[n]\, d\mathbf{r} - \mu_e \int n(\mathbf{r})\, d\mathbf{r}, \tag{16.40}$$

$$\Omega_{ei} = -\int \frac{\bar{Z} n(\mathbf{r})\rho(\mathbf{r}')}{|\mathbf{r} - \mathbf{r}'|}\, d\mathbf{r}\, d\mathbf{r}' + \int F^{ei}_{c}[n, \rho]\, d\mathbf{r} \tag{16.41}$$

and

$$\Omega_i = \bar{Z} \int \frac{Z}{r} \rho(\mathbf{r})\, d\mathbf{r} + \frac{\bar{Z}^2}{2} \int \frac{\rho(\mathbf{r})\rho(\mathbf{r}')}{|\mathbf{r} - \mathbf{r}'|} d\mathbf{r}\, d\mathbf{r}' + \int F_c^i[\rho]\, d\mathbf{r} - \mu_i \int \rho(\mathbf{r})\, d\mathbf{r}. \tag{16.42}$$

These equations contain the mean nuclear charge of the ions, denoted by $\bar{Z} = Z - n_a$, where n_a is the number of bound electrons calculated from the effective Schrödinger equation of the one-body form of density functional theory [see Kohn and Sham (1965), at zero temperature]. μ_e and μ_i define, as in Chapter 14, the electron and ion chemical potentials. The potential Ω_e in (16.40) contains an electron-exchange-correlation term F_{xc}^e. Similarly, the ion grand-canonical potential Ω_i contains the ion-ion correlation contribution F_c^i. This can be evaluated, following DPA, from the formalism of the hypernetted chain (HNC) theory, whereas F_c^{ei}, which appears in Ω_{ei}, was taken as negligible.

In order to maintain electroneutrality, the actual functional minimized by Dharma-wardana and Perrot (1982) was of the form

$$\Omega^*[n, \rho] = \Omega[n, \rho] - \lambda\left(\bar{Z} \int \rho(\mathbf{r})\, d\mathbf{r} - \int n(\mathbf{r})\, d\mathbf{r}\right), \tag{16.43}$$

λ being a Lagrange multiplier.

The self-consistent solutions of the equivalent of the Kohn-Sham (1965)

Figure 16.4. Shows proton-proton pair correlation functions in hydrogen plasma for $r_s = 1$. Different curves correspond to different values of coupling parameter Γ (after Dharma-wardana, Perrot, and Aers, 1983).

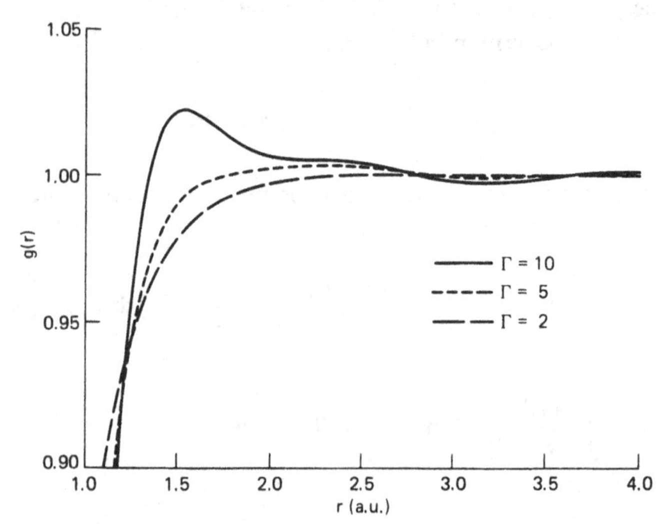

equations [see also Slater (1951)] allow one to obtain density distributions. In particular, if one writes the density distribution $\rho(\mathbf{r})$ as

$$\rho(r) = g_{ii}(r)\bar{\rho}, \tag{16.44}$$

then one can obtain the ion-ion pair function $g_{ii}(r)$ from $\rho(r)$.

Figure 16.4, taken from the work of DPA, shows proton-proton pair-correlation functions thus obtained from self-consistent density functional theory. These correspond to a mean interelectronic distance r_s equal to unity. The three curves correspond to a different coupling parameter $\Gamma = \bar{Z}e^2/r_s k_B T$.

16.2.2. *Effective proton-proton potentials*

Given the appropriate ion-ion correlation function $g_{ii}(r)$, the object now is to use the analogue of the inversion procedure set out in Chapter 5 to obtain the effective proton-proton potential, denoted below by $V_{ii}(r)$, treating the protons as an effective one-component classical system of ions. Then one has [see also Balescu (1975) and DeWitt (1978)]

$$g_{ii}(r) = \exp[-\beta V_{ii}(r) + N(r) + B(r)]: \qquad \beta = (k_B T)^{-1} \tag{16.45}$$

together with the Ornstein-Zernike direct correlation relationship (2.5). In (16.45), $N(r)$ and $B(r)$ are referred to as the nodal and bridge contributions [see also (5.13)] to the potential of mean force, respectively. Here

$$N(r) = h(r) - c(r), \tag{16.46}$$

with h the total correlation function $g(r) - 1$ and $c(r)$ as usual the direct correlation function. As discussed in Chapters 5 and 6, the choice $B(r) = 0$ is made in the HNC approximation. This is a moderate coupling approximation and seems to give reasonable results in the range in which potential energy divided by kinetic energy is less than about 20. One can then write

$$\beta[V_{ii}(r)]_{\text{HNC}} = g(r) - 1 - c(r) - \ln g(r) \tag{16.47}$$

as the HNC approximation to the pair potential.

As DPA note, in the case of hydrogen and other low-Z plasmas, the coupling parameter remains well within the range of the HNC approximation as far as laboratory or fusion plasmas are concerned.

In the case of liquid metals and high-Z-systems, it is no longer possible to drop the bridge term. A very clear-cut example of the inadequacy of the HNC inversion is that given by Taylor and Watts (1981) in connection with liquid Li and Al.

As an example of the results obtained by DPA, Figure 16.5 shows the proton-proton pair potential for $r_s = 1$, $k_B T = 0.1$ au (i.e., $\Gamma = 10$), as calculated from density functional theory, from Thomas-Fermi theory, and from the random phase approximation. The proton-proton potential obtained from the density functional theory allows the ions to approach more closely than in the RPA or TF models.

The point should be stressed, in conclusion, that such effective proton-proton potentials are obtained based on a one-centre density functional calculation followed by HNC inversion. As DPA point out, these results should be comparable to a pair potential that would be obtained from the problem of electrons scattering off two centres fixed in an electron gas. However, even within the over-simplified Thomas-Fermi approximation, such two-centre problems are numerically formidable [see Alfred and March (1957) at $T = 0$]. The DPA approach should provide a useful tool for the generation of ion-ion pair potentials for systems at finite temperatures and arbitrary degeneracies (see also Appendix 16.1).

Figure 16.5. Effective proton-proton potential $v_{ii}(r)$ for $r_s = 1$, $k_B T = 0.1$ a.u. (i.e., coupling parameter $\Gamma = 10$). The density functional theory result is compared with Thomas-Fermi (dashed curve) and RPA (dash-dot) predictions (after Dharmawardana, Perrot, and Aers, 1983).

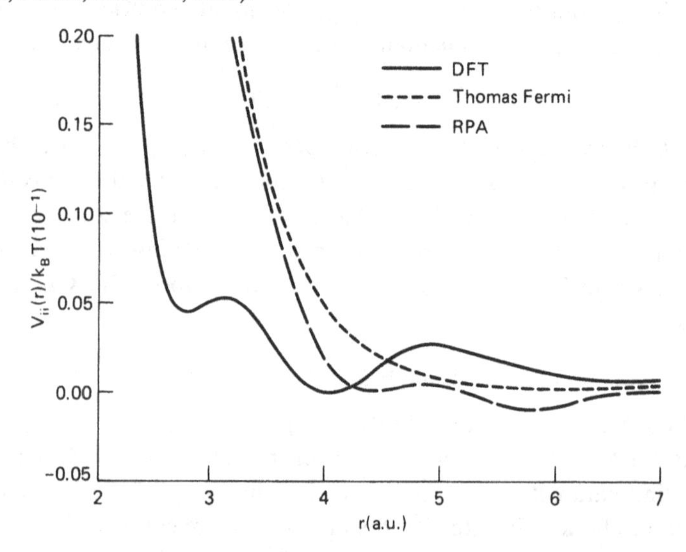

16.3. Hydrogen-helium mixtures and constitution of giant planets

The study of mixtures of hydrogen and helium is relevant to the understanding of the atmospheres of the giant planets, which are largely composed of these elements. Here, in this section, the prediction of fluid-fluid and fluid-solid equilibria will be considered in the molecular system He-H up to 1 megabar.

In addition to the interest for the giant planets, the system helium-hydrogen is of considerable significance from a first-principles standpoint. In particular, the behaviour of these mixtures provides a test for theoretical models of phase equilibria in mixtures at high pressures. From both quantum-mechanical calculations and experiment, the H_2-H_2 repulsion is found to be much softer than the intermolecular repulsion of other substances (Ree and Bender, 1979; 1980). To describe the thermodynamic properties of mixtures containing hydrogen, one has to use different stiffnesses for the intermolecular repulsion of the different constituent species in the mixture. This means, as has been emphasized by Ree (1983), that the van der Waals one-fluid model (vdW1f], which is very useful for Lennard-Jones (LJ) potentials is decidedly less useful to describe the He-H_2 systems. Another reason that may be cited for the interest in the mixtures of He and H_2 is just because the molecules are light, and therefore quantum effects become significant at sufficiently low temperatures.

Young, McMahan, and Ross (1981) have shown that an exponential six (exp-6) intermolecular potential together with a modified variational theory (Ross, 1979) yields good agreement with experimental p-V-T and melting curve data for He^4 up to 120 kbar. Ross, Ree and Young (1983) determined an effective exp-6 potential for H_2 and D_2, which agreed with available shock-wave data up to 760 kbar.

Ree (1983) extended the van der Waals model to systems with an exp-6 pair potential by deriving a set of mixing rules for the effective one-component potential parameters. By comparison of theory for the system He-H_2 with Monte Carlo computer calculations for a number of densities and temperatures, the optimum set was obtained.

The phase behaviour of He-H_2 up to 75 kbar and 360 K has been studied by van den Bergh et al. (1987). In particular, these workers determined experimentally the fluid-fluid coexistence surface and the critical line of the system, as well as the solid-fluid surface and the three-phase line. Loubeyre et al. (1987) also observed fluid-fluid equilibria in this system up to 80 kbar.

Subsequently, van den Bergh and Schouten (1988) have used the preceding model of Ree to calculate the fluid-fluid isotherms of He-H_2 with the modified variational theory. These workers found a parameter set for the unlike intermolecular potential that yielded good agreement with experiment at 9 kbar and 100 K as well as with their own experimental results up to 75 kbar and 360 K. They also employed the same set of parameters to calculate the fluid-solid isotherms and the three-phase line. Subsequently this parameter set was used to predict phase equilibria in He-H_2 up to 2500 K and 1 Mbar. Some details of this work will now be discussed.

16.3.1. Helmholtz free energy

PURE FLUIDS

The Helmholtz free energy will be written as

$$F = \tfrac{3}{2}Nk_B T + F_{\text{int}} + F_{\text{pot}} - Nk_B T \ln V + C \tag{16.48}$$

where V is the molar volume. F_{int} is the contribution from rotational and vibrational degrees of freedom, whereas F_{pot} is that from the intermolecular potential energy of the molecules.

In the case of helium, evidently $F_{\text{int}} = 0$; it is, of course, nonzero for H_2 and D_2. In the calculation of fluid-fluid equilibria in mixtures, only that part of F_{int} that is nonlinearly dependent on the composition gives a contribution to the excess Gibbs free energy (compare Chapter 13). This term was omitted by van den Bergh and Schouten, since they believe the effect on the calculated equilibria to be small (see Loubeyre et al., 1987).

Following Ree (1983) these workers used the variational perturbation theory as developed by Mansoori and Canfield (1969), Rasaiah and Stell (1970) and Ross (1979). The potential energy term to first order in the perturbation is

$$F_{\text{pot}} = F_{\text{hs}}(\eta) + A(\eta)Nk_B T + F_{\text{pert}} + F_{\text{qm}} \tag{16.49}$$

with η, as usual, denoting the packing fraction $\eta = \pi d^3 \rho/6$, ρ being the number density and d the hard sphere diameter of the reference system.

In (16.49), $F_{\text{hs}}(\eta)$ is the excess free energy of the hard sphere reference system (see Section 3.1), taken from the semiempirical hard sphere excess free energy derived by Carnahan and Starling (1969; see also Section 3.1):

$$F_{\text{hs}} = Nk_B T \frac{4\eta - 3\eta^2}{(1-\eta)^2}. \tag{16.50}$$

This expression represents an accurate approximation to the excess hard sphere free energy.

The factor $A(\eta) = -(\eta^{4/2} + \eta^2 + \eta/2)$ has been introduced (Ross, 1979) to compensate for the softness of the repulsive part of the potential in comparison with the hard sphere reference system. This, in fact, makes the reference system resemble more nearly that of an inverse -12 system.

The interaction energy of the assembly is given by

$$F_{\text{pert}} = \frac{\rho N}{2} \int_d^\infty \phi(r) g_{\text{hs}}(r, \eta)\, \mathrm{d}\mathbf{r} \tag{16.51}$$

with $\phi(r)$ the intermolecular potential energy and $g_{\text{hs}}(r, \eta)$ the hard sphere radial distribution function.

The term F_{qm} in (16.49) incorporates the first-order quantum correction (high-order terms being neglected) in the Wigner-Kirkwood expansion (Wigner, 1932; Kirkwood, 1933). This first-order term is given by

$$F_{\text{qm}} = \frac{\beta h^2 \rho N}{96\pi^2 m} \int_d^\infty \nabla^2 \phi(r) g_{\text{hs}}(r, \eta)\, \mathrm{d}\mathbf{r}. \tag{16.52}$$

Van den Bergh and Schouten (1988) used the representation of $g(r, \eta)$ given by Smith and Henderson (1970) in evaluating F_{pert} and F_{qm} in (16.51) and (16.52).

One should note here that using these equations for F_{pert} and F_{qm} implies that the H_2-H_2 potential is spherically symmetric over the whole pressure range under consideration (van den Bergh and Schouten, 1988). It is argued by reference to the work of Fischer and Lago (1983) that the error introduced by this assumption in calculating the free energy is small. The assumption seems to be supported by experimental data on the vibrational frequency of H_2. The shift in the vibrational frequency of H_2 is less than 3% at pressures up to 600 kbar at 300 K (Sharma, Mao, and Bell, 1980). Moreover it has been argued by Silvera and Wijngaarden (1981) that ortho D_2 remains spherically symmetric at low (< 5 K) temperatures up to 278 kbar.

The potential used in (16.51) and (16.52) for F_{pert} and F_{qm} to calculate the phase equilibria by van den Bergh and Schouten was of the form

$$\phi(r) = \frac{\varepsilon}{\alpha - 6} \left\{ 6 \exp\left[\alpha \left(1 - \frac{r}{r^*} \right) \right] - \alpha \left(\frac{r^*}{r} \right)^6 \right\}. \tag{16.53}$$

The parameter ε represents the attractive well depth, r^* measures the range of interaction ($\phi(r^*) = -\varepsilon$), whereas the parameter α regulates the stiffness of the repulsion. The parameters are listed in Table 16.1. In this table, the parameters for the He-He potential were taken from the work of Young et al. (1981). These parameters reproduce the experimental pressure-volume and melting-line data accurately up to pressures of 120 kbar. The parameter set for pure hydrogen in Table 16.1 was used by Ross, Ree, and Young (1983) to reproduce shock-wave data (see Chapter 15) on liquid H_2 up to 900 kbar and 7000 K.

Van den Bergh and Schouten use the so-called extended van der Waals one-fluid model developed by Ree (1983) to calculate fluid-fluid coexistence

Table 16.1. *Set of parameters in potential of (16.53), used by van den Bergh and Schouten (1988) to calculate phase equilibria of* H-He *mixtures.*

	ε/k_B (K)	r^* (Å)	α
He-He	10.8	2.9673	13.1
H_2-H_2	36.4	3.43	11.1
He-H_2	17.3	3.28	12.49

Figure 16.6. Gibbs free energy of mixing G^m for four different pressures plotted as a function of composition for $T = 2000$ K for H-He mixture (after Van den Bergh and Schouten, 1988).

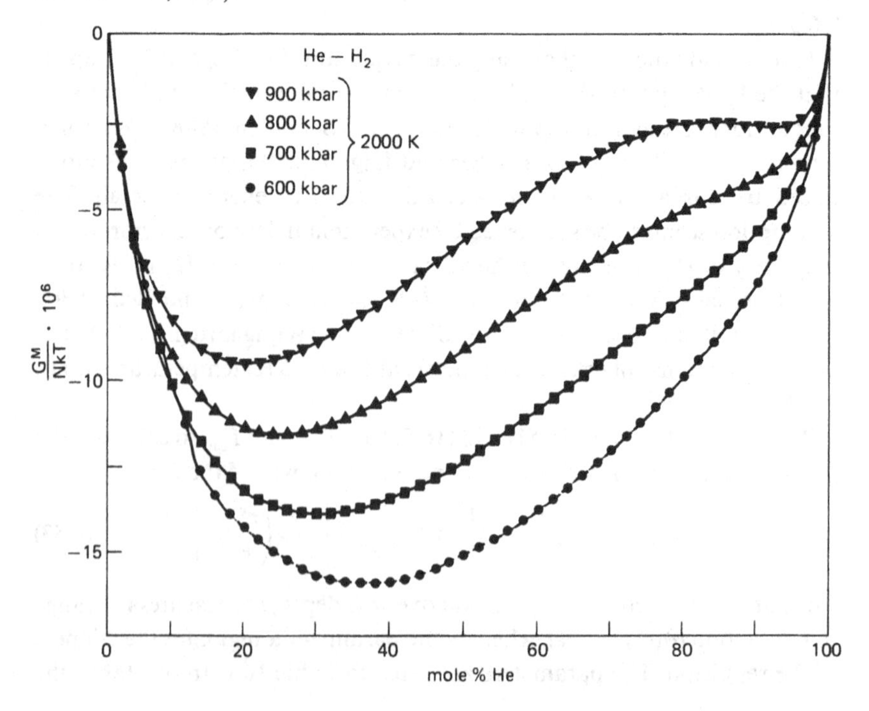

surfaces of He-H_2 mixtures up to 1 Mbar. As an illustration of their results, Figure 16.6 gives an example of the Gibbs free energy of mixing G^m plotted as a function of composition for $T = 2000$ K.

Figure 16.7 shows calculated fluid-fluid and solid-fluid isotherms at 500, 700, and 1000 K. Dashed-dotted lines in this figure represent solid-fluid-fluid three-phase equilibrium.

Figure 16.7. Shows calculated fluid-fluid and solid-fluid isotherms at 500, 700 and 1000 K. Dashed-dotted lines represent solid-fluid-fluid three-phase equilibrium (after Van den Bergh and Schouten, 1988).

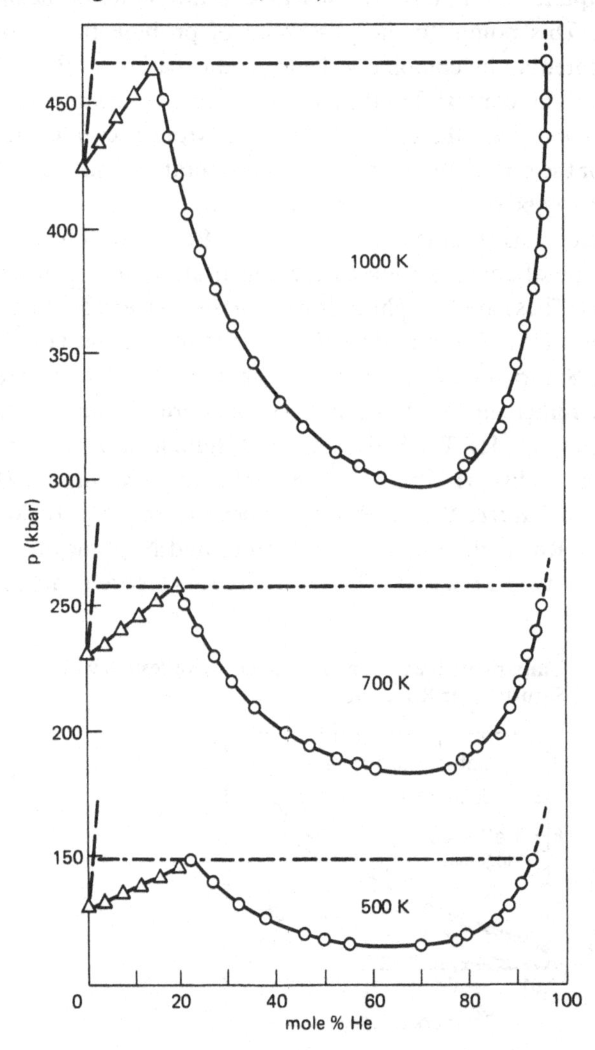

16.3.2. Summary and possible relevance to giant planets

As far as the conditions met in the giant planets are concerned, the preceding results are over a quite restrictive range of thermodynamic states. Nevertheless, this account has been included to demonstrate how it has proved possible to give a theoretical description of the fluid-fluid behaviour of the system He-H_2.

It is worth stressing the good agreement with experimental results that occurs in the preceding work both at 100 K and at 300 K. The calculated fluid-fluid equilibrium is, however, sensitive to the choice of the molecular parameters. This points to the possibility of probing the unlike intermolecular potential by combining theory and experiment. But it also has to be added in this context that the parameters for the like interactions also considerably influence the calculated results. More accurate equation of state data for the pure components at high temperature and pressure would therefore clearly be of considerable value here.

To conclude this discussion, let us follow Ross (1985) in relating the preceding type of consideration to the constitution and structure of the giant planets. These are thought to consist of three layers; an outer layer of H_2 and nonmetallic He; a middle layer either of metallic H and He for Jupiter and Saturn or of ammonia, methane, and water (referred to as "ices") for Uranus and Neptune, and a rocky core of iron, nickel, silicon and magnesium oxides. The hydrogen and helium in Jupiter are subjected to a range of conditions from 1 bar and 160 K to 45 Mbar and 24,000 K. In the case of Saturn, the latter two values are roughly 10 Mbar and 10,000 K (see Ross, 1985). The ices in Uranus and Neptune are subjected to pressures between 0.2 and 0.7 Mbar. The rocky core components are

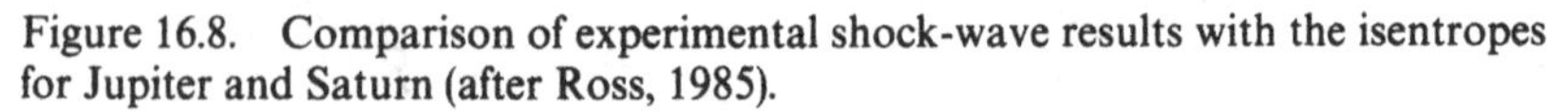

Figure 16.8. Comparison of experimental shock-wave results with the isentropes for Jupiter and Saturn (after Ross, 1985).

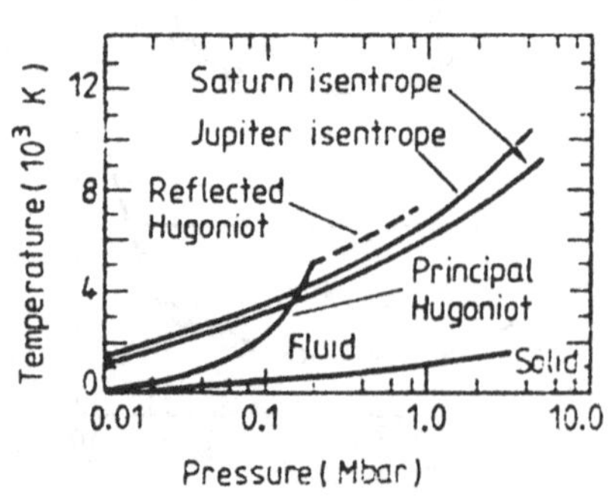

compressed to roughly 10 Mbar in Uranus and Neptune and as high as 100 Mbar at the centre of Jupiter.

Equation-of-state data relevant to the modeling of the giant planets has been obtained from shock-wave experiments (see Chapter 15). A comparison of these results with the calculations made for Jupiter and Saturn is shown in Figure 16.8.

Of the above planets, Ross (1985) notes that Jupiter is the best understood. DeMarcus (1958) had already noted that the general structure of the planet is determined primarily by the equation of state of hydrogen. Although, at the time of writing, Saturn is considered to have a composition similar to Jupiter, Stevenson and Salpeter (1976) have addressed an earlier difficulty in modeling this planet by proposing a phase separation in the He/metallic H layer. At a given pressure Saturn has a lower temperature than Jupiter, and if the critical temperature of the mixture were intermediate, then precipitation of He droplets would occur only on Saturn. This led to a number of theoretical studies for the solubility of fully pressure-ionized H-He mixtures that confirmed, in general, the existence of a miscibility gap in this system [a possible objection is that He may not be pressure-ionized below 35 Mbar; see Young, McMahan, and Ross (1981); also Ross (1985)]. The highest pressure to which He is subjected in Saturn is only 10 Mbar.

The continuing interest of the molecular-metallic transition in hydrogen and of the theory of H-He mixtures should be abundantly clear from the preceding discussion.

Appendix 2.1

Fluctuation theory derivation of $S(0)$ in terms of compressibility

Let us consider an open region, i.e., one in which particles can come and go freely, drawn in a system of infinite extent. What will now be shown is that the fluctuation in the number of particles in this region is given by the volume integral of $g(r) - 1$, which is specifically the isothermal compressibility of the liquid. Another interesting example of such a relation between fluctuations and thermodynamic quantities yields the specific heat c_v; this is discussed in Appendix A5.4.

One reason for the interest in the above relation between the volume integral of the radial distribution function—or, equivalently, from (2.4), the long wavelength limit of the structure factor $S(k)$—and the compressibility (first derived by Ornstein and Zernike) is because of the difficulty of extending diffraction experiments to very small scattering angles.

Let us consider a member of the grand canonical ensemble in which the open region, of volume V, contains exactly N particles. For a specified configuration of the particles, $\mathbf{R}_i$, say, the singlet density $\rho(\mathbf{r}_1)$ at point $\mathbf{r}_1$ in this region, and the density $\rho^{(2)}(\mathbf{r}_1, \mathbf{r}_2)$ of pairs of particles at points $\mathbf{r}_1$ and $\mathbf{r}_2$ are given, respectively, by

$$\rho(\mathbf{r}_1) = \sum_{i=1}^{N} \delta(\mathbf{R}_i - \mathbf{r}_1) \tag{A2.1.1}$$

and

$$\rho^{(2)}(\mathbf{r}_1, \mathbf{r}_2) = \sum_{i=j=1}^{N} \delta(\mathbf{R}_i - \mathbf{r}_1)\delta(\mathbf{R}_j - \mathbf{r}_2). \tag{A2.1.2}$$

It follows directly from these definitions that

$$\int_V d\mathbf{r}_1\, \rho(\mathbf{r}_1) = N \tag{A2.1.3}$$

$$\int_V \int_V d\mathbf{r}_1\, d\mathbf{r}_2\, \rho^{(2)}(\mathbf{r}_1, \mathbf{r}_2) = N^2 - N. \tag{A2.1.4}$$

The distribution functions for single particles and for pairs of particles are obtained by averaging the respective densities over phase space and over all numbers N of particles with the probability distribution of the grand-canonical ensemble

$$W_{GC} = \exp\left(\frac{\Omega + N\mu - H_N}{k_B T}\right) \tag{A2.1.5}$$

where $\Omega = -pV$, μ is the chemical potential, and H_N is the Hamiltonian of the set of N particles. In a fluid the averaged densities have the form $\langle \rho(\mathbf{r}_1)\rangle = \langle N\rangle/V = \rho$ and $\langle \rho^{(2)}(\mathbf{r}_1, \mathbf{r}_2)\rangle = \rho^2 g(\mathbf{r}_{12})$, where ρ is the bulk number density of the fluid and $g(r_{12})$ is the radial distribution function, dependent only on the scalar distance r_{12}. By taking the average of (A2.1.4), it follows that

$$\lim_{k\to 0} S(k) = 1 + \rho \int \mathrm{d}V[g(r_{12}) - 1] = \frac{\langle N^2\rangle - (\langle N\rangle)^2}{\langle N\rangle}, \tag{A2.1.6}$$

which is the desired relation between the structure factor in the long wavelength limit and the fluctuations in the particle number.

Next, one recalls from fluctuation theory that the particle number fluctuations are related to the thermodynamic properties of the system. The grand partition function is

$$\exp\left(\frac{-\Omega}{k_B T}\right) = \sum_{N=0}^{\infty} \exp\frac{[N\mu - F(N, T, V)]}{k_B T} \tag{A2.1.7}$$

where $F(N, T, V)$ is the Helmholtz free energy of a member of the grand ensemble containing N particles. By differentiating Ω and the average number of particles,

$$\langle N\rangle = \sum_{N=0}^{\infty} \frac{N \exp[\Omega + N\mu - F(N, T, V)}{k_B T}, \tag{A2.1.8}$$

with respect to the chemical potential, one finds

$$\left(\frac{\partial \Omega}{\partial \mu}\right)_{T,V} = -\langle N\rangle \tag{A2.1.9}$$

and

$$\left(\frac{\partial N}{\partial \mu}\right)_{T,V} = \frac{1}{k_B T}\left[\langle N\rangle \left(\frac{\partial \Omega}{\partial \mu}\right)_{T,V} + \langle N^2\rangle\right]. \tag{A2.1.10}$$

Hence it follows that

$$\langle N^2\rangle - (\langle N\rangle)^2 = k_B T \left(\frac{\partial N}{\partial \mu}\right)_{T,V}. \tag{A2.1.11}$$

Finally, by making use of the thermodynamic identity

$$\left(\frac{\partial \mu}{\partial N}\right)_{T,V} = -\frac{1}{\rho^2}\left(\frac{\partial p}{\partial V}\right)_{T,N} = \frac{1}{\rho^2 V K_T}, \qquad \text{(A2.1.12)}$$

where K_T is the isothermal compressibility, one arrives at the desired relation (2.5).

Appendix 3.1
Percus-Yevick hard sphere solution for direct correlation function

In Chapter 3, brief reference was made to the collective coordinate theory of Percus and Yevick. Here a summary will be given of the solution of that approximate structural theory for a dense hard sphere liquid (Wertheim, 1963; Thiele, 1963). In accordance with experience from the intimate relation between the direct correlation function $c(r)$ and the pair interaction $\phi(r)$ from the Percus-Yevick equation, the direct correlation function in the present model becomes zero outside the hard sphere diameter σ. Study of the virial expansion (Nijboer and van Hove, 1952; see also Ashcroft and March, 1967) suggests that $c(r)$ is a polynomial of order r^3 inside the hard sphere diameter; this can be verified by substitution in the Percus-Yevick equation. If one works with the dimensionless variables

$$\eta = \tfrac{1}{6}\pi\sigma^3\rho \tag{A3.1.1}$$

with ρ the number density and

$$x = \frac{r}{\sigma} \tag{A3.1.2}$$

then the direct correlation function $c(x)$ has the form

$$c(x) = \begin{cases} \alpha + \beta x + \gamma x^3, & x < 1 \\ 0, & x > 1 \end{cases} \tag{A3.1.3}$$

exactly as anticipated from the behaviour of the virial coefficients. Here, in (A3.1.3), α, β, and γ are functions of the packing fraction η (the fraction of the total fluid volume occupied by the hard spheres). The defining equations for α, β, and γ are, in fact,

$$\left.\begin{aligned} (1-\eta)^4\alpha &= -(1+2\eta)^2 \\ (1-\eta)^4\beta &= 6\eta(1+2\eta)^2 \\ (1-\eta)^4\gamma &= -\tfrac{1}{2}\eta(1+2\eta)^2 \end{aligned}\right\}. \tag{A3.1.4}$$

If one uses (A3.1.3) for $c(x)$ in conjunction with the density fluctuation formula (2.5), one regains (3.15).

Bhatia and March (1984a), in their discussion of vacancy formation energy in relation to liquid structure (Section 4.4), pointed out that the preceding theory leads to a direct relation between the Ornstein-Zernike direct correlation function $c(r)$ at $r = 0$ and its Fourier transform $c(q)$ evaluated at $q = 0$. This relation is readily shown from (A3.1.3) to have the form

$$c_{\mathrm{HS}}^{\mathrm{PY}}(q = 0) = 1 + c_{\mathrm{HS}}^{\mathrm{PY}}(r = 0). \tag{A3.1.5}$$

Bernasconi and March (1986) have tested this relation on some 15 liquid metals using experimental diffraction data. Their results are recorded in Table A3.1.1. It will be seen that for most cases, the ratio $c(r = 0)/c(q = 0)$ recorded there is near unity, in accordance with (A3.1.5) because, near freezing, both c's in that equation have magnitudes large compared with unity. However, bearing in mind the uncertainties that exist in the diffraction data, three metals clearly deviate from the prediction (A3.1.5), namely, the polyvalent metals Ga, Pb and Sn. More surprisingly, so does

Table A3.1.1. *Values of Ornstein-Zernike function at $r = 0$ and at $q = 0$ for liquid metals near freezing point. Ratio $c(r = 0)/c(q = 0)$ is near to unity according to Percus-Yevick theory for hard spheres.*

Liquid metal	$-c(r = 0)$	$-c(q = 0)$	Ratio $\dfrac{c(r = 0)}{c(q = 0)}$
Na	43	41	1.0
K	42	40	1.0
Rb	45	42	1.1
Cs	50	38	1.3
Cu	60	47	1.3
Ag	51	53	1.0
Au	35	38	0.9
Mg	31	39	0.8
Al	45	54	0.8
Ga	34	200	0.2
Pb	44	110	0.4
Sn	40	140	0.3
Fe	46	48	1.0
Ni	41	50	0.8
Co	35	50	0.7

Note: For comparison, liquid argon values are:

	33	17	1.9

the liquid insulator Ar, whose values are also recorded in the footnote to the table. Some conclusions as to the schematic form of $c(r)$ are thereby drawn for these three metals Ga, Pb and Sn by Bernasconi and March; the interested reader is referred to their paper for further details.

Appendix 3.2
Weeks-Chandler-Andersen (WCA) approximation to structure factor

The object of this appendix is to summarize the Weeks-Chandler-Andersen (WCA) approximation to the calculation of liquid structure that is referred to in Chapter 3. The following account is based on the presentation of Young (1987).

In the WCA perturbation method, the potentials of Figure 3.2 are separated (in a somewhat arbitrary way) into core and tail contributions. Actually Weeks, Chandler, and Andersen (1971) studied conventional potentials, with well-defined principal minima as in diagram (i) of Figure 3.2. They used the separation

$$v = v_{\text{core}} + v_{\text{tail}} \tag{A3.2.1}$$

where

$$v_{\text{core}} = \begin{cases} v - v_{\min}, & r < r_0 \\ 0, & r > r_0 \end{cases} \tag{A3.2.2}$$

and

$$v_{\text{tail}} = \begin{cases} v_{\min}, & r < r_0 \\ v, & r > r_0. \end{cases} \tag{A3.2.3}$$

The lack of a pronounced principle minimum (at or around the nearest-neighbour configuration) need not be an insurmountable obstacle; (A3.2.2) and (A3.2.3) can be applied to all the forms depicted in Figure 3.2 when the appropriate identifications are made. The approach of WCA is then to solve the core-only problem first and next to incorporate the tail using perturbation theory.

One possible way to solve the core-only problem is to replace the cores by an optimum hard sphere choice, using the Gibbs-Bogoliubov method of Section 3.2. Under such circumstances, an approximate analytical solution can be found (Jacobs and Andersen, 1975; Telo da Gama and Evans, 1980). Thereby, some basic justification for the result (3.26) is obtained.

Weeks, Chandler, and Andersen, however, incorporated some description of the departure of the core from hard-sphere form into the structure factor. Their result was

$$S_{\text{core}}(q) = S_{\text{hs}}(q) + b(q) \tag{A3.2.4}$$

where $b(q)$ (the "blip function") vanishes at $q = 0$. Actually, at small values of q [see Young (1987)] this equation leads to a spurious bump in the structure factor, and it is better to write (Meyer et al., 1981, 1984) at least in this region:

$$\tilde{c}_{\text{core}}(q) = \tilde{c}_{\text{hs}}(q) + b(q), \tag{A3.2.5}$$

which implies, using (2.5), that $S_{\text{core}}^{-1} = S_{\text{hs}}^{-1}(q) - b(q)$.

The hard sphere formalism applies in an unmodified form only at $q = 0$ (because $b(0) = 0$), whereas at higher values of q, $b(q)$ provides modifications due to core softening. The diameter in the WCA perturbation treatment is, therefore, really applicable only to the gentler [phonon-type: see Young (1987)] collisions at $q \approx 0$; it follows, therefore, that the diameter is larger than its all-purpose Gibbs-Bogoliubov counterpart. This matter has been quantified by Meyer et al. (1981, 1984), who found that in the WCA treatment it is better to replace (3.5) by

$$v(\sigma) - v_{\text{min}} \approx k_{\text{B}}T. \tag{A3.2.6}$$

This is solved to give $\eta = (\pi/6)\rho\sigma^3 \approx 0.50$ near melting, but in view of what has been said, one ought really not to call it a packing fraction.

Finally, one includes the tail by perturbation theory. There are numerous ways of achieving this, with correspondingly varying degrees of accuracy. The simplest of these is probably the random-phase approximation (RPA), used in a different context in Appendix 6.1. This leads to

$$\tilde{c}(q) = \tilde{c}_{\text{core}}(q) - \frac{\tilde{v}_{\text{tail}}(q)}{k_{\text{B}}T}. \tag{A3.2.7}$$

The justification of this equation is that it leads to the correct asymptotic result (3.10), whereas at small r it is dominated by the core, as it should be. These arguments lead, with a little manipulation, to (3.26).

Appendix 5.1
Pressure dependence of pair function related to three-particle correlations

In this appendix, a brief summary of the relation between the pressure dependence of the pair correlation function $g(r)$ and the three-particle correlation function $g^{(3)}$ entering the important force equation (5.2) will be given. The following argument is essentially a version of the work of Schofield (1966).

The starting point is usefully taken as the density dependence of the pair function $g(r)$. In the grand ensemble the dependence of $g(r)$ on density at constant temperature is only in the fugacity z, and one can write

$$\left(\frac{\partial[\rho^2 g(r)]}{\partial\rho}\right)_T = V\left(\frac{\partial[\rho^2 g(r)]}{\partial\langle N\rangle}\right)_T = V\left(\frac{\partial[\rho^2 g(r)]}{\partial z}\right)_T \bigg/ \left(\frac{\partial\langle N\rangle}{\partial z}\right)_T. \quad \text{(A5.1.1)}$$

From the grand partition function Ξ and N-body potential Φ_N, one can then show that

$$z\left(\frac{\partial[\rho^2 g(r)]}{\partial z}\right)_T = \Xi^{-1}\sum_{N=2}^{\infty}(N-\langle N\rangle)\frac{z^N}{(N-2)!}\int\cdots\int d\mathbf{R}_3\cdots d\mathbf{R}_N$$
$$\times \exp\left[\frac{-\Phi_N(\mathbf{r}_1,\ldots,\mathbf{R}_N)}{k_B T}\right]$$
$$= \langle N\rho^{(2)}(\mathbf{r}_1,\mathbf{r}_2)\rangle - \langle N\rangle\langle\rho^{(2)}(\mathbf{r}_1,\mathbf{r}_2)\rangle, \quad \text{(A5.1.2)}$$

which evidently represents the fluctuation in the product of N with the pair density denoted by $\rho^{(2)}$. After some manipulation, (A5.1.2) can be rewritten in terms of the three-particle correlation function $g^{(3)}$ as

$$z\left(\frac{\partial[\rho^2 g(r)]}{\partial z}\right)_T = 2\rho^2 g(r) + \rho^3\int d\mathbf{r}_3[g^{(3)}(\mathbf{r}_1,\mathbf{r}_2,\mathbf{r}_3) - g(r)]. \quad \text{(A5.1.3)}$$

But it can also be shown that

$$\frac{z}{V}\left(\frac{\partial\langle N\rangle}{\partial z}\right)_T = \frac{z}{V}\left(\frac{\partial N}{\partial\mu}\right)_T\left(\frac{\partial\mu}{\partial z}\right)_T = \frac{k_B T}{V}\left(\frac{\partial N}{\partial\mu}\right)_T = \rho k_B T\left(\frac{\partial\rho}{\partial p}\right)_T, \quad \text{(A5.1.4)}$$

and hence one reaches the result

$$\left(\frac{\partial[\rho^2 g(r)]}{\partial p}\right)_T = \frac{\left(\frac{\partial[\rho^2 g(r)]}{\partial \rho}\right)_T}{\left(\frac{\partial p}{\partial \rho}\right)_T}$$

$$= (k_B T)^{-1}\left\{2\rho g(r) + \rho^2 \int d\mathbf{r}_3[g^{(3)}(\mathbf{r}_1, \mathbf{r}_2, \mathbf{r}_3) - g(r)]\right\}. \tag{A5.1.5}$$

This expression can be reorganized into its most transparent form

$$k_B T\left(\frac{\partial g(r)}{\partial p}\right)_T = \int d\mathbf{r}_3[g^{(3)}(\mathbf{r}_1, \mathbf{r}_2, \mathbf{r}_3) - g(r)g(r_{23}) - g(r)g(r_{31}) + g(r)]. \tag{A5.1.6}$$

Thus, it can be seen that the experimental study of the pressure dependence of the pair correlation function gives information on the three-body correlations, integrated over all values of the coordinates of the third particle. The appearance of the various pair functions in the square bracket in (A3.1.6) implies that only the behaviour of the triplet function over the configurations in which the third particle is close to the other two particles is really relevant here.

A similar discussion can be given concerning the temperature dependence of correlation functions [see Schofield (1966) and March and Tosi (1976)], but further details are not pursued here. Finally, it is to be noted that three- and four-particle correlation functions enter the theory of the specific heats of simple liquids (see Appendix 5.4).

Appendix 5.2

Conditions to be satisfied by thermodynamically consistent structural theories

The purpose of this appendix is to set down the way in which thermodynamic consistency can be achieved, at least in principle, in structural theories. The first part of the argument follows the approach of Kumar, March, and Wasserman (1982). For simplicity the following argument will be presented for density-independent pair potentials. Then the virial expression for the pressure is given by (3.3), which will be conveniently written in the short-hand notation

$$\frac{p}{nk_{\mathrm{B}}T} = 1 - nI. \tag{A5.2.1}$$

This equation, when combined with (3.3), evidently defines I.

Next, let us note the exact expressions for the compressibility, namely,

$$n\frac{\partial p}{\partial n} = nk_{\mathrm{B}}T\left(1 - n\int c(r)\,\mathrm{d}\mathbf{r}\right) = 2p - nk_{\mathrm{B}}T - k_{\mathrm{B}}Tn^3\frac{\partial I}{\partial n}, \tag{A5.2.2}$$

the last part of (A5.2.2) following by differentiation of (A5.2.1) with respect to n.

This is the place to stress that a sufficient condition for thermodynamic consistency can be obtained by substituting p in (A5.2.2) from (A5.2.1) to obtain

$$-n\int c(r)\,\mathrm{d}\mathbf{r} = -\frac{\partial}{\partial n}\left[\int \mathrm{d}\mathbf{r}\, rg\frac{\mathrm{d}\phi}{\mathrm{d}r}\frac{n^2}{6k_{\mathrm{B}}T}\right] \tag{A5.2.3}$$

or, equivalently,

$$-r^2c(r)n = \phi(r)\frac{\partial^2}{\partial n\,\partial r}\left[\frac{gr^3n^2}{6k_{\mathrm{B}}T}\right] + F(r, n, T), \tag{A5.2.4}$$

where the function F must satisfy

$$\int_0^\infty F\,\mathrm{d}r = 0. \tag{A5.2.5}$$

A5.2.1. Example of two-dimensional one-component plasma

To explore the way (A5.2.4) works out in practice, Senatore, Rashid, and March (1986) have applied it to the two-dimensional one-component plasma. To do so, they note first the generalization to d dimensions of the "potential" part $c_p(r)$ as

$$c_p(r) = -\frac{\phi(r)}{k_B T}\frac{1}{2\,d\rho\, r^{d-1}}\frac{\partial^2}{\partial\rho\,\partial r}[\rho^2 r^d g(r)], \qquad \text{(A5.2.6)}$$

which is valid for potentials falling off faster than r^{-d}.

For one coupling strength $\Gamma = 2$, the two-dimensional OCP model can be solved for the pair function $g(r)$, to obtain

$$g(r) = 1 - \exp(-\pi\rho r^2). \qquad \text{(A5.2.7)}$$

The direct correlation function can be constructed, and Figure A5.2.1

Figure A5.2.1. Direct correlation function $c(r)$ for a two-dimensional one-component plasma for the special case when the coupling parameter $\Gamma = 2$. Solid curve, $c(x)$; dot-dash curve, $c_p(x)$ from (A5.2.6). Dashed curve shows $c_c = c - c_p$ (after Rashid, Senatore, and March, 1986; a somewhat similar plot is made by Hernando, 1986).

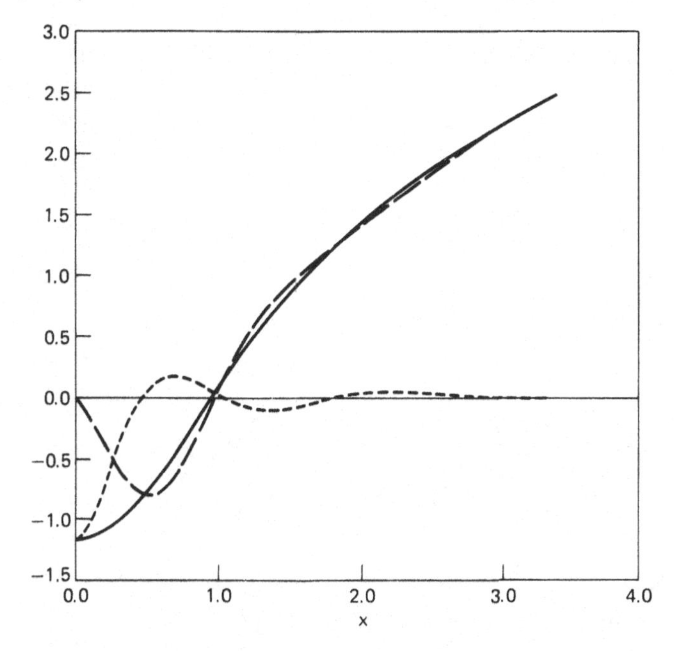

shows the form of $c(r)$ as well as the two contributions c_p and $c - c_p \equiv c_c$ to the direct correlation function obtained from (A5.2.7). The first potential contribution is dominant outside an initial core region. However, the other term, arising from the function F in (A5.2.4), makes the whole contribution at $r = 0$.

Appendix 5.3
Gaussian core model and Kirkwood decoupling of triplet correlations

The Gaussian core model of Stillinger (1976) has the total interaction potential

$$\Phi(\mathbf{r}_1, \ldots, \mathbf{r}_N) = \lambda \sum \exp(-r_{ij}^2) \qquad \text{(A5.3.1)}$$

where the positions of the N particles are denoted by $\mathbf{r}_i$; $i = 1, \ldots, N$, λ is a coupling constant and r_{ij} is the scalar distance between particles i and j. When $\lambda > 0$, all particles repel, and at moderate densities many of the properties of this model resemble those for models with more realistic pair interactions. In particular, the Gaussian core model exhibits a conventional freezing transition as the temperature is lowered into the face-centred or body-centred cubic structure at low or high densities, respectively (Stillinger, 1976; Stillinger and Weber, 1978, 1979; Stillinger, 1979). Exact molecular dynamics simulations have been carried out for this model with $\lambda > 0$ (Stillinger and Weber, 1980; 1981).

If the number N of particles is large, changing the sign of λ produces an instability whereby the particles mutually attract and collapse into a tightly bound and compact aggregate; we return to this state later.

One virtue of this model is that exact high temperature series can be exhibited for its thermodynamics and its distribution functions in the large-system limit (Stillinger, 1979). Here the natural expansion parameter is $\beta\lambda$, with $\beta = (k_B T)^{-1}$, and owing to the collapse instability that occurs at $\beta\lambda = 0$ in the large N limit, these are asymptotic series with vanishing radii of convergence. In the case of the pair correlation function, one has

$$g(r, \beta\lambda) = 1 + \sum_{n=1}^{\infty} (-\beta\lambda)^n f_n(r) \qquad \text{(A5.3.2)}$$

where f_1 through f_4 are given by Root, Stillinger and Washington (1988). For any density ρ, one expects to be able to find pair correlation functions from any reasonable approximate theory at least for small positive values of $\beta\lambda$. However, no such solutions should exist for negative $\beta\lambda$ if the approximate theory is powerful enough to reflect the collapse referred to before, even if only qualitatively.

Root, Stillinger and Washington test the Born-Green and Percus-Yevick theories (see Chapter 5) in this particular respect.

A5.3.1. Kirkwood superposition

Kirkwood superposition on the potential of mean force leads to the product form (3.5) for $g^{(3)}$ in terms of $g(r)$. Inserting in the resulting Born-Green approximate theory the result $\phi(r) = \lambda \exp(-r^2)$ appropriate to the Gaussian core model, one finds

$$\frac{d \ln g(r_{12})}{dr_{12}} = 2\beta\lambda \exp(-r_{12}^2) + \left(\frac{2\pi\rho\beta\lambda}{r_{12}^2}\right) \int_0^\infty dr_{13} \int_{|r_{12}-r_{13}|}^{r_{12}+r_{13}} dr_{23}\, r_{13} r_{23} \times (r_{12}^2 + r_{13}^2 - r_{23}^2) \exp(-r_{13}^2) g(r_{13}) g(r_{23}). \tag{A5.3.3}$$

Series solution in $\beta\lambda$ can be deduced, and the structure is that of (A5.3.2). Indeed, f_1 through f_3 are exact; the first discrepancy (affecting just one contribution) arises in f_4 from the work of Root, Stillinger, and Washington.

A5.3.2. Percus-Yevick collective coordinate theory

Root, Stillinger, and Weber do the same calculation for the Percus-Yevick approximation and again find f_1 through f_3 are exact; but f_4 is not. f_4 differs from the exact result for this model by a single missing term.

A5.3.3. Numerical solutions

Root, Stillinger, and Washington then obtain numerical solutions for the pair function $g(r)$ for the Born-Green and Percus-Yevick equations. This they do, first of all for positive values of $\beta\lambda$, followed by a continuous change of this quantity toward and into negative values, to see to what extent "collapse" is represented. These workers find from both these approximate methods that solutions for $g(r)$ continue smoothly for fixed density from positive $\beta\lambda$ through the origin and into the negative regime of $\beta\lambda$. Thus, the attractive interactions present when $\beta\lambda < 0$ now fail to produce collapse in these approximate methods but instead are found to lead to a modest local density enhancement. As Root, Stillinger, and Washington point out, the fact that these methods smoothly continue from the positive to the negative $\beta\lambda$ axis indicates that the corresponding series have nonzero radii of convergence.

A5.3.4. Some exact results for completely aggregated state

To conclude this appendix, we shall summarize some exact results obtained by Root, Stillinger, and Washington for the completely aggregated state with attractive interactions. First, the pair function is simply as if the potential of mean force were an attractive parabola; i.e.,

$$g(r_{12}) = V\left(\frac{|\beta\lambda|N}{2\pi}\right)^{3/2} \exp\left(-\frac{1}{2}|\beta\lambda|Nr_{12}^2\right). \tag{A5.3.4}$$

The important three-particle correlation function $g^{(3)}$ can also be calculated analytically as

$$g^{(3)}(\mathbf{r}_1, \mathbf{r}_2, \mathbf{r}_3) = V^2\left(\frac{|\beta\lambda|N}{3^{1/2}\pi}\right)^3 \exp\left[\left(-\frac{|\beta\lambda|N}{3}\right)(r_{12}^2 + r_{23}^2 + r_{31}^2)\right]. \tag{A5.3.5}$$

These workers define the Kirkwood superposition correction factor K through

$$K(r_{12}, r_{23}, r_{31}) = \frac{g^{(3)}(r_{12}, r_{23}, r_{31})}{g(r_{12})g(r_{23})g(r_{31})}, \tag{A5.3.6}$$

which is readily written explicitly from (A5.3.4) and (A5.3.5). Appeal is made in the main text to these results for the completely aggregated state.

Appendix 5.4
Specific heats of liquids in terms of higher-order correlation functions

Bratby, Gaskell and March (1970) have given a method by which the difference in specific heats of a simple liquid can be expressed in terms of the three-particle distribution function. Their starting point is the usual thermodynamic relation

$$c_p - c_v = -\frac{T}{N}\left(\frac{\partial p}{\partial T}\right)_V^2 \left(\frac{\partial V}{\partial p}\right)_T. \tag{A5.4.1}$$

This can be rewritten in the form

$$c_p - c_v = \frac{1}{\rho^2 k_B}\left(\frac{\partial p}{\partial T}\right)_V^2 \left[-\rho k_B T \frac{1}{V}\left(\frac{\partial V}{Vp}\right)_T\right]. \tag{A5.4.2}$$

Using (2.5) in the equivalent form (plus $(\partial E/\partial V)_T = T(\partial p/\partial T)_V - p$ (Zemansky, 1951)):

$$S(0) = -\rho k_B T \frac{1}{V}\left(\frac{\partial V}{\partial p}\right)_T, \tag{A5.4.3}$$

one finds for the specific heat difference the result

$$c_p - c_v = S(0)\left\{\frac{p}{\rho k_B T} + \frac{1}{\rho k_B T}\left(\frac{\partial E}{\partial V}\right)_T\right\}^2 k_B \tag{A5.4.4}$$

where p is the pressure and E is the internal energy. Though it is worth noting here that for hard spheres, $(\partial E/\partial V)_T = 0$, and that then (A5.4.4) yields an exact result for this model, for real, dense liquids the term $(\partial E/\partial V)_T$ is in fact the dominant one.

As seen in Chapter 3, a useful form of $(\partial E/\partial V)_T$ can be obtained from liquid state theory when the pair potential $\phi(r)$ is independent of density:

$$\left(\frac{\partial E}{\partial V}\right)_T = -\frac{\rho^2}{2}\int \frac{\partial}{\partial \rho}[\rho g(r)]\phi(r)\,d\mathbf{r}. \tag{A5.4.5}$$

Using the result of Appendix 5.1 for the density dependence of the radial distribution function $g(r)$ in terms of the three-particle correlation function g_3 (see Chapter 5), one obtains

$$\left(\frac{\partial E}{\partial V}\right)_{\mathrm{T}} = -\frac{\rho^2}{2S(0)}\left[\int g(r)\phi(r)\,\mathrm{d}\mathbf{r} + \rho\int \{g_3(\mathbf{r},\mathbf{s}) - g(r)g(s)\}\phi(r)\,\mathrm{d}\mathbf{r}\,\mathrm{d}\mathbf{s}\right]. \tag{A5.4.6}$$

Combining (A5.4.4) and (A5.4.6), one can write

$$\frac{c_{\mathrm{p}} - c_{\mathrm{v}}}{S(0)} = \left\{1 - \frac{2\pi\rho}{3k_{\mathrm{B}}T}\int g(r)r^3\frac{\partial\phi}{\partial r}\,\mathrm{d}r - \frac{\rho}{2k_{\mathrm{B}}TS(0)}\right.$$
$$\left.\times\left[\int g(r)\phi(r)\,\mathrm{d}\mathbf{r} + \rho\int \{g_3(\mathbf{r},\mathbf{s}) - g(r)g(s)\}\phi(r)\,\mathrm{d}\mathbf{r}\,\mathrm{d}\mathbf{s}\right]\right\}, \tag{A5.4.7}$$

where the virial result for the pressure has been employed.

Schofield (1966) had earlier used fluctuation theory (see Appendix 5.1) to calculate c_{v}, with the result

$$c_{\mathrm{v}} = \frac{3}{2}k_{\mathrm{B}} + \frac{1}{k_{\mathrm{B}}T^2}\left[\frac{1}{2}\rho\int g(r)\phi^2(r)\,\mathrm{d}\mathbf{r} + \rho^2\int g_3(\mathbf{r},\mathbf{s})\phi(r)\phi(s)\,\mathrm{d}\mathbf{r}\,\mathrm{d}\mathbf{s}\right.$$
$$+\frac{1}{4}\rho^3\int \{g_4(\mathbf{r},\mathbf{s},\mathbf{t}) - g(r)g(|\mathbf{t}-\mathbf{s}|)\}\phi(r)\phi(|\mathbf{t}-\mathbf{s}|)\,\mathrm{d}\mathbf{r}\,\mathrm{d}\mathbf{s}\,\mathrm{d}\mathbf{t}$$
$$\left.-\frac{\left[\rho\int g(r)\phi(r)\,\mathrm{d}\mathbf{r} + \frac{1}{2}\rho^2\int \{g_3(\mathbf{r},\mathbf{s}) - g(\mathbf{r})\}\phi(r)\,\mathrm{d}\mathbf{r}\,\mathrm{d}\mathbf{s}\right]^2}{S(0)}\right]. \tag{A5.4.8}$$

As Bratby, Gaskell and March (1970) noted, $c_{\mathrm{p}} - c_{\mathrm{v}}$, which involves only the three-particle correlation function, is basically simpler than c_{p} or c_{v} separately, which also involve the four-particle correlation function g_4. Equation (A5.4.7) and Schofield's result (A5.4.8) are the main results of this appendix. For cautionary remarks on the difficulties, in practice, of direct use of such formulae in numerical estimates, the reader is referred to the discussion of Bratby, Gaskell and March (1970).

Appendix 5.5

Inversion of measured structure, constrained by pseudopotential theory, to extract ion-ion interaction

The use of pair force laws to predict liquid structure is exemplified by the calculations of Swamy (1986) on liquid Na and Al. Using the hypernetted chain ((5.5) with $E \to c$), and also the method of Machin-Woodhead and Chihara (MWC): see Chihara (1984; and other references there), Swamy has explored the results of using various oscillatory potentials to assess the applicability of these integral equations for such force laws. The results thereby obtained have been compared with molecular-dynamics simulation. His studies indicate that the HNC equation underestimates the main peak in $S(k)$. The MWC method seems to give good results for $S(k)$ according to Swamy's studies for shorter-range potentials, but when oscillatory effects are included, it is also deficient near the first peak of $S(k)$.

While difficulties remain with integral-equation treatments of liquid structure for such direct calculations of $S(k)$ from a given potential, it has been known for a long time that the inverse problem of extracting $\phi(r)$ from an experimentally determined $S(k)$ (Johnson and March, 1963) is a much more stringent probe of an approximate integral equation. This point has been made quite clear in the work of Levesque, Weis and Reatto (1985; see also Reatto, 1988), which is summarized in Section 5.4.

In this appendix, procedures will be described by which the inversion of liquid structure data can yield pair potentials when these are constrained by requirements imposed by the form of pseudopotential theory, already discussed in Chapters 3 and 4.

Dharma-wardana and Aers (1983, 1986) proposed a method based only on the use of experimental structure-factor data $S(k)$ over the measured range of k, no use being made of extrapolation of these data. Thus a procedure was devised whereby a calculated $S(k)$ gave the best fit to an experimental $S(k)$ provided in some limited range $k_{min} < k < k_{max}$. The calculated $S(k)$ in their method was also required to reproduce accurately the compressibility, via $S(0)$, from (2.5).

The next important constraint was the use of a parametrized form of the pair potential, guided by the pseudopotential theory as described in Chapters 3 and 4. Thus the method used can be described as a parametrized modified hypernetted chain (MHNC) fit to $S(k)$ and not a direct inversion of $S(k)$.

As an example of the procedure of Dharma-wardana and Aers (1983, 1986), the results of their approach as applied to computer simulation data for Al at 1050 K is outlined next. These simulation data (Levesque, Weis and Reatto, 1985) were fitted in the range $1.8 < kr_0 < 19$, where $r_0 = 3.1268$ atomic units is the ion-sphere radius. The compressibility was taken to be 0.0195.

The important constraint imposed on the trial potential in this work was to write the (assumed) Fourier transform of the effective pair potential $\phi(r)$ as

$$\phi(k) = \frac{z^2 4\pi}{k^2} - \chi(k) V_{ie}^2(k) \qquad \text{(A5.5.1)}$$

with the electron-ion pseudopotential $V_{ie}(k)$ chosen to be a simple two-parameter form (well depth A_0 and cutoff radius R_0 being the explicit parameters). The electron-gas response function, denoted by $\chi(k)$ in (A5.5.1) was taken to be of the form given by the local density approximation of the density functional method (see, for example, Callaway and March, 1984). In the work of Dharma-wardana and Aers, this contained an effective mass

Figure A5.5.1. Potential $\phi(r)$ as recovered from computer simulation data on a model of liquid Al. This is shown by the boxes (MHNC fit). The solid line is the original potential used (Dagens et al., 1975). The inversion result of Levesque et al. (1985) is shown by the triangles (after Dharma-wardana and Aers, 1986; see also March, 1987).

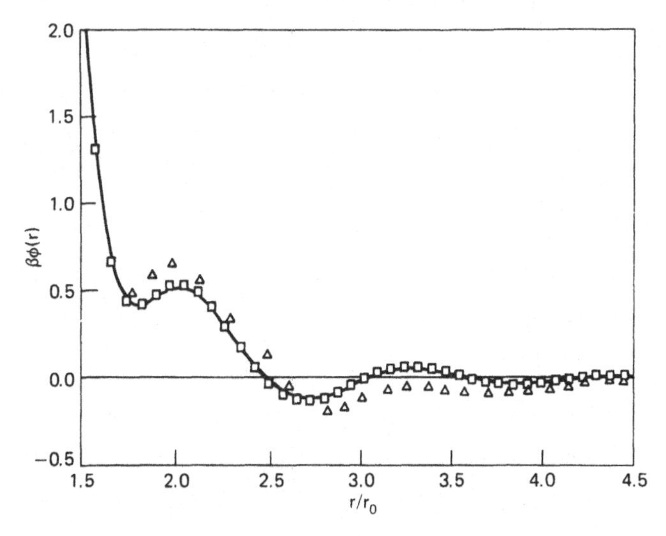

parameter equal to r_s^*/r_s, where r_s, as usual, is the electron sphere radius, $r_0/z^{1/3}$, the valence z being 3 for Al. Thus the $S(k)$ data were fitted with A_0, R_0 and r_s^* as parameters, with one parameter appearing in the bridge function $B(r)$ of MHNC (see Rosenfeld and Ashcroft, 1979).

In connection with the bridge function $B(r)$, they used the hard sphere parametrization employed by Rosenfeld and Ashcroft (1979; see also Verlet and Weis, 1972). Here V_0 is simply z/r_0. In Figure A5.5.1, the potential thereby extracted is compared with the original potential of Dagens, Rasolt and Taylor (1975) and also with an alternative form of inversion, due to Levesque et al. (see Section 5.4). The Dagens et al. potential is recovered to quite high accuracy, using $A_0 V_0 = 0.21797$, $R_0/r_0 = 0.37848$, $r_s^*/r_s = 0.99312$, $V_0 = 0.96132$ a.u, and $\eta = 0.44328$.

Appendix 6.1

Vacancy formation energy evaluated in a hot (model) crystal

The object of this appendix is to show how the formula (6.28) leads to a relationship between E_v and the thermal energy $k_B T_m$ associated with the melting temperature T_m. The following argument is due to Bhatia and March (1984a), and their model is appropriate primarily to hot crystals built from rare gas atoms [see also March (1989)].

For a given effective interaction $\phi(r)$ and radial distribution function $g(r)$ as near as possible to the freezing point, the vacancy formation energy in a hot close-packed crystal, where relaxation of the atoms neighbouring the vacancy can be neglected, is

$$E_v = -\frac{\rho}{2}\int g\phi \, d\mathbf{r} - k_B T. \tag{A6.1.1}$$

For argon, for instance, Bhatia and March have now argued that since $g = 0$ within the core diameter σ and $c = 0$ for hard spheres in the Percus-Yevick approximation (Appendix 3.1) outside the core, then following the study of Woodhead-Galloway, Gaskell and March (1968), where (see the so-called mean spherical approximation, MSA)

$$c(r) = -\frac{\phi_{lr}(r)}{k_B T}, \qquad r > \sigma \tag{A6.1.2}$$

(ϕ_{lr} denotes the long-range part of the pair potential), (A6.1.1) can be rewritten in the form

$$2E_v = \left\{\rho k_B T\left[\int g(r)c(r)\, d\mathbf{r} - \frac{2}{\rho}\right]\right\}_{T_m}. \tag{A6.1.3}$$

One must caution that (A6.1.3), appropriate for argon, will not apply as it stands to metals, the main concern here.

However, for considered rare gases, (A6.1.3) can be rewritten in terms of the total correlation function $h(r) = g(r) - 1$ as

$$\frac{E_v}{k_B T_m} = \frac{1}{2}\rho\int h(r)c(r)\, d\mathbf{r} - \frac{1}{2}c(q = 0) - 1. \tag{A6.1.4}$$

Utilizing now the convolution relation (2.7) between $h(r)$ and $c(r)$ at the point $r = 0$ and the fact that in dense liquids $g(r = 0) = 0$, one readily finds that

$$-1 = c(r = 0) + \rho \int h(r)c(r)\,\mathrm{d}\mathbf{r}. \tag{A6.1.5}$$

Using (A6.1.5) to eliminate the integral in (A6.1.4), one reaches the formula of Bhatia and March (1984) for the vacancy formation energy E_v of hot crystals composed of rare gas atoms; namely,

$$-\frac{E_v}{k_B T_m} = \frac{1}{2}[c(r = 0) - \tilde{c}(q = 0) + 3]_{T_m}, \tag{A6.1.6}$$

where it seems most natural to evaluate at the melting temperature $T = T_m$. The vacancy formation energy in units of $k_B T_m$ is therefore determined in this simple model entirely by the direct correlation function $c(r)$ of the liquid just above the freezing point. Because $c(r = 0)$ is a large negative number ~ -40 and $\tilde{c}(q = 0)$ is approximately the same (e.g., in a hard sphere model they differ by unity, as shown by Bhatia and March, 1984a), it is evident why the ratio $E_v/k_B T_m$ is a large number, found empirically to be of the order of 10. March (1987b, 1988) (see also the discussion in the main text) was concerned with the relation of E_v to departures from Joule's law in the liquid. Although the resulting argument is presumably more accurate than that leading to (A6.1.6), the gist of the results is similar. Equation (A6.1.6) has the merit of relating, as in the freezing theory given in Chapter 6, a characteristic of the hot solid, namely, its vacancy energy, directly to the bulk liquid through the Ornstein-Zernike direct correlation function $c(r)$ at the melting temperature T_m. It can hardly be in question that this puts the well-established correlation found empirically between E_v and T_m on a quite firm statistical thermodynamic footing.

Appendix 6.2
Vacancy formation energy related to Debye temperature

Let us begin with a model applicable in a sense that will be explained further to simple close-packed metals. Here, it will be assumed that the vacancy can be treated as a perturbation on the free conduction electron gas.

The simplest approach is to use the Thomas-Fermi relation between density and potential [see March and Deb (1987)]. Since, however, it is assumed that the vacancy can be treated in perturbation theory, one can linearize this relation, thus obtaining for the displaced electron density $\rho(r) - \rho_0$ around the vacant site, ρ_0 being the mean conduction electron density before the vacancy is introduced,

$$\rho(r) - \rho_0 = \frac{q^2}{4\pi} V(r), \tag{A6.2.1}$$

where q^{-1} is the so-called Thomas-Fermi screening length, which is given by

$$q^2 = \frac{4k_f}{\pi a_0}. \tag{A6.2.2}$$

In a self-consistent manner, one can now calculate the perturbing potential energy due to the screened vacancy, denoted by $V(r)$, by combining (A6.2.1) with Poisson's equation to find (Mott, 1936)

$$\nabla^2 V = q^2 V. \tag{A6.2.3}$$

In a simple metal, the screening length q^{-1} is readily evaluated from mean density $\rho_0 = k_f^3/3\pi^2$, with k_f as usual the Fermi wave number, a_0 the Bohr radius, and q^{-1} is typically of the order of 1 Å.

In a metal of valence Z, one can model the vacancy, before screening, as a point charge Ze, and screening converts the bare Coulomb potential energy Ze^2/r felt by an electron at distance r from the vacancy to the screened Coulomb form

$$V(r) = \frac{Ze^2}{r} \exp(-qr), \tag{A6.2.4}$$

which is readily verified to satisfy (A6.2.3) and the appropriate boundary conditions that V tends to Ze^2/r as r tends to zero and $V(r)$ tends to zero as r tends to infinity.

While (A6.2.4) is useful for a number of purposes, it is not a sufficiently accurate approximation in general because the resulting potential varies rapidly in space, whereas the Thomas-Fermi theory, as emphasized in Chapter 2, is based on local free-electron relations and thus is quantitatively valid only for relatively slow spatial variations in density and potential.

March and Murray (1960) proposed the appropriate generalization of the linear equation (A6.2.3) that accounts for the fact that there are diffraction effects when the conduction electron waves are scattered from the point defect. Their equation, which is discussed in Chapter 4, replaces (A6.2.3) by

$$\nabla^2 V(r) = \frac{2k_f^2}{\pi} \int V(\mathbf{r}') \frac{j_1(2k_f|\mathbf{r} - \mathbf{r}'|)}{|\mathbf{r} - \mathbf{r}'|^2} \, d\mathbf{r}', \qquad \text{(A6.2.5)}$$

which would reduce, in fact, to (A6.2.3) if it were permissible to replace $V(\mathbf{r}')$ inside the integral in (A6.2.5) by $V(\mathbf{r})$. This would obviously be valid only for a suitably slowly varying potential in $\mathbf{r}$ space.

Although the precise solution of (A6.2.5) can be calculated later, let us proceed to estimate the vacancy formation energy, E_v say, by considering the change in the sum of the one-electron energy levels brought about by the self-consistent perturbing potential energy V. Since one is assuming a free electron gas model, the unperturbed conduction electron wave functions are simply $\mathscr{V}^{-1/2} \exp(i\mathbf{k} \cdot \mathbf{r})$, and the energy change $\Delta\varepsilon_\mathbf{k}$ of the state of wave vector $\mathbf{k}$ is evidently, from first-order perturbation theory,

$$\Delta\varepsilon_\mathbf{k} = \mathscr{V}^{-1} \int \exp(-i\mathbf{k} \cdot \mathbf{r}) V(\mathbf{r}) \exp(i\mathbf{k} \cdot \mathbf{r}) \, d\mathbf{r}, \qquad \text{(A6.2.6)}$$

which is clearly independent of $\mathbf{k}$. Summing over the occupied states, replacing the summation by an integration with the usual factor for the density of states, leads to the change in the one-electron eigenvalue sum as

$$\Delta E = \rho_0 \int V(\mathbf{r}) \, d\mathbf{r}. \qquad \text{(A6.2.7)}$$

But now, by integration of (A6.2.5) over $\mathbf{r}$ and using the perfect screening of the vacancy, which assures one that the displaced charge $\rho(r) - \rho_0$ must integrate precisely to Ze, one finds the desired result

$$\Delta E = (\tfrac{2}{3}) Z E_f \qquad \text{(A6.2.8)}$$

with $E_f = \hbar^2 k_f^2/2m$ written for the Fermi energy.

There is an important correction to be made to (A6.2.8) before using it as an estimate of the vacancy formation energy for small Z. This is because, when an atom is removed and placed on the surface, the crystal volume is increased by one atomic volume in the present model of no relaxation. This lowers the kinetic energy of the conduction electrons by an amount that is easily calculated (Fumi, 1955) to be $(\frac{2}{5})ZE_f$, and subtracting this from (A6.2.8) yields the approximation

$$E_v = (\tfrac{4}{15})ZE_f. \tag{A6.2.9}$$

Of course, this perturbative argument could, at best, be applicable only for small Z, as already mentioned, i.e., for Cu. This is clearly demonstrated in Figure A6.2.1, where experimental values for E_v, divided for convenience by the valence times the free electron Fermi energy, are plotted against Z. The constant $\frac{4}{15}$ fits reasonably as the Z tends to zero limit of this plot, but this figure shows that (A6.2.9), although valid for sufficiently small Z, must not be used for other than the monovalent metals.

However, while (A6.2.9) makes it plain that the Fermi energy is, in practice, an inappropriate unit in terms of which to measure E_v for poly-

Figure A6.2.1. Plot of experimental values of vacancy formation energy E_v, conveniently expressed in units of the valence Z times the free electron Fermi energy E_f, versus valence Z.

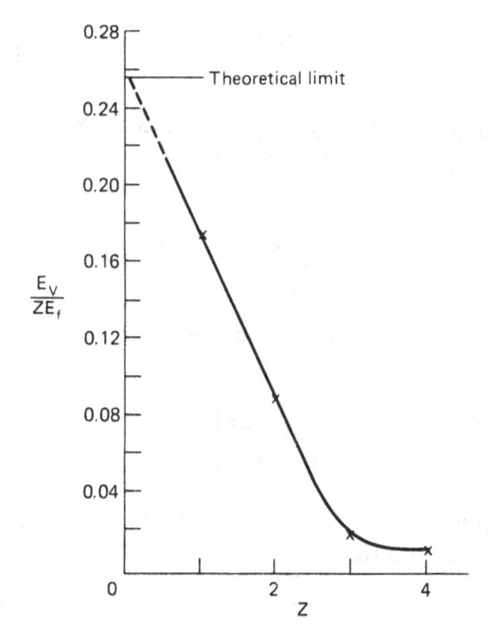

valent metals, let us now combine this equation with the so-called Bohm-Staver formula for the velocity of sound, v_s, in a metal.

This formula may be derived as follows. One writes a formula for the plasma frequency of the ions, taken to have mass M and carrying charge Ze. If the ionic density is denoted by ρ_i, then the plasma frequency of the ions is given by

$$\omega_{\text{plasma}}^{\text{ions}} = \left\{\frac{4\pi\rho_i(Ze)^2}{M}\right\}^{1/2} \tag{A6.2.10}$$

from the Langmuir formula.

But now one notes that:

1. $\rho_i Z$ is the conduction electron density ρ_0, and
2. Equation (A6.2.10) represents an optical, not an acoustic, mode.

The reason for this is clear from the discussion of the screening of the vacancy argument, now applied to the screening of the ions. The preceding argument has failed to allow for the attractive power of the ions in piling up the hitherto uniform conduction electron density around them. To describe this pileup, let us return to (A6.2.4) and note that in Fourier transform (**k** space), the bare Coulomb potential $4\pi Ze^2/k^2$ due to an ion gets screened according to

$$\frac{4\pi Ze^2}{k^2} \to \frac{4\pi Ze^2}{k^2 + q^2}, \tag{A6.2.11}$$

the last expression simply representing the screened Coulomb potential in **k** space. Thus, in the long wavelength limit appropriate to sound waves, i.e., as $k \to 0$,

$$Ze^2 \to \frac{Ze^2k^2}{q^2} \tag{A6.2.12}$$

represents the effect of electrons screening the ions, and rewriting (A6.2.10) leads to

$$\left\{\frac{4\pi\rho_0 Ze}{M}\right\}^{1/2} \to \left\{\frac{4\pi\rho_0 Ze}{M}\right\}^{1/2}\frac{k}{q}, \tag{A6.2.13}$$

which converts an "optical mode" into an acoustic mode to yield a dispersion relation as **k** tends to zero:

$$\omega = v_s k, \tag{A6.2.14}$$

where the velocity of sound v_s is given by

$$v_s = \left\{\frac{Zm}{3M}\right\}^{1/2} v_f \tag{A6.2.15}$$

with v_f denoting the Fermi velocity. Equation (A6.2.15) is known as the Bohm-Staver formula (see Chapter 14).

6.2.1. Relation to Debye temperature

Using elementary Debye theory for an isotropic solid, as set out for example in the book by Mott and Jones (1936), the Debye temperature θ is given in terms of v_s by

$$\theta = \frac{v_s}{\Omega^{1/3}} \frac{h}{k_B} \left(\frac{3}{4\pi}\right)^{1/3} \tag{A6.2.16}$$

where Ω is the atomic volume. Eliminating ZE_f between the formulae for E_v and v_s yields both

$$E_v \propto Mv_s^2 \tag{A6.2.17}$$

and, following March (1966):

$$\theta = \frac{h}{k_B} \left\{\frac{3}{4\pi\Omega}\right\}^{1/3} \left\{\frac{2E_v}{3\alpha M}\right\}^{1/2} \tag{A6.2.18}$$

where we have written $E_v = \alpha ZE_f$. For the predicted value $\alpha = \frac{4}{15}$, the formula (A6.2.18) is not quite quantitatively in agreement with the empirical relation proposed by Mukherjee (1965), but changing α from $\frac{4}{15}$ to $\frac{1}{6}$ brings (A6.2.18) into agreement with Mukherjee's empirical formula.

The important conclusion here is that the vacancy energy in close-packed metals is most fundamentally related to a phonon energy, either through (A6.2.17) or (A6.2.18). By eliminating ZE_f, the obvious weaknesses of using perturbation theory based on a free-electron gas model are avoided; it is to be emphasized that (A6.2.17) and (A6.2.18) are quantitatively useful formulae for the vacancy formation energy.

Although the preceding argument is based on electron theory, with neglect of ionic relaxation, it can be shown that similar formulae are, in fact, also regained from a wholly different starting point, appropriate to open body-centred cubic metals.

Thus, Flores and March (1981) have set up a theory showing the way in which the long-range ionic displacements around a vacancy come from both the elastic, long-wavelength limit and from effect of the topology of the Fermi surface, known as the Kohn (1959) anomaly.

However, when attention is focused on the alkali metals Na and K, with almost spherical Fermi surfaces, it can be reasonably assumed that the elastic displacements dominate. A model of "complete relaxation" around a vacancy is then shown to relate formation energy E_v, bulk modulus B, and atomic volume Ω through

$$E_v = \text{constant } B\Omega, \tag{A6.2.19}$$

which is equivalent to (A6.2.17).

Flores and March emphasize that, in contrast to close-packed metals, where electron redistribution around the vacancy plays an important role, in open body-centred cubic metals it is to be expected that relaxation will be of considerable importance. They therefore made a study of two aspects of ionic relaxation in these open structures:

1. The long-range ionic displacements
2. The relevance of local relaxation in calculating vacancy formation energies.

The interested reader should consult their paper for details.

Appendix 7.1
Inverse transport theory for noninteracting electrons

In the body of the text of Chapter 7, use was made of the inverse transport theory of noninteracting electrons. The following treatment leads back to the results of Rousseau, Stoddart, and March (RSM) (1972), with a polarization denominator contribution relating to the role of bound states in electronic transport. To date, the theory has not been derived for interacting electrons (however, see Chapter 11).

Following McCaskill and March (1984), independent electrons moving in a total scattering potential V will be analyzed. The inverse transport theory is known from the work of Leung and March (1977) not to be exact at finite temperature as a result of a distribution of drift velocities rather than a single value $\bar{v}$. In the following, the finite temperature expressions will be retained, but the analysis is exact only in the limit as the Fermi function $f(E)$ approaches the step function characteristic of Fermi statistics at $T = 0$.

The structure of this appendix is as follows. Immediately below, a force balance is set up in the asymptotic time scale to be defined. It will be demonstrated that this leads to a polarization denominator in the inverse transport theory of RSM. One then accounts specifically for bound states without any a priori assumptions, showing thereby how their influence is calculated within the theory. Following this, a discussion using time-dependent techniques will be given. In conclusion, the relation to the decomposition of the forces in electromigration theory will be briefly referred to (see Appendix 13.8).

It will first be shown that the balance of force in the direct response to an external field, as in the Greenwood (1958) calculation of the current, does not describe a transport situation; an additional force is needed. The inverse transport theory meets this requirement, if one also retains the terms resulting from the direct linear response. One of these then contributes a polarization contribution to the denominator, thereby transcending the work of RSM.

Using the gauge defined by vector potential $\mathbf{A} = (-cFt, 0, 0)$, where F is the electric field strength and c is the velocity of light, and periodic boundary conditions, McCaskill and March (1984) obtain, for the linear response to the equilibrium density matrix $f(H)$:

$$\rho = f(H) + g, \tag{A7.1.1}$$

the off-diagonal matrix with elements

$$g_{\rm nm} = i\hbar eFv_{\rm nm}\frac{f_n - f_m}{(E_{\rm n} - E_{\rm m})^2}\left[1 - \exp(E_{\rm n} - E_{\rm m})\frac{t}{i\hbar}\right] \tag{A7.1.2}$$

in the adiabatic basis defined by

$$H(t)|n(t)\rangle = E_{\rm n}(t)|n(t)\rangle \tag{A7.1.3}$$

for the perturbed Hamiltonian

$$H(t) = \frac{1}{2m}\left[\left(\frac{\hbar}{i}\frac{\partial}{\partial x} + eFt\right)^2 + \left(\frac{\hbar}{i}\frac{\partial}{\partial y}\right)^2 + \left(\frac{\hbar}{i}\frac{\partial}{\partial z}\right)^2\right] + V(x, y, z). \tag{A7.1.4}$$

Here e is the electronic charge, V is the electronic potential in the condensed matter, $f(E)$ is the Fermi function as above, and v is the velocity operator $(\hbar/mi)\,\partial/\partial x$.

The force on the electrons, after perturbation by the electric field, may be directly computed as

$$F_{x_{\rm dir}}(t) = -2\,{\rm Re}\left\{{\rm Tr}\left(\frac{\partial V}{\partial x}g(t)\right)\right\}. \tag{A7.1.5}$$

Inserting (A7.1.2) and using the Fourier-Laplace transform to take the $t \to \infty$ limit, one obtains

$$\lim_{t\to\infty} F_{x_{\rm dir}} = 2\pi\hbar eF\sum_{m\neq n}{\rm Re}(V'_{\rm nm}V_{\rm mn})\left(\frac{\partial f}{\partial E}\right)_{E_n}\delta(E_{\rm n} - E_{\rm m})$$
$$- 2\hbar eF\sum_{m\neq n}{\rm Im}(V'_{\rm nm}v_{\rm mn})\frac{f_n - f_m}{E_n - E_m}\frac{P}{E_n - E_m}, \tag{A7.1.6}$$

where P denotes the principal part in integrals. Only on the time scale $\hbar/k_{\rm B}T \ll t \ll \hbar/\Delta E$ does the quasi-delta function in the first term provide a time-independent result, and the evaluation of the second term requires similar care. Here ΔE is a characteristic energy spacing, as discussed at a similar stage of the current calculation by Greenwood (1958).

Using this treatment, the resistivity R is calculated by McCaskill and March (1984) to have the RSM form $N/(1 + D)$, where N is the numerator of RSM. The correction D in the denominator is given by

$$D = \frac{2\pi\hbar}{N_e} \sum_{m \neq n} \text{Re}\left\{\left(\frac{\partial V}{\partial x}\right)_{nm} v_{mn}\right\}\left(\frac{\partial f}{\partial E}\right)_{E_n} \delta(E_n - E_m)$$

$$+ \frac{2\hbar}{N_e} \sum_{m \neq n} \text{Im}\left\{\left(\frac{\partial V}{\partial x}\right)_{mn} V_{mn}\right\} \frac{f_n - f_m}{E_n - E_m} \frac{P}{E_n - E_m}. \qquad \text{(A7.1.7)}$$

If there exists a time scale that separates the polarization behaviour of localized states from those delocalized over the entire system,

$$\frac{\hbar}{\Delta E_b} \ll t \ll \frac{\hbar}{\Delta E_d}, \qquad \text{(A7.1.8)}$$

where ΔE_b, ΔE_d are characteristic energy differences between nearest levels for bound and delocalized states, then the denominator yields a constant value, which is calculated by McCaskill and March as

$$D = -\frac{N_b}{N_e}, \qquad \text{(A7.1.9)}$$

i.e., the ratio of the number of bound electrons to the total number of electrons. The effect of the denominator when (A7.1.9) holds is to yield a resistivity that is just that of the numerator N but with N_e modified to $N_e - N_b$. However, in the asymptotic time scale, the bound electrons do not drift with respect to the potential V, and hence the density ρ_e should also be modified to $(N_e - N_b)/\mathscr{V}$, where $\mathscr{V}$ is the volume. This provides a total correction $(N_e/N_e - N_b)^2$ to the RSM expression for the resistivity. The bound electrons make no contribution to the numerator, which would vanish with box boundary conditions.

The clear separation of the behaviour of bound and delocalized electrons is no longer possible in the vicinity of the metal-insulator transition in particular, where the polarization time for some marginally bound state may be long. In any case, in the absence of an a priori estimate of N_b, the resistivity may still be calculated as

$$R = \frac{N}{(1 + D)^2} \qquad \text{(A7.1.10)}$$

with D given by (A7.1.7) and N by the RSM numerator, but with ρ_e the total electron density. Equation (A7.1.10) describes the structure of a transport theory independently of assumptions about the number of electrons participating. The preceding argument avoids the difficulties associated with the definition of such a description, at least at a formal level.

In concluding this appendix, it is of interest to stress, following McCaskill and March (1984), that the decomposition of the force that occurs in inverse transport theory outlined above has a parallel history in the theory of electromigration (see Appendix 13.8). In particular, the discussion there has revolved around the ambiguity of a division of forces into electron wind and direct field parts. Here, the numerator plays the role of the "ergodic" electron wind contribution and the polarization denominator that of the direct field force. The degree to which polarization can take place on a physical time scale (asymptotic time domain) is limited for nearly free electrons (and the numerator provides the dominant force) but absolute for bound electrons and significant for strong scattering disordered condensed matter.

Appendix 8.1
Method of fluctuating hydrodynamics

In this appendix, the method of Bedeaux and Mazur (1973, 1974; see also Pomeau and Résibois, 1975) is summarized. Their starting point is the continuity equation for the number density $n_1(r,t)$ of tagged particles in a fluid:

$$\frac{\partial n_1(r,t)}{\partial t} = -\operatorname{div} J(r,t). \tag{A8.1.1}$$

The current here is the sum of a diffusion current $-D_0\nabla n_1$ and a convective current $v(r,t)n_1(r,t)$, plus a fluctuating current J_R:

$$J(r,t) = -D_0\nabla n_1(r,t) + v(r,t)n_1(r,t) + J_R. \tag{A8.1.2}$$

The term J_R is the source of the thermal fluctuations of n_1; yet it is neglected below. Moreover D_0 is a "bare diffusion coefficient," which is not to be equated with the experimentally determined D.

The important term in (A8.1.2) is the convective contribution, which involves a coupling between n_1 and the fluid velocity $v(r,t)$; however, the simplicity of the approach presented here stems from the fact that the fluid is discernable from the tagged particles. The velocity field $v(r,t)$ can thus consistently be taken as known. Following Bedeaux and Mazur (see also Pomeau and Résibois), it will be chosen as a Gaussian stationary random function obeying the usual linearized fluctuating hydrodynamics.

Taking the Fourier-Laplace transform of (A8.1.1) and (A8.1.2) and neglecting J_R, one finds

$$(i\omega + D_0 q^2)n_{1;\mathbf{q},\omega} = iq\cdot A n_{1;\mathbf{q},\omega} + n_{1,\mathbf{q}}(t=0) \tag{A8.1.3}$$

where A is a vector operator that depends linearly on $v_{\mathbf{q},\omega}$ and acts on any function $\phi_{\mathbf{q},\omega}$ as:

$$A\phi_{\mathbf{q},\omega} = \frac{1}{(2\pi)^4}\int \mathrm{d}^3q'\,\mathrm{d}\omega'\, v_{\mathbf{q}-\mathbf{q}',\omega'}\phi_{\mathbf{q}',\omega}. \tag{A8.1.4}$$

Writing

$$G_0 = -(i\omega + D_0 q^2)^{-1} \tag{A8.1.5}$$

and

$$n^0_{1;\mathbf{q},\omega} = iG_0 n_{1,\mathbf{q}}(t=0) \tag{A8.1.6}$$

one finds from (A8.1.3)

$$n_{1;\mathbf{q},\omega} = (l + iG_0 q \cdot A)^{-1} n^0_{1:\mathbf{q},\omega}. \tag{A8.1.7}$$

One can also obtain for the current the result

$$J_{\mathbf{q},\omega} = (A + iqD_0)(1 + iG_0 q \cdot A) n^0_{\mathbf{q},\omega}. \tag{A8.1.8}$$

The aim is now to relate the mean values of $J_{\mathbf{q},\omega}$ and $n_{1;\mathbf{q},\omega}$ for a given $n^0_{1:\mathbf{q},\omega}$, the fluctuating quantity being A. This will then provide a relation between $\langle J_{\mathbf{q},\omega}\rangle$ and $\langle n_{1;\mathbf{q},\omega}\rangle$ and hence a value of the macroscopic ("dressed") diffusion coefficient, which will account for fluctuations of the fluid velocity. One finds then, after some algebra:

$$D = D_0 + q^{-2} \frac{\langle iq \cdot A(1 + iG_0 q \cdot A)^{-1}\rangle}{\langle (1 + iG_0 q \cdot A)^{-1}\rangle}. \tag{A8.1.9}$$

It is to be noted that one is led in this way quite naturally to a q and ω dependent diffusion coefficient, without recourse to Green-Kubo limit formulae.

To reveal the $\omega^{1/2}$ dependence at small ω, one now formally expands the right-hand side of (A8.1.9) in increasing powers of A:

$$D = D_0 + q^{-2}\langle q \cdot AG_0 q \cdot A\rangle + \text{high-order terms.} \tag{A8.1.10}$$

It is now to be noted that the average involved in this expression is very close to a mode-mode contribution, since A is proportional to the fluctuating velocity field. From (A8.1.4):

$$\langle q \cdot AG_0 q \cdot A\rangle = \frac{1}{(2\pi)^4} \int \mathrm{d}^3 q' \int \mathrm{d}\omega'$$
$$\times \langle q \cdot v_{\mathbf{q}-\mathbf{q}',\omega-\omega'} G_0(\omega', q')(q - q') \cdot v_{\mathbf{q}-\mathbf{q}',\omega-\omega'}\rangle. \tag{A8.1.11}$$

From space-time translational invariance

$$\langle v_{\mathbf{q}-\mathbf{q}',\omega-\omega}, v_{\mathbf{q}'-\mathbf{q}'',\omega'-\omega''}\rangle = (2\pi)^4 \delta(\omega - \omega'')\delta(q - q'')S_v(q - q'; \omega - \omega'), \tag{A8.1.12}$$

where from the Wiener-Khintchin theorem $S_{v;\mathbf{q},\omega}$ is the Fourier-Laplace transform of the space-time correlation of the velocity field fluctuations. From linearized hydrodynamics, these fluctuations decay according to the sound modes and to a vorticity mode. Retaining only the latter (Pomeau and Résibois, 1975), one has

$$S_{\mathbf{v};\mathbf{q},\omega} = \left(1 - \frac{\mathbf{q}\mathbf{q}}{q^2}\right)\frac{k_B T}{\rho}\frac{1}{i\omega + \nu q^2}. \tag{A8.1.13}$$

Inserting this expression into (A8.1.12), one finds a contribution to D of mode-mode form. Expanding around $\omega = 0$, one finds almost immediately a $\omega^{1/2}$ term in the small ω expansion of D. The coefficient agrees with the result (8.13) except that the bare self-diffusion coefficient D_0 appears instead of the physical (dressed or renormalized) transport coefficient D. However, as Pomeau and Résibois stress, this difference is actually due to the neglect of the higher-order terms in (A8.1.10). When these are properly accounted for, one recovers the physical diffusion constant in the final result. Due to the simplicity of their method, Bedeaux and Mazur (1973, 1974) were also able to obtain results at small but finite q. The interested reader should consult their paper for the details.

Appendix 8.2
Asymptotic behaviour of other Green-Kubo time correlation functions

Let us write the thermal conductivity κ in the form

$$\kappa = \frac{1}{T}\int_0^\infty \kappa(t)\,\mathrm{d}t. \tag{A8.2.1}$$

Then, with details set out, for instance, by Pomeau and Résibois (1975), one finds in the long-time limit $t \to \infty$; with $v = \eta/\rho$ and $\rho = Mn$:

$$\kappa(t) \simeq \frac{k_\mathrm{B}T}{3}\left(\frac{1}{4\pi t}\right)^{3/2}\left[\frac{2C_\mathrm{p}T}{(v + \kappa/nC_\mathrm{p})^{3/2}} + \frac{Mc^2}{(2\Gamma)^{3/2}}\right]. \tag{A8.2.2}$$

Similarly, for the shear viscosity, with $2\Gamma = \rho^{-1}(\frac{4}{3}\eta + \zeta + \kappa(\gamma - 1)/nC_\mathrm{p}$:

$$\eta(t) \simeq \frac{k_\mathrm{B}T}{15}\left(\frac{1}{8\pi t}\right)^{3/2}\left(\frac{7}{v^{3/2}} + \frac{1}{\Gamma^{3/2}}\right). \tag{A8.2.3}$$

One sees again from these results that the $t^{-3/2}$ behaviour is recovered. The only differences from the self-diffusion case discussed earlier at some length are:

1. More complicated mode couplings, related to the different symmetry properties of the flows.
2. The appearance of thermodynamic quantities such as the specific heat c_p and the sound velocity c. These are consequences of the thermodynamic derivatives that appear in expansions of, for example, the Fourier transform $h_\mathrm{q}(t)$ of the local enthalpy density:

$$h_\mathrm{q}(t) = \left.\frac{\partial h}{\partial n}\right|_\mathrm{T} n_\mathrm{q} + \left.\frac{\partial h}{\partial T}\right|_\mathrm{n} T_\mathrm{q} + 0(n_\mathrm{q}^2, T_\mathrm{q}^2, n_\mathrm{q}T_\mathrm{q}). \tag{A8.2.4}$$

It should be noted that no such thermodynamic derivatives enter (A8.2.3), which has a particularly simple structure. In contrast to this, the formula for the bulk viscosity involves higher-order derivatives because, by analogy with (A8.2.4), one now has to expand to second order in the conserved variables n_q and T_q. The result is (Pomeau, 1972; Ernst, Hauge and Van

Leeuwen, 1970, 1971) that $\zeta(t)$ can be usefully separated as a sum of three parts:

$$\zeta(t) \simeq \zeta^v(t) + \zeta^\theta(t) + \zeta^s(t) \tag{A8.2.5}$$

where as t tends to infinity,

$$\zeta^v(t) = 4(k_B T)^2 \left(\frac{1}{8\pi t}\right)^{3/2} \left[\frac{1}{3\nu^{3/2}} + \frac{\alpha}{2}(\gamma - 1)\right] \tag{A8.2.6}$$

$$\zeta^\theta(t) = \frac{\alpha^2}{2}(\gamma - 1)^2 (k_B T)^2 \left(\frac{nc_p}{8\pi t\kappa}\right)^{3/2} \left[\frac{T}{c_p}\frac{\partial c_p}{\partial T}\bigg|_p - n\alpha T\frac{\partial c_p}{\partial P}\bigg|_T\right]^2 \tag{A8.2.7}$$

and

$$\zeta^s(t) = \frac{(k_B T)^2}{2}\left(\frac{1}{8\pi\Gamma t}\right)^{3/2} \left[\frac{1}{3} - \frac{\partial p}{\partial \varepsilon}\bigg|_n - \frac{n}{c}\frac{\partial c}{\partial n}\bigg|_s\right]. \tag{A8.2.8}$$

Here α is the derivative $(n/T)\partial T/\partial n|_p$; $p = p(n, \varepsilon)$; ε the energy density.

Similar formulae have also been obtained for binary nonreacting mixtures (see Appendix 13.5).

Appendix 8.3
Dynamics of $S(k, \omega)$ included through self-function $S_s(k, \omega)$

The origins of this appendix go back to Vineyard (1958), who proposed the approximation (8.20) for the dynamical structure factor $S(k, \omega)$. This is still useful for some purposes but has also severe limitations, one of which being that, if $S_s(k, \omega)$ possesses a Green-Kubo limit according to (8.6), then the Vineyard form (8.20) for $S(k, \omega)$ does not.

Following Gyorffy and March (1969, 1971), the purpose of this appendix is to show that Vineyard's philosophy of including the dynamics in $S(k, \omega)$ solely through $S_s(k, \omega)$ is indeed justified. However, in principle, static structural correlations enter at all orders—i.e., n-particle correlations in principle—and not just two-particle static correlations, as assumed in writing (8.20).

One works with the density-density response function $\chi(q, \omega)$, which may be defined as follows. If one switches on to the liquid an interaction Hamiltonian of the form

$$H_{\text{int}}(t) = -\int d\mathbf{r}\, V_{\text{ext}}(\mathbf{r}, t)\rho(\mathbf{r}), \tag{A8.3.1}$$

where

$$\rho(\mathbf{r}) = \sum_i \delta(\mathbf{r} - \mathbf{r}_i), \tag{A8.3.2}$$

then the corresponding density change $\delta\rho$ induced by $V_{\text{ext}}(\mathbf{r}, t)$ may be written in terms of the double Fourier transform $V_{\text{ext}}(q, \omega)$ with respect to r and t:

$$\delta\rho(q, \omega) = \chi(q, \omega) V_{\text{ext}}(q, \omega). \tag{A8.3.3}$$

In the classical limit, the fluctuation-dissipation theorem allows one to write the Laplace transforms of $\chi(q, t)$ and $\chi_s(q, t)$, with transformed variable (t to p):

$$\chi(q, p) = -\beta \int_0^\infty dt \exp(-pt) \frac{\partial F(q, t)}{\partial t} \tag{A8.3.4}$$

and

$$\chi_s(q,p) = -\beta \int_0^\infty dt \exp(-pt) \frac{\partial F_s(q,t)}{\partial t} \tag{A8.3.5}$$

in terms of F and F_s. These are defined in terms of the density fluctuations by

$$F(q,t) = \langle \rho_{\mathbf{q}}(t) \rho_{-\mathbf{q}}(0) \rangle \tag{A8.3.6}$$

and

$$F_s(q,t) = \langle \rho_{\mathbf{q}}^s(t) \rho_{-\mathbf{q}}^s(0) \rangle \tag{A8.3.7}$$

where $\rho_{\mathbf{q}} = \sum_i \exp(i\mathbf{q} \cdot \mathbf{R}_i)$ and $\rho_{\mathbf{q}}^s = \exp(i\mathbf{q} \cdot \mathbf{R}_s)$, with $\mathbf{R}_i$, as usual, denoting the position vector of the ith particle.

The argument of Gyorffy and March (1971) then centres on the fact that $\chi_s(q,p)$ is a monotonically decreasing function of p with increasing p and that it therefore has an inverse that is single-valued. Hence it follows formally that $\chi(q,p) = \chi(q, \chi_s(q,p))$. This is their main result from the point of view of the Vineyard philosophy. From a practical point of view, these workers argue that χ can now be expanded around $\chi_s = 0$ to yield

$$\chi = \sum_{n=1}^{\infty} a_n \chi_s^n(q,p). \tag{A8.3.8}$$

They then demonstrate that the coefficients a_n can be written in terms of the moments of $S(q,\omega)$, denoted by α_n, and those of $S_s(q,\omega)$, written γ_n. In particular (Gyorffy and March, 1971):

$$a_1 = 1; \qquad a_2 = \frac{\alpha_2 - \gamma_2}{\gamma_1^2}, \ldots. \tag{A8.3.9}$$

A8.3.1. Approximate theories

To conclude this appendix, let us turn from the preceding exact results to approximate theories. Table A8.3.1 records three such theories. The theory of Hubbard and Beeby (1969) has been treated at some length in the text; the result recorded in the table for that theory has, in fact, employed the hydrodynamic form of $\chi_s(q,p)$, obtained from (8.4) as

$$\chi_s(q,p) = \frac{\beta D q^2}{(p + Dq^2)}. \tag{A8.3.10}$$

Table A8.3.1. *Summary of three approximate theories relating dynamical structure factor to self-function via static correlation functions.*

Vineyard (1958; see (8.20))	$\chi(q,p) = S(q)\chi_s(q,p)$
Kerr (1968)	$\chi(q,p) = \dfrac{\chi_s(q,p)}{1 - \beta\tilde{c}(q)\chi_s(q,p)}$
Hubbard-Beeby (1969; see (8.70))	$\chi(q,p) = -\dfrac{pq^2}{M}\dfrac{(\chi_s/Dq^2\beta)^2}{1 + \omega_q^2(\chi_s/Dq^2\beta)^2}$

Note: Here, as usual, $\tilde{c}(q)$ is the Ornstein-Zernike direct correlation function, related to the structure factor $S(q)$ by $\tilde{c}(q) = (S(q) - 1)/S(q)$.

The third theory entered in Table A8.3.1 is due to Kerr (1968). Only the Hubbard-Beeby theory, of the three, has a Green-Kubo limit for $S(k,\omega)$, and that is not accurate quantitatively [see Gyorffy and March (1969, 1971); also March and Paranjape (1987)].

Appendix 8.4

Fourth moment theorem for dynamical structure factor

The intermediate scattering function $F_s(k, t)$ is the Fourier transform on ω of the self function $S_s(k, \omega)$. At short times, $F_s(k, t)$ can be expanded in a power series; the objective here is to extract the coefficient of t^4, which is the fourth moment of $S_s(k, \omega)$. After a straightforward calculation, one finds the real part (the imaginary part is easily shown to average to zero) as

$$[\tfrac{1}{24}k^4\langle\{\dot{x}_1(0)\}^4\rangle - \tfrac{1}{6}k^2\langle\dot{x}_1(0)\dddot{x}_1(0)\rangle - \tfrac{1}{8}k^2\langle\{\ddot{x}_1(0)\}^2\rangle].$$

As a consequence of the fact that for a fluid, the various time-dependent correlation functions must be independent of the origin of time, it is readily shown that $\langle\dot{x}_1(0)\dddot{x}_1(0)\rangle = -\langle\{\ddot{x}_1(0)\}^2\rangle$, and then it follows straightforwardly that

$$\left.\frac{\partial^4 F_s(k, t)}{\partial t^4}\right|_{t=0} = k^4\langle\{\dot{x}_1(0)\}^4\rangle + k^2\langle\{\ddot{x}_1(0)\}^2\rangle. \qquad \text{(A8.4.1)}$$

The first term can immediately be calculated to yield $3k^4(k_B T)^2/m^2$, whereas in the second one can use the Newtonian equation of motion

$$m\ddot{x}_1 = -\frac{\partial \Phi_N}{\partial x_1}. \qquad \text{(A8.4.2)}$$

Carrying out the averaging with respect to $\exp(-\Phi_N/k_B T)$, one notes that

$$\frac{\partial}{\partial x_1}\left\{\exp\left(-\frac{\Phi_N}{k_B T}\right)\right\} = -\frac{1}{k_B T}\frac{\partial \Phi_N}{\partial x_1}\exp\left(-\frac{\Phi_N}{k_B T}\right) \qquad \text{(A8.4.3)}$$

and using this to integrate by parts, one obtains

$$\left.\frac{\partial^4 F_s(k, t)}{\partial t^4}\right|_{t=0} = \frac{3k^4(k_B T)^2}{m^2} + \frac{\rho k^2 k_B T}{m^2}\int d\mathbf{r}\, g(r)\frac{\partial^2 \Phi}{\partial x^2}. \qquad \text{(A8.4.4)}$$

Hence one can get the fourth moment of $S_s(k, \omega)$ in the form

$$\int_{-\infty}^{\infty}\frac{d\omega}{2\pi}\omega^4 S_s(k, \omega) = \frac{3k^4(k_B T)^2}{m^2} + \frac{\rho k^2 k_B T}{3m^2}\int d\mathbf{r}\, g(r)\nabla^2\Phi(r). \qquad \text{(A8.4.5)}$$

A closely related calculation to this one leads to the fourth moment of $S(k, \omega)$ in terms of the fair potential Φ as:

$$\int_{-\infty}^{\infty} \frac{d\omega}{2\pi} \omega^4 S(k, \omega) = \frac{3k^4(k_B T)^2}{m^2} + \frac{\rho k^2 k_B T}{m^2} \int d\mathbf{r}\, g(r)(1 - \cos kx) \frac{\partial^2 \Phi}{\partial x^2}. \tag{A8.4.6}$$

This is the result used, for instance, in checking the Hubbard-Beeby theory of Section 8.6 to be correct for the fourth moment.

Appendix 8.5
One-dimensional barrier crossing: Kramers' theory

The idea behind Kramers' work is that a particle that is trapped in a potential well, essentially through Brownian motion, can escape over a potential barrier and that such a description affords a useful model for elucidating the applicability of the transition state method for calculating rate processes.

More specifically, Kramers (1940) studied the problem of a particle moving in an external field of force but in addition subjected to the irregular forces of a surrounding medium in thermal equilibrium (Brownian motion). The idea is then to study a particle that is initially trapped in a potential well but that can escape after a time by passing over a potential barrier. Essentially, Kramers then calculates the probability of escape and, in particular, the way it depends on the temperature and viscosity of the medium. A comparison proves possible with the results of the "transition state" method. The essence of the calculation is to set up and discuss the equation of diffusion obeyed by a density distribution of particles in phase space.

A8.5.1. One-dimensional model

Consider a particle of unit mass moving in one dimension and acted upon by an external field of force $K(q)$ and a stochastic force $X(t)$ due to the surrounding medium. Its equation of motion has the form

$$\dot{p} = K(q) + X(t):\ \dot{q} = p. \tag{A8.5.1}$$

Now consider the diffusion equation for an ensemble of particles with density $\rho(p, q)$ in the phase space. Kramers shows that this obeys an equation of the Fokker-Planck type, specifically,

$$\frac{\partial \rho}{\partial t} = -K(q)\frac{\partial \rho}{\partial p} - p\frac{\partial \rho}{\partial q} - \frac{\partial}{\partial p}(\mu_1 \rho) + \frac{1}{2}\frac{\partial^2}{\partial p^2}(\mu_2 \rho) + \cdots. \tag{A8.5.2}$$

He investigates the simple case, which is that due initially to Einstein, where

$$\mu_1 = -\eta p; \qquad \mu_2 = 2\eta T, \tag{A8.5.3}$$

η being the viscosity (friction constant). Then the preceding equation reads

$$\frac{\partial \rho}{\partial t} = -K(q)\frac{\partial \rho}{\partial q} - p\frac{\partial \rho}{\partial q} + \eta\frac{\partial}{\partial p}\left(p\rho + T\frac{\partial \rho}{\partial p}\right). \tag{A8.5.4}$$

A8.5.2. Limiting cases

Large friction constant

In this limit, the effect of the Brownian forces on the velocity of the particle is much larger than that of the external force $K(q)$. Assuming that K does not change very much over a distance $T^{1/2}/\eta$, it can be anticipated that starting from an arbitrary initial ρ distribution, a Maxwell velocity distribution will be established in a time $\sim\eta^{-1}$ for each value of q:

$$\rho(q, p, t) = \sigma(q, t)\exp\left\{\frac{-p^2}{2T}\right\}. \tag{A8.5.5}$$

Subsequently, a slow diffusion of the density σ in q-space will take place, which can be expected to be governed by the Smoluchowski diffusion equation

$$\frac{\partial \sigma}{\partial t} = -\frac{\partial}{\partial q}\left(\frac{K}{\eta}\sigma - \frac{T}{\eta}\frac{\partial \sigma}{\partial q}\right), \tag{A8.5.6}$$

where T/η represents the diffusion constant.*

A stationary diffusion current then obeys

$$\omega = \frac{K}{\eta}\sigma - \frac{T}{\eta}\frac{\partial \sigma}{\partial q} = \text{constant} \tag{A8.5.7}$$

and, since this can also be written in the form

$$\omega = -\frac{T}{\eta}\exp\left(\frac{-U}{T}\right)\frac{\partial}{\partial q}(\sigma e^{U/T}), \tag{A8.5.8}$$

one obtains, after integrating between two points A and B on the q coordinate:

* The detail of how to get the Smoluchowski equation from the Fokker-Planck form is discussed fully in Kramers' (1940) paper.

$$\omega = \frac{T\sigma e^{U/T}|_B^A}{\int_A^B \eta e^{U/T}\,dq}. \tag{A8.5.9}$$

This result can be used to obtain an expression for the escape of a particle from a potential well over a potential barrier.

Small viscosity

For small viscosity, one must work with the area inside a curve of constant energy, namely,

$$I(E) = \oint p\,dq, \tag{A8.5.10}$$

and one obtains instead of (A8.5.10) for ω the result in this weak friction limit

$$\omega = \frac{\eta T\rho e^{E/T}|_B^A}{\int_A^B \frac{1}{I} e^{E/T}\,dE}. \tag{A8.5.11}$$

A8.5.3. Escape over a potential barrier

Consider, following Kramers, the potential function U to have the form shown in Figure A8.5.1. Initially, the particle is trapped at A. The height Q

Figure A8.5.1. Schematic form of potential energy function U versus configurational coordinate q.

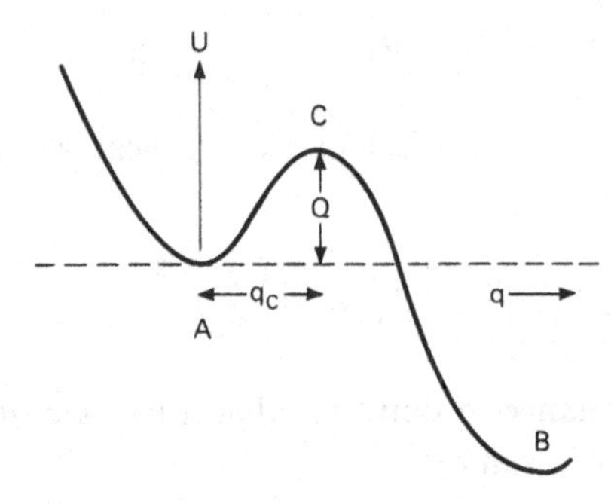

of the potential barrier is assumed large compared with T. The discussion of the escape from A over the hump C to B now follows Christiansen's (1936) treatment of a chemical reaction as a diffusion problem. The ensemble in phase space is thought of as illustrating the phases of a large number of similar particles, each in its own field U.

Qualitatively, before giving the quantitative results, at A the particle can be in a bound state. B obviously corresponds to another state of binding but of lower energy. If the system of particles were in thermodynamic equilibrium, the ensemble density would be proportional to $e^{-E/T}$, and the net number of particles passing from A to B would vanish. If, however, initially the number of particles bound at A is larger than would correspond to thermal equilibrium with the number at B, a diffusion process will begin, tending toward the establishment of equilibrium. The process will be slow if $Q \gg T$. One can then expect that, at any moment, it can be compared with a stationary diffusion process. Even if the particles in the well A do not initially satisfy a Boltzmann distribution, such a distribution near A (and also near B) will be realized a long time before an appreciable number of particles have escaped. Thus, the quasi-stationary diffusion will correspond to a flow from a quasi-infinite reservoir of particles with a Boltzmann distribution at A to the region B.

Kramers solves this quasi-stationary diffusion in phase space in the limiting case of large and small η, for which it reduces to a one-dimensional problem.

Large η

Application of (A8.5.9) now yields

$$\omega = \frac{T}{\eta}\sigma_A\left\{\int_A^B \mathrm{e}^{U/T}\,\mathrm{d}q\right\}^{-1} : \sigma_A = (\sigma \mathrm{e}^{U/T})_{\text{near } A}. \tag{A8.5.12}$$

The number n_A of particles near A can be calculated, if U near A is assumed harmonic, with the result

$$n_A = \int_{-\infty}^{\infty} \sigma_A \mathrm{e}^{-(2\pi\omega)^2 q^2/2T}\,\mathrm{d}q = \frac{\sigma_A}{\omega}\left\{\frac{T}{2\pi}\right\}^{1/2}. \tag{A8.5.13}$$

The ratio $r = \omega/n_A$ is the chance in unit time that a particle originally trapped in A escapes to B. It is given by

$$r = \frac{\omega}{n_A} = \frac{\omega}{\eta}(2\pi T)^{1/2}\left\{\int_A^B e^{U/T}\,dq\right\}^{-1}. \tag{A8.5.14}$$

The main contribution to the integral turns out to come from a small region near C. If one writes there

$$U_{\text{near C}} = Q - \tfrac{1}{2}(2\pi\omega')^2(q - q_c)^2 \tag{A8.5.15}$$

then one finds the result

$$r = \frac{2\pi\omega\omega'}{\eta}e^{-Q/T}. \tag{A8.5.16}$$

If one takes a potential asymmetrical about C, then Kramers shows that

$$r = \frac{2\pi\omega^2}{\eta}\left(\frac{\pi Q}{T}\right)^{1/2}e^{-Q/T}. \tag{A8.5.17}$$

More general treatment

If one again assumes, near C, that (A8.5.15) holds, the corresponding diffusion process in phase space can be described exactly for any value of η. If one considers the stationary situation again, then

$$0 = -(2\pi\omega')^2 q'\frac{\partial \rho}{\partial p} - p\frac{\partial \rho}{\partial q'} + \eta\frac{\partial}{\partial p}\left(p\rho + T\frac{\partial \rho}{\partial p}\right) \tag{A8.5.18}$$

where $q' \equiv q - q_c$. Making the substitution

$$\rho = \delta e^{-[p^2 - (2\pi\omega')^2 q^2]/2T}, \tag{A8.5.19}$$

we get

$$0 = -(2\pi\omega')^2 q'\frac{\partial \delta}{\partial p} - p\frac{\partial \delta}{\partial q'} - \eta p\frac{\partial \delta}{\partial p} + \eta T\frac{\partial^2 \delta}{\partial p^2}. \tag{A8.5.20}$$

The solution δ = constant corresponds to thermal equilibrium. Equation (A8.5.20) also has a solution where δ is a function of a linear combination μ of p and q',

$$\delta = \delta(u)\colon u = p - aq', \tag{A8.5.21}$$

the condition determining a yielding

$$a = \frac{\eta}{2} \pm \left\{\frac{\eta^2}{4} + (2\pi\omega')^2\right\}^{1/2}. \tag{A8.5.22}$$

The solution for δ is then found to be

$$\delta = K \int^{u} e^{-(a-\eta)u^2/2\eta T} \, du. \tag{A8.5.23}$$

If the plus sign in a is chosen, then $a - \eta$ will be positive, and the solution represents diffusion as desired.

The number ω of particles passing the point C in unit time will be obtained by integrating ρp over p from $-\infty$ to $+\infty$ for $q' = 0$, while the number of particles trapped near A, n_A, can likewise be calculated.

The possibility of escape is, therefore,

$$r = \frac{\omega}{n_A} = \frac{\omega}{2\pi\omega'}\left(\left\{\frac{\eta^2}{4} + (2\pi\omega')^2\right\}^{1/2} - \frac{\eta}{2}\right) e^{-Q/T}. \tag{A8.5.24}$$

For $\eta/2 \gg 2\pi\omega'$, this reduces to the previous result $r \sim (2\pi\omega\omega'/\eta)e^{-Q/T}$, whereas for small viscosity $\eta/2 \ll 2\pi\omega'$, it reduces to

$$r = \omega e^{-Q/T} = r_{\text{transition state method}}. \tag{A8.5.25}$$

Thus, the probability of escape corresponds exactly to the value that would be given by the transition state method (Pelzer and Wigner, 1932; Evans and Polanyi, 1935).

According to this method, one considers the particles near A to be in perfect temperature equilibrium with those near B, so that one has thermal equilibrium also at C, and then one calculates the number of particles ω, which in unit time pass the transition point C from left to right. This number is of course equal to that passing C from the right to the left and is given by

$$\omega = \int_{(\mathring{c})}^{\infty} \rho_0 p \, dp = e^{-Q/T} \int_0^{\infty} p e^{-p^2/2T} \, dp = T e^{-Q/T} \tag{A8.5.26}$$

where

$$\rho_0 = e^{-E/T} \tag{A8.5.27}$$

represents the Boltzmann-Gibbs distribution in phase space. The number n_A of particles trapped near A can also be calculated and leads to r given by (A8.5.25).

Small viscosity

The preceding argument from the transition state method is now not correct, since the effect of the Brownian motion is so small that the subsequent delivery of particles with energy Q (or more) near C may be insufficient. The rate of escape is now determined by a diffusion process along the energy coordinate. Kramers calculates this, to obtain the probability of escape as

$$r = \frac{\omega}{n_A} = \eta \frac{Q}{T} e^{-Q/T}. \tag{A8.5.28}$$

If the U function to the left of C could be represented exactly by $\frac{1}{2}(2\pi\omega)q^2$, the error in this approximation for r would be only of the relative order of magnitude T/Q.

The validity of (A8.5.28) implies that η is small compared with ω, i.e., the motion of the oscillator is of periodic damped form. A periodic damping would mean $\eta = 4\pi\omega$.

It is relevant to add here that the work of Visscher (1976) for the case of small η shows the rate of escape to have a largely Arrhenius exponential dependence on barrier height but to be surprisingly insensitive to (1) barrier shape and (2) injection energy distribution.

Appendix 8.6

Mode-coupling and velocity field methods

In this appendix, a summary will be given of mode-coupling and velocity field (Gaskell and Miller, 1978) methods. The following presentation is based on that of Munakata (1983).

It is convenient to begin the discussion with the velocity autocorrelation function of a one-component liquid. If $n_0(\mathbf{k}) = \exp(i\mathbf{k}\cdot\mathbf{r}_0)$ characterizes the density of a tagged particle with position $\mathbf{r}_0$ and momentum $\mathbf{p}_0$, then one conveniently uses the correlation function

$$\phi_s(k,t) = \langle n_0(-\mathbf{k})n_0(\mathbf{k},t)\rangle \equiv \langle n_0(\mathbf{k})|n_0(\mathbf{k},t)\rangle \qquad \text{(A8.6.1)}$$

and its Laplace transform

$$\phi_s(k,z) = i\int_0^\infty \mathrm{d}t\,\exp(izt)\phi_s(k,t), \qquad \text{(A8.6.2)}$$

where as usual $\langle A\rangle$ represents the canonical ensemble average of A. The velocity autocorrelation function is defined by (see also main text):

$$\phi(t) = \frac{\langle \mathbf{p}_0(t)\cdot\mathbf{p}_0\rangle}{\langle \mathbf{p}_0\rangle^2}. \qquad \text{(A8.6.3)}$$

A8.6.1. Mode-coupling method

Using the memory function formalism (Mori, 1965a, b), (A8.6.2) can be expressed in the form

$$\phi_s(k,z) = -\frac{1}{z + k^2 D(k,z)} \qquad \text{(A8.6.4)}$$

where the generalized diffusion constant $D(k,z)$ is

$$D(k,z) = -(3M^2)^{-1}\sum_\alpha \langle f^\alpha(\mathbf{k})|(z+QL)^{-1}|f^\alpha(\mathbf{k})\rangle \qquad \text{(A8.6.5)}$$

with L and $P = 1 - Q$ denoting the Liouville operator $\mathring{A} = \mathrm{d}A/\mathrm{d}t = iLA$

and the projection operator on to the variable $n_0(k)$, respectively. The random force $f^{\alpha}(\mathbf{k})$, α denoting a Cartesian component, is

$$f^{\alpha}(\mathbf{k}) = p_0^{\alpha} \exp(i\mathbf{k}\cdot\mathbf{r}_0). \qquad \text{(A8.6.6)}$$

The velocity autocorrelation function (VAF) is related to $D(k, z)$ by

$$\phi''(\omega) = \frac{1}{2}\int_{-\infty}^{\infty} \mathrm{d}t \exp(i\omega t)\phi(t) = \left(\frac{k_B T}{M}\right)^{-1} D''(k = 0, \omega) \qquad \text{(A8.6.7)}$$

where

$$D(k, z = \omega + iO) \equiv D'(k, \omega) + iD''(k, \omega). \qquad \text{(A8.6.8)}$$

As an important two-mode (bilinear) variable, one takes (see Munakata, 1983)

$$B_M^{\alpha}(\mathbf{k}, \mathbf{q}) = j_M^{\alpha}(\mathbf{q}) n_0(\mathbf{k} - \mathbf{q}) \qquad \text{(A8.6.9)}$$

with $j_M(q)$ the mass current given by, in volume V of liquid,

$$j_M(q) = V^{-1/2} \sum_i P_i \exp(i\mathbf{q}\cdot\mathbf{r}_i). \qquad \text{(A8.6.10)}$$

Denoting the projection operator on to the two-mode variable (A8.6.9) by P_2, one now approximates the random force, defined in (A8.6.6), by

$$|f^{\alpha}(k)\rangle \simeq P_2 |f^{\alpha}(k)\rangle = \sum_{\mathbf{q}} (nV^{1/2})^{-1} |B_M^{\alpha}(\mathbf{k}, \mathbf{q})\rangle \qquad \text{(A8.6.11)}$$

with number density of atoms $n = N/V$. Using a decoupling approximation (Munakata, 1983), it follows from (A8.6.5) and (A8.6.11) that

$$3M^2 D(k, z) = \left(\frac{Mk_B T}{nV}\right) i \int_0^{\infty} \mathrm{d}t \exp(izt)$$

$$\times \sum_{\mathbf{q}} [\phi_M^L(q, t) + 2\phi_M^T(q, t)]\phi_s(|\mathbf{k} - \mathbf{q}|, t). \quad \text{(A8.6.12)}$$

Here the normalized mass-current correlation functions $\phi_M^L(q, t)$ and $\phi_M^T(q, t)$ have been introduced via

$$\langle j_M^{\alpha}(\mathbf{q}) j_M^{\beta}(q, t)\rangle = Mnk_B T\left[\phi_M^L(q, t)\frac{q^{\alpha}q^{\beta}}{q^2} + \phi_M^T(q, t)\left(\delta_{\alpha\beta} - \frac{q^{\alpha}q^{\beta}}{q^2}\right)\right].$$

(A8.6.13)

Thus, from (A8.6.7) and (A8.6.12), it follows that the mode-coupling method just outlined leads to

$$\phi_{MC}(t) = (3V)^{-1} \sum_q n^{-1} [\phi_M^L(q, t) + 2\phi_M^T(q, t)]\phi_s(q, t). \qquad \text{(A8.6.14)}$$

This equation has been applied to the VAF of liquid argon (Munakata and Igarashi, 1977, 1978) and gives the long-time tail correctly (see also Chapter 7 and Appendices 8.1 and 8.2).

A8.6.2. Velocity field method

It is now of some interest to compare the preceding result (A8.6.14) with that obtained by the velocity field method of Gaskell and Miller (1978; see also Munakata, 1983). Following these workers one uses a microscopic velocity field (VF), denoted by $v(\mathbf{r}, t)$, such that

$$v(r, t) = \sum_i \frac{P_i(t) f(|\mathbf{r} - \mathbf{r}_i(t)|)}{M} \tag{A8.6.15}$$

where the "form" factor $f(r)$ becomes zero for $r > \sigma$ with σ a measure of the atomic sphere radius. From $v(\mathbf{r}_i, t) = \mathbf{p}_i(t)/M = \mathbf{v}_i(t)$ and a hydrodynamic condition, one sees that $f(r = 0) = 1$ and

$$\int \mathbf{dr}\, f(r) = \frac{1}{n}. \tag{A8.6.16}$$

From (A8.6.15) and the condition $f(r = 0) = 1$, one finds, using again a decoupling approximation,

$$\langle \mathbf{v}_0(t) \cdot \mathbf{v}_0 \rangle \simeq \int \mathbf{dr}\, \langle \mathbf{v}(r + r_0, t) \cdot \mathbf{v}_0 \rangle \phi_s(r, t) \tag{A8.6.17}$$

with

$$\phi_s(r, t) = \langle \delta\{\mathbf{r} - [\mathbf{r}_0(t) - \mathbf{r}_0]\} \rangle = V^{-1} \sum_{\mathbf{q}} \phi_s(q, t) \exp(-i\mathbf{q} \cdot \mathbf{r}). \tag{A8.6.18}$$

Since

$$v(\mathbf{k}, t) = \int \mathbf{dr}\, \mathbf{v}(\mathbf{r}, t) \exp(i\mathbf{k} \cdot \mathbf{r}) = f(k) \left(\frac{V^{1/2}}{M} \right) j_{\mathrm{M}}(\mathbf{k}, t), \tag{A8.6.19}$$

one can rewrite (A8.6.17) as

$$\phi_{\mathrm{VF}}(t) = (3V)^{-1} \sum_{\mathbf{q}} f(q) [\phi_{\mathrm{M}}^{L}(\mathbf{q}, t) + 2\phi_{\mathrm{M}}^{T}(q, t)] \phi_s(q, t). \tag{A8.6.20}$$

Equation (A8.6.20) has been applied with some success to the VAF's of several types of simple liquids (see Gaskell, 1982). One notes that n^{-1} on the right-hand side of (A8.6.14) is replaced by $f(q)$ in (A8.6.20).

Appendix 9.1
Ornstein-Zernike treatment of critical correlations

The original argument of Ornstein and Zernike (1918) for the form of the pair function $g(r)$ near the liquid-gas critical point started from the convolution definition (2.7) of the direct correlation function $c(r)$. These workers then made two assumptions:

1. That $c(r)$ is short-ranged compared with the total correlation function $h(r) = g(r) - 1$. This is only true near the critical point (see (3.10)).
2. That $h(\mathbf{r}')$ in the convolution integral in (2.7) can be expanded in a Taylor series about the point $\mathbf{r}$.

From 2, the first term, $h(r)$, in this expansion gives a contribution $h(r)\rho \int c(r)\,d\mathbf{r}$ to the convolution, the term grad h integrates to zero, and the term proportional to $|\mathbf{r}' - \mathbf{r}|^2$ evidently contributes

$$\text{constant} \times \nabla^2 h \int cr^2\,d\mathbf{r}.$$

There being no reason why $\int cr^2\,d\mathbf{r}$ should vanish, one then finds the following differential equation for $h(r)$:

$$\nabla^2 h = \text{constant} \times [1 - \tilde{c}(0)]h = \kappa^2 h. \tag{A9.1.1}$$

Since $S(0)$ tends to infinity at T_c (from (2.5)) and $S(0) = [1 - \tilde{c}(0)]^{-1}$, it follows from the definition of κ in (A9.1.1) that this quantity tends to zero as T tends to T_c. The solution of (A9.1.1), which decays to zero at infinity, is evidently

$$h = \text{constant}\,\frac{\exp(-\kappa r)}{r}. \tag{A9.1.2}$$

In the remainder of this appendix and then in the more general statistical mechanical context of Appendix 9.2, it will be noted that modern critical point theory, based on the renormalization group [see Wilson (1972) and Wilson and Kogut (1974)], requires some small corrections to (A9.1.2). These involve the additional critical exponent η in the long-range form at

T_c of $h(r)$, namely, $h(r) \sim \text{constant}/r^{1+\eta}$, where η is as introduced in the main text. However, it is worth noting here that structural theories of liquids, as discussed in Chapter 5, have been applied to this general area, though not to an as yet final calculation of η [see, for example, Choy and Mayer (1967), Kayser and Raveché (1982) and Senatore and March (1984)].

The theory of renormalization (with $\kappa = 0$) starts from a result obtained by Gell-Mann and Low (1951). In the present context, this result can be expressed by writing the structure factor $S(k)$ in the form [compare (9.16)]

$$S(k) = k^{-2} s\left(\frac{k}{\Lambda}\right), \tag{A9.1.3}$$

and by noting that there is an arbitrariness of scale. This arbitrariness can be transferred to a choice of reference momentum λ, with $\lambda \ll \Lambda$, the quantity Λ being a cutoff momentum. The renormalization approach essentially relates $s(k/\Lambda)$ to $s(\lambda/\Lambda)$. Since s is dimensionless, its k dependence can involve only the remaining length in the problem, the cutoff Λ^{-1}.

The burden of the argument is, of course, then to determine the form of s. It can be shown (see March and Tosi, 1976, p. 242) that

$$S(k) \propto k^{-2}\left(\frac{k}{\Lambda}\right)^{\eta} \tag{A9.1.4}$$

with η greater than zero.

A generalization of this argument can be effected for $T \neq T_c$, i.e., $\kappa \neq 0$. The interested reader may refer to Pfeuty and Toulouse (1977) for a full account of renormalization methods in phase transitions. An introductory discussion follows in Appendix 9.2.

Appendix 9.2

Homogeneity, scaling, and an introduction to renormalization group method

Prior to renormalization group theory, the so-called homogeneity hypothesis (Widom, 1965; Griffiths, 1965), that the critical correlation functions—and hence the free energy—were homogeneous functions of the basic variables, which allowed the various critical exponents to be related, had been proposed. Here, we do not distinguish between the two possibly different exponents on either side of the transition. The relations thereby obtained (see (A9.2.6) for definition of α),

$$\alpha + 2\beta + \gamma = 2 \tag{A9.2.1}$$

$$\gamma = \beta(\delta - 1) \tag{A9.2.2}$$

$$\gamma = (2 - \eta)\nu \tag{A9.2.3}$$

$$\alpha = 2 - \nu d, \tag{A9.2.4}$$

are satisfied by the Onsager solution for the two-dimensional ($d = 2$) Ising model [see, for example, Callaway (1976)], and [except for (A9.2.4)] by the mean-field exponents. Series expansion results give very sensitive tests of these relations between exponents.

A second advance came through the scaling approach of Kadanoff (1966), which for a simple magnetic model showed how homogeneity could be understood by considering the explicit transformations of the basic parameters (temperature and field) under a length scaling. At one and the same time, this illustrated the importance of dimensionality, and the irrelevance of lattice structure, spin magnitude, and range and strength of the forces. The following argument, developing this, is based on the account of Stinchcombe (1984).

The basic idea is that at the transition, provided it is second order, the correlation length ξ diverges. Near the transition it is therefore the largest and, presumably, the only characteristic length associated with critical phenomena. Then all features on the length scale of the atomic spacing, say, become irrelevant. But, as discussed in the text (see (9.14)) the correlation length diverges in a temperature-dependent fashion. This temperature dependence of ξ then leads to the temperature dependence of all the singular properties in the critical regime.

Consider a dilatation (i.e., increase of length scale) of the system by an arbitrary factor b. On the new length scale, the correlation length appears to have been reduced by the same factor: $\xi \to \xi' = \xi/b$. Now consider a second quantity, say the (extensive) free energy, with scaling $F \to F' = b^{-d}F$. But F' should be the same function of ξ' as F is of ξ if the correlation length is the only relevant length. Therefore, it follows that $F(\xi) = F'(\xi') = b^{-d}F(\xi/b)$. Since b is arbitrary, it can be put equal to ξ at this point to find

$$F(\xi) = \xi^{-d}F(1) \propto |T - \mathrm{T_c}|^{\nu d}, \tag{A9.2.5}$$

which yields by thermodynamics the singular part of the specific heat C,

$$C \propto |T - T_c|^{-\alpha}, \tag{A9.2.6}$$

and hence (A9.2.4).

The preceding discussion, essentially along the lines laid down by Kadanoff (1966), identifies the homogeneity parameter b with the dilatation. The treatment also demonstrates that instead of obtaining just exponent relationships, the exponents themselves could be found if it were known how the parameters describing the assembly (temperature, field, etc.) transformed under the dilatation of the system. As an example, suppose that under the dilatation by b, the scale of temperature (measured from the critical temperature T_c) transforms as $\delta T = T - T_c \to \delta T' = \lambda_b \delta T$, where λ_b is a b-dependent pure number. Then employing (9.14):

$$\frac{1}{b} = \frac{\xi'}{\xi} = \frac{(\delta T')^{-\nu}}{(\delta T)^{-\nu}} = \lambda_b^{-\nu}, \tag{A9.2.7}$$

and hence

$$\nu = \frac{\ln b}{\ln \lambda_b}. \tag{A9.2.8}$$

A9.2.1. Renormalization group method

The achievement of the renormalization group methods (Wilson, 1972; Wilson and Fisher, 1972; Wilson, 1974; Niemeyer and Van Leeuwen, 1974) is that it not only ties these ideas together beautifully but, at the same time, provides methods for obtaining the transformation of parameters under a dilatation. Hence the critical exponents can be estimated, and one can attempt to give an answer to the question of universality. The brief discussion of this method given next again follows the account of Stinchcombe (1984) very closely.

The idea is to set up explicitly a transformation of basic parameters such that under a dilatation, the system appears exactly as it did before the dilatation, in the sense of preserving the same absolute correlation length, etc. This is normally achieved by removing degrees of freedom and finding the transformation of parameters that leaves unchanged the Boltzmann probabilities for the remaining degrees of freedom. As Stinchcombe points out, the original, and most controlled, methods of doing this use field theory methods in momentum space, where the degrees of freedom removed are the short-wavelength components of the fields. If d_2 and d_1 are upper and lower critical dimensionalities, respectively, and n denotes the number of components of the order parameter, then perturbation graphs are usually selected by classifying their order with respect to $\varepsilon = d_2 - d$ or sometimes with respect to $d - d_1$, or $1/n$. In the first case, the perturbation analysis (ε-expansion) is an expansion from the mean-field (Gaussian) results. (The $1/n$ expansion (Ma, 1973) is from the spherical-model results, which are exact in the limit, as n tends to infinity.)

The second principal method of explicitly constructing the renormalization group transformation is the direct-position space method, as set out by Niemeyer and Van Leeuwen (1976). This is based on grouping degrees of freedom (block spin methods) or on deleting degrees of freedom (Barber, 1975; Kadanoff and Houghton, 1975: (e.g., decimation methods removing every other spin or atom).

Let us suppose that some such method has been applied to construct a transformation. If μ denotes all the parameters (temperature, field, etc.), the resulting transformation can be formally written as $\mu \to \mu' = R_b^\mu$, where R_b is the operator giving the transformation resulting from dilatation by the factor b. At the second-order transition under consideration, the only relevant length ξ diverges, and the system becomes length-invariant and should not notice the dilatation. One way this situation may occur is through the basic parameters μ having particular values given by μ^*, where $\mu^* = R_b\mu^*$ so that they do not change under the dilatation; μ^* is one of the equation's so-called fixed points in the multidimensional parameter space. Another way by which the system can become critical, thereby displaying divergent ξ, is to have μ one of, for example, a "critical surface" of points such that repeated applications of R_b to any point on the surface takes the point into a fixed point. The transformation of points near the fixed point then determines the asymptotic critical behaviour. Following Stinchcombe, this concept of the transformations effecting "flow" in the parameter space is illustrated in Figure A9.2.1. It provides an explanation of universality and of crossover. Universality is the irrelevance of many parameters (such as spin and strength of forces) which merely scale away as the system approaches the fixed point, whose character only basically depends on the spatial dimensionality d and the number n of components of the order parameter. Crossover is the change of critical behaviour from that characteristic of one canonical system to that of another as the point flows between the fixed points associated with the two characteristic behaviours. (Thus a weakly anisotropic Heisenberg system crosses over ultimately to the critical behaviour of an Ising model as the flow goes from the vicinity of the isotropic Heisenberg fixed point ($n = 3$) to the Ising fixed point ($n = 1$).)

Following again the account of Stinchcombe (1984), let us now consider the transformation in more detail and relate thereby to the scaling treatment above. Near the fixed point, it is useful to work with $\delta\mu = \mu - \mu^*$ and $\delta\mu' = \mu' - \mu^*$, which are related through $\delta\mu' = R_b^L \delta\mu$. Two successive dilatations being equivalent to a single dilatation, one obtains the (semi-) group property

$$R_b^L R_{b'}^L = R_{bb'}^L. \tag{A9.2.9}$$

This implies that the eigenvalues λ_b and R_b^L satisfy a similar equation for any b, b' so that $\lambda_b = b^y$, where y is independent of b. In a simple one-parameter case (where typically $\delta\mu = T - T_c$)

$$\lambda_b = \frac{dR(\mu^*)}{d\mu^*}, \tag{A9.2.10}$$

Figure A9.2.1. Illustration of concept of the transformations effecting "flow" in the parameter space.

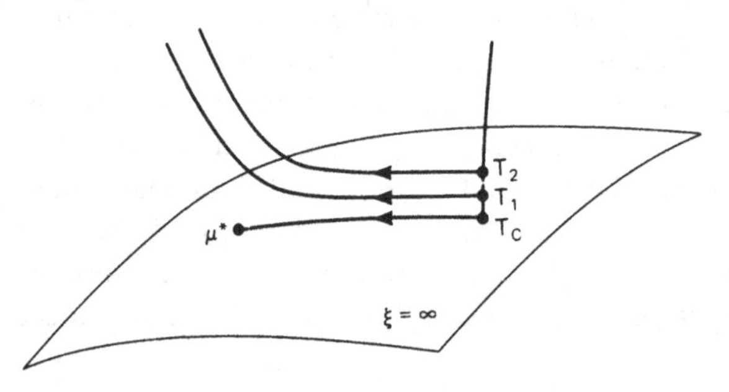

and arguments as before can be used to obtain

$$\xi \propto (\delta\mu)^{-\nu} \tag{A9.2.11}$$

where ν is as in (A9.2.8), which one has therefore rederived.

The eigenvalues of the linearized transformations therefore give the critical exponents. Further, whether a parameter is relevant or irrelevant (scales away under the transformation) depends on whether its associated eigenvalue is greater or less than 1. The critical surface referred to earlier is the surface through the fixed point on which all relevant parameters have their fixed-point value and so includes all points flowing into the fixed point.

Thus the transformation of parameters under a change-of-length scale determines the critical behaviour of a system. Stinchcombe (1984) supplements the preceding account with elegant examples of renormalization group transformations; the interested reader should refer to his review for details.

Appendix 9.3

Compressibility ratios and thermal pressure coefficients of simple monatomic liquids from model equations of state

This appendix will be concerned mainly with the compressibility ratio Z_c defined by

$$Z_c = \frac{p_c V_c}{RT_c} \tag{A9.3.1}$$

where p_c, V_c, and T_c are the usual critical values of these thermodynamic quantities and the thermal pressure coefficient to be defined below. Simple two-parameter equations of state such as those of van der Waals and of Dieterici (see Guggenheim, 1949) lead to specific values of Z_c (see Section 9.4), whereas Z_c varies for the fluid alkali metals from a value somewhat less than 0.1 for Li to 0.21 for Cs. As Chapman and March (1987) emphasize, it is therefore necessary to transcend such two-parameter equations of state; they propose to do this via (9.18).

They then study the so-called thermal pressure coefficient $\gamma_V = (\partial p/\partial T)_V$ from (9.18), which is given by

$$\frac{1}{\gamma_V} = \left(\frac{V - \alpha V_c}{R}\right) g(T^*, V^*) \tag{A9.3.2}$$

where $T^* = T/T_c, \ldots,$ and

$$g(T^*, V^*) = \left[\frac{\partial}{\partial T^*}(T^* f_2)\right]^{-1}. \tag{A9.3.3}$$

In particular, $g = 1$ for the original van der Waals equation. For liquid alkali metals, Chapman and March plot experimental data for $1/\gamma_V$ versus molar volume for points along the high density side of the liquid-vapour coexistence curve. The near linearity of each set of data provides ample support for the form of (A9.3.2) with $g(T^*, V^*)$ only a slowly varying function of its arguments. By extrapolation in the direction of low volume and temperature, it is found that $g \simeq 1.2$ in each case, whereas values of α lie in the range 0.12–0.16. However, when these findings for the alkalis are compared with liquid argon, significantly greater curvature of $(\gamma_V)^{-1}$ versus molar volume plot is found, and the deviation of g from unity is more substantial.

Appendix 9.4
Mode coupling applied to critical behaviour

In Appendix 8.6, mode-coupling theory was summarized. Here, in this appendix, the equations used by Olchowy and Sengers (1988) in their critical point studies, which are derived by the mode-coupling method, will first be recorded. The solutions of these equations by these workers lead to results of the character discussed in the main text.

The mode-coupling theory of dynamic critical phenomena (Kawasaki, 1976) yields two coupled integral equations for the critical parts of the thermal diffusivity D_T and the viscosity η:

$$\Delta D_T(q) = \frac{k_B T}{(2\pi)^3 \rho} \int^{q_0} \mathbf{dk} \frac{c_p(|\mathbf{q} - \mathbf{k}|)}{c_p(q)} \frac{\sin^2 \theta}{k^2 \eta(k)/\rho + |\mathbf{q} - \mathbf{k}|^2 D_T(|\mathbf{q} - \mathbf{k}|)} \tag{A9.4.1}$$

and

$$\Delta \eta(q) = \frac{1}{2q^2} \frac{k_B T}{(2\pi)^3} \int^{q_0} \mathbf{dk}\, c_p(k) c_p(|\mathbf{q} - \mathbf{k}|) \left(\frac{1}{c_p(k)} - \frac{1}{c_p(|\mathbf{q} - \mathbf{k}|)} \right)^2 \times \frac{k^2 \sin^2 \theta \sin^2 \phi}{k^2 D_T(k) + (\mathbf{q} - \mathbf{k})^2 D_T(|\mathbf{q} - \mathbf{k}|)}. \tag{A9.4.2}$$

Here θ and ϕ are the polar and azimuthal angles of $\mathbf{k}$ in a coordinate system with the polar axis in the direction of the wave vector $\mathbf{q}$. A cutoff wave number q_0 has been retained in (A9.4.1) for the long-range fluctuations.

Olchowy and Sengers solve by assuming: (1) $\eta(k)$ independent of k, (2) D_T negligible in (A9.4.1), and (3) $c_p \gg c_v$ and c_p proportional to the (symmetrized) compressibility $(\partial \rho / \partial \mu)_T$ with μ the chemical potential. The q-dependence of the compressibility is taken to be of Ornstein-Zernike form $[1 + q^2 \xi^2]^{-1}$ with ξ the correlation length.

Following the preceding summary of the practical route followed by Olchowy and Sengers, it is of interest here to note, somewhat more generally, that mode-mode coupling has been used with considerable success to discuss critical dynamics. The account below follows that of Pomeau and Résibois (1975). To illustrate the way mode coupling plays a prominent role in this area, consider, following the discussion in Appendix 8.5, the correction to the viscosity of, for instance, a van der Waals fluid due to the coupling between two heat modes, denoted below by $\Delta\eta$. Following Appendix 8.5 one finds, with $\gamma(y) = C_p(y)/C_v(y)$:

$$\Delta\eta \alpha \gamma \int_0^\infty \mathrm{d}y[n\chi_T(y)]^2 \left(\frac{\gamma(y)-1}{\gamma(y)}\right)^2 \left(ny\frac{\partial V_y^L}{\partial y}\right)^2 \frac{C_p(y)}{K^R}, \qquad \text{(A9.4.3)}$$

where the pair interaction V has a short-range repulsive part V^R and a long-range small attractive part $\gamma^3 V^L(\gamma r)$.

It is easily demonstrated that, when the temperature is decreased, one finally reaches a temperature T_c, such that, for a well-defined density n_c, the zero wave-number compressibility $\chi_T(0)$ given by

$$\chi_T \to \chi_T(q\gamma^{-1}) = \left[\frac{\partial p}{\partial n}\bigg|_T^R + nV_{q\gamma^{-1}}^L\right], \qquad \text{(A9.4.4)}$$

the superscript R here referring to the short-range reference system, becomes singular. One then has

$$-n_c V_0^L = \frac{\partial p}{\partial n_c}\bigg|_{T_c}^R, \qquad \text{(A9.4.5)}$$

which defines the critical point in this model.

Of course, it is known that the equilibrium γ-expansion fails to converge close to the critical point (Hemmer 1964); moreover (A9.4.4) was derived under the assumption that all thermodynamical coefficients were well behaved; no longer true near the critical point (n_c, T_c). If, nevertheless, following Pomeau and Résibois, one ignores these difficulties provisionally, then one can investigate the consequences of (A9.4.5) on (A9.4.3). Consider then a temperature T such that

$$0 < \frac{T - T_c}{T_c} \equiv \varepsilon < 1$$

and examine, in the integral (A9.4.3), the region where $y \ll 1$. Expanding the denominator in

$$\chi_T(y) = \frac{n^{-1}}{\partial p/\partial n|_T^R + nV_y^L} \qquad \text{(A9.4.6)}$$

with the help of

$$\frac{\partial p}{\partial n_c}\bigg|_T^R \equiv \frac{\partial p}{\partial n_c}\bigg|_{T_c}^R + \alpha\varepsilon\ldots, \qquad V_y^L = V_0^L + y^2\frac{1}{2}\frac{\partial^2 V^L}{\partial y^2}\bigg|_0 + \ldots, \qquad \text{(A9.4.7)}$$

one reaches the well-known Ornstein-Zernike equation (see Appendix 9.1)

$$\chi_T(y) \alpha \frac{1}{\xi_\gamma^{-2} + y^2} \tag{A9.4.8}$$

where

$$\xi_\gamma \equiv \left(\sqrt{\frac{\alpha\varepsilon}{\frac{1}{2}n\partial^2 V^L/\partial y^2 | y = 0}}\right) \propto \sqrt{\varepsilon^{-1}} \tag{A9.4.9}$$

defines the correlation length in units of γ^{-1}; the usual correlation length is simply

$$\xi = \xi_\gamma \gamma^{-1}.$$

ξ and ξ_γ diverge at the critical point. As discussed by Pomeau and Résibois, one can estimate that $\gamma(y) \sim 1$ and $y\partial V^L | \partial y \sim y^2$. Hence the contribution to (A9.4.3) coming from the region $y \ll 1$ (or $q \ll \gamma$) is of the order

$$\Delta\eta \sim \gamma \int_0^\infty \frac{\mathrm{d}y\, y^4}{(y^2 + \xi_\gamma^{-2})^2} \frac{C_p(y)}{K^R} \sim \frac{C_p(0)\xi^{-1}\gamma}{K^R}, \tag{A9.4.10}$$

with K^R the contribution to the thermal conductivity from the short-range repulsion V^R. For γ small but finite, this so-called correction becomes infinite as T_c is approached and dominates the short-range term η^R. Similarly, one can show that $\delta K = K - K^R$ is not small when T_c is approached.

It can be said that these results indicate that the preceding simple treatment is inadequate close to T_c. However, qualitatively it is not difficult to "repair" this inadequacy: One must simply put the full transport coefficients in the mode-mode term instead of their value in the reference fluid. Then (A9.4.10) becomes a self-consistency condition

$$\eta K \sim \xi^{-1} C_p(0) \tag{A9.4.11}$$

where the factor γ, a small but finite constant, has been dropped. If one uses $C_p(0) \sim \xi^2$, one then finds

$$\eta K \sim \xi, \tag{A9.4.12}$$

which indicates a necessary divergence either in K or in η. The more proper treatment of Kadanoff and Swift (1968) produces similar results to these. These authors built up a formalism that introduces mode-mode coupling effects at the level of the formal solution of the Liouville equation. A second ingredient of their treatment is the model-independent definition of wavenumber-dependent thermodynamic coefficients. As an example, one can take a formula like

$$nk_B T \chi_T = \lim_{q \to 0} \langle \tilde{n}_q \tilde{n}_{-q} \rangle \tag{A9.4.13}$$

and extend it to define a wavenumber-dependent compressibility

$$nk_B T\chi_T(q) = \langle \tilde{n}_q \tilde{n}_{-q} \rangle. \tag{A9.4.14}$$

In general, the right-hand side of such an expression is hard to evaluate explicitly; yet close to T_c, when $\xi \gg$ any molecular length, the idea of *static scaling* can be used to motivate the assumption that the generalized compressibility, which generally depends separately on the parameters q and ξ^{-1}, is in fact a homogeneous function:

$$\chi_T(q, \xi) = q^u f(q\xi), \tag{A9.4.15}$$

where $f(X)$ is an unknown but well-defined function that tends to a constant as $X \to 0$. Moreover, the exponent u can be determined from purely thermodynamic properties because, for $q \ll \xi^{-1}$,

$$\chi_T(q, \xi) \alpha \xi^u, \tag{A9.4.16}$$

as a consequence of the homogeneity of the function f.

With these ideas, the Kadanoff-Swift theory proceeds along lines parallel to the examples developed above. Points where differences appear are as follows:

1. Because the wavenumber q is not necessarily the smallest parameter in the problem (very close to T_c, one may well have a regime $q\xi \gg 1$), the transport modes cannot be written in the simple form

$$\lambda_\alpha^q = q^2 \mu_\alpha(\xi), \tag{A9.4.17}$$

where μ_α solely depends on temperature, but one should retain the complete q and dependence of μ_α:

$$\lambda_\alpha(q, \omega) = q^2 \mu_\alpha(\xi, q, \omega). \tag{A9.4.18}$$

2. Because of the absence of a small parameter in the problem, one is not allowed to retain only two mode terms. Indeed, there are rather clear indications that many mode terms contribute equally to the diverging transport coefficients. Fortunately, simple dimensional arguments show that this does not affect the nature of the divergence (nor the value of the dynamical critical exponents) found by taking only two- and three-mode coupling.
3. As illustrated above by (A9.4.11), this divergence is determined by solving self-consistently homogeneous mode-mode coupling equations.

With these points considered, the mode-mode coupling technique can be applied straightforwardly at least in principle. In practice, the difficulty is the large number of equations (corresponding to the various transport coefficients and to the various frequency regimes) that have to be solved simultaneously. These equations will not be written explicitly here. It must be noted, though, that in a model-independent calculation, one should take the actual, i.e., not van der Waals, critical behaviour for the static quantities; roughly one has

$$\xi \alpha \varepsilon^{-2/3}; \quad C_p \sim \varepsilon^{-4/3}. \tag{A9.4.19}$$

The theory referred to above has been remarkably confirmed by experiment [see, for example, Stanley (1974)].

Appendix 9.5
Proof of Wiedemann-Franz law up to metal-insulator transition for Fermi liquid model

Castellani et al. (1987) have studied the heat diffusion and the thermal conductivity of an interacting disordered electron liquid in the metallic regime close to the metal-insulator transition. The heat-diffusion constant provides a direct measurement of the *dressed*, or *quasi-particle* diffusion constant, which scales differently from the charge diffusion constant. These workers demonstrate, within the framework of their Fermi liquid model, that the thermal conductivity scales like the electrical conductivity, thereby establishing in this model the validity of the Wiedemann-Franz law (7.1) up to the metal-insulator transition.

Various workers have studied transport coefficients and thermodynamic susceptibilities of interacting disordered electrons close to the metal-insulator transition. Analogies with Fermi liquid behaviour have been exploited by Castellani and DiCastro (1985), and the scaling parameters of the metal-insulator transition originally due to Finkelstein (1983) have been related to the spin susceptibility, specific heat, and compressibility of the disordered interacting system [see also Castellani et al. (1984, 1986)]. The transport coefficients, the corresponding susceptibility, and the associated Fermi liquid parameters exhibit the structure:

$$\sigma_\rho = D_\rho \frac{dn}{d\mu}, \qquad D_\rho = \frac{D}{(dn/d\mu)/\rho}, \tag{A9.5.1}$$

$$\sigma_s = D_s \chi, \qquad D_s = \frac{D}{\chi/\chi_0} \tag{A9.5.2}$$

with σ_ρ, σ_s, D_ρ, and D_s the charge and spin conductivities and diffusion constants, respectively. $\mathrm{d}n/\mathrm{d}\mu$ and χ are the charge and spin susceptibilities, whereas ρ and χ_0 are the corresponding bare noninteracting values.

The following discussion appeals to the physical picture of the metal-insulator transition based on the excitation of quasi-particles in a disordered system (see Castellani et al., 1987). The quasi-particles are characterized by a density of states z, a singlet and a triplet short-range scattering amplitude $\gamma_s = a^2\Gamma_s z$, and $\gamma_t = a^2\Gamma_t z$, where a is the spectral weight of the quasi-particles and the quasi-particle diffusion constant $D_Q = D/z$. Assuming this picture, the specific heat at constant volume should be given by $C_v = \rho z T$, a relation first suggested by Castellani and DiCastro (1986). Equations (A9.5.1) and (A9.5.2) suggest the following expressions for the heat-diffusion constant and the thermal conductivity:

$$\kappa = \sigma_H = C_v D_H; \qquad D_H = \frac{D}{(C_v/C_v^0)} = D_Q \tag{A9.5.3}$$

where $C_v^0 = (\pi^2/3)\rho T$ is the bare noninteracting specific heat. Equation (A9.5.3), which relates the quasi-particle diffusion constant to an observable quantity, as well as the general structure of (A9.5.1)–(A9.5.3) follows very naturally, as Castellani et al. (1987) emphasize, if one assumes that transport is entirely due to quasi-particles.

For this purpose, let us consider first a phenomenological transport equation for the quasi-particle distribution function $n_\sigma(\varepsilon_n, r)$. ε_n here denotes the energy of a single isolated quasi-particle and plays the role of $k^2/2m^*$ in conventional Fermi liquid theory [see, for example, Jones and March (1973)]. The energy as a functional of $n_\sigma(\varepsilon_n, r)$ is given by

$$E = \sum_\sigma \sum_n \int n_\sigma(\varepsilon_n, r)\varepsilon_n \mathrm{d}^d r + \frac{1}{2}\sum_{\sigma\sigma'} \int \mathrm{d}^d r \int \mathrm{d}^d r' \, \delta N_\sigma(r) f_{\sigma\sigma'} \delta N_{\sigma'}(r'). \tag{A9.5.4}$$

Here $f_{\sigma\sigma'}$ is the Landau interaction function, which in a strongly disordered system does not depend on angle, since at distances larger than the mean free path, only s-wave scattering is important. $N_\sigma(r)$ is the total density per spin: $N_\sigma(r) = \sum_n n_\sigma(\varepsilon_n, r)$. For long-wave disturbances, one can write a kinetic equation for diffusing quasi-particles, namely,

$$0 = \frac{\partial n_\sigma}{\partial t} - D_Q \nabla^2 n_\sigma + \left(\frac{\partial n_\sigma}{\partial \varepsilon}\right)[-D_Q \nabla^2][\phi_\sigma + \sum f_{\sigma\sigma'} \delta N_{\sigma'}]. \tag{A9.5.5}$$

In this equation, ϕ_σ is an external potential, which is to be put equal to zero in the discussion of thermal properties. There is here a relation to—but also important differences from—earlier work of McMillan (1985).

It is now straightforward to show (Castellani et al., 1987) that (A9.5.1)–(A9.5.3) follow from (A9.5.5) with $\mathrm{d}n/\mathrm{d}\mu$ and D written in terms of the Landau parameters in the customary manner (Jones and March, 1973; see also Appendix 11.1). In particular, just as in the conventional case, the term $\sum_{\sigma'} f_{\sigma\sigma'} \delta N_{\sigma'}$ is of order T^2 for a thermal disturbance:

$$\delta n_\sigma = -\left(\frac{\partial n}{\partial \varepsilon}\right)(\varepsilon - \mu)\frac{\delta T}{T} \tag{A9.5.6}$$

so that the heat diffusion constant is given by $D_{\mathrm{H}} = D_Q$, without any additional Fermi liquid corrections.

Relations (A9.5.3) and (A9.5.1) are important in the present context because they demonstrate that the Wiedemann-Franz law (7.1) holds up to the metal-insulator transition in the presence of the interactions, within the model framework used by Castellani et al. (1987). This law, which was proved by Chester and Thellung (1961) to be valid for noninteracting electrons in a disordered medium, depends only on the Fermi liquid nature of the ground state and is independent of the scaling of the Fermi liquid parameters. The identity $D_{\mathrm{H}} = D_{\mathrm{Q}}$ gives a direct experimental route to the quasi-particle diffusion constant if and when the model is appropriate.

Castellani et al. (1987) give a microscopic derivation of (A9.5.3). Their derivation validates the more qualitative Fermi liquid considerations, and, in particular, their many-body technique treats correctly the energy vertex in the presence of disorder. The reader is referred to their work for the detailed diagrammatic techniques employed. But it is to be added that in the presence of electron-electron interactions, the heat-diffusion constant D_{H} differs from the charge diffusion constant D_ρ. $D_{\mathrm{H}}/D_\rho \sim 1/z$ and thus will scale differently if $z \to 0$ or $z \to \infty$. Nevertheless, the singular corrections to the specific heat introduce additional renormalization to the thermal conductivity. The net result in this model is then the validity of the Wiedemann-Franz law (7.1) up to the metal-insulator transition. In fact, (A9.5.1) and (A9.5.3) yield (7.1) when the relations

$$\sigma = e^2\left(\frac{\partial n}{\partial \mu}\right)D_\rho = e^2\rho D \tag{A9.5.7}$$

are used.

While in no way denying the considerable interest of the above theory, Chapman and March (1988; see also Section 11.12) have cautioned that there is evidence that, in the fluid alkalis, to which one might consider applying the preceding description, the Fermi liquid picture does not hold up to the metal-insulator transition. It remains to be seen, therefore, whether there are other liquid-metallic systems where the above model will come into its own.

Appendix 10.1
Plasmon properties as function of phenomenological relaxation time

For reasons outlined in the text, direct use will be made here of Mermin's formula for $\varepsilon(q, \omega)$, in which a relaxation time τ is introduced. Because one is concerned with high-frequency (plasmon) properties, the limitation of the Mermin formula referred to in the text—that it does not reflect the blurring of the Fermi surface in the static limit $\omega = 0$—is evidently not serious for the present application.

Denoting now the free-electron Lindhard dielectric function (see also Section 4.2) without Fermi surface blurring by $\varepsilon^\circ(q, \omega)$, Mermin's result takes the form

$$\varepsilon(q, \omega) = 1 + \frac{(1 + i/\omega\tau)[\varepsilon^\circ(q, \omega + i/\tau) - 1]}{1 + \{(i/\omega\tau)[\varepsilon^\circ(q, \omega + i/\tau) - 1]/[\varepsilon^\circ(q, 0) - 1]\}}. \qquad \text{(A10.1.1)}$$

Because the plasmon is essentially a long-wavelength electronic excitation, one need specify $\varepsilon_1(q, \omega)$ for small q only. To do so, let us return to (10.55) and (10.56), substitute $\varepsilon_k = \hbar^2 k^2/2m$ and $f_{\mathbf{k}}$ by the usual Fermi step function (i.e., $T = 0$ and no blurring due to disorder scattering). In the small q limit, exploiting the fact that the derivative of the Fermi function is a delta function at the Fermi surface, one obtains, with $\mu = \mathbf{q} \cdot \mathbf{k}/qk$,

$$H^\circ(q, \omega) = \frac{1}{2\pi^2} \int_{-1}^{1} \mathrm{d}\mu \frac{k_F \dfrac{m}{\hbar^2} [2k_F q\mu + q^2]}{2k_F q\mu + q^2 - \dfrac{2m}{\hbar}\omega}. \qquad \text{(A10.1.2)}$$

Clearly $\varepsilon^\circ(q, \omega + i/\tau)$ required in (A10.1.1) is to be obtained from (10.56) by putting $H = H^\circ(q, \omega + i/\tau)$.

With this form for $\varepsilon^\circ(q, \omega + i/\tau)$, (A10.1.1) has been used to plot the real and imaginary parts of $\varepsilon(q, \omega)$ as a function of ω for fixed q in (A10.1.1), and these are shown in Figure A10.1.1 for a value of τ appropriate to liquid Na.

From the real part of this plot, it turns out that the plasmon dispersion, corresponding to $\varepsilon_1(q, \omega) = 0$, is negligibly affected by τ and indeed by a shorter relaxation time by an order of magnitude. Thus, (A10.1.1) shows no changes of significance from the usual Lindhard result for the plasmon dispersion.

Figure A10.1.1. Real and imaginary parts of the dielectric function $\varepsilon(q, \omega)$ defined in (10.3.1) plotted for fixed q (10^7 m^{-1}) as a function of ω. These are shown for τ appropriate to liquid metal Na (1.6×16^{-14} s). Note the tail on the imaginary part due to finite relaxation time τ. For $\tau \to \infty$, ε_2 is zero for $\omega > v_F q$, where v_F is the Fermi velocity (after March and Paranjape, 1987).

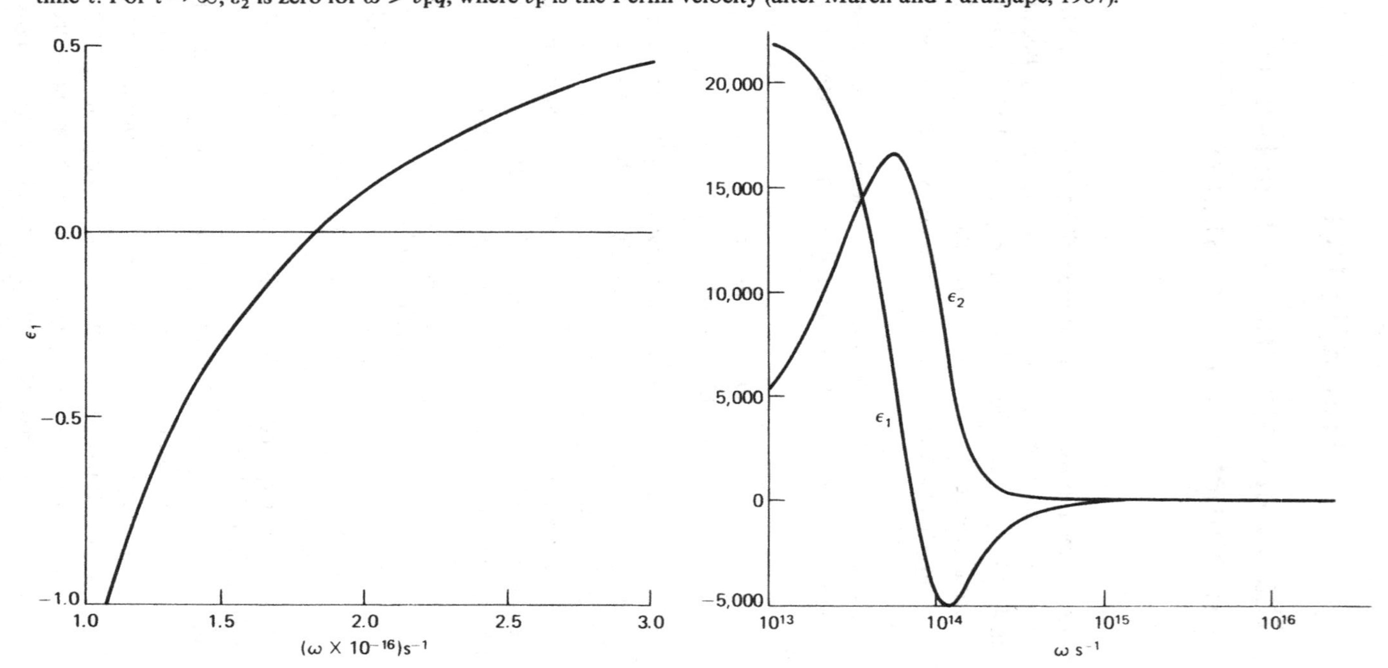

However, as Figure A10.1.1 shows, the effect on ε_2 of τ is qualitative. Whereas with $\tau \to \infty$, ε_2 vanishes for $\omega > v_F q$, v_F being the Fermi velocity, there is now a "tail" on ε_2 for ω greater than this value.

Therefore, March and Paranjape (1987) have also studied the energy-loss

Figure A10.1.2. Energy loss function $\mathrm{Im}(1/\varepsilon(q,\omega))$ plotted as a function of ω for fixed q ($q = 10^8$ and 10^9 m^{-1}). Note the peaks at the plasmon frequency ω_p have nonzero width because of the finite relaxation time τ. As $\tau \to \infty$, one recovers a delta function peak at ω_p.

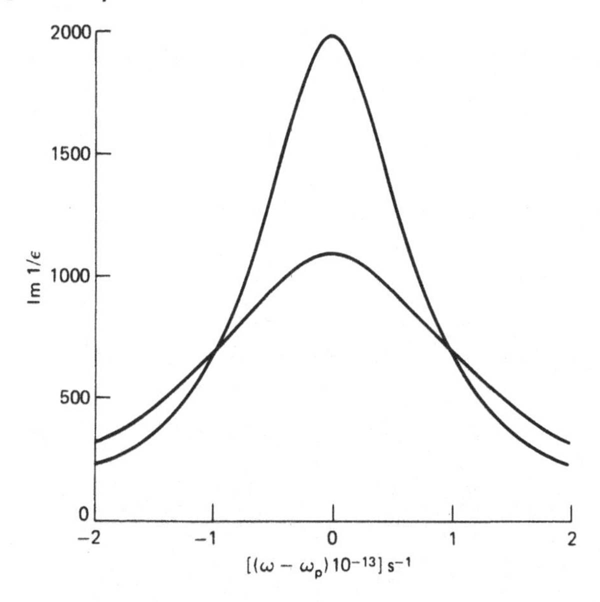

Figure A10.1.3. Total width at half maximum peak height from Figure A10.1.2 plotted as a function of relaxation time τ.

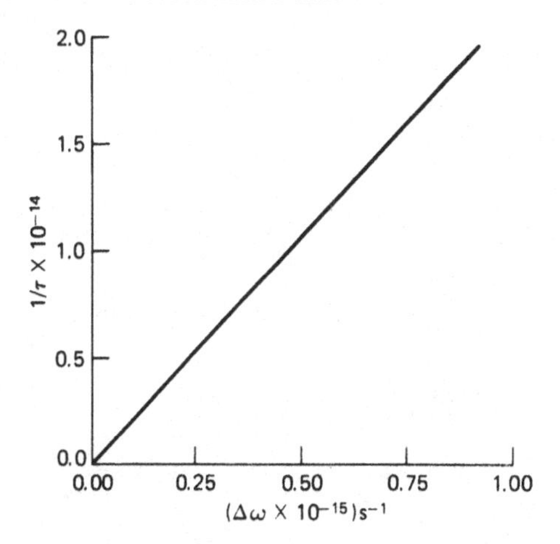

function $\mathrm{Im}(1/\varepsilon(q,\omega))$ from the result (A10.1.1). This function is plotted in Figure A10.1.2 for $\tau = 10^{-14}$ s. It will be seen that although, as already mentioned, the plasma frequency is not affected by finite τ of physical magnitudes, the delta function peak in the loss function at $\omega_p(q)$ is broadened because of the collisions built in through finite τ. The total width of the peak at half maximum height is plotted against $1/\tau$ in Figure A10.1.3. To date, for pure liquid metals the writer knows of no data directly comparable with the damping of the plasmon shown in the figure.

Appendix 11.1
Heavy Fermion theory

Here a brief summary will be given of the background in heavy Fermion theory to the application of it by Chapman and March (1988) to the magnetic susceptibility of liquid Cs along the coexistence curve. The following summary is based on the review by Fulde (1988). He notes that as an alternative to the term *heavy Fermions*, one can speak of metals with strongly correlated, yet delocalized, electrons.

This general area originated from a study of systems containing ions with f electrons. These might be Ce ions with $4f$ states, or U or Np ions with $5f$ electrons. The low-temperature specific heat of heavy Fermion systems shows that there are many more low-lying excitations present in them than in an ordinary metal. One important finding is that there appears to be a one-to-one correspondence between the low-lying excitations in heavy-Fermion systems and those of a nearly free electron gas when the parameters of the latter (e.g., the mass or the magnetic moment of the electrons) are properly renormalized. One example to cite is that, at sufficiently low temperatures, the specific heat is approximately of the form $c = \gamma T + \cdots$ [Fulde (1988) gives the example of $CeAl_3$ in his Figure 1] and the magnetic susceptibility χ is temperature-independent [compare (11.1)], as his Figure 2 shows for the same material. The postulated mutual correspondence of excitations then enables the Fermi liquid framework set up by Landau to be used to treat heavy Fermion systems.

The origin of the name *heavy Fermions* resides in the observations that the enormously large specific heat coefficient γ, which is of order J mol^{-1} K^{-2} instead of order mJ mol^{-1} K^{-2} as in ordinary metals, implies an effective mass m^* of the quasi-particles that is often several hundred times the free electron mass.

A11.1.1. Quasi-particle states

The assumption just cited is that there is a one-to-one correspondence between the low-lying excitations (quasi-particles) of a system of strongly correlated electrons and of a (nearly) free electron gas.

A characteristic feature of Landau's description of the quasi-particles is that their energy dispersion $\tilde{E}_\sigma(k)$ is dependent on the number of other quasi-particles that are present. The quasi-particle distribution is denoted by $\delta n_\sigma(k)$. According to Landau, one can write

$$\tilde{E}_\sigma(\mathbf{k}) = E(\mathbf{k}) + \sum_{k'\sigma'} f_{\sigma\sigma'}(\mathbf{k}, \mathbf{k}')\delta n_{\sigma'}(\mathbf{k}') \tag{A11.1.1}$$

where f characterizes the quasi-particle interactions. Therefore, the problem of describing quasi-particles can be separated into two parts:

1. The determination of $E(\mathbf{k})$
2. The determination of the quasi-particle interactions, which are conveniently expressed in terms of Landau's interaction parameters (called F_l^s and F_l^a)

As to (1), the excitation energies of a homogeneous electron gas are determined from

$$E(\mathbf{k}) = \frac{k^2}{2m} + \sum (\mathbf{k}, E(\mathbf{k}) - \mu \tag{A11.1.2}$$

where $\sum(\mathbf{k}, \omega)$ is the wavenumber and frequency-dependent self-energy of the electrons.

However, in heavy Fermion systems, the self-energy is known to contain a large term linear in frequency, i.e.,

$$\sum (\mathbf{k}, \omega) = \alpha\omega + \cdots, \qquad \alpha \text{ large}. \tag{A11.1.3}$$

This term gives rise to a large effective mass because

$$\frac{m^*}{m} = \frac{(1 - \partial\sum/\partial\omega)_{\omega=E(k_f)}}{1 + \left(\dfrac{m}{k_f}\right)\left(\dfrac{\partial\sum}{\partial k}\right)_{k=k_f}}. \tag{A11.1.4}$$

Turning to point (2), direct interactions between quasi-particles play an important role in heavy Fermion systems. They strongly influence the normal state properties of such assemblies (examples being the static thermodynamic quantities such as compressibility and spin susceptibility).

For homogeneous systems, $f_{\sigma\sigma'}(\mathbf{k}, \mathbf{k}')$ can depend only on the angle between $\mathbf{k}$ and $\mathbf{k}'$ and on $\sigma \cdot \sigma'$. Therefore, one writes

$$f(\theta) = f^s(\theta) + \sigma \cdot \sigma' f^a(\theta). \tag{A11.1.5}$$

The Landau parameters mentioned above, F_l^s and F_l^a, are obtained when f^s and f^a are expanded in Legendre polynomials as

$$f^{s(a)}(\theta) = \frac{1}{2N^*(0)} \sum_{l=0}^{\infty} F_l^{s(a)} P_l(\cos\theta). \tag{A11.1.6}$$

Here $N^*(0)$ is the density of states per spin direction at the Fermi surface. Often only the parameters $F_0^{s(a)}$ and $F_1^{s(a)}$ are assumed to differ from zero. Their number is sufficiently small to allow them to be related without difficulty to experimentally observable quantities.

The most important Landau parameter in heavy Fermion theory is F_0^s. It enters into the electronic density susceptibility χ_c (or the compressibility) through (see, for example, Jones and March, 1973)

$$\frac{\chi_c}{\chi_c^0} = \frac{m^*/m}{1 + F_0^s} \tag{A11.1.7}$$

where χ_c^0 is the susceptibility in the noninteracting case. Without the quasi-particle interactions, the electronic compressibility would be enhanced by a factor m^*/m compared to that of ordinary metals. This is so because of the large density of states $N^*(0)$ that the metal has at the Fermi level. In contrast, a large Landau parameter F_0^s is an important feature of heavy Fermion systems [see Fulde et al. (1988)].

Rice et al. (1985) have generalized to elevated temperature some aspects of heavy Fermion theory, their equations being cited in the main text in relation to the magnetic susceptibility of liquid Cs along the coexistence curve.

Appendix 13.1
Conformal solution theory: thermodynamics and structure

The idea is to relate the thermodynamic properties of a multicomponent solution to those of a reference species, called L_0. The basic equation derived by Longuet-Higgins (1951) for the molar Gibbs free energy in terms of the mole fractions x_r and x_s of species L_r and L_s in the solution is

$$G = \sum_r x_r[G_r + \mathrm{RT}\ln x_r] + \sum_{r<s} x_r x_s E_0 d_{rs}. \tag{A13.1.1}$$

Here G_r denotes the molar free energy of species L_r at given temperature and pressure, whereas E_0 is the molar configurational energy of the reference species and is explicitly given by $\mathrm{RT} - Q_0$, where Q_0 is the latent heat of vaporization of L_0 at temperature T and pressure p. The quantity d_{rs} denotes an interaction parameter for each pair of components.

The dependence of volume V on concentration follows from (A13.1.1) as

$$V = \sum_r x_r V_r + V_0(p\beta_0 - T\alpha_0) \sum_{r<s} x_r x_s d_{rs} \tag{A13.1.2}$$

where α_0 and β_0 are, respectively, the thermal expansion coefficient and the isothermal compressibility of the reference liquid L_0.

These forms, (A13.1.1) and (A13.1.2), achieve the desired aim of expressing the thermodynamic properties of a conformal solution in terms of those of its components, together with a single interaction parameter for each pair of components. Generalizations of this theory have been effected by Byers Brown (see Rowlinson, 1969 for a summary and further references).

A13.1.1. Structure factors of conformal solutions

In Section 13.2, the thermodynamics of mixtures was discussed on the basis of conformal solution theory, use being made of the long-wavelength limit $S_{cc}(0)$ of the concentration-concentration structure $S_{cc}(k)$. This latter quantity has been calculated from conformal solution theory by Parrinello, Tosi, and March (1974); their results are summarized next.

The specific assumption made in conformal solution theory (Longuet-Higgins, 1951) is that the pair potentials $\phi_{\alpha\beta}(r)$ between the different atoms in the mixture are related to the pair potential $\phi(r)$ in the reference liquid by

$$\phi_{\alpha\beta}(r) = A_{\alpha\beta}\phi(\lambda_{\alpha\beta}r). \tag{A13.1.3}$$

In the lowest-order theory, the further assumption made is that the A's and λ's deviate only slightly from unity and that $\lambda_{12} = \frac{1}{2}(\lambda_{11} + \lambda_{12})$. These assumptions allow one to express the thermodynamics of the binary mixture in terms of the product of the parameter $d_{12} = 2A_{12} - A_{11} - A_{22}$ and of the quantity E_0 in (A13.1.1), E_0 being simply given by

$$E_0 = \tfrac{1}{2}\rho \int \mathrm{d}\mathbf{r}\, g(r)\phi(r), \tag{A13.1.4}$$

the interchange energy w in Section 13.2 being then $w = d_{12}E_0$.

The calculation of the pair correlation functions in the mixture can be tackled by perturbation theory (Mo et al., 1974; Parrinello, Tosi and March, 1974). The simplest result of Parrinello, Tosi, and March is for the concentration-concentration pair function $g_{cc}(r)$ in the mixture, this being given in terms of the properties $g(r)$ and $\phi(r)$ of the underlying reference liquid by

$$g_{cc}(r) = c_1^2 c_2^2 \beta d_{12} g(r)\phi(r) \tag{A13.1.5}$$

or in k space,

$$S_{cc}(k) = c_1 c_2[1 + c_1 c_2 \beta d_{12}(\phi g)_k] \tag{A13.1.6}$$

where

$$(\phi g)_k = \rho \int \mathrm{d}\mathbf{r}\, \exp(i\mathbf{k}\cdot\mathbf{r})\phi(r)g(r). \tag{A13.1.7}$$

In the long-wavelength limit, this gives back the conformal solution value of $S_{cc}(0)$ to first order in w.

Calculations of $S_{cc}(k)$ have been presented by Johnson et al. (1974), g and ϕ being chosen to reflect whether the mixture was insulating (like Ar-Kr) or metallic (like Na-K).

Appendix 13.2

Results for concentration fluctuations from quasi-chemical approximation

The purpose of this appendix is to outline the main result obtained from the quasi-chemical approximation for the concentration fluctuations $S_{cc}(0)$ and the short-range order parameter α_1. The following presentation is based on that of Singh (1987).

The derivation is quite similar to that given in Section 13.8, except that the grand partition function Ξ for a cluster of two lattice sites now reads

$$\Xi_1'^{(2)} = \xi_A^2 \phi_A^{2(z-1)} \rho_{AA} + \xi_B^2 \phi_B^{2(z-1)} \rho_{BB} + 2\xi_A \xi_B \phi_A^{z-1} \phi_B^{z-1} \rho_{AB} \qquad \text{(A13.2.1)}$$

where

$$\rho_{ij} = e^{-(\varepsilon_{ij} + P_{ij}\Delta\varepsilon_{ij})/k_B T}, \qquad i, j = \mathrm{A, B}. \qquad \text{(A13.2.2)}$$

ε_{ij} in (A13.2.2) denotes the energy of the free ij bond, and $\Delta\varepsilon_{ij}$ is the change in the energy if the ij bond is in the complex $A_\mu B_\nu$. P_{ij} denotes the probability that the bond in the cluster is part of the complex, that is

$$P_{AB} = c^{\mu-1}(1-c)^{\nu-1}[2 - c^{\mu-1}(1-c)^{\nu-1}] \qquad \text{(A13.2.3)}$$

$$P_{AA} = c^{\mu-2}(1-c)^{\nu}[2 - c^{\mu-2}(1-c)^{\nu}], \qquad \mu \geqslant 2 \qquad \text{(A13.2.4)}$$

$$P_{BB} = c^{\mu}(1-c)^{\nu-2}[2 - c^{\mu}(1-c)^{\nu-2}], \qquad \nu \geqslant 2. \qquad \text{(A13.2.5)}$$

For $\mu = \nu = 1$, P_{AA} and P_{BB} are zero. It has been noted by Bhatia and Singh (1982) that equations (A13.2.3)–(A13.2.5) represent a rather simplistic approximation to P_{ij} and that, as a consequence, the performance of the model for extremely interactive systems, $G_M/Nk_BT \leqslant -3.0$, may not be fully adequate.

As in Section 13.8, one is then led to the result

$$\sigma = \frac{(2c - 1 + \beta)}{2c\eta}, \qquad \text{(A13.2.6)}$$

which is similar to (13.50). β is the same as in Section 3.1 whereas η is now different and is to be found from

$$\eta^2 = \exp\left(\frac{2\omega}{zk_B T}\right) \exp\left(\frac{2P_{AB}\Delta\varepsilon_{AB} - P_{AA}\Delta\varepsilon_{AA} - P_{BB}\Delta\varepsilon_{BB}}{k_B T}\right). \qquad \text{(A13.2.7)}$$

A13.2.1. Thermodynamic expressions

The ratio of the two activity coefficients, $\gamma = \gamma_A/\gamma_B$, follows as

$$\ln\gamma = z\ln\sigma + \left(\frac{z}{2k_B T}\right)(P_{AA}\Delta\varepsilon_{AA} - P_{BB}\Delta\varepsilon_{BB}) + \mathscr{I} \qquad \text{(A13.2.8)}$$

where $\mathscr{I}$ is a constant independent of concentration that may depend on temperature and pressure. In this expression $z\ln\sigma$ stands for

$$z\ln\sigma = \frac{1}{2}z\ln\frac{(1-c)(\beta+2c-1)}{(c)(\beta-2c+1)}. \qquad \text{(A13.2.9)}$$

The excess free energy of mixing,

$$\frac{G_M^{xs}}{Nk_B T} = \int_0^c \ln\gamma\,dc, \qquad \text{(A13.2.10)}$$

can therefore be expressed as

$$\frac{G_M^{xs}}{Nk_B T} = z\int_0^c [\ln\sigma + (2k_B T)^{-1}(P_{AA}\Delta\varepsilon_{AA} - P_{BB}\Delta\varepsilon_{BB})]\,dc + \mathscr{I}c. \qquad \text{(A13.2.11)}$$

The constant $\mathscr{I}$ is determined from the requirement that $G_M^{xs} = 0$ at $c = 1$. The desired expression for $S_{cc}(0)$, from which Figure 13.4 was obtained, is

$$S_{cc} = c(1-c)\left\{1 + \frac{1}{2}z\left(\frac{1}{\beta} - 1\right) + \frac{zc(1-c)}{2\beta k_B T}[2(1-2c)P'_{AB}\Delta\varepsilon_{AB} + (\beta - 1 + 2c)P'_{AA}\Delta\varepsilon_{AA} - (\beta + 1 - 2c)P'_{BB}\Delta\varepsilon_{BB}]\right\}^{-1} \qquad \text{(A13.2.12)}$$

where the prime on P indicates a first derivative with respect to c. The expression (13.56) is adequate for the SRO parameter α_1 for compound-forming systems, except that the exchange parameter η, which is required in the calculation, is now to be evaluated from (A13.2.7).

A13.2.2. Simplification of quasi-chemical results for weakly interacting compound-forming alloys

To conclude this appendix, let us note simplifications of the preceding formulae that occur if the interaction parameters w and $\Delta\varepsilon_{ij}$ are assumed to be small. This is true for Ag-Al considered in Figure 13.5. In this case, η^2 in (A13.2.7) can be expressed to linear terms in these parameters as

$$\eta^2 \simeq 1 + \frac{1}{zk_B T}(2w + 2P_{AB}\Delta w_{AB} - P_{AA}\Delta w_{AA} - P_{BB}\Delta w_{BB}) \quad (A13.2.13)$$

where $\Delta w_{ij} = z\Delta\varepsilon_{ij}$ and P_{ij} have the same meaning as in (A13.2.3)–(A13.2.5). Also, β now becomes

$$\beta \simeq 1 + \frac{2c(1-c)}{zk_B T}[2w + 2P_{AB}\Delta w_{AB} - P_{AA}\Delta w_{AA} - P_{BB}\Delta w_{BB}]. \quad (A13.2.14)$$

Therefore, (A13.2.8) for the activity coefficient simplifies to

$$\ln\gamma = \frac{1}{k_B T}[(1-2c)(w + P_{AB}\Delta w_{AB} + cP_{AA}\Delta w_{AA} - (1-c)P_{BB}\Delta w_{BB}] + \mathscr{I}. \quad (A13.2.15)$$

The constant $\mathscr{I}$ appearing in (13.2.15) can readily be determined by setting $\int_0^1 \ln\gamma \, dc = 0$, which yields

$$\begin{aligned}\mathscr{I} = {} & \frac{\Delta\omega_{AB}}{k_B T}[2\mathscr{B}(\mu+1,\nu) - 2\mathscr{B}(\mu,\nu+1) + \mathscr{B}(2\mu-1,2\nu) - \mathscr{B}(2\mu,2\nu-1)] \\ & + \frac{\Delta\omega_{AA}}{k_B T}[\mathscr{B}(2\mu-2,2\nu+1) - 2\mathscr{B}(\mu,\nu+1)] \\ & + \frac{\Delta\omega_{BB}}{k_B T}[2\mathscr{B}(\mu+1,\nu) - \mathscr{B}(2\mu+1,2\nu-2)] \quad (A13.2.16)\end{aligned}$$

where $\mathscr{B}(m,n)$ is the beta function related to the gamma function, as usual. These relations can now be used to obtain, for G_M^{xs}, the result

$$G_M^{xs} = N[c(1-c)w + \phi_{AB}\Delta w_{AB} + \phi_{AA}\Delta w_{AA} + \phi_{BB}\Delta w_{BB}] \quad (A13.2.17)$$

where the ϕ_{ij}'s are simple polynomials in c, which can be written explicitly for given values of μ and ν. For some explicit expressions for particular pairs μ and ν, the reader may consult Singh (1987). Finally, one can obtain the concentration fluctuations from (A13.2.17) as

$$S_{cc}(0) = \frac{c(1-c)}{1 + \frac{c(1-c)}{k_B T}[-2w + \phi''_{AB}\Delta w_{AB} + \phi''_{AA}\Delta w_{AA} + \phi''_{BB}\Delta w_{BB}]} \quad (A13.2.18)$$

where $\phi''_{ij} = d^2\phi_{ij}/dc^2 (i,j = A,B)$. The position of the asymmetry in $S_{cc}(0)$ is, in fact, controlled by these functions ϕ''_{ij}. The expression for the short-range order parameter α_1 is also simplified considerably, but the result will not be given here [see Singh (1987)].

Appendix 13.3
Density profiles, direct correlation functions, and surface tension of liquid mixtures

The purpose of this appendix is to present the statistical mechanical equations relating density profiles, partial direct correlation functions, and the surface tension of liquid mixtures. The line of argument employed here follows that given by Bhatia, March and Tosi (1980).

Their argument generalizes that of Lovett, Buff and Mou (1976) to multicomponent mixtures. Let $u_i(r)$ denote the dimensionless one-body potential per particle for species i,

$$u_i(r) = \beta(\mu_i - U_i(r)); \qquad \beta = (k_B T)^{-1}. \tag{A13.3.1}$$

Here $U_i(r)$ is the external potential for species i, whereas μ_i is its chemical potential.

The system is considered to be open and at constant V and T. The single-particle densities are then

$$\rho_i(r) = \langle \hat{\rho}_i(r) \rangle \tag{A13.3.2}$$

where $\langle \cdot \rangle$ denotes the ensemble average, as usual.

Given all the $u_i(r)$, the various $\rho_j(r)$ are uniquely determined, and vice versa at given volume and temperature. Hence the quantities $u_i(r)$ can be regarded as functionals of the various $\rho_j(r)$ and vice versa. One then has

$$\frac{\delta \rho_i(r)}{\delta u_j(r')} = \langle \hat{\rho}_i(r)\hat{\rho}_j(r) \rangle - \langle \hat{\rho}_i(r) \rangle \langle \hat{\rho}_j(r') \rangle \tag{A13.3.3}$$

and

$$\frac{\delta u_i(r)}{\delta \rho_j(r')} = \delta_{ij}\frac{\delta(r - r')}{\rho_i(r)} - c_{ij}(r, r') \equiv K_{ij}(r, r'), \tag{A13.3.4}$$

$c_{ij}(r, r')$ being the direct correlation functions for the mixture. If $u_i(r)$ is written as

$$u_i(r) = \ln(\rho_i(r)\Lambda_i^3) - C_i(r) \tag{A13.3.5}$$

where $\Lambda_i = h(2\pi m_i k_B T)^{-1/2}$, then

$$c_{ij}(r, r') = \frac{\delta C_i(r)}{\delta \rho_j(r')}. \tag{A13.3.6}$$

Denoting the functional dependence of the quantity u_i by

$$u_i(r_1, [\rho_1(r), \rho_2(r), \ldots, \rho_\nu(r)]) \equiv u_i(r_1), \tag{A13.3.7}$$

then translational invariance implies

$$u_i(r_1, [\rho_1(r + \delta), \ldots, \rho_\nu(r + \delta)]) \equiv u_i(r_1 + \delta). \tag{A13.3.8}$$

Hence one finds

$$u_i(r_1 + \delta) - u_i(r_1) = \sum_{j=1}^{\nu} \int \mathbf{dr} \frac{\delta u_i(r_1)}{\delta \rho_j(r)} (\rho_j(r + \delta) - \rho_j(r)) + \cdots. \tag{A13.3.9}$$

Proceeding to the limit $\delta \to 0$ yields

$$\nabla u_i(r_1) = \sum_{j=1}^{\nu} \int \mathbf{dr} \frac{\delta u_i(r_1)}{\delta \rho_j(r)} \nabla \rho_j(r) = \sum_{j=1}^{\nu} \int \mathbf{dr}\, K_{ij}(r_1, r) \nabla \rho_j(r) \tag{A13.3.10}$$

where in (A13.3.10), use has been made of (A13.3.4).

If the external potential is now reduced to zero, i.e., $\nabla u_i(r) = 0$, then (A13.3.10) becomes

$$\sum_{j=1}^{\nu} \int \mathbf{dr}\, K_{ij}(r_1, r) \nabla \rho_j(r) = 0 \tag{A13.3.11}$$

or using (A13.3.4):

$$\frac{\nabla \rho_i(r_1)}{\rho_i(r_1)} = \sum_{j=1}^{\nu} \int \mathbf{dr}\, c_{ij}(r_1, r) \nabla \rho_j(r). \tag{A13.3.12}$$

These, then, are the basic equations determining the density profiles; the partial direct correlation functions c_{ij} being those in the presence of the surface.

A13.3.1. Surface tension of mixtures

Knowing the density profile and direct correlation function in a one-component liquid, the surface can be calculated from a formula associated with the names of Yvon and of Triezenberg and Zwanzig (TZ) (1972). The equivalence of this formula, for a liquid with pair interaction potentials, to the well-known theory of Kirkwood and Buff (1949) has been established by Schofield (1979). The generalization of the TZ argument to liquid mixtures is outlined next (Bhatia, March and Tosi, 1980).

To calculate the surface tension, consider that the equilibrium profiles ρ_i have gradient along the x axis. Then the total energy depends on the cross-sectional area of the fluid perpendicular to the x axis. Attention will now be focused on that surface, which satisfies the Gibbs equimolar criteria with respect to the species i. The origin of the Gibbs surface will be located at $(0, y, z)$. Then if one takes the system in the form of a cylinder of basal area a_0, one can write, for extension a from the Gibbs surface in phase I and extension b similarly in phase II,

$$a_0 \int_{-a}^{b} \rho_i(x)\,\mathrm{d}x = a_0(a\rho_i^{\mathrm{I}} + b\rho_i^{\mathrm{II}}) = N_i, \qquad \text{(A13.3.13)}$$

where ρ_i^{I} and ρ_i^{II} are the densities of species i in the bulk phases I and II.

Let us next write r for the two-dimensional vector $\mathbf{r}(y, z)$. Then if there is a small fluctuation $\Delta\rho_j(r, x)$ in the various $\rho_j(x)$, $j = 1 \ldots v$ the change in the Helmholtz free energy F is

$$\Delta F = \tfrac{1}{2}k_{\mathrm{B}}T \sum_{ij} \int \Delta\rho_i(r_1 x_1)\Delta\rho_j(r_2 x_2)K_{ij}(r_1 x_1, r_2 x_2)\,\mathrm{d}^2 r_1\,\mathrm{d}^2 r_2\,\mathrm{d}x_1\,\mathrm{d}x_2. \qquad \text{(A13.3.14)}$$

Fourier-analyzing $\Delta\rho_j$ and remembering that $K_{ij}(r_1 x_1, r_2 x_2) \equiv K_{ij}(0, x_1, r_2 - r_1, x_2)$ which has the Fourier transform

$$K_{ij}(q, x_1 x_2) = \int K_{ij}(0, x, r_2 - r_1, x_2)\exp(i\mathbf{q}\cdot\mathbf{r}_2 - \mathbf{r}_1)\,\mathrm{d}^2(r_2 - r_1) \qquad \text{(A13.3.15)}$$

one finds from a direct generalization of the TZ argument that

$$K_{ij}(q, x_1 x_2) = \delta_{ij}\frac{\delta(x_1 - x_2)}{\rho_i(x_1)} - \hat{c}_{ij}(q, x_1 x_2) \qquad \text{(A13.3.16)}$$

with

$$\hat{c}_{ij}(q, x_1 x_2) = \int c_{ij}(0, x_1, r, x_2)\exp(i\mathbf{q}\cdot\mathbf{r})\,\mathrm{d}^2 r. \qquad \text{(A13.3.17)}$$

For small q

$$K_{ij}(q, x_1 x_2) = K_{ij}^{(0)}(x_1 x_2) + q^2 K_{ij}^{(2)}(x_1 x_2) + \cdots \qquad \text{(A13.3.18)}$$

$$K_{ij}^{(0)} = \delta_{ij}\frac{\delta(x_1 - x_2)}{\rho_i(x_1)} - \int c_{ij}(r_1 = 0, x_1, r x_2)\,\mathrm{d}^2 r \qquad \text{(A13.3.19)}$$

$$K_{ij}^{(2)}(x_1x_2) = \tfrac{1}{4}\int c_{ij}(r = 0, x_1, r, x_2)r^2\,\mathrm{d}^2r \tag{A13.3.20}$$

and, after some analysis set out by Bhatia, March and Tosi (1980) in detail, the change in free energy ΔF can be written (now omitting the sum over q) as

$$\Delta F = \frac{1}{2}a_0 k_B T|x_0(q)|^2\left\{\sum_{ij}\int\frac{\mathrm{d}\rho_i}{\mathrm{d}x_1}\frac{\mathrm{d}\rho_j}{\mathrm{d}x_2}K_{ij}^{(0)}(x_1x_2)\,\mathrm{d}x_1\,\mathrm{d}x_2 + \sum_{ij}q^2\int\frac{\mathrm{d}\rho_i}{\mathrm{d}x_1}\frac{\mathrm{d}\rho_j}{\mathrm{d}x_2}K_{ij}^{(2)}(x_1x_2)\,\mathrm{d}x_1\,\mathrm{d}x_2 + \cdots\right\}. \tag{A13.3.21}$$

Here, $x_0(r)$ is the shift in the Gibbs surface at $\mathbf{r}$ due to the fluctuation $\Delta\rho_i(r, x)$, and the area of the new surface is, for a given q,

$$a = a_0 + \tfrac{1}{2}a_0 q^2|x_0(q)|^2. \tag{A13.3.22}$$

In the limit $q \to 0$, $\Delta a \to 0$ and ΔF tends to zero, yielding conditions that the first summations in the curly brackets in (A13.3.21) must be zero. Sufficient conditions for this to be true turn out to be just (A13.3.12) determining the density profiles.

The final step is to note that the surface tension σ is

$$\sigma = \frac{\Delta F}{(a - a_0)}, \tag{A13.3.23}$$

and hence one finds

$$\sigma = k_B T\sum_{ij}\int\frac{\mathrm{d}\rho_i}{\mathrm{d}x_1}\frac{\mathrm{d}\rho_j}{\mathrm{d}x_2}K_{ij}^{(2)}(x_1x_2)\,\mathrm{d}x_1\,\mathrm{d}x_2, \tag{A13.3.24}$$

which is the desired generalization of the TZ formula to multicomponent alloys. The same result follows in fact, by generalizing the one-component treatment based on the pressure difference across a curved surface, but the details are not given here.

Bhatia, March and Tosi (1980) also consider the density gradient form of the density profile equations, but again their paper should be consulted for the details.

Appendix 13.4

Relation of surface segregation phenomenology to first-principles statistical mechanics

To supplement the derivation of the formula (13.40) for the surface tension of a liquid binary alloy given in the text, it is of interest to summarize the way in which that argument is related to first-principles statistical mechanics. The treatment in this appendix is closely related to that of Bhatia, March and Sutton (1978).

Following Fleming, Yang and Gibbs (1976), one can expand the free-energy density $\psi(x)$ for a liquid binary alloy to lowest order in the density gradients as

$$\psi(x) = \psi[\rho_1(x), \rho_2(x)] + \tfrac{1}{2}\sum_{ij} A_{ij}[\rho_1(x), \rho_2(x)]\rho_i'(x)\rho_j'(x). \quad \text{(A13.4.1)}$$

In (A13.4.1), $\rho_1(x)$ and $\rho_2(x)$ are the density profiles through the surface of components 1 and 2 of the mixture. The first term is the local density approximation, while the correction terms account in lowest order for the inhomogeneity through the surface region. The derivative $\partial\rho_i/\partial x$, $i = 1, 2$, is denoted by ρ_i'. Introducing Lagrange multipliers μ_1 and μ_2 in the customary way, one can write Euler equations for the variation problem posed by minimization of (A13.4.1), namely,

$$\left.\begin{aligned} \mu_1 &= \mu_1[\rho_1(x), \rho_2(x)] + F_1 \\ \mu_2 &= \mu_2[\rho_1(x), \rho_2(x)] + F_2 \end{aligned}\right\} \quad \text{(A13.4.2)}$$

with

$$\mu_i[\rho_1(x), \rho_2(x)] = \frac{\partial\psi[\rho_1(x), \rho_2(x)]}{\partial\rho_i(x)} \quad \text{(A13.4.3)}$$

$$F_1 = (\rho_1')^2\left(-\frac{1}{2}\frac{\partial A_{11}}{\partial\rho_1}\right) - \rho_1'\rho_2'\frac{\partial A_{11}}{\partial\rho_2} + (\rho_2')^2\left(\frac{1}{2}\frac{\partial A_{22}}{\partial\rho_1} - \frac{\partial A_{12}}{\partial\rho_2}\right)$$
$$- (\rho_1''A_{11} + \rho_2''A_{12}), \quad \text{(A13.4.4)}$$

and

$$F_2 = (\rho_1')^2\left(\frac{1}{2}\frac{\partial A_{11}}{\partial \rho_2} - \frac{\partial A_{12}}{\partial \rho_1}\right) - \rho_1'\rho_2'\frac{\partial A_{22}}{\partial \rho_1} - \frac{1}{2}(\rho_2')^2\frac{\partial A_{22}}{\partial \rho_2}$$
$$- (\rho_1'' A_{12} + \rho_2'' A_{22}). \qquad \text{(A13.4.5)}$$

Following the work of Fleming, Yang and Gibbs (1976) for the one-component case, the surface tension for the binary system comes out to be

$$\sigma = \int\left[\sum_{ij} A_{ij}\rho_i''(x)\rho_j'(x)\right]\mathrm{d}x, \qquad \text{(A13.4.6)}$$

which is, in fact, the gradient expansion approximation to formula (9.40).

Employing the Euler equations and introducing the equilibrium pressure p, one can reexpress the surface tension in a more useful form, analogous to that given by Bhatia and March (1978; see also Section 12.4) for the one-component case

$$\sigma = 2\int\{p - \mu_1\rho_1(x) - \mu_2\rho_2(x) + \psi[\rho_1(x), \rho_2(x)]\}\,\mathrm{d}x. \qquad \text{(A13.4.7)}$$

Following the same procedure as for the pure liquid, (A13.4.7) can now be developed to second order in $\rho_i(x) - \rho_i$, with the result

$$\sigma \sim \int\left\{\frac{\partial \mu_1}{\partial \rho_i}[\rho_1(x) - \rho_1]^2 + \left(\frac{\partial \mu_1}{\partial \rho_2} + \frac{\partial \mu_2}{\partial \rho_1}\right)[\rho_1(x) - \rho_1][\rho_2(x) - \rho_2]\right.$$
$$\left. + \frac{\partial \mu_2}{\partial \rho_2}[\rho_2(x) - \rho_2]^2 + \cdots\right\}\mathrm{d}x. \qquad \text{(A13.4.8)}$$

Unless one introduces further assumptions or models, this is as far along the route of gradient expansions as one can go. The connection of (A13.4.8) with the phenomenology of Bhatia and March (1978) considered in Section 12.5 will therefore be the main motivation for the rest of this appendix. If one now assumes that $F_1 = F_2$, then it follows from (A13.4.2) and (A13.4.3) that

$$\mu_1 - \mu_1[\rho_1(x), \rho_2(x)] = \mu_2 - \mu_2[\rho_1(x), \rho_2(x)] \qquad \text{(A13.4.9)}$$

or, to lowest order, that

$$\frac{\partial \mu_1}{\partial \rho_1}[\rho_1(x) - \rho_1] + \frac{\partial \mu_1}{\partial \rho_2}[\rho_2(x) - \rho_2] = \frac{\partial \mu_2}{\partial \rho_1}[\rho_1(x) - \rho_1]$$
$$+ \frac{\partial \mu_2}{\partial \rho_2}[\rho_2(x) - \rho_2]. \qquad \text{(A13.4.10)}$$

Using (A13.4.10) in (A13.4.8) to express this in terms of $\rho(x) - \rho$ only, where $\rho(x) = \rho_1(x) + \rho_2(x)$, gives back the phenomenological formula (12.40).

The condition (A13.4.9) that the local chemical potential difference be constant throughout the interface has already been considered in detail by Fleming, Yang and Gibbs (1976). To achieve (A13.4.9), which in the present context, of course, is the basic approximation to (A13.4.2)–(A13.4.3) required to regain the phenomenological formula (12.40), one would have to apply a fictitious external field that acts differentially on the two components, as the discussion of Fleming, Yang and Gibbs (1976) shows. As mentioned earlier, the problem can then be reduced to terms involving the total particle density through the interface.

Further refinement of the theory beyond the approximate formula (12.40) requires the use of (A13.4.6) for σ in conjunction with the full Euler equations (A13.4.2). At the time of writing, to make progress one would have to invoke models, e.g., the conformal-solution treatment of Bhatia, March and Tosi (1980), to estimate the difference $F_1 - F_2$ neglected in (A13.4.9). Because this same assumption has been shown to underlie (12.40), it would be too much to expect this formula to be quantitative over an extended range of concentration, as already mentioned in the text.

Appendix 13.5
Long-time behaviour of correlation functions in binary alloys

The long-time tail of the velocity-autocorrelation function in a monatomic liquid was referred to in Section 8.5. For concentration diffusion in a mixture of particles of species 1 and 2, one finds (see Pomeau, 1972):

$$D(t) \underset{t\to\infty}{=} \frac{3}{\mu_\gamma}(k_B T)^2 \left(\frac{\rho}{4\pi t}\right)^{3/2} (\eta_+ - \eta_-)^{-1}$$

$$\sum_{a=\pm} a(\eta + \eta_a)^{-3/2} \left[\frac{\kappa + DT^2\mu_T^2 + 2D_1 T\mu_T}{Tc_p} - \eta_a\right], \quad \text{(A13.5.1)}$$

where $D(t)$ is the appropriate flux of concentration correlation function.

In this asymptotic formula for $D(t)$, the thermohydrodynamic parameters that appear are defined as follows. Let $\mu = \mu_1 - \mu_2$ be the difference between the dynamical potential per unit mass of species 1 and 2. With γ_1 the mass concentration of species 1, then $\mu_\gamma = \partial\mu/\partial\gamma_1|_{T,p}$ and $\mu_T = \partial\mu/\partial T|_{p,\gamma_1}$. Furthermore, D_1 is the thermodiffusion coefficient, which is defined such that the linearized hydrodynamic equation for γ_1 is

$$\frac{\partial \gamma_1}{\partial t} = D\Delta\mu + \frac{D_1}{T}\Delta T. \quad \text{(A13.5.2)}$$

The quantities η_{+-} in (A13.5.1) are the roots of the equation

$$\eta_\alpha^2 + \eta_\alpha\left[\left(\frac{K}{T} + 2D_1\mu_T + DT\mu_T^2\right)\Big/ c_p + D\mu_\gamma\right] + \frac{\mu_\gamma}{Tc_p}(KD - D_1^2) = 0, \quad \text{(A13.5.3)}$$

while here c_p is the heat capacity per unit mass. It is to be noted that one can recover from (A13.5.1) the asymptotic form appropriate to a one-component liquid (see Section 8.5):

$$D(t) \simeq \frac{2k_B T}{3\rho} \frac{1}{[4\pi(v + D)t]^{3/2}}, \quad \text{(A13.5.4)}$$

with ν the kinematic viscosity η/ρ, because, in the limit of low concentration:

$$DT^2\mu_T^2 \simeq D_1 T\mu_T \simeq 0, \qquad \eta_\pm = \frac{K}{Tc_p} \tag{A13.5.5}$$

whereas $\mu\gamma D$ becomes the self-diffusion coefficient defined from the Einstein formula.

Appendix 13.6
Hydrodynamic correlation functions in a binary alloy

The linearized hydrodynamic equations for a two-component neutral fluid comprise (Landau and Lifshitz, 1959):

1. The mass continuity equation

$$\frac{\partial m(\mathbf{r},t)}{\partial t} + \nabla \cdot \mathbf{p}(\mathbf{r},t) = 0, \tag{A13.6.1}$$

2. The Navier-Stokes equation (see also Chapter 14), from which one retains only the longitudinal part to obtain

$$\left(\frac{\partial}{\partial t} - D_{\mathrm{L}}\nabla^2\right)\mathbf{p}(\mathbf{r},t) = -\nabla P(\mathbf{r},t) \tag{A13.6.2}$$

with $P(\mathbf{r},t)$ the local pressure,

3. The heat diffusion equation

$$\frac{\partial Q(\mathbf{r},t)}{\partial t} = x\nabla^2 T(\mathbf{r},t) - M\rho\left[k_{\mathrm{T}}\left(\frac{\partial \mu}{\partial x}\right)_{\mathrm{p,T}} - T\left(\frac{\partial \mu}{\partial T}\right)_{\mathrm{p,x}}\right]\nabla\cdot\mathbf{j}(\mathbf{r},t), \tag{A13.6.3}$$

and

4. The continuity equation for the mass concentration $x(\mathbf{r},t)$:

$$\frac{\partial x(\mathbf{r},t)}{\partial t} + \nabla\cdot\mathbf{j}(\mathbf{r},t) = 0. \tag{A13.6.4}$$

Equations (A13.6.1) and (A13.6.2) are the same as those for a one-component fluid.

D_{L} denotes $(\frac{4}{3}\eta + \zeta)/M\rho$ and $M = c_1 M_1 + c_2 M_2$. The heat diffusion equation contains, instead, an additional term determined by the diffusion flux $j(\mathbf{r},t)$, with a coefficient involving the thermal diffusion ratio, k_{T}, and thermodynamic derivatives of $\mu = (\mu_1/M_1) - (\mu_2/M_2)$, the difference in chemical potential per unit mass of the two components. Finally, the continuity equation for the mass concentration leads to the diffusion equation when one writes the diffusion flux as

$$\mathbf{j}(\mathbf{r},t) = -D\left[\nabla x(\mathbf{r},t) + \frac{k_{\mathrm{T}}}{T}\nabla T(\mathbf{r},t) + \frac{k_{\mathrm{P}}}{P}\nabla P(\mathbf{r},t)\right] \tag{A13.6.5}$$

in terms of the interdiffusion coefficient D, the thermal diffusion ratio k_T, and the thermodynamic quantity $k_p = p_0(\partial\mu/\partial P)_{T,x}/(\partial\mu/\partial x)_{p,T}$.

The solution of the preceding equations to determine the time-dependent correlation functions in the mixture, up to terms linear in the transport coefficients, has been given by Cohen, Sutherland and Deutch (1971). Formally, one can rewrite the hydrodynamic equations as

$$\frac{\partial N_i(\mathbf{r}, t)}{\partial t} = -\sum_j \int d\mathbf{r}' \, L_{ij}(\mathbf{r} - \mathbf{r}') N_j(\mathbf{r}', t), \qquad \text{(A13.6.6)}$$

where $L_{ij}(\mathbf{r} - \mathbf{r}')$ is a suitable matrix and $N_i(\mathbf{r}, t)$ are a set of four dynamic variables, which are conveniently chosen as the velocity potential $\psi(\mathbf{r}, t) = (1/M)\nabla \cdot \mathbf{p}(\mathbf{r}, t)$, the pressure $P(\mathbf{r}, t)$, the mass concentration $x(\mathbf{r}, t)$, and the quantity $\phi(\mathbf{r}, t) = T(\mathbf{r}, t) - (T\alpha_T/\rho M C_p)P(\mathbf{r}, t)$, where α_T is the thermal expansion coefficient and C_p is, as usual, the heat capacity. The formal solution of this set of equations has the form

$$\langle \tilde{N}_i(\mathbf{k}, z) N_j(-\mathbf{k}) \rangle = [\det M(\mathbf{k}, z)]^{-1} \sum_l P_{il}(\mathbf{k}, z) \langle N_l(\mathbf{k}) N_j(-\mathbf{k}) \rangle \qquad \text{(A13.6.7)}$$

where $\tilde{N}_i(\mathbf{k}, z)$ is the Laplace-Fourier transform of $N_i(\mathbf{r}, t)$, whereas $M_{ij}(\mathbf{k}, z) = z\delta_{ij} + L(\mathbf{k})$ and $P_{ij}(\mathbf{k}, z)$ are suitable algebraic functions. The static correlation functions in the thermodynamic limit that enter the right-hand side of (A13.6.7) are determined by fluctuation theory, which for the preceding choice of dynamical variables gives the entropy change per unit mass in a fluctuation as

$$\Delta s = -\frac{1}{2T_0}\left[\frac{C_p}{T}(\delta\phi)^2 + \frac{K_T}{M\rho\gamma}(\delta P)^2 + \left(\frac{\partial\mu}{\partial x}\right)_{p,T}(\delta x)^2\right]. \qquad \text{(A13.6.8)}$$

Explicitly one finds for the concentration-concentration correlation function (Bhatia, Thornton and March, 1974) the expression

$$S_{CC}(\mathbf{k}, \omega) = \frac{Nk_B T}{Z}\left\{\frac{2A_7 X k^2}{\omega^2 + X^2 k^4} + \frac{2A_8 Y k^2}{\omega^2 + Y^2 k^4}\right\}, \qquad \text{(A13.6.9)}$$

where

$$X = \tfrac{1}{2}\{\chi + \mathcal{D} + [(\chi + \mathcal{D})^2 - 4\chi D]^{1/2}\} \qquad \text{(A13.6.10)}$$

$$Y = \tfrac{1}{2}\{\chi + \mathcal{D} - [(\chi + \mathcal{D})^2 - 4\chi D]^{1/2}\} \qquad \text{(A13.6.11)}$$

$$A_7 = \frac{Y - D}{Y - X} \qquad \text{(A13.6.12)}$$

$$A_8 = \frac{X - D}{X - Y} \qquad \text{(A13.6.13)}$$

and

$$\chi = V\kappa/C_p, \mathscr{D} = D[1 + Zk_T^2/TC_p],$$

and

$$Z = (\partial^2 G/\partial c_2^2)_{p,T,N}, Z_x = (\partial^2 G/\partial x^2)_{p,T,N}.$$

Similarly, the number-concentration correlation function is given by

$$S_{NC}(k,\omega) = Nk_B T\left\{\frac{2A_4 Xk^2}{\omega^2 + X^2k^4} + \frac{2A_5 Yk^2}{\omega^2 + Y^2k^4}\right.$$
$$\left. + A_6\frac{k}{c_0}\left[\frac{\omega + c_0 k}{(\omega + c_0 k)^2 + \Gamma^2 k^4} - \frac{\omega - c_0 k}{(\omega - c_0 k)^2 + \Gamma^2 k^4}\right]\right\}, \tag{A13.6.14}$$

where the new parameters are the adiabatic speed of sound, $c_0 = (\gamma/M\rho K_T)^{3/2}$ and

$$\Gamma = \frac{1}{2}\left[D_L + (\gamma - 1)\chi + \frac{\gamma DVZ}{K_T}\Sigma^2\right] \tag{A13.6.15}$$

$$A_4 = (Y - X)^{-1}\left[(D - Y)\frac{\delta}{Z} + \frac{Dk_T\alpha_T}{C_p}\right] \tag{A13.6.16}$$

$$A_5 = (X - Y)^{-1}\left[(D - X)\frac{\delta}{Z} + \frac{Dk_T\alpha_T}{C_p}\right] \tag{A13.6.17}$$

$$A_6 = -D\Sigma \tag{A13.6.18}$$

with $\Sigma = (\delta - \delta_m)/Z + k_T\alpha_T/C_p$, $\delta = (1/V)(\partial V/\partial c)_{p,T,N}$ and $\delta_m = (M_1 - M_2)/M$.

Finally the number-number correlation function is given by

$$S_{NN}(k,\omega) = \frac{Nk_B TK_T}{V\gamma}\left\{\frac{2A_1 Xk^2}{\omega^2 + X^2k^4} + \frac{2A_2 Yk^2}{\omega^2 + Y^2k^4}\right.$$
$$+ \left[\frac{\Gamma k^2}{(\omega + c_0 k)^2 + \Gamma^2 k^4} + \frac{\Gamma k^2}{(\omega - c_0 k)^2 + \Gamma^2 k^4}\right]$$
$$\left. + A_3\frac{k}{c_0}\left[\frac{\omega + c_0 k}{(\omega + c_0 k)^2 + \Gamma^2 k^4} - \frac{\omega - c_0 k}{(\omega - c_0 k)^2 + \Gamma^2 k^4}\right]\right\}, \tag{A13.6.19}$$

where the additional parameters are

$$A_1 = \frac{1-\gamma}{X-Y}\left[D - X - \frac{2Dk_T\delta}{T\alpha_T} + (Y - D)\frac{\delta^2 C_p}{ZT\alpha_T^2}\right]$$

$$A_2 = \frac{1-\gamma}{Y-X}\left[D - Y - \frac{2Dk_T\delta}{T\alpha_T} + (X - D)\frac{\delta^2 C_p}{ZT\alpha_T^2}\right] \qquad \text{(A13.6.20)}$$

and

$$A_3 = (3\Gamma - D_L) + \frac{2V\gamma D}{K_T}\delta_m\Sigma. \qquad \text{(A13.6.21)}$$

Equations (A13.6.9), (A13.6.14), and (A13.6.19) for the number-concentration dynamical structure factors constitute the main results of this appendix.

Appendix 13.7
Metallic binary liquid-glass transition

Many complications in the metal physics description of binary alloys have been referred to as arising from the concentration dependence of the force fields. Thus a simpler approach seems called for in discussing the metallic binary liquid-glass transition.

Cohen and Turnbull (1959) [see also Turnbull and Cohen (1961) and Cohen and Grest (1979)] proposed a free-volume model in order to examine the thermodynamic and diffusive behaviour in the vicinity of the glass transition. In this model [see Li, Moore, and Wang (1988a, b)]:

1. An atom in the supercooled liquid or glass, for the most part, vibrates in a cage formed by its surrounding atoms, and
2. The atom inside the cage may escape to a void and diffuse from its original position, when it gains sufficient activation energy to overcome the barrier between its cage and the void.

The void referred to is defined as having a free volume greater than an atomic volume and is adjacent to the cage.

Point (1) has been established to be valid from a computer simulation of the static and dynamic properties for a Lennard-Jones (LJ) system (Kimura and Yonezawa, 1983). However, the same computer study implies that point (2) may not be actually applicable to the atomic mean square displacements.

Because of the fundamental differences between a LJ system and a metal, Li, Moore, and Wang (1988a,b) have made similar computer studies of metallic binary systems, which can become metallic glasses by a rapid quench from the melt not only in computer experiments but also in the laboratory (in which LJ systems such as argon never become glassy). In this way, Li, Moore, and Wang (1988a,b) demonstrate how well such static and dynamic properties as those appearing in the metallic liquid-glass transition can be theoretically described by incorporating the cage concept. Their Monte Carlo and analytic study (Li, Moore and Wang, 1988a) of the transition for the metallic binary alloy $Ca_{0.7}Mg_{0.3}$, which becomes a stable glass even at room temperature, when quenched from the melt in the laboratory, has been extended (Li, Moore and Wang, 1988b) to calculate (1) the angular distribution functions required to determine the structure of the cages and to provide a model for the atomic motion in the cage and (2) the atomic mean square displacements (MSD).

A13.7.1. Cage structure and angular distribution functions

The cage structure is determined by angular distribution functions (ADF), which in lowest order require three particles. Li, Moore, and Wang proceed by letting one of the particles be at the centre of a sphere of radius r. Then, for any pair of the other particles lying on this sphere, the probability of finding the angle to be θ between their position vectors relative to the sphere centre is $A(\cos\theta, r)\sin\theta$ [see also Jacobeus et al. (1980)]; that is

$$A(x;r) = \frac{1}{N_c}\sum_{i<j}\Theta(r - r_{ci})\Theta(r - r_{cj})\delta\left(x - \frac{\mathbf{r}_{ci}\cdot\mathbf{r}_{cj}}{r_{ci}r_{cj}}\right) \qquad \text{(A13.7.1)}$$

where $x = \cos\theta$ and N_c is defined by

$$N_c = \sum_{i<j}\Theta(r - r_{ci})\Theta(r - r_{cj}). \qquad \text{(A13.7.2)}$$

Here Θ is a step function and $\mathbf{r}_{ci} = \mathbf{r}_i - \mathbf{r}_c$, with $\mathbf{r}_i$ and $\mathbf{r}_c$, respectively, the position vectors of the ith particle and the particle at the centre of the sphere. Starting from the three-particle correlation function (see Chapter 5), (A13.7.1) can be written in an integral form (Jacobeus et al., 1980). Li, Moore, and Wang rewrite (A13.7.1) in an equivalent form better suited to suppress statistical errors, but the interested reader should refer to their paper for these technical details. The angular distribution functions Li, Moore, and Wang thereby display confirm the cage concept rather strikingly.

A13.7.2. Mean square displacements

Using the framework developed by Wong and Chester (1987), the mean square displacement (MSD) of the particles in a Monte Carlo study can be written

$$u^2(I) = \frac{1}{N}\sum_{i=1}^{N}[\mathbf{r}_i(I) - \mathbf{r}_i(0)]^2 \qquad \text{(A13.7.3)}$$

where I stands for the number of iterations in the Monte Carlo simulation, N is the number of particles, $\mathbf{r}_i(0)$ is the position vector of the ith particle for the system in its initial configuration, and $\mathbf{r}_i(I)$ is its position vector after I iterations. Figure A13.7.1 shows values of u^2 for both liquid and glass states. Li, Moore, and Wang note the following points from their computer studies of the CaMg metallic alloy:

1. The MSD is greater for the Mg component than for the Ca component at each temperature they considered. The reason is that the two components have different ion-core radii.
2. For temperatures below the glass transition temperature T_g, MSD versus I curves have similar shapes to the MSD-t (time) curves from molecular dynamics calculations for an LJ system, as expected from the fact that the particles in any glass are quite localized. This similarity of shape implies a linear I versus t relation.
3. For temperatures above T_g, the MSD-I curves differ significantly from the corresponding MSD-t plots. This is consistent with the significant differences in

Figure A13.7.1. Mean square displacement u^2 for both liquid and glassy states [after Li et al. (1988)].

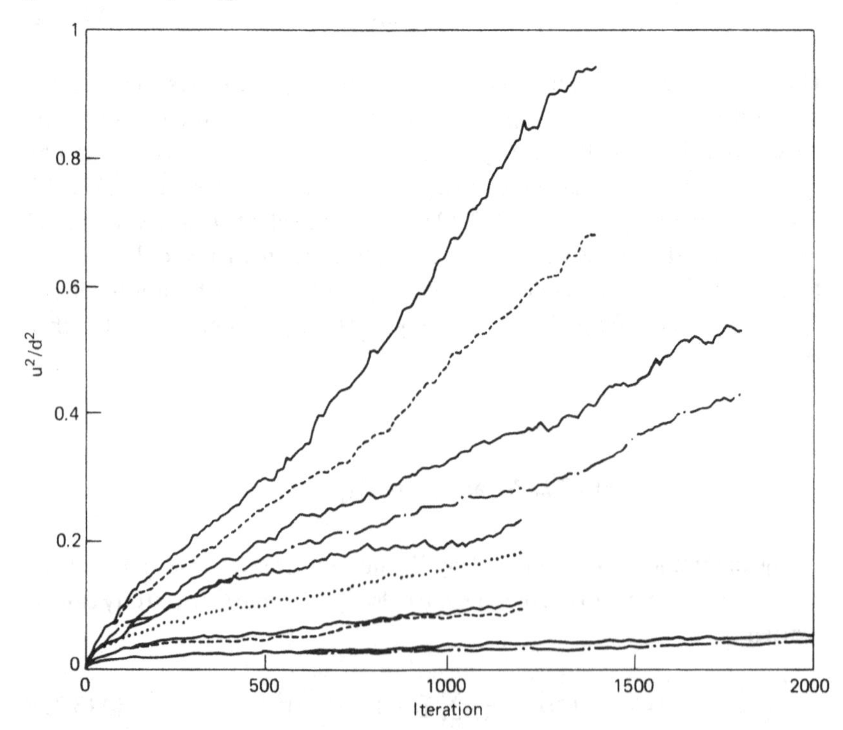

the atomic excitations determined experimentally between LJ type systems and metallic liquids [see, for instance, Bratby, Gaskell and March, (1970) and Gray, Yokoyama and Young (1980)].

4. For any of the temperatures considered, the MSD increased with I. The initial curvature in the MSD-I curve (Figure A13.7.1) is consistent with the generalized Einstein model in which the atom in a cage oscillates about a centre, which is itself undergoing Brownian motion (Li, Moore, and Wang, 1988b; see also Egelstaff, 1967).
5. In the later linear stage (corresponding to $I > 500$), the slopes of the Mg and Ca MSD versus I curves are about the same for $T < T_g$ but differ for $T > T_g$. The $T < T_g$ behaviour also follows from the generalized Einstein model. The case $T > T_g$ is interpreted by Li, Moore, and Wang (1988b) as follows. At high temperatures, the cage breaks immediately after the early state due to the strong thermal effects on it and, as a result the atom, which was initially in the cage, is then described by the Langevin model of diffusion (see Chapter 7; see also Gerl, 1985).

The main conclusion is therefore that the cage concept (see also Chapter 7, Section 5) is valid for both liquid and glassy states of metallic systems such as this CaMg alloy. The lifetime of the cage, however, depends strongly on the temperature T. As T decreases from the melting temperature T_m, the cages, each enclosing an atom, become more stable, so that the atoms in a glass are well localized. The mean-square-displacement calculations support the validity of point (1); however Li, Moore, and Wang (1988b) express the view that point (2) is probably not applicable to atomic diffusion in metallic systems. But it is striking, of course, that the Monte Carlo simulations of $Ca_{0.7}Mg_{0.3}$, consisting of 1000 particles, are—for temperatures ranging from the melting temperature T_m down to temperatures significantly below the liquid-glass transition—consistent with an Einstein model of atomic motions. The cage concept is valid over the entire temperature range.

As the next step in this appendix, let us return to the discussion of the concentration fluctuations $S_{CC}(0)$ in relation to the glassy state.

Role of $S_{CC}(0)$ in metallic glass formation

Here, the discussion of concentration fluctuations in Section 13.6 will be related to metallic glass formation. Many of the compound-forming binary molten alloys are good glass formers, such as AlLa, CaMg, MgZn, and

CuTi. RamachandraRao, Singh, and Lele (1984) have utilized the knowledge of $S_{CC}(0)$ to explain the glass formation in such systems. These authors discussed the possible correlations among $S_{CC}(0)$, the equilibrium phase diagram, the existence of chemical complexes in the liquid state (see Section 7.3), and glass formation. They pointed out that in the glass-forming composition range, which usually lies far away from the stoichiometric

Figure A13.7.2. The phase diagram and concentration fluctuations $S_{CC}(0)$ for AlLa at 1693 K. $S_{CC}(0)$ has been found from activity data (Knonenko et al, 1978; (after Singh, 1987).

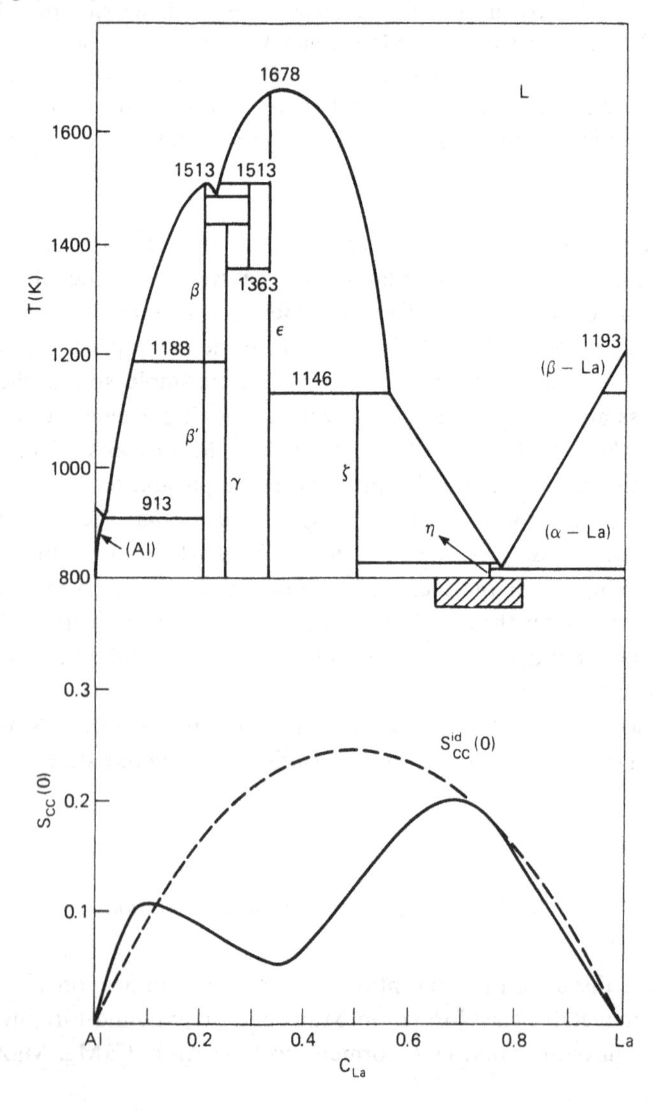

composition, $S_{CC}(0)$ attains ideal values; that is to say, the liquid exhibits ideal behaviour for this composition (as an example, see Figure A13.7.2). Consequently, in the framework of the complex-formation model, one infers that the unassociated species (A and B atoms) and the complex ($A_\mu B_\nu$) mix randomly. Such random mixing can presumably hinder nucleation and helps glass formation. It may, however, be stated that the phenomenon of glass formation in molten alloys is very involved and may depend on many other factors (for reviews, see Chen (1980) and Cahn (1980)). Nonetheless it seems that the ideal value of $S_{CC}(0)$ may not be sufficient but rather is a necessary condition for obtaining glass from molten alloys.

As was mentioned in the text, knowledge of $S_{CC}(0)$ can also be used to assess the nature and the strength of the chemical complexes ($A_\mu B_\nu$) if they exist in the liquid alloys. At the stoichiometric compositions, $S_{CC}(0)$ deviates considerably from the ideal value $S_{CC}^{id}(0)$ and shows a dip. It tends to zero for strongly interacting systems such as AlLa and BiMg and takes intermediate values for weakly interacting systems such as CaMg and MgZn. The position of the dip in the $S_{CC}(0)$-c plot yields information on the nature of the complex and its depth signifies strength. Matsunaga et al. (1983) and, subsequently, Saboungi, Herron, and Kumar (1985) have used it for such analysis in NaPb molten alloys. $S_{CC}(0)$ for NaPb obtained from observed activity data (Saboungi et al., 1985) is shown in Figure A13.7.3.

Figure A13.7.3. Concentration fluctuations $S_{CC}(0)$ for a NaPb melt at 723 K, obtained from activity data (Saboungi, Herron, and Kumar, 1985; after Singh, 1987).

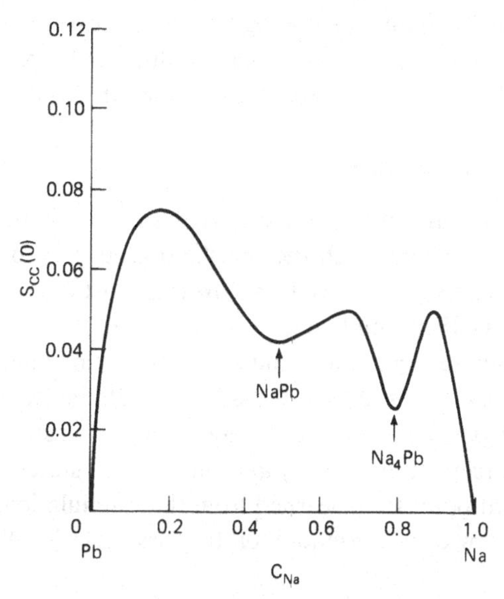

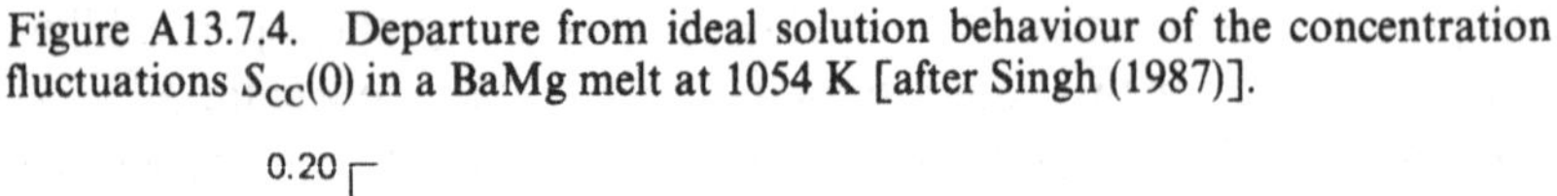
Figure A13.7.4. Departure from ideal solution behaviour of the concentration fluctuations $S_{CC}(0)$ in a BaMg melt at 1054 K [after Singh (1987)].

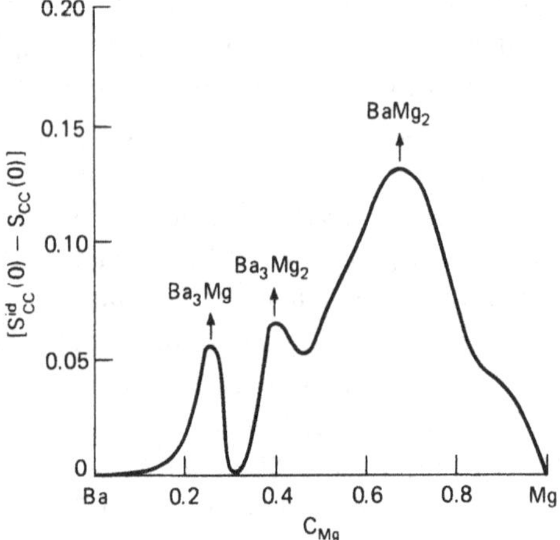

It indicates that the nature of complexes in NaPb melt are NaPb and Na_4Pb, the latter being dominant. Particularly in weakly interacting systems, the plot of $[S^{id}_{CC}(0) - S_{CC}(0)]$ is more informative. This is shown in Figure A13.7.4 for the BaMg melt, where Ba_3Mg, Ba_3Mg_2 and $BaMg_2$ are found to be the appropriate complex compositions.

In the light of the discussion on $S_{CC}(0)$ and α_1 (see Chapter 13) it is worth mentioning that in the range of the glass-forming composition, the contribution to the short-range order parameter α_1 due to mixing of unassociated species and the complex should be negligibly small. Any remnants of short-range order at this composition might be due to the preferential orientation of the unlike atoms in whatever complexes are available there.

Binary Mixture of Soft Spheres

Mountain and Thirumalai (1987) have reported simulation results for binary mixtures of soft spheres with the specific objective of identifying possible physical mechanisms that result in slow transport in supercooled liquids and glasses. As they point out, it is difficult to draw definitive conclusions from computer simulation studies involving small numbers of particles and short time scales. Nevertheless, the results indicate several interesting aspects of glass formation that can be used to construct model dynamics for viscous relaxation in glassy and supercooled states.

The most important point to be learned from these simulations is that the changes associated with the formation of the glass occur locally rather than collectively and hence there is a need for an "order parameter," to characterize the local order. For the specific model adopted by Mountain and Thirumalai (1987), the mixture consists of N_1 soft spheres of type 1 with mass M_1 and diameter σ_{11} and N_2 soft spheres with mass M_2 and diameter σ_{22}. The cross-interaction diameters are assumed additive: $\sigma_{12} = \frac{1}{2}(\sigma_{11} + \sigma_{22})$ and $N_1 + N_2$ was always 500 in their molecular-dynamiccs simulations. The structural results obtained from the pair and triplet correlation functions indicate that, for r^{-12} soft spheres, the local order develops into a smeared-out *fcc* arrangement of atoms in the first shell of neighbours around a given atom and that this order is sufficiently distorted that it does not extend to the second-neighbour distance in the crystal. This leads to a length, of the order $1.5\sigma_{11}$, which characterizes the local order.

Another important feature of the glassy state is the freezing in of the local environment for each atom. Mountain and Thirumalai (1987) suggest that the orientation of near-neighbor bonds might be a useful means of characterizing local order.

One of the objectives of the work of Mountain and Thirumalai (1987) was to identify possible "local modes" involving few particles and to relate the relaxation of these modes to the glass transition. The physical motivation for such a picture stems from the notion that activated processes, which dominate structural relaxation for temperatures close to the glass-transition temperature, can involve only rearrangement of particles over a small length scale. However, the identification of such an order parameter does not appear to be unique, but the reader should refer to Mountain and Thirumalai (1987) for further details.

Appendix 13.8

Haeffner effect, electromigration, and thermal transport

Isotope effects are known to exist in liquid metals from a variety of experiments; in particular, those using the light isotopes Li^6 and Li^7. Some striking—and surprising—regularities exist more generally, especially the effect discovered by Haeffner (1953). Here, in an applied electric field, the light isotope in the isotopic liquid metal mixture is found, invariably, to move toward the anode. No known exceptions to this rule exist. The problem of electromigration is closely related, but presumably the understanding of the Haeffner effect is an essential prerequisite to an understanding of this phenomenon. For a review of electromigration, the article by Huntington (1973; see also Jones, 1980) may be consulted.

The facts, and some basic phenomenology, have been presented by Ginoza and March (1985) for (1) the Haeffner effect just discussed, (2) self- and mutual diffusion, and (3) shear viscosity.

Here, the aim is to present the theory underlying the Haeffner effect, in lowest-order Born approximation, in a form that is directly related to electrical resistivity. Some attempts to transcend this approximation will then be briefly discussed.

A13.8.1. Haeffner effect

As mentioned already, it is useful to regard the Haeffner effect as a special case of the more general electromigration problem in liquid metal alloys. This is the effect found in a number of binary systems, where the constituent ions drift in opposite directions under the influence of an applied dc electric field (see also Tyrrell and Harris, 1984). This effect, known alternatively as electrotransport or electrodiffusion, has been generally explained in terms of a competitive transfer of momentum from the conduction electron current to the two species of ions in the alloy, as discussed, for example, by Landauer and Woo (1974).

The force acting on each ion is assumed to be the sum of two components:

1. A direct force due to the applied electric field itself, which tends to drive the ion towards the cathode, and
2. An "electron drag" component arising from the scattering of conduction electrons by the ion, which acts to pull the ions towards the anode, that is, in the same direction as the electron current. The ionic species which experiences the greater electron drag will then presumably migrate toward the anode, whereas the other, in order not to build up density gradient, will migrate in the opposite direction.

As Stroud (1976) points out, one should be cautious because, to date, the attempts at this problem have been largely focused on calculating a force rather than a current. The ionic current induced in the liquid binary alloys under discussion depends on an ionic mobility as well as a driving force. If the local fluctuations in the drag force are not too large it should be possible to obtain an ionic mass current by simply multiplying the average driving force by an average mobility. Stroud's work, however, suggests that these fluctuations are substantial. They should therefore be taken into account in a quantitative theory of electromigration.

To press the preceding points in the alloy, before specializing to the Haeffner effect, Stroud, in his work on the average driving force for electromigration in liquid metal alloys, concludes that there is a correlation between the calculated driving force acting on a solute ion and its assumed hard sphere diameter. He concludes that "in virtually every case it is found that the effective valence of the solute atom becomes more negative as the assumed hard-sphere diameter decreases, i.e., the smaller the atom, the more likely it is to migrate to the anode."

However, he then goes on to point out that the correlation between the driving force and hard sphere diameter reflects a sensitivity of the driving force to the structure factors, i.e., to the local environment of the solute ion. This sensitivity presumably arises from quantum-mechanical interference between the electron wave scattered off an ion on which a force is being exerted and the wave scattered off a neighbouring ion. The recoil of the ion induced by this scattered wave gives rise to the electron drag force. Although such sensitivity to local environment is to be expected, the reason for the direction of the correlation, i.e., large drag forces corresponding to small diameter, is still difficult to pinpoint. Stroud (1976) observes that since this environment is likely to fluctuate in a liquid, this sensitivity suggests that the driving forces may also experience substantial local fluctuations, which are likely to correlate with local fluctuations in solute mobility. Such correlated fluctuations have long been thought necessary to account for electric field induced isotope separations, i.e., the Haeffner effect (see also the Born approximation treatment of Section A13.8.3).

A13.8.2. Thermotransport

It is relevant here to add some comments on thermotransport. The process in which a concentration gradient is induced by a temperature gradient in solids, liquids, or gases has been termed either thermotransport, thermomigration, thermal diffusion, or the Ludwig-Soret effect. The phenomenon, while well studied in many materials, as reviewed by Grew (1969) and by Huntington (1973), has been less extensively studied to date in liquid metals.

In a typical experimental setup, a liquid with atomic fraction x_1 of component 1 is contained in a vertical capillary several centimeters long. It is then subjected to a temperature gradient of 1 to 5°C/mm for an appropriate time t inversely proportional to the diffusion constant D, so that a steady-state concentration profile is developed. For liquid metals, it is usual to express the results in terms of the net heat of transport Q^*, which for dilute alloys is given by

$$\frac{d(\ln x_1)}{d(1/T)} = \frac{Q^*}{R}. \tag{A13.8.1}$$

Evidently Q^* is then obtained directly from a plot of $\ln x_1$ versus $1/T$.

It is worthy of note at this stage that Gonzalez and Oriani (1965) have pointed out that a strong correlation exists between Q^* in solid metal systems and the effective valence measured in electrotransport experiments. This is expected, according to the arguments of Gerl (1971), if the electron contribution to Q^* is dominant. The correlation turns out to be particularly good for dilute liquid metals, in which the component which migrates to the hot end in thermotransport is the one that goes to the anode in electrotransport. Particularly important in the present context is that it holds for the isotopes of Li, since the migration of a Li^6 species to the hot end is consistent with the measured Haeffner effect, according to Ott and Lunden (1964) and to Verhoeven (1969).

Even with the large positive temperature gradients that are generally used for these experiments, Rigney (1977) has emphasized that convection can develop during thermotransport if the component that migrates to the hot end has a higher density than the other component. The work of Bhat and Swalin (1972) should also be referred to in this same connection.

A13.8.3. Microscopic theory of Haeffner effect: the Born approximation

As a start on the theory of the Haeffner effect, it seemed natural to begin at the level of Born approximation. It is not claimed that such a level of approximation is adequate for a quantitative theory of the Haeffner effect, but it does serve to establish a framework for further studies of the problem. The treatment below is due to Parrinello, Tosi and March (1975).

Consider an external, static electric field $\mathbf{E} = (E, 0, 0)$ applied to the liquid metal in the positive direction of the x-axis and that a steady current of electrons is thereby established. The system may be described by the Hamiltonian:

$$\left.\begin{aligned} \mathscr{H} &= \mathscr{H}_0 + H_{\mathrm{ei}} \\ \mathscr{H}_0 &= \mathscr{H}_{\mathrm{e}} + \mathscr{H}_{\mathrm{i}} \end{aligned}\right\}, \tag{A.13.8.2}$$

where $\mathscr{H}_{\mathrm{i}}$ is the effective Hamiltonian of the two kinds of ionic components and $\mathscr{H}_{\mathrm{e}}$ is the effective Hamiltonian of the electron component. The most general expression for H_{ei} may be written

$$H_{\mathrm{ei}} = \frac{1}{V}\sum_{\mathbf{k}}\int \mathrm{d}\omega \sum_{\mathbf{p}\sigma}\sum_{\mathbf{p}'} \delta_{\mathbf{p}-\mathbf{p}',\hbar\mathbf{k}}\, \delta\left(\frac{\{\varepsilon_p - \varepsilon_{p'}\}}{\hbar} - \omega\right)$$
$$\times \sum_{\alpha=1,2} \langle \mathbf{p}'|U_{\mathrm{e}\alpha}(\mathbf{k},\omega)|\mathbf{p}\rangle \rho_\alpha^+(\mathbf{k}) a^+_{\mathbf{p}'\sigma} a_{\mathbf{p}\sigma} \tag{A13.8.3}$$

where $a^+_{\mathbf{p}\sigma}$ and $a_{\mathbf{p}\sigma}$ are creation and annihilation operators of a $\mathbf{p}\sigma$-electron, respectively:

$$\rho_\alpha(\mathbf{k}) = \sum_{i=1}^{N_\alpha} \exp(-i\mathbf{k}\cdot\mathbf{R}_i^{(\alpha)}), \tag{A13.8.4}$$

$\mathbf{R}_i^{(\alpha)}$ being the position vector of the ith ion of the α-ion component. In (A13.8.3), $\langle \mathbf{p}'|U_{\mathrm{e}\alpha}(\mathbf{k},\omega)|\mathbf{p}\rangle$ denotes the interaction matrix element corresponding to the electron-α-type ion interaction. The remaining factors mean energy-momentum conservation in the process of interaction (see Ginoza and March (1985). One may assume that $\mathscr{H}_{\mathrm{e}}$ describes the assembly of effectively free electrons and then

$$\frac{\mathrm{Tr}[\mathrm{e}^{-\beta\mathscr{H}_{\mathrm{e}}} a^+_{\mathbf{p}\sigma} a_{\mathbf{p}\sigma}]}{\mathrm{Tr}[\mathrm{e}^{-\beta\mathscr{H}_{\mathrm{e}}}]} = \{\exp[\beta(\varepsilon_{\mathbf{p}+\tau e\mathbf{E}} - \mu)] + 1\}^{-1} \equiv f(\mathbf{p} + \tau e\mathbf{E}) \tag{A13.8.5}$$

where as usual $\beta = (k_{\mathrm{B}}T)^{-1}$, whereas $\varepsilon_{\mathbf{p}} = p^2/2m$, τ is the electronic relaxation time, and μ is the chemical potential.

The "electron-wind" force, $\mathbf{f}_\alpha$, acting on the α-ion component may be calculated as the average of the force operator $\sum_{j=1}^{N_\alpha}[-\nabla_{\mathbf{R}_j}^{(\alpha)} H_{\mathrm{ei}}]/N_\alpha$. The method has similarities to that of Rousseau, Stoddart, and March (1972; see also Appendix 7.1). In the case of weak electron-ion interaction, however, one can obtain the same expression for $\mathbf{f}_\alpha$ directly from the work of Baym (1964). In Born approximation for the treatment of electron inelastic scattering, the rate at which momentum $\hbar k$ transfers from the electrons to the α-ion component is

$$W_\alpha(\mathbf{k}) = \frac{2\pi}{V^2\hbar^2}\int d\omega \sum_{\mathbf{p},\mathbf{p}'} \delta_{\mathbf{p}-\mathbf{p}',\hbar\mathbf{k}}\delta(\{\varepsilon_p - \varepsilon_{p'}\}/\hbar - \omega)$$
$$\times \sum_\beta \langle \mathbf{p}'|U_{e\alpha}(\mathbf{k},\omega)|\mathbf{p}\rangle\langle \mathbf{p}|U_{e\beta}(-\mathbf{k},-\omega)|\mathbf{p}'\rangle\{N_\beta/N_\alpha\}^{1/2}S_{\beta\alpha}(\mathbf{k},\omega)$$
$$\times [f(\mathbf{p}+\tau e\mathbf{E})(1 - f(\mathbf{p}'+\tau e\mathbf{E})$$
$$- \exp(-\beta\hbar\omega)f(\mathbf{p}'+\tau e\mathbf{E})(1 - f(\mathbf{p}+\tau e\mathbf{E}))] \tag{A13.8.6}$$

where

$$S_{\beta\alpha}(\mathbf{k},\omega) = \int \frac{dt}{2\pi}\exp(i\omega t)\langle \rho_\beta(\mathbf{k},t)\rho_\alpha^+(\mathbf{k})\rangle/\{N_\beta N_\alpha\}^{1/2}, \tag{A13.8.7}$$

$\rho(\mathbf{k},t)$ and $\langle\cdot\rangle$ meaning the Heisenberg representation and ensemble average with respect to $\mathscr{H}_i$, respectively. The expression for $\mathbf{f}_\alpha$ is then given by

$$\mathbf{f}_\alpha = \sum_{\mathbf{k}} \hbar\mathbf{k}W_\alpha(\mathbf{k}). \tag{A13.8.8}$$

In the case of a weak electric field, the force $\mathbf{f}_\alpha$ is proportional to $\tau e\mathbf{E}$. Let us define a quantity τ_α by

$$\mathbf{f}_\alpha = -\frac{\tau}{\tau_\alpha}z_\alpha e\mathbf{E}. \tag{A13.8.9}$$

From (A13.8.6), (A13.8.8), and (A13.8.9), it follows that

$$\frac{1}{\tau_\alpha} = \frac{\pi\beta}{mz_\alpha V^2}\sum_{\mathbf{k}} k_x^2 \sum_{\substack{\mathbf{p}\alpha\\ \mathbf{p}'}} \int d\omega\delta_{\mathbf{p}-\mathbf{p}',\hbar\mathbf{k}}\delta\left(\frac{[\varepsilon_p - \varepsilon_{p'}]}{\hbar} - \omega\right)f(\mathbf{p})(1 - f(\mathbf{p}'))$$
$$\times \sum_{\beta=1,2}\langle \mathbf{p}'|U_{e\alpha}(\mathbf{k},\omega)|\mathbf{p}\rangle\langle \mathbf{p}'|U_{e\beta}(\mathbf{k},\omega)|\mathbf{p}\rangle^* \sqrt{\frac{c_\beta}{c_\alpha}}S_{\beta\alpha}(\mathbf{k},\omega). \tag{A13.8.10}$$

In the case when the matrix element of the interaction simplifies to

$$\langle \mathbf{p}'|U_{e\alpha}(\mathbf{k},\omega)|\mathbf{p}\rangle \equiv V_{e\alpha}(k), \tag{A13.8.11}$$

one finds

$$\frac{1}{\tau_\alpha} = \frac{m}{12\pi^3\hbar^3 z_\alpha}\int_0^{2k_f} dk\,k^3 \sum_{\beta=1,2} V_{e\beta}^*(k)V_{e\alpha}(k)\sqrt{\frac{c_\beta}{c_\alpha}}\int_{-\infty}^{\infty}\frac{S_{\beta\alpha}(\mathbf{k},\omega)\beta\hbar\omega}{\exp(\beta\hbar\omega) - 1}d\omega. \tag{A13.8.12}$$

In the limit of the one-component ion, this expression reduces to that for the relaxation time given in the work of Baym (1964).

The total force acting on the α-type ion is the sum of the direct term $z_\alpha eE$ and the "electron drag" term f_α. The resultant force is conveniently expressed as $z_\alpha^* e\mathbf{E}$, with z_α^*, as usual, denoting the effective valency:

$$z_\alpha^* = z_\alpha\left(1 - \frac{\tau}{\tau_\alpha}\right). \tag{A13.8.13}$$

Now the force-balance equation in the electron component is

$$-Ne\mathbf{E} - \sum_\alpha N_\alpha \mathbf{f}_\alpha = 0 \qquad \left(N \equiv \sum_\alpha z_\alpha N_\alpha\right), \tag{A13.8.14}$$

and by making use of (A13.8.9) one finds

$$\tau = \frac{\bar{z}}{\sum_\beta c_\beta \dfrac{z_\beta}{\tau_\beta}}, \qquad \left(\bar{z} \equiv \sum_\alpha c_\alpha z_\alpha\right). \tag{A13.8.15}$$

In fact, the unknown quantity τ could be taken to be defined by this equation. From (A13.8.13) and (A13.8.15), one obtains

$$\begin{aligned} z_\alpha^* &= z_\alpha\left(\frac{1 - \dfrac{\bar{z}}{\tau_\alpha}}{\left(\sum_\beta \dfrac{z_\beta}{\tau_\beta} c_\beta\right)}\right) \\ &= \frac{z_\alpha - \bar{z}\rho_\alpha}{\sum_\beta c_\beta \rho_\beta}, \end{aligned} \tag{A13.8.16}$$

where a *partial electrical resistivity* ρ_α is defined by

$$\rho_\alpha \equiv \frac{m}{ne^2}\frac{z_\alpha}{\tau_\alpha}. \tag{A13.8.17}$$

It should be noted that

$$\sum_\alpha c_\alpha z_\alpha^* = 0, \tag{A13.8.18}$$

which is another expression for the force balance equation in the electron component. From (A13.8.16) and (A13.8.18), one obtains (A13.8.29) given later.

The responses of the ion components in the system defined by (A13.8.2) may be approximately treated in such a manner that this system has applied to it a "fictitious" external electric field $z_\alpha^* e\mathbf{E}$. Let us then consider a system defined by the following Hamiltonian:

$$H_i^{\text{eff}} = H_i + H' \tag{A13.8.19}$$

where $H_i \equiv [\mathscr{H}_i]_{\mathbf{E}=0}$ and

$$H' = -\sum_{\beta=1,2} N_\beta z_\beta^* eEx_\beta, \tag{A13.8.20}$$

E and x_β being, respectively, the x component of $\mathbf{E}$ and $\sum_{j=1}^{N_\beta} \mathbf{R}_j^{(\beta)}/N_\beta$. There exist the following force-balance equations:

$$-\sum_\beta \sum_{j=1}^{N_\beta} \nabla_{\mathbf{R}_i^{(\beta)}} H_i = 0 \qquad \text{(internal force balance)}$$

$$-\sum_\beta \sum_{j=1}^{N_\beta} \nabla_{\mathbf{R}_j^{(\beta)}} H^1 = (N_1 + N_2) \sum_\beta c_\beta z_\beta^* e\mathbf{E}$$

$$= 0 \qquad \text{(external force balance equation).} \tag{A13.8.21}$$

One now introduces a mean ion-velocity v_α as

$$v_\alpha = \operatorname{Tr} \widehat{w_i} \dot{x}_\alpha$$

where $\widehat{w_i}$ is the statistical density matrix corresponding to (A.13.8.19) and $\dot{x}_\alpha = dx_\alpha/dt$. From the preceding force balance equations and the theorem of Ehrenfest, one obtains (see Ginoza and March, 1985) the total mass-flow conservation law (equal to the total momentum conservation law) as

$$\sum_\beta c_\beta M_\beta v_\beta = 0. \tag{A.13.8.22}$$

Now one may assume that H' is sufficiently weak and is introduced adiabatically in the remote past so that one can obtain by linear response theory the result

$$v_\alpha = \sum_{\beta=1,2} \frac{z_\beta^* eE}{n_\alpha} \lim_{\omega \to 0} \lim_{k_x \to 0} \frac{i\omega}{k_x^2} \tilde{\chi}_{\alpha\beta}(|k_x|, \omega), \tag{A.13.8.23}$$

where

$$\tilde{\chi}_{\alpha\beta}(k, \omega) = \int_0^\infty dt \exp(i\omega t) \frac{1}{i\hbar V} \frac{1}{\operatorname{Tr} e^{-\beta H_i}} \operatorname{Tr} e^{-\beta H_i} [\rho_\alpha(\mathbf{k}, t), \rho_\beta^+(\mathbf{k})]. \tag{A13.8.24}$$

From (A13.8.22), and (A13.8.23)

$$\operatorname{Im} \tilde{\chi}_{\alpha\beta}(k, \omega) = -(\pi/\hbar)[1 - \exp(-\beta\hbar\omega)](n_\alpha n_\beta)^{1/2} [S_{\alpha\beta}(k, \omega)]_{E=0}, \tag{A13.8.25}$$

one finds

$$v_1 - v_2 = Bc_1 z_1^* eE \tag{A13.8.26}$$

where

$$B = \frac{\pi\beta}{(c_1c_2)^2} \lim_{\omega \to 0} \lim_{k_x \to 0} \frac{\omega^2}{k_x^2} S_{CC}(|k_x|, \omega) \tag{A13.8.27}$$

while

$$S_{CC}(k,\omega) = c_1c_2\{c_2S_{11}(\mathbf{k},\omega) + c_1S_{22}(\mathbf{k},\omega) - 2(c_1c_2)^{1/2}S_{12}(\mathbf{k},\omega)\}_{E=0}. \tag{A.13.8.28}$$

Having set out the formal theory at the level of Born approximation, let us summarize the results to which one is led (see also Ginoza and March (1985)):

1. The effective valency, z_α^*, can be calculated in terms of the (concentration-dependent) resistivities of isotopes 1 and 2, denoted by ρ_1 and ρ_2, respectively, from the following equation:

$$c_1 z_1^* = -c_2 z_2^* = \frac{c_1c_2z_1z_2}{c_1\rho_1 + c_2\rho_2}\left(\frac{\rho_2}{z_2} - \frac{\rho_1}{z_1}\right). \tag{A13.8.29}$$

2. The difference between the ion mean-velocities is given by (A13.8.26).
3. Ion mean-velocities must satisfy the total ion-mass flow conservation

$$c_1M_1v_1 + c_2M_2v_2 = 0 \quad \text{or} \quad v_\alpha = X_{\bar{\alpha}}(v_\alpha - v_{\bar{\alpha}}) \tag{A13.8.30}$$

where

$$X_\alpha = \frac{M_\alpha c_\alpha}{M_1c_1 + M_2c_2}. \tag{A13.8.31}$$

From (1)–(3), one obtains:

$$\text{if} \quad \frac{\rho_1}{z_1} > \frac{\rho_2}{z_2}, \quad c_1z_1^* < 0 \qquad \text{and then} \qquad v_2 > 0, \quad v_1 < 0 \tag{A.13.8.32}$$

where the electric field $\mathbf{E}$ is in the positive direction of the x axis.

Typically, a current of 10^4 A cm^{-2} is employed for 2000 h to establish an isotopic concentration gradient in pure liquid Hg, and the difference between ionic drift velocities required to separate the isotopes on the scale of 1 to 10 cm (Faber, 1972) is about 10^{-8} to 10^{-9} cm s^{-1}. Therefore, in order that this theory is acceptable quantitatively, it must satisfy

1. If $M_1 < M_2$, $\rho_1 > \rho_2$, and
2. $|v_1 - v_2| \sim 10^{-8}$ to 10^{-9} cm s^{-1}.

If the origin of the Haeffner effect is to be understood by the electron-ion interaction (A13.8.11) treated as weak, it must arise from quantum effects (see also Omini (1986)), since $V_{e1}(k) = V_{e2}(k)$ for isotopes, and the classical ionic structure factors then cancel exactly in the difference $\rho_1 - \rho_2$. Other possibilities as to the origin of the Haeffner effect may be summarized as follows:

1. It has been implicitly assumed that the electron-ion interaction is independent of ionic mass. Otherwise, this mass dependence might be the origin of the effect. This possibility is discussed in terms of some models by Ginoza and March (1985).
2. The dynamical electron-ion interaction might be the origin of the effect.

The matter remains open at the time of writing.

A13.8.4. Response of liquid metal to relative forces, with application to electromigration

In this part of the appendix, following work by Sorbello (1978, 1988), by Jones (1980) and others, a discussion will be presented concerning the motion of the particles of a liquid under the influence of driving forces depending on the relative positions of pairs of particles. In this situation (see Jones and Barker, 1987) it can be shown that generalized reciprocal relations exist that relate the response to velocity distributions around a diffusing particle. Jones and Barker (1987) apply such a theory to the electromigration of a K impurity in Na, where, hitherto, the configuration-dependence of the driving forces has been ignored. By simple approximations, these workers can nevertheless include in the calculation the recognition that solvent ions in the vicinity of the impurity will suffer different driving forces from those in the bulk liquid. In turn, this will directly influence the impurity migration. The calculations of Jones and Barker (1987) suggest that this is of considerable importance. These workers also indicate how results relating diffusion coefficients to current-current correlation functions, of a type obtained by a velocity-field approach (see Appendix 8.5), may be obtained from the same formalism.

Electromigration of K *in* Na

To illustrate the above approach, let us note again that electromigration experiments are analyzed under the assumption that the driving force $\bar{\mathbf{F}}$ is written as $eZ^*\mathbf{E}$, where Z^* is an effective valence for the impurity. The drift velocity is then $\mu\bar{\mathbf{F}}$, where the mobility μ may be obtained in terms of the diffusion coefficient using the Nernst-Einstein relation. Jones and Barker (1987) then write down an expression for Z^*, taking account of the configuration dependence of the actual driving forces. They write

$$Z^* = Z_d^* + Z_{\text{ind}}^* \tag{A13.8.33}$$

where Z_d^* is the usual expression for the effective valence, so that $Z_d^*\mathbf{E}$ is the electrostatic force on the ion, namely, $eZ\mathbf{E}$, with Z the actual valence, plus the force due to electron scattering averaged over configuration. This is then corrected by adding the contribution $Z_{\text{ind}}^*\mathbf{E}$. To obtain this, Jones and Barker invoke the Kirkwood decoupling approximation at a stage in their calculation. Furthermore, they use a model in which the three independent pair correlation functions are approximated by a single function. Then if $\mathbf{E}$ is the α direction (the direction of velocity $\mathbf{u}$), Jones and Barker write*

$$Z_{\text{ind}}^* = n\int d\mathbf{r}\, g(\mathbf{r}) V_\beta(\hat{\mathbf{u}}, \mathbf{r})\left[f_{\alpha\beta}^{0i}(\mathbf{r}) + n\int d\mathbf{r}'\, f_{\alpha\beta}^{ij}(\mathbf{r}) g(\mathbf{r}')[g(|\mathbf{r}-\mathbf{r}'|) - 1\right]^*. \tag{A13.8.34}$$

The results of Jones and Barker show that while the details of the calculation are model-sensitive, the important point is that Z_{ind}^* in (A13.8.33) is comparable with Z_d^*. Furthermore, the inclusion of configuration dependence through Z_{ind}^* is important and can even influence the direction of electromigration, the impurity K moving in the same direction as the conduction electrons according to experiment in Na-K.

Clearly, a lot of uncertainties remain in this area, and important progress in the theory can be expected here.

* Here

$$f_{\alpha\beta}^{ij} = \frac{e\tau}{m}\sum_{|\mathbf{k}|<2k_f}\frac{k_\alpha k_\beta}{k}v_i^*(\mathbf{k})v_j(\mathbf{k})\exp(i\mathbf{k}\cdot\mathbf{r}_{ij}),$$

where τ is the relaxation time, $v_i(k)$ is the scattering form factor for ion i and r_{ij} is the position of i relative to j.

Appendix 13.9
Theory of disorder localization of noninteracting electrons

Although this volume is basically about metallic conductors, it is relevant to a discussion of electron states in disordered materials to give here some general background, plus relevant references, to the theory of localization of noninteracting electrons in a random potential.

The pioneering paper in this area was that of Anderson (1958) on the absence of diffusion in certain random lattices. But there was subsequently a lot of effort on this problem of noninteracting electrons in a static disordered potential before there was agreement on, at least, the answers to some important questions in this general area.

One might, at first sight, think it fruitful to compare the theory of electrons in disordered systems with the Bloch wave theory of the behaviour of electrons in a regular lattice. As Thouless (1979) emphasizes in his survey article, the theory of electrons in disordered systems is much more closely analogous to and owes much more to the theory of critical phenomena (see Chapter 9). In a sense then, developments in electron states in disordered systems awaited the synthesis of the theory of critical behaviour.

Following Thouless (1979), it is useful to group approaches into a number of areas, the first category being perturbative methods.

A13.9.1. Perturbative methods

Anderson's original paper (1958) was based on the application of perturbation theory to a system that was strongly localized by a lot of disorder. In some ways, it resembles the Ursell-Mayer approach to phase transitions by examining the convergence of a high temperature perturbation series. Anderson demonstrated convincingly that states are localized by strong disorder and that such states are quite distinct from the extended states familiar from the theory of metal crystals, so that something like a phase transition must occur. The theory, however, predicted a much narrower range of stability for localized states than numerical simulations showed.

The opposite approach, from the limit of weak disorder, proved in some respects more difficult. Edwards and other later workers developed functional integral techniques based on treating the disorder as a perturbation of free electrons. This certainly led to the anticipation of some of the techniques used in subsequent field theory approaches.

Later various workers calculated conductivity in the weak scattering limit of perturbation theory. This involved, among other things, a reinterpretation of the older results of Langer and Neal (1966) and was one of the vital ingredients of the scaling theory of localization. Among other results, it predicts that there is no metallic conduction in two dimensions in the absence of magnetic effects.

A13.9.2. Scaling and conductance

Abrahams et al. (1979) used the idea that consideration of blocks of material leads to a rescaled Anderson model, in which the coupling between sites is replaced by a coupling between blocks and the only parameter of importance is the ratio of the coupling between these blocks to the spacing of energy levels within the blocks, the details being given by Thouless (1979) in the volume on ill-condensed matter. Further, the square of this ratio is proportional to the conductance, as also reviewed by Thouless (1979). Since localization is understood in the strong disorder limit and the conductance can be calculated in the weak disorder case, a sensible interpolation between the two proves possible.

In one dimension, scaling theory gives the well-known result that all states are localized by disorder, as first established by Borland (1960). In two dimensions, the same scaling theory says that all states are also localized. In three dimensions, it predicted no minimum metallic conductivity, since as the conductance on small-length scales approaches the critical value, the resistivity on large-length scales increases without limit.

Finite size-scaling calculations have also been carried out. The idea of a finite size-scaling theory is that a strip of wire of finite cross section is made long enough that one-dimensional localization dominates. The ratio of the one-dimensional localization length l_M to the diameter M is a function, at least in the critical region, of the ratio of some characteristic length l_∞ to M. The length l_∞ is the bulk (or two-dimensional) localization length for the wire (or strip) where there is bulk (or two-dimensional) localization, and it is related to the conductance σ by $l_\infty = e^2/h\sigma$ in the extended regime. It is therefore given by the length scale at which coupling between the blocks is comparable with the energy spacing. In two dimensions, Thouless notes that l_∞ seems to go exponentially to infinity as the disorder goes to zero, but in three dimensions it diverges at a finite value of l_M/M. If the relation between l_M/M and conductance is known in one dimension, then the behaviour of l_M/M as a function of disorder can be related to the behaviour of the conductance as a function of disorder. This work allows scaling theory to be extended to shapes that are not simply cubes or squares and allows anisotropy to be included.

A13.9.3. Spectral properties and localization

Although it is clear that the spectrum averaged over an infinite system or over an ensemble does not give adequate information about localization, there is an intimate connection between the local, nonaveraged properties of the spectrum, as given, for example, by the spectral representation of a diagonal element of the one-particle Green function and localization. If the spectrum is continuous in some energy range, states in that energy range are not normalized and are therefore extended, whereas a point spectrum corresponds to localized states, as discussed in detail by Kunz and Souillard (1979) for instance. If, as is generally believed, metallic conductivity and diffusion are associated with an absolutely continuous spectrum, what transport properties are associated with the singular continuous spectrum, where the density of states is zero almost everywhere but there are no delta functions? Thouless suggests that scaling properties then might imply that the AC conductivity $\sigma(\omega)$ may go to zero linearly with ω instead of going quadratically to zero, as with localized states, or to a constant, as with extended states.

To summarize, there is, by now, considerable understanding of localization of independent electrons and of the transition from the localized to the extended regime.

Appendix 14.1
Phonon-plasmon model

On the assumption that, in the long-wavelength limit, single-particle and multiparticle excitations are less important than collective modes, which are taken to be (see also Chapter 16, Section 1):

1. Plasmonlike, with dispersion relation $\omega = \omega_p(q)$, and
2. Phononlike, with $\omega = v_s(q)q$,

one can write down approximate expressions for the partial dissipation functions of the liquid metal (for ion-ion, ion-electron, and electron-electron terms as (see March, 1974):

$$\chi''_{ij}(q,\omega) = -\frac{\pi}{2}\{n_i n_j\}^{1/2}\{\alpha_{ij}(q)[\delta(\omega - \omega_{\mathrm{p}}(q)) - \delta(\omega + \omega_{\mathrm{p}}(q))]$$

$$+ \beta_{ij}(q)[\delta(\omega - qv_{\mathrm{s}}(q)) - \delta(\omega + qv_{\mathrm{s}}(q))]\}. \qquad \text{(A14.1.1)}$$

It is to be noted at this point that the partial structure factors $S_{ij}(q)$, related to the usual pair functions $g_{ij}(\mathbf{r}_1 - \mathbf{r}_2)$ by

$$S_{ij}(q) = \delta_{ij} + (n_i n_j)^{1/2}\int \mathrm{d}(\mathbf{r}_1 - \mathbf{r}_2)[g_{ij}(\mathbf{r}_1 - \mathbf{r}_2) - 1]\exp(i\mathbf{q}\cdot\mathbf{r}_1 - \mathbf{r}_2) \qquad \text{(A14.1.2)}$$

are to be obtained from χ'' using the relation

$$S_{ij}(q) = -\int \frac{\mathrm{d}\omega}{2\pi}\chi''_{ij}(q,\omega)\coth\left(\frac{\hbar\omega}{2k_B T}\right)(n_i n_j)^{-1/2}. \qquad \text{(A14.1.3)}$$

To satisfy the first moment theorem on χ''_{ij} in the long-wavelength limit, one must choose $\alpha_{ij}(q) \propto q^2$ and $\beta_{ij}(q) \propto q$ in (A14.1.1). It is then easy to see that the static structure factor $S_{ii}(q)$ at $q = 0$ is entirely dominated by the phonon mode, and the compressibility sum rule gives the coefficients β_{ij} in $\beta_{ij}(q) \to \beta_{ij}q$ for small q as

$$\beta_{\mathrm{ii}} = Z^{-1}\beta_{\mathrm{ee}} = Z^{-1/2}\beta_{\mathrm{ie}} = \hbar n_{\mathrm{i}} K_{\mathrm{T}} v_{\mathrm{s}} \qquad \text{(A14.1.4)}$$

with v_{s} the velocity of sound.

In Chapter 8, since it was seen that the ratio of the specific heats was $\gamma \sim 1.2$–1.3 in many liquid metals at their melting points, the fact that the sound velocity entering above is identified with the isothermal sound velocity is not very important in practice.

Next, let us consider the coefficients s, s_i, and s_e defined by $\alpha_{ij}(q) \to sq^2$, $\alpha_{ie}(q) \to s_i q^2$, and $\alpha_{ee}(q) \to s_e q^2$ as $q \to 0$. These coefficients involve both the plasmon mode and the phonons (Tosi and March, 1973b; March, 1974). More specifically, one has

$$Zs_e^{\text{plasmon}} = \frac{\hbar}{2m\omega_p} \frac{n_i M}{\mathcal{M}} \tag{A14.1.5}$$

$$s_i^{\text{plasmon}} = \frac{\hbar}{2M\omega_p} \frac{n_e m}{\mathcal{M}} \tag{A14.1.6}$$

and

$$s^{\text{plasmon}} = -\frac{\hbar}{2m\omega_p} \frac{n_i m}{\mathcal{M}}. \tag{A14.1.7}$$

Here m and M are electron and ion masses, respectively, whereas $\mathcal{M}$ is defined by

$$\mathcal{M} = n_i M + n_e m. \tag{A14.1.8}$$

Also, ω_p is the reduced plasma frequency given by

$$\omega_p^2 = 4\pi n_i Z^2 e^2 [M^{-1} + (Zm)^{-1}]. \tag{A14.1.9}$$

In practice, because of the large difference between electron and ion masses, ω_p can be equated to the electronic plasma frequency in the homogeneous electron gas and $Zs_e^{\text{plasmon}} \sim h/2m\omega_p$, whereas s_i^{plasmon} and s^{plasmon} are extremely small.

Appendix 14.2
Response functions for mass densities

An alternative description of the two-component system to that derived in the main text is afforded by defining correlation functions for mass and charge [compare Bhatia, Thornton and March (1974)]. Thus one writes

$$M_{\mathbf{k}}(t) = m\rho_{e\mathbf{k}}(t) + M\rho_{i\mathbf{k}}(t) \tag{A14.2.1}$$

and

$$X_{\mathbf{k}}(t) = \tfrac{1}{2}[M\rho_{i\mathbf{k}}(t) - m\rho_{e\mathbf{k}}(t)]. \tag{A14.2.2}$$

Now one defines correlation functions through

$$S_{\mathrm{MM}}(q,\omega) = \left(\frac{1}{\mathscr{M}}\right)\int_{-\infty}^{\infty} dt \exp(-i\omega t)\langle M_{\mathbf{k}}^{+}(0); M_{\mathbf{k}}(t)\rangle \tag{A14.2.3}$$

and

$$S_{\mathrm{XX}}(q,\omega) = \left(\frac{1}{\mathscr{M}}\right)\int_{-\infty}^{\infty} dt \exp(-i\omega t)\langle X_{\mathbf{k}}^{+}(0); X_{\mathbf{k}}(t)\rangle \tag{A14.2.4}$$

etc. In (A14.2.3) and (A14.2.4), $\mathscr{M}$ is simply the total mass $mn_{e} + Mn_{i}$. Similarly, one defines corresponding response functions

$$\chi_{\mathrm{MM}}(q,\omega) = \int_{0}^{\infty} dt \exp(-i\omega t)\left(\frac{1}{i\hbar}\right)\langle[M_{\mathbf{q}}^{+}(0), M_{\mathbf{q}}(t)]\rangle$$

$$= \beta\int_{0}^{\infty} dt \exp(-i\omega t)\langle \dot{M}_{\mathbf{q}}^{+}(0); M_{\mathbf{q}}(t)\rangle, \tag{A14.2.5}$$

etc. One can then calculate moments using the following relations:

$$i\hbar\dot{\rho}_{e\mathbf{q}}^{+} = [\rho_{e\mathbf{q}}^{+}, H] = -\left(\frac{\hbar}{m}\right)\sum_{i}\exp(i\mathbf{q}\cdot\mathbf{r}_{i})\left(\mathbf{q}\cdot\mathbf{p}_{i} + \frac{1}{2}\hbar q^{2}\right) \tag{A14.2.6}$$

and

$$i\hbar\dot{\rho}_{i\mathbf{q}}^{+} = [\rho_{i\mathbf{q}}^{+}, H] = -\left(\frac{\hbar}{M}\right)\sum_{j}\exp(i\mathbf{q}\cdot\mathbf{R}_{j})\left(\mathbf{q}\cdot\mathbf{P}_{j} + \frac{1}{2}\hbar q^{2}\right). \tag{A14.2.7}$$

One then finds

$$\left(\frac{1}{\pi}\right)\int_{-\infty}^{\infty} \mathrm{d}\omega\, \omega\chi''_{\mathrm{MM}}(q,\omega) = \left(\frac{1}{i\hbar}\right)\langle[M_{\mathbf{q}}(0), \dot{M}_{\mathbf{q}}^{+}(0)]\rangle$$

$$= q^2\mathcal{M} \quad \text{(A14.2.8)}$$

$$\left(\frac{1}{\pi}\right)\int_{-\infty}^{\infty} \mathrm{d}\omega\, \omega\chi''_{\mathrm{MX}}(q,\omega) = \left(\frac{1}{i\hbar}\right)\langle[X_{\mathbf{q}}(0), \dot{M}_{\mathbf{q}}^{+}(0)]\rangle$$

$$= 0 \quad \text{(A14.2.9)}$$

and

$$\left(\frac{1}{\pi}\right)\int_{-\infty}^{\infty} \mathrm{d}\omega\, \omega\chi''_{XX}(q,\omega) = \left(\frac{1}{i\hbar}\right)\langle[X_{\mathbf{q}}(0), \dot{X}_{\mathbf{q}}^{+}(0)]\rangle = \frac{q^2\mathcal{M}}{4}. \quad \text{(A14.2.10)}$$

To pass to the classical limit, one uses the relations

$$\tilde{S}_{\rho\rho}(q,\omega) = \left(\frac{1}{\mathcal{M}}\right)\int_{-\infty}^{\infty} \mathrm{d}t \exp(-i\omega t)\left\langle\frac{1}{2}\{\rho_{\mathbf{q}}^{+}(0), \rho_{\mathbf{q}}(t)\}\right\rangle$$

$$\rightarrow \left(\frac{2}{\beta\omega}\right)\left(\frac{\chi''_{\rho\rho}(q,\omega)}{\mathcal{M}}\right), \quad \text{(A14.2.11)}$$

etc., when one finds

$$\left(\frac{1}{2\pi}\right)\int_{-\infty}^{\infty} \tilde{S}_{\mathrm{MM}}(q,\omega)\omega^2\, \mathrm{d}\omega = \frac{q^2}{\beta} \quad \text{(A14.2.12)}$$

and

$$\left(\frac{1}{2\pi}\right)\int_{-\infty}^{\infty} \tilde{S}_{\mathrm{XX}}(q,\omega)\omega^2\, \mathrm{d}\omega = \frac{1}{4}\left(\frac{q^2}{\beta}\right). \quad \text{(A14.2.13)}$$

Equations (A14.2.12) and (A14.2.13) are precisely the results of Bhatia et al. (1974) for a classical binary alloy.

Appendix 14.3
Quantum hydrodynamic limit of two-component theory

Because of the interest in the anomalous isotopic mass scaling of the shear viscosities of the pure isotopes Li^6 and Li^7 (Ban et al., 1962; Omini, 1986), let us approach the theory of atomic transport coefficients already introduced in Chapter 7 from the basic conservation equations of the two-component theory that were derived in the text.

First, one defines the following quantities characteristic of the liquid metal as a whole:

$$\mathcal{M}(\mathbf{R},t) = mn_e(\mathbf{R},t) + Mn_i(\mathbf{R},t) \tag{A14.3.1}$$

$$\mathbf{J}_M(\mathbf{R},t) = m\mathbf{j}(\mathbf{R},t) + M\mathbf{J}(\mathbf{R},t) \tag{A14.3.2}$$

and

$$Q(\mathbf{R},t) = Zn_i(\mathbf{R},t) - n_e(\mathbf{R},t). \tag{A14.3.3}$$

Adding the continuity equations for electrons and ions, one evidently finds

$$\frac{\partial \mathcal{M}(\mathbf{R},t)}{\partial t} + \nabla_{\mathbf{R}} \cdot \mathbf{J}_M(\mathbf{R},t) = 0. \tag{A14.3.4}$$

From the equations for $\partial \mathbf{j}/\partial t$ and $\partial \mathbf{J}/\partial t$, one obtains

$$\begin{aligned}
&\frac{\partial \mathbf{J}_M(\mathbf{R},t)}{\partial t} + \nabla_{\mathbf{R}} \cdot [\pi(\mathbf{R},t) + \Pi(\mathbf{R},t)] \\
&\quad = -\int d\mathbf{x}\, \nabla_{\mathbf{R}} v(\mathbf{R}-\mathbf{x}) \langle Q(\mathbf{R},t) Q(\mathbf{x},t)\rangle_c \\
&\qquad - \int d\mathbf{x}\, \nabla_{\mathbf{R}} v^s(\mathbf{R}-\mathbf{x}) \langle \rho_i(\mathbf{R},t)\rho_i(\mathbf{x},t)\rangle \\
&\qquad - \int d\mathbf{x}\, \nabla_{\mathbf{R}} v^s_{ie}(\mathbf{R}-\mathbf{x}) \langle \rho_e(\mathbf{R},t)\rho_i(\mathbf{x},t) + \rho_i(\mathbf{R},t)\rho_e(\mathbf{x},t)\rangle.
\end{aligned} \tag{A14.3.5}$$

It is (A14.3.5) that will lead to the analogue of the Navier-Stokes equation for the liquid metal as a whole.

In particular, the correlation function $\langle \rho_e(\mathbf{R},t)\rho_i(\mathbf{x},t) + \rho_i(\mathbf{R},t)\rho_e(\mathbf{x},t)\rangle$ has precisely the symmetry required to yield a "stress tensor" in the hydrodynamic limit.

It is to be noted here that the right-hand side of (A14.3.5), giving the internal force driving the current, has a rather direct physical interpretation. For structureless ions, the force is due only to correlations in the charge fluctuations, that is, the first term on the right-hand side. For ions with core electrons, (A14.3.5) shows that there are additional forces due to the short-range ion-ion and short-range electron-ion interactions.

Having interpreted some aspects of (A14.3.5), it is now useful to cast it into linearized form and to introduce nonequilibrium pair-distribution functions (March and Tosi, 1973). Then one can show that the equation takes the form

$$\begin{aligned}\frac{\mathrm{D}\mathbf{J}_{\mathrm{M}}(\mathbf{R},t)}{\mathrm{D}t} &= -\nabla_{\mathbf{R}}\cdot[\pi^0 + \Pi^0] - \int \mathrm{d}\mathbf{x}''\,\nabla_{\mathbf{R}}v(\mathbf{R}-\mathbf{x}'')\langle\rho_e(\mathbf{R},t)\rho_e(\mathbf{x}'',t)\rangle \\ &\quad - \int \mathrm{d}\mathbf{x}''\,\nabla_{\mathbf{R}}v(\mathbf{R}-\mathbf{x}'')\langle\rho_i(\mathbf{R},t)\rho_i(\mathbf{x}'',t)\rangle \\ &\quad - \int \mathrm{d}\mathbf{x}''\,\nabla_{\mathbf{R}}v_{ie}(\mathbf{R}-\mathbf{x}'')\langle\rho_i(\mathbf{R},t)\rho_e(\mathbf{x}'',t) + \rho_e(\mathbf{R},t)\rho_i(\mathbf{x}'',t)\rangle \\ &= -\nabla_{\mathbf{R}}\cdot[\pi^0 + \Pi^0] - \int \mathrm{d}\mathbf{x}''\,\nabla_{\mathbf{R}}v(\mathbf{R}-\mathbf{x}'')g_{ee}(\mathbf{R},\mathbf{x}'',t) \\ &\quad - \int \mathrm{d}\mathbf{x}''\,\nabla_{\mathbf{R}}v(\mathbf{R}-\mathbf{x}'')g_{ii}(\mathbf{R},\mathbf{x}'',t) \\ &\quad - 2\int \mathrm{d}\mathbf{x}''\,\nabla_{\mathbf{R}}v_{ie}(\mathbf{R}-\mathbf{x}'')g_{ie}(\mathbf{R},\mathbf{x}'',t) \end{aligned} \tag{A14.3.6}$$

where g_{ee}, g_{ii}, and g_{ie} are the nonequilibrium pair functions of electron-electron, ion-ion and ion-electron correlations. No divergence difficulties arise in this equation, as can be seen by invoking (A14.3.5).

The final step in converting (A14.3.6) into a hydrodynamic equation is to expand all the g's around their equilibrium values and then to carry out gradient expansions. Thus, for example, one writes the deviation of g_{ii} from equilibrium as δg_{ii} and then splits this into two parts. One of these, δg_{ii}^{ρ}, is dependent on density and the second δg_{ii}^{v} on velocity v. One can then write for the change in $g_{ii}(r) \equiv g_{ii}(|\mathbf{x}-\mathbf{y}|)$ from equilibrium,

$$\delta g_{ii} = \delta g_{ii}^{\rho} + \delta g_{ii}^{v}, \tag{A14.3.7}$$

where

$$\delta g_{\text{ii}}^{\rho} = \tfrac{1}{2}[\delta n_{\text{i}}(\mathbf{x}, t) + \delta n_{\text{i}}(\mathbf{y}, t)]\Gamma_1^i(r) + \tfrac{1}{2}[\delta n_{\text{e}}(\mathbf{x}, t) + \delta n_{\text{e}}(\mathbf{y}, t)]\Gamma_1^e(r) + \cdots \tag{A14.3.8}$$

with a similar equation for $\delta g_{\text{ii}}^{\text{v}}$, following Fröhlich (1967).

With corresponding expansions of g_{ie} and g_{ee}, the analogue of the Navier-Stokes equation can be written, and the quantity $(\frac{4}{3}\eta + \zeta)$ can be related to combinations of quantities such as $\int \mathbf{dr}(\partial V/\partial r)\Gamma_1^i(\mathbf{r})$, as is clear from (A14.3.6) and (A14.3.8). There are a formidable number of functional derivatives (the Γ's, etc.) of the pair functions entering this theory. Fortunately, the fact that part of δg^{ρ} may be combined with the kinetic terms to yield a pressure term in analogy with that in the classical Navier-Stokes equation can be demonstrated (March and Tosi, 1973, their Appendix 4).

Appendix 14.4
Evaluation of transport coefficients

Following Tosi, Parrinello, and March (1974), the full expression for (14.11) is

$$
\begin{aligned}
\delta\langle\varrho_j(\mathbf{R},t)\varrho_l(\mathbf{x},t)\rangle_c = {} & [v_l(\mathbf{x},t) - v_j(\mathbf{R},t)]\sigma_{lj}(\mathbf{r}) + [(\hat{r}\cdot\nabla)v_l(\mathbf{x},t) \\
& + (\hat{r}\cdot\nabla)v_j(\mathbf{R},t)]\sigma_{lj}^{(1)}(\mathbf{r}) + [(\hat{q}\cdot\nabla)v_l(\mathbf{x},t) \\
& + (\hat{q}\cdot\nabla)v_j(\mathbf{R},t)]\sigma_{lj}^{(2)}(\mathbf{r}) + [(\hat{r}\cdot\nabla)^2 v_l(\mathbf{x},t) \\
& - (\hat{r}\cdot\nabla)^2 v_j(\mathbf{R},t)]\sigma_{lj}^{(3)}(\mathbf{r}) + [(\hat{q}\cdot\nabla)(\hat{r}\cdot\nabla)v_l(\mathbf{x},t) \\
& - (\hat{q}\cdot\nabla)(\hat{r}\cdot\nabla)v_j(\mathbf{R},t)]\sigma_{lj}^{(4)}(\mathbf{r}) + [(\hat{q}\cdot\nabla)^2 v_l(\mathbf{x},t) \\
& - (\hat{q}\cdot\nabla)^2 v_j(\mathbf{R},t)]\sigma_{lj}^{(5)}(\mathbf{r}) \\
= {} & [v_l(\mathbf{R},t) - v_j(\mathbf{R},t)]\sigma_{lj}(\mathbf{r}) + (\hat{r}\cdot\nabla)v_l(\mathbf{R},t)[r\sigma_{lj}(\mathbf{r}) \\
& + \sigma_{lj}^{(1)}(\mathbf{r})] + (\hat{r}\cdot\nabla)v_j(\mathbf{R},t)\sigma_{lj}^{(1)}(\mathbf{r}) \\
& + (\hat{r}\cdot\nabla)^2 v_l(\mathbf{R},t)[\tfrac{1}{2}r^2\sigma_{lj}(\mathbf{r}) + r\sigma_{lj}^{(1)}(\mathbf{r}) + \sigma_{lj}^{(3)}(\mathbf{r})] \\
& - (\hat{r}\cdot\nabla)^2 v_j(\mathbf{R},t)\sigma_{lj}^{(3)}(\mathbf{r}) + (\hat{q}\cdot\nabla)[v_l(\mathbf{R},t) \\
& + v_j(\mathbf{R},t)]\sigma_{lj}^{(2)}(\mathbf{r}) + (\hat{q}\cdot\nabla)(\hat{r}\cdot\nabla)v_l(\mathbf{R},t)(r\sigma_{lj}^{(2)}(\mathbf{r}) \\
& + \sigma_{lj}^{(4)}(\mathbf{r})] - (\hat{q}\cdot\nabla)(\hat{r}\cdot\nabla)v_j(\mathbf{R},t)\sigma_{lj}^{(4)}(\mathbf{r}) \\
& + (\hat{q}\cdot\nabla)^2[v_l(\mathbf{R},t) - v_j(\mathbf{R},t)]\sigma_{lj}^{(5)}(\mathbf{r}) + \cdots, \qquad \text{(A14.4.1)}
\end{aligned}
$$

where $v_l(\mathbf{R},t) = (1/n_l)J_l(\mathbf{R},t)$ and

$$\sigma_{lj}(\mathbf{r}) = -\sigma_{jl}(\mathbf{r}) = \frac{\partial\langle\varrho_j(\mathbf{R})\varrho_l(\mathbf{x})\rangle_c}{\partial v_l}, \qquad \text{(A14.4.2)}$$

with analogous definitions for the other σ's. The symmetry properties of these quantities, which may be written as $\sigma_{lj}^{(n)} = (-1)^n\sigma_{jl}^{(n)}$, have been used in (A14.4.1). Equation (14.13) follows at once, with γ given by the expression (14.14) and

$$\eta_{jl} = \eta_{lj} = -\int d\mathbf{r}\{\tfrac{3}{5}[\tfrac{1}{2}r^2\sigma_{lj}(\mathbf{r}) + r\sigma_{lj}^{(1)}(\mathbf{r})](\hat{q}\cdot\nabla)v_{jl}(\mathbf{r}) + \tfrac{1}{3}r\sigma_{lj}^{(2)}(\mathbf{r})v'_{jl}(\mathbf{r})\}, \tag{A14.4.3}$$

$$\eta_1 = -\int d\mathbf{r}\{[\tfrac{3}{5}\sigma_{\mathrm{ei}}^{(3)}(\mathbf{r}) + \sigma_{\mathrm{ei}}^{(5)}(\mathbf{r})](\hat{q}\cdot\nabla)v_{\mathrm{ie}}(\mathbf{r}) + \tfrac{1}{3}\sigma_{\mathrm{ei}}^{(4)}(\mathbf{r})v'_{\mathrm{ie}}(\mathbf{r})\}. \tag{A14.4.4}$$

This completes the derivation of (14.13).

Appendix 14.5

Electron-ion structure factor in a nonequilibrium situation

The decoupling techniques used to evaluate functional derivatives such as that entering (14.10) have been developed in detail in the book by Kadanoff and Baym (1962). The electron-ion structure factor in a nonequilibrium situation, to lowest order in the electron-ion interaction, can be written

$$
\begin{aligned}
&\langle \varrho_{\mathrm{i}}(\mathbf{R},t)\varrho_{\mathrm{e}}(\mathbf{x},t)\rangle_c \\
&= i \iint \mathrm{d}\mathbf{y}\,\mathrm{d}\mathbf{y}'\, v_{\mathrm{ie}}(\mathbf{y}-\mathbf{y}') \int_{-\infty}^{t} \mathrm{d}\tau [\langle \varrho_{\mathrm{i}}(\mathbf{R},t)\varrho_{\mathrm{i}}(\mathbf{y},\tau)\rangle_c \langle \varrho_{\mathrm{e}}(\mathbf{x},t)\varrho_{\mathrm{e}}(\mathbf{y}',\tau)\rangle_c \\
&\quad - \langle \varrho_{\mathrm{i}}(\mathbf{y},\tau)\varrho_{\mathrm{i}}(\mathbf{R},t)\rangle_c \langle \varrho_{\mathrm{e}}(\mathbf{y}',\tau)\varrho_{\mathrm{e}}(\mathbf{x},t)\rangle_c],
\end{aligned}
\tag{A14.5.1}
$$

which in Fourier transform becomes

$$
\begin{aligned}
&\mathrm{F.T.}\{\langle \varrho_{\mathrm{i}}(\mathbf{R},t)\varrho_{\mathrm{e}}(\mathbf{x},t)\rangle_c\}_{\mathbf{k}} \\
&= -v_{\mathrm{ie}}(\mathbf{k}) \iint_{-\infty}^{\infty} \frac{\mathrm{d}\omega}{2\pi}\frac{\mathrm{d}\omega'}{2\pi} \frac{S'_{\mathrm{ii}}(\mathbf{k},\omega)S'_{\mathrm{ee}}(-\mathbf{k},\omega') - S'_{\mathrm{ii}}(-\mathbf{k},\omega)S'_{\mathrm{ee}}(\mathbf{k},\omega')}{\omega+\omega'+i\varepsilon}
\end{aligned}
\tag{A14.5.2}
$$

where $S'_{\mathrm{ii}}(\mathbf{k},\omega)$ and $S'_{\mathrm{ee}}(\mathbf{k},\omega)$ are the dynamic structure factors of ions and electrons in the nonequilibrium state. In the Born-Oppenheimer approximation, it may be assumed that the equilibration rates for the ionic system are much more rapid than the electron-ion scattering rates; namely, one can take $S'_{\mathrm{ii}}(\mathbf{k},\omega)$ in the preceding equation as the equilibrium structure factor for the ions and thus allow the functional derivative in (14.10) to operate only on $S'_{\mathrm{ee}}(\mathbf{k},\omega)$. The imaginary part of (A14.5.2), which alone is of interest in (14.10) for reasons of symmetry, can then be written

$$
\begin{aligned}
\mathrm{Im}[\mathrm{F.T.}\{\langle \varrho_{\mathrm{i}}(\mathbf{R},t)\varrho_{\mathrm{e}}(\mathbf{x},t)\rangle_c\}_{\mathbf{k}}] &= \tfrac{1}{2}v_{\mathrm{ie}}(\mathbf{k}) \int_{-\infty}^{\infty} \frac{\mathrm{d}\omega}{2\pi} [S_{\mathrm{ii}}(\mathbf{k},\omega)S'_{\mathrm{ee}}(-\mathbf{k},-\omega) \\
&\quad - S_{\mathrm{ii}}(-\mathbf{k},-\omega)S'_{\mathrm{ee}}(\mathbf{k},\omega)].
\end{aligned}
\tag{A14.5.3}
$$

This completes the derivation of (14.11).

Appendix 14.6
Relations between long-wavelength limit structure factors in binary metallic alloys

Suppose that ions of types A and B have valences z and Z, respectively. Six distinct correlation functions, S_{AA}, S_{BB}, S_{ee} and cross-correlation functions S_{Ae}, S_{Be}, and S_{AB}, are required to describe the mixture of A, B, and electrons e. Let us write down the condition of perfect screening (compare the nondegenerate case of Debye-Hückel theory with the present (totally degenerate) electron assembly) of an ion A, ion B, and an electron e in terms of the pair-correlation functions $g_{AA}(r)$, etc.

To do so (March, Tosi, and Bhatia 1973), one views the surrounding charge density first from an A ion, then a B ion, etc. For the A ion, the positive charges ze are distributed at distance r from the ion one has chosen as origin, with charge density $zen_A g_{AA}(r)$, ions with charge Ze with density $Zen_B g_{AB}(r)$, and electrons with charge density $-en_e g_{Ae}(r)$, n_A, n_B, and n_e being the number densities of the three species. The total charge density thus seen from A must evidently integrate to $-ze$ for perfect screening out of long-range electric fields required in a conducting medium. Thus one finds

$$-ze = zen_A \int (g_{AA}(r) - 1)\,\mathrm{d}\mathbf{r} + Zen_B \int (g_{AB}(r) - 1)\,\mathrm{d}\mathbf{r} - en_e \int (g_{Ae}(r) - 1)\,\mathrm{d}\mathbf{r} \tag{A14.6.1}$$

where one has subtracted unity from each of the g's for convergence at infinity. This can be done trivially, as shown, since

$$zen_A + Zen_B = en_e. \tag{A14.6.2}$$

Similarly, by sitting on a B ion, one finds

$$-Ze = Zen_B \int (g_{BB}(r) - 1)\,\mathrm{d}\mathbf{r} + zen_A \int (g_{AB}(r) - 1)\,\mathrm{d}\mathbf{r} - en_e \int (g_{Be}(r) - 1)\,\mathrm{d}\mathbf{r} \tag{A14.6.3}$$

and for an electron,

$$+e = -en_e \int (g_{ee}(r) - 1)\,d\mathbf{r} + zen_A \int (g_{Ae}(r) - 1)\,d\mathbf{r}$$
$$+ Zen_B \int (g_{Be}(r) - 1)\,d\mathbf{r}. \qquad \text{(A14.6.4)}$$

If one defines the structure factors S_{AA}, etc., in the long-wavelength limit by

$$S_{ij} = \delta_{ij} + (n_i n_j)^{1/2} \int (g_{ij}(r) - 1)\,d\mathbf{r} \qquad \text{(A14.6.5)}$$

then (A7.3.1), (A7.3.2), and (A7.3.4) may be rewritten as

$$S_{ee} = z^2\left(\frac{n_A}{n_e}\right)S_{AA} + 2zZ\frac{(n_A n_B)^{1/2}}{n_e}S_{AB} + Z^2\left(\frac{n_B}{n_e}\right)S_{BB} \qquad \text{(A14.6.6)}$$

$$S_{Ae} = \left(\frac{n_A}{n_e}\right)^{1/2}\left\{zS_{AA} + ZS_{AB}\left(\frac{n_B}{n_A}\right)^{1/2}\right\} \qquad \text{(A14.6.7)}$$

and

$$S_{Be} = \left(\frac{n_B}{n_e}\right)^{1/2}\left\{ZS_{BB} + z\left(\frac{n_A}{n_B}\right)^{1/2}S_{AB}\right\}. \qquad \text{(A14.6.8)}$$

One next uses (A14.6.6)–(A14.6.8) to express S_{ee}, S_{Ae}, and S_{Be} in terms of the ionic partial structure factors a_{AA}, a_{BB}, and a_{AB} (in the long-wavelength limit again, of course) used, for example, in the experimental study of McAlister and Turner (1972). With $i, j = A, B$:

$$a_{ij} = 1 + (n_A + n_B) \int (g_{ij}(r) - 1)\,d\mathbf{r}. \qquad \text{(A14.6.9)}$$

In order to write the resulting expressions compactly, it will be convenient to employ the notation metal $A = 1$, metal $B = 2$, $c_1 = n_1/(n_1 + n_2)$, $c_2 = 1 - c_1$, $z = z_1$, $Z = z_2$, and $\bar{z} = c_1 z_1 + c_2 z_2$. One then finds

$$S_{ee} = (\bar{z})^{-1}\{c_1^2 z_1^2 a_{11} + c_2^2 z_2^2 a_{22} + 2c_1 c_2 z_1 z_2 a_{12} + c_1 c_2 (z_1 - z_2)^2\}, \qquad \text{(A14.6.10)}$$

$$S_{1e} = (c_1 \bar{z})^{-1/2}\{z_1 c_1^2 a_{11} + z_2 c_1 c_2 a_{12} + c_1 c_2 (z_1 - z_2)\}, \qquad \text{(A14.6.11)}$$

and

$$S_{2e} = (c_2 \bar{z})^{-1/2}\{z_2 c_2^2 a_{22} + z_1 c_1 c_2 a_{12} - c_1 c_2 (z_1 - z_2)\}. \qquad \text{(A14.6.12)}$$

These results reduce for pure liquid metals to those given by Watabe and Hasegawa (1972) and Chihara (1972).

From (A14.6.10)–(A14.6.12), plus the data for a_{ij} given by McAlister and Turner (1972), March, Tosi, and Bhatia (1973) have plotted S_{ee}, etc., as functions of concentration for the liquid Na-K system ($z_1 = z_2 = 1$); the interested reader should refer to their Figure 1 and to a subsequent review by Tamaki (1987).

Appendix 16.1

Integral equations for correlations in liquid metals, especially hydrogen

When liquid metals are treated as a mixture of ions and electrons, as in Chapter 14, the density-functional method can yield formally exact expressions for the radial distribution functions concerning an ion in the forms of the density distributions $n_i(r|I)$ ($i = I$ and e) around a fixed ion in the liquid metal (see Chihara, 1984 and other references there):

$$g_{\rm II}(r) = \frac{n_{\rm I}(r|I)}{n_0^I} = \exp\{-\beta v_{\rm II}^{\rm eff}(r)\}, \qquad \text{(A16.1.1)}$$

$$g_{\rm eI}(r) = \frac{n_{\rm e}(r|I)}{n_0^e} = \frac{n_{\rm e}^0(r|v_{\rm eI}^{\rm eff})}{n_0^e}. \qquad \text{(A16.1.2)}$$

Here, $n_{\rm e}^0(r|U)$ means the density distribution of the noninteracting electron gas under the presence of an external potential $U(r)$, and it is calculated by solving the wave equation

$$\left\{-\frac{\hbar^2}{2m}\nabla^2 + U(r)\right\}\phi_i(r) = \varepsilon_i\phi_i(r), \qquad \text{(A16.1.3)}$$

as

$$n_{\rm e}^0(r|I) \equiv \sum_i [\exp\{\beta(\varepsilon_i - \mu_0^e)\} + 1]^{-1}|\phi_i(r)|^2 \qquad \text{(A16.1.4)}$$

with μ_0^e the chemical potential of the noninteracting system. The effective interactions $v_{ij}^{\rm eff}(r)$ involved in (A16.1.1) and (A16.1.2) are described by terms of the interaction part of the intrinsic free energy $\mathscr{F}_{\rm int}$ of this system:

$$v_{ij}^{\rm eff}(r) = v_{ij}(r) + \frac{\delta\mathscr{F}_{\rm int}}{\delta n_i(r|j)} - \mu_i^{\rm int} \qquad \text{(A16.1.5)}$$

with $\mu_i^{\rm int}$ the interaction part of the chemical potential of i species. These interactions can be rewritten as

$$\begin{aligned} v_{ij}^{\rm eff}(r) &= v_{ij}(r) + \sum_\ell \int \frac{\delta^2\mathscr{F}_{\rm int}}{\delta n_i(r)\delta n_\ell(r')}\bigg|_0 \delta n_\ell(r'|j)\,{\rm d}r' - \frac{B_{ij}(r)}{\beta} \\ &= v_{ij}(r) - \frac{\Gamma_{ij}(r)}{\beta} - \frac{B_{ij}(r)}{\beta} \end{aligned} \qquad \text{(A16.1.6)}$$

with

$$\Gamma_{ij}(r) \equiv \sum_{\ell} \int C_{i\ell}(|r - r'|)\{n_\ell(r'|j) - n_0^\ell\}\, dr' \qquad (A16.1.7)$$

in terms of the bridge functions $B_{ij}(r)$ and the direct correlation functions (DCF) defined by

$$C_{ij}(|r - r'|) \equiv -\beta \frac{\delta^2 \mathscr{F}_{\text{int}}}{\delta n_i(r)\delta n_j(r')}\bigg|_0. \qquad (A16.1.8)$$

This definition (Chihara, 1984) of the DCF leads to expressions of the partial structure factors $S_{II}(Q)$, $S_{eI}(Q)$ and the density response function of electrons χ_Q^{ee} by the DCF's $C_{ij}(Q)$ and the density response function χ_Q^0 of the noninteracting system in the forms:

$$S_{II}(Q) = \frac{1 - n_0^I C_{ee}(Q)\chi_Q^0}{D(Q)} \qquad (A16.1.9)$$

$$S_{eI}(Q) = \frac{(n_0^e n_0^I)^{1/2} C_{eI}(Q)\chi_Q^0}{D(Q)} = \frac{(n_0^e n_0^I)^{1/2} C_{eI}(Q)\chi_Q^0}{1 - n_0^e C_{ee}(Q)\chi_Q^0} S_{II}(Q), \qquad (A16.1.10)$$

$$\chi_Q^{ee} = \frac{1 - n_0^I C_{II}(Q)}{D(Q)} = \frac{\{1 - n_0^I C_{II}(Q)\}}{\{1 - n_0^e C_{ee}(Q)\chi_Q^0\}} S_{II}(Q) \qquad (A16.1.11)$$

with

$$D(Q) = \{1 - n_0^I C_{II}(Q)\}\{1 - n_0^e C_{ee}(Q)\chi_Q^0\} - n_0^e n_0^I |C_{eI}(Q)|^2 \chi_Q^0. \qquad (A16.1.12)$$

Here, we have used the fact that the density response functions χ_Q^{iI} concerning ions become identical with the structure factors $S_{iI}(Q)$, since the ions in a liquid metal can be treated as classical particles. The Ornstein-Zernike (OZ) relations for the mixture are obtained by the inverse Fourier transforms of the preceding equations as

$$g_{II}(r) - 1 = C_{II}(r) + \Gamma_{II}(r), \qquad (A16.1.13)$$

$$g_{eI}(r) - 1 = \hat{B}\cdot C_{eI}(r) + \hat{B}\cdot\Gamma_{eI}(r), \qquad (A16.1.14)$$

$$\frac{n_e(r|e)}{n_0^e} - 1 = \hat{B}\cdot C_{ee}(r) + \hat{B}\cdot\Gamma_{ee}(r), \qquad (A16.1.15)$$

where $\hat{B}$ denotes an operator defined by

$$\mathscr{F}_Q[\hat{B}^\alpha \cdot f(r)] \equiv (\chi_Q^0)^\alpha \cdot \mathscr{F}_Q[f(r)] \equiv (\chi_Q^0)^\alpha \cdot \int e^{iQ\cdot r} f(r)\, dr, \qquad (A16.1.16)$$

for an arbitrary real number α. Here, it should be remarked that the OZ relation (A16.1.15) for the electron density distribution $n_e(r|e)$ around a "fixed" electron in the mixture is derived on the basis of the relation:

$$\mathscr{F}_Q[n_e(r|e) - n_0^e] = \frac{\chi_Q^{ee}}{\chi_Q^0} - 1, \tag{A16.1.17}$$

which has been derived from a certain ansatz (see Chihara, 1982). In the above, it should be noted that (A16.1.1)–(A16.1.14) are formal, but exact, expressions.

At this point, one can derive many kinds of integral equations for the RDFs in a liquid metal from the above equations by introducing various approximations. In the first place, the hypernetted-chain equation (see Chapter 5) is obtained by the neglect of the bridge functions $B_{ij}(r)$ in (A16.1.6); this equation is written with the aid of the OZ relations (A16.1.13)–(A16.1.15) as follows.

I. *The quantal* HNC (QHNC) *equation.*

$$C_{II}(r) = \exp\{-\beta v_{II}(r) + \Gamma_{II}(r)\} - 1 - \Gamma_{II}(r) \tag{A16.1.18}$$

$$C_{eI}(r) = \hat{B}^{-1} \cdot \left[\frac{n_e^0(r|v_{eI} - \Gamma_{eI}/\beta)}{n_0^e} - 1 \right] - \Gamma_{eI}(r) \tag{A16.1.19}$$

$$C_{ee}(r) = \hat{B}^{-1} \cdot \left[\frac{n_e^0(r|v_{ee} - \Gamma_{ee}/\beta)}{n_0^e} - 1 \right] - \Gamma_{ee}(r). \tag{A16.1.20}$$

In the HNC approximation, the DCF's $C_{ij}(r)$ are all determined self-consistently from (A16.1.18)–(A16.1.20). In contrast to this approach, if the DCF's $C_{ij}(Q)$ are given by some other method, (A16.1.9)–(A16.1.11) can provide the structure factors, that is, the RDFs. If one can approximate $C_{II}(Q)$ in the mixture as the DCF $C^{OCP}(Q)$ of the one-component plasma (OCP) and $C_{ee}(Q)$ as $C^{jell}(Q)$ of the jellium model, where the ions in a liquid metal are approximated by a uniform background of positive charge, one can then obtain the electron-ion plasma (EIP) model with the use of the pseudopotential $w(Q)$ in the following form.

II. *The* EIP *model.*

$$C_{II}(Q) \simeq C^{OCP}(Q), \tag{A16.1.21}$$

$$C_{eI}(Q) \simeq -\beta v_{eI}(Q) = -\beta w(Q), \tag{A16.1.22}$$

$$C_{ee}(Q) \simeq C_{ee}^{jell}(Q) \equiv \frac{(\chi_Q^{jell})^{-1} - (\chi_Q^0)^{-1}}{n_0^e}, \tag{A16.1.23}$$

where χ_Q^{jell} is the density response function of the jellium model. Also, this model can be represented by using the local-field corrections $G_{ij}(Q)$ defined by

$$C_{ij}(Q) \equiv -\beta v_{ij}(Q)\{1 - G_{ij}(Q)\}, \tag{A16.1.24}$$

in the forms

$$G_{\text{II}}(Q) \simeq G^{\text{OCP}}(Q), \tag{A16.1.25}$$

$$G_{\text{eI}}(Q) \simeq 0, \tag{A16.1.26}$$

$$G_{\text{ee}}(Q) \simeq G_{\text{ee}}^{\text{jell}}(Q). \tag{A16.1.27}$$

By mixed use of the approximations in the QHNC equation and in the EIP model, various kinds of integral equations are derived as follows (see Chihara, 1984).

III. *The conventional pseudopotential theory.* The combination of (A16.1.18), (A16.1.26), and (A16.1.27) constitutes a set of integral equations for $g_{\text{II}}(r)$ and $g_{\text{eI}}(r)$. It should be noted that this set of integral equations is equivalent to the HNC equation for the one-component fluid interacting via an effective ion-ion interaction

$$v^{\text{eff}}(Q) \equiv v_{\text{II}}(Q) - n_0^e |w(Q)|^2 \beta \chi_Q^{\text{jell}}. \tag{A16.1.28}$$

IV. *The nonlinear pseudopotential theory.* If the exchange-correlation effect $G_{\text{ee}}(Q)$ of electrons in a liquid metal is approximated as $G_{\text{ee}}^{\text{jell}}(Q)$ in the jellium that stands for the ions, a set of equations, (A16.1.18), (A16.1.19), and (A16.1.27), can determine the RDFs, $g_{\text{II}}(r)$ and $g_{\text{eI}}(r)$, and a "nonlinear" pseudopotential $w(Q)\{1 - G_{\text{eI}}(Q)\}$. This set of integral equations can be proved to be equivalent to the HNC equation for the one-component fluid with an effective interparticle potential:

$$v^{\text{eff}}(Q) \equiv v_{\text{II}}(Q) - n_0^e |w(Q)\{1 - G_{\text{eI}}(Q)\}|^2 \beta \chi_Q^{\text{jell}}. \tag{A16.1.29}$$

With the aid of the OZ relations, the effective interactions in these approximations may be written in the alternative forms

$$\begin{aligned} v_{\text{II}}^{\text{eff}}(r) = v_{\text{II}}(r) + V_{\text{p}}(r) &- \frac{1}{\beta}\int [\hat{B}^{-1}\cdot\{g_{\text{eI}}(|r - r'|) - 1\} \\ &+ \beta v_{\text{eI}}^{\text{eff}}(|r - r'|)] n_0^I \{g_{\text{eI}}(r') - 1\}\,\mathrm{d}r' \\ &- \frac{1}{\beta}\int \{h_{\text{II}}(r') + \beta v_{\text{II}}^{\text{eff}}(r')\} n_0^I h_{\text{II}}(|r - r'|)\,\mathrm{d}r' \end{aligned} \tag{A16.1.30}$$

with

$$h_{\text{II}}(r) \equiv g_{\text{II}}(r) - 1, \tag{A16.1.31}$$

$$V_{\text{p}}(r) \equiv \int v_{\text{II}}(|r - r'|)\{n_I(r'|I) - n_e(r'|I)\}\,\mathrm{d}r' \tag{A16.1.32}$$

and

$$v_{\mathrm{eI}}^{\mathrm{eff}} = v_{\mathrm{eI}}(r) - V_{\mathrm{p}}(r) - \frac{1}{\beta}\int [\hat{B}^{-1}\cdot\{g_{\mathrm{eI}}(|r-r'|) - 1\}$$

$$+ \beta v_{\mathrm{eI}}^{\mathrm{eff}}(|r-r'|)] n_0^I h_{\mathrm{II}}(r')\,\mathrm{d}r'$$

$$-\frac{1}{\beta}\int C_{\mathrm{ee}}^{\mathrm{XC,jell}}(|r-r'|)\{n_{\mathrm{e}}(r'|I) - n_0^e\}\,\mathrm{d}r' \qquad \text{(A16.1.33)}$$

with

$$C_{\mathrm{ee}}^{\mathrm{XC,jell}}(Q) \equiv C_{\mathrm{ee}}^{\mathrm{jell}}(Q) + \beta v_{\mathrm{ee}}(Q) = \beta v_{\mathrm{ee}}(Q) G_{\mathrm{ee}}^{\mathrm{jell}}(Q). \qquad \text{(A16.1.34)}$$

If one uses the local-density approximation $\mu_{\mathrm{jell}}(n_{\mathrm{e}}(r|I))$ for the exchange-correlation effect represented by the fourth term on the right-hand side of (A16.1.33), one obtains the following.

V. *An extension of the* DWP *equation.* A combination of (A16.1.1) and (A16.1.2) with effective interactions and

$$v_{\mathrm{eI}}^{\mathrm{eff}}(r) = v_{\mathrm{eI}}(r) - V_{\mathrm{p}}(r) - \frac{1}{\beta}\int [\hat{B}^{-1}\cdot\{g_{\mathrm{eI}}(|r-r'|) - 1\}$$

$$+ \beta v_{\mathrm{eI}}^{\mathrm{eff}}(|r-r'|)] n_0^I h_{\mathrm{II}}(r')\,\mathrm{d}r'$$

$$+ \mu_{\mathrm{XC}}^{\mathrm{jell}}(n_{\mathrm{e}}(r|I)) - \mu_{\mathrm{XC}}^{\mathrm{jell}}(n_0^e) \qquad \text{(A16.1.35)}$$

extends the DWP equation (see Section 16.3) to take account of the ion-electron correlation. That is, the neglect of the third terms on the right-hand side of (A16.1.30) and (A16.1.35) yields the following.

VI. *The* DWP *equation.* Chihara (1984) numerically solved five integral equations [I, III, IV, V, and VI] for the case of liquid-metallic hydrogens of density $n_0 \equiv n_0^e = n_0^I$. It should be noted that the equation introduced by Ichimaru's group is essentially equal to the conventional pseudopotential theory III. The nonlinear effect in the electron-proton RDF g_{ep}, which is neglected in their equation, can be examined by comparing the results of III and IV. Furthermore, the comparison of V and VI exhibits the significance of the third terms on the right-hand side of (A16.1.30) and (A16.1.35), which were discarded in DWP's treatment. The QHNC equation I, compared with III–VI, shows the effect of ion-configuration on the exchange-correlation potential, since in the other four equations it is approximated by that calculated from the jellium model, where the ions are replaced by the uniform background of positive charge, in contrast to I, where the ions are treated as a component of the system on an equal footing with the electrons.

The numerical procedure to solve the QHNC equation was described in detail in Chihara (1984). Also, the other four integral equations can be solved by the use of the fast Fourier transform algorithm in a similar manner to the case of the QHNC equation. In this calculation, the electrons were assumed to be fully degenerate, since the calculation was performed for a sufficiently high-density region. The local-density approximation for the exchange-correlation potential used in V and VI, was taken to be of the form proposed by Gunnarsson and Lundqvist (1980):

$$\mu_{XC}(r_s) = -\frac{2}{\pi\alpha r_s}\left\{1 + 0.0545 r_s \cdot \ln\left(1 + \frac{11.4}{r_s}\right)\right\} Ry. \quad \text{(A16.1.36)}$$

In III and IV, one uses the Hubbard-Geldart-Vosko form [see March and Deb (1987)] of the LFC, which is written as

$$G_{ee}^{\text{jell}}(q) = \frac{q^2}{2q^2 + 4g} \quad \text{(A16.1.37a)}$$

with

$$g = \frac{1}{1 + 0.0155\alpha\pi r_s}, \quad \text{(A16.1.37b)}$$

and $q \equiv Q/Q_F$ and $\alpha \equiv (4/9\pi)^{1/3}$.

States of liquid-metallic hydrogen can be specified by the two parameters $\Gamma \equiv \beta e^2/a$ and $r_s \equiv a/a_B$, with $a \equiv (3/4\pi n_0)^{1/3}$ and a_B the Bohr radius. When the density of the system is as high as $r_s < 1$, all five equations give almost the same results for the proton-proton and electron-proton RDFs, g_{pp}, and g_{ep}, except for g_{ep} of III. The integral equations, I, IV and V, yield g_{pp}'s that cannot be discriminated from each other. The situation is similar in the case of g_{ep} obtained from IV and V.

It should be mentioned that the effective proton-proton interaction $v^{\text{eff}}(r)$, calculated from the DWP equation, is significantly different from those of other equations at large distances. As r_s increases up to, for example, 1, these integral equations give different proton-proton RDFs, although the integral equations IV and V yield the same $g_{pp}(r)$. Furthermore, in this region, the conventional pseudopotential method III cannot give a good electron-proton RDF.

To summarize, the density-functional formalism leads to several integral equations for the pair correlations in a liquid metal as a coupled electron-ion mixture with the aid of the hypernetted-chain (HNC) approximation and the information on electrons contained in the jellium model; the quantal HNC equation, the integral equations of Dharma-wardana and Perrot, and of Ichimaru and coworkers. The latter two integral equations have been extended (Chihara, 1984) to include the effect of the electron-ion correlation, which was neglected in their treatments. By solving these five integral equations numerically for the case of liquid-metallic hydrogen, the range of validity of their approximations has been investigated. Furthermore, the approximation of using the exchange-correlation effect in the jellium model in place of that for electrons coupled with the protons has been examined by Chihara on the basis of the quantal HNC equation.

Appendix 16.2
Quantum Monte Carlo calculations of ground state of solid hydrogen

Quantum Monte Carlo calculations of some properties of bulk hydrogen at $T = 0$ have been carried out by Ceperley and Alder (1987). The only approximations invoked are:

1. Restriction to finite systems (64 to 432 atoms),
2. Use of fixed node approximation to treat Fermi statistics (see, for example, the review by March and Senatore, 1989), and
3. Finite length of Monte Carlo runs.

The Born-Oppenheimer approximation is avoided in their work by solving the quantum many-body problem simultaneously for both electron and proton degrees of freedom. Using different trial functions and several different crystalline structures, the transition between those molecular and atomic phases explored was determined to occur at 3.0 ± 0.4 Mbar. The transition to a rotationally ordered molecular phase was found at around 1.0 Mbar.

Here, because the main interest is with the metallic phase, it should be noted that extrapolation to the thermodynamic limit is much more difficult than for the molecular phase because of electron delocalization and thus sensitivity to boundary effects.

Ceperley and Alder (1987) utilized Fermi liquid theory (see Fulde, 1987; also Appendix 11.1) to write for N atoms:

$$E_N = E^{\infty} + c_1(r_s)\frac{(T_N - T_{\infty})}{r_s^2} + \frac{c_2(r_s)}{Nr_s} \qquad \text{(A16.2.1)}$$

where c_1 and c_2 are functions of the density to be determined from the simulation, whereas T_N denotes the kinetic energy of the ideal gas at $r_s = 1$. c_1 and c_2 are written as in (A16.2.1) so as to be roughly independent of r_s in the high-density limit r_s tends to zero.

Their main results are summarized in Table A16.2.1 and in Figure A16.2.1. These lead to the conclusion that hydrogen changes from a face-

Table A16.2.1. *Quantum Monte Carlo results for solid hydrogen at $T = 0$ [after Ceperley and Alder (1987)].*

r_s	Crystal	N	$T\,(\times 10^{-3})$	$-E_{var}$	$-E^N_{mix}$	$-E^\infty_{mix}$	P (Mbar)
			Static lattice of protons				
1.0	*fcc*	108	1.0	0.635(2)	0.670(1)	0.727	19.5
1.13	*fcc*	108	1.6	0.816(1)	0.849(1)	0.893	8.4
1.13	*bcc*	54	1.2	0.949(2)	0.983(3)	0.891	8.5
1.13	*bcc*	128	1.6	0.835(2)	0.876(3)	0.897	8.3
1.30	*fcc*	108	2.0	0.935(1)	0.970(2)	1.002	2.7
1.31	*bcc*	128	1.2	0.957(1)	0.988(1)	1.002	2.6
1.31	*fcc*	256	0.8	0.957(1)	0.992(1)	1.002	2.6
1.45	*bcc*	54	3.4	1.065(1)	1.094(1)	1.035	0.88
1.45	*fcc*	108	1.5	0.982(1)	1.011(1)	1.039	0.85
1.61	*fcc*	108	2.8	0.999(1)	1.033(1)	1.052	0.07
1.77	*fcc*	108	3.2	0.998(1)	1.033(1)	1.048	−0.18
2.00	*fcc*	108	4.0	0.980(1)	1.021(1)	1.033	−0.29
			Dynamic lattice of protons				
1.13	*bcc*	54	0.6	0.782(1)	0.813(1)	0.856	9.13
1.31	*sc*	64	5.0	0.967(1)	0.985(1)	0.962	2.84
1.31	*fcc*	108	0.8	0.913(1)	0.942(1)	0.973	3.05
1.45	*bcc*	54	2.3	1.050(1)	1.071(1)	1.012	1.17
1.45	*fcc*	108	1.2	0.960(1)	0.990(1)	1.016	1.18
1.61	*bcc*	54	1.2	1.057(1)	1.081(1)	1.032	0.25
1.77	*fcc*	108	0.8	0.985(1)	1.019(1)	1.035	−0.03

centred cubic molecular phase to a cubic atomic crystal at a pressure of 3.0 Mbar, as mentioned in general terms before. The atomic phase is stable for $r_s < 1.30$ (1.65 cc/mole) and the assumed molecular phase is the stable state for $r_s > 1.39$ (2.01 cc/mole). Thus the relative volume change is about 20%. Many previous estimates of this transition density had been made; some of them are similar to the quantum Monte Carlo results.

Ceperley and Alder also give the pair correlation functions in the atomic simple cubic phase at the lowest stable atomic density. The electron-electron pair function shown in Figure A16.2.2 is similar to that of an electron gas at the same density.

Finally, these workers also studied the dielectric response of hydrogen. It has been suggested that molecular hydrogen could become a metal at

Figure A16.2.1. Energy of ground state of metallic hydrogen in atomic phase as a function of mean interelectronic spacing r_s in Bohr radii. *Upper line:* fit to Monte Carlo results for finite mass proton lattice of hydrogen. *Lower line:* for static lattice (protons of assumed infinite mass) for either fcc or bcc structure. *Open triangles:* pertubation theory results. *Open circles*: variational correlated theory. *Open squares:* density functional method. [After Ceperley and Alder (1987).]

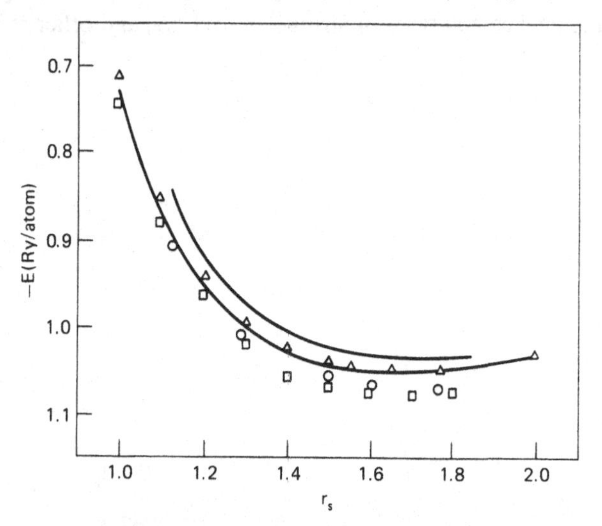

Figure A16.2.2. Electron-electron pair correlation function in ground state of atomic phase of hydrogen [after Ceperley and Alder (1987)].

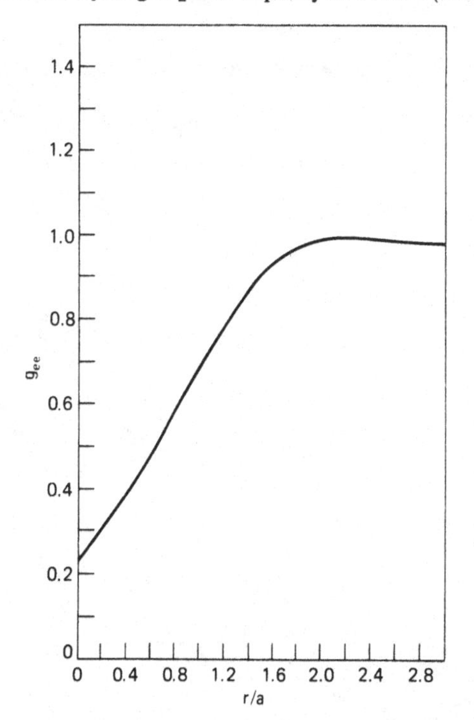

high pressure before undergoing a transition to an atomic phase. This is because, as the density increases, its energy bands broaden and there is an indirect band gap closure, leading to a transition to a metallic state. Such a phenomenon has been observed in iodine at high pressure [see the quantum chemical calculations of Siringo et al. (1989) and other references there].

References

Abraham, F. F. (1980) *J. Chem. Phys.* **72,** 359.

Abrahams, E., Anderson, P. W., Licciardello, D. C., and Ramakrishnan, T. V. (1979) *Phys. Rev. Lett.* **42,** 673.

Adamson, A. W. (1960) *Physical chemistry of surfaces* (Interscience: New York) p. 73.

Alder, B. J., and Wainwright, T. E. (1967) *Phys. Rev. Lett.* **18,** 988; (1957) *J. Chem. Phys.* **27,** 1208.

Alekseev, V. A., and Iakubov, I. T. (1983) *Phys. Rev.* **96,** 1.

Alfred, L. C. R., and March, N. H. (1957) *Phil. Mag.* **2,** 985.

Aliawadi, N. K., Rahman, A. and Zwanzig, R. (1971) *Phys. Rev.* **A4,** 1616.

Alonso, J. A., and March, N. H. (1985) *Surf. Sci.* **160,** 509.

Alonso, J. A., and March, N. H. (1989) *Phys. Chem. Liquids* **20,** 235.

Alonso, J. A., and March, N. H. (1989) *Electrons in metals and alloys* (Academic: London).

Altshuler, L. V. (1965) *Sov. Phys. Usp.* **8,** 52.

Amar, J. G., and Mountain, R. D. (1987) *J. Chem. Phys.* **86,** 2236.

Anderson, P. W. (1958) *Phys. Rev.* **109,** 1492.

Anderson, P. W. (1961) *Phys. Rev.* **124,** 41.

Andrade, E. N. da C. (1934) *Phil. Mag.* **17,** 497.

Angus, W. R. (1932) *Proc. Roy. Soc.* **A136,** 569.

Asano, S., and Yonezawa, F. (1980) *J. Phys.* **F10,** 75: see also (1977) 3rd Int. Conf. Liquid Metals: Bristol.

Asay, J. R., and Hayes, D. B. (1975) *J. Appl. Phys.* **46,** 4789.

Ascarelli, P., and Harrison, R. J. (1969) *Phys. Rev. Lett.* **22,** 385.

Ashcroft, N. W., and Langreth, D. C. (1967) *Phys. Rev.* **155,** 682; *Phys. Rev.* **159,** 500.

Ashcroft, N. W., and March, N. H. (1967) *Proc. Roy. Soc.* **A297,** 336.

Ashcroft, N. W., and Hammerberg, J. (1974) *Phys. Rev.* **B9,** 409.

Baer, Y., and Myers, H. P. (1977) *Solid State Commun.* **21,** 833.

Balescu, R. (1975) *Equilibrium and nonequilibrium statistical mechanics* (Wiley: New York).

Ballentine, L. (1975) *Adv. Chem. Physics* **31,** 263: (1966) *Can. J. Phys.* **44,** 2533.

Baltensperger, W. (1966) *Phys. Kondens. Mater.* **5,** 341.

Balucani, U., Vallauri, R., and Gaskell, T. (1985) *J. Phys.* **C18,** 3133.

Balucani, U., Vallauri, R., and Gaskell, T. (1987) *Phys. Rev. A* **35,** 4263.

Balucani, U. et al. (1988) *Phys. Rev.* **A37,** 3386.

Ban, N. T., Randall, C. M., and Montgomery, D. J. (1962) *Phys. Rev.* **128,** 6.

Bansil, A. et al. (1976) in *Liquid Metals* (IOP Conf. Series **30:** Bristol) p. 313.

Barber, M. N. (1975) *J. Phys.* **C8,** L203.

Bardeen, J. (1956), *Handbuch der Physik, Berlin* **15,** 274.

Barker, M. I., March, N. H., Johnson, M. W., and Page, D. I. (1973) 2nd Int. Conf. Liquid Metals (Taylor and Francis: London) p. 99.

Barrat, J. L., Hansen, J. P., and Pastore, G. (1987), *Phys. Rev. Lett.* **58,** 2075.

Baus, M. and Colot, J. L. (1985) *Mol. Phys.* **55,** 653.

Baym, G. (1964) *Phys. Rev.* **135,** A1691.

Bedeaux, R., and Mazur, P. (1973) *Phys. Lett.* **43A,** 401.

Bedeaux, R., and Mazur, P. (1974) *Physica* **76,** 247.
Benedek G. B., and Kushida, T. (1958) *J. Phys. Chem. Solids* **5,** 241.
Bernasconi, J. M., and March, N. H. (1986) *Phys. Chem. Liquids*, **15,** 169.
Bernasconi, J. M., March, N. H., and Tosi, M. P. (1986) *Phys. Chem. Liquids* **16,** 39.
Berry, M. V., Durrans, R. F., and Evans, R. (1972) *J. Phys.* **A5,** 166.
Bethe, H. A. (1935) *Proc. Roy. Soc.* **A150,** 552.
Bhat, B. N., and Swalin, R. A. (1972) *Acta Metall.* **20,** 1387.
Bhatia, A. B. (1977) *Conf. Series Inst. Phys.* **30,** 21.
Bhatia, A. B., and Hargrove, W. H. (1974) *Phys. Rev.* **B10,** 3186.
Bhatia, A. B., Hargrove, W. H., and March, N. H. (1973) *J. Phys.* **C6**, 621.
Bhatia, A. B., Hargrove, W. H., and Thornton, D. E. (1974) *Phys. Rev.* **B9,** 435.
Bhatia, A. B., and Krishnan, K. S. (1948) *Proc. Roy. Soc.* **A194,** 185.
Bhatia, A. B., and Krishnan, K. S. (1945) *Nature* **156,** 503.
Bhatia, A. B., and March, N. H. (1972) *Phys. Lett.* **41A,** 397.
Bhatia, A. B., and March, N. H. (1975) *J. Phys.* **F5,** 1100: *Phys. Lett.* **51A,** 401.
Bhatia, A. B., and March, N. H. (1978) *J. Chem. Phys.* **68**, 1999, 4651.
Bhatia, A. B., and March, N. H. (1979) *Phys. Chem. Liquids* **9,** 1.
Bhatia, A. B., and March, N. H. (1984a) *J. Chem. Phys.* **80,** 2076.
Bhatia, A. B., and March, N. H. (1984b) *Phys. Chem. Liquids* **13**, 313.
Bhatia, A. B., March, N. H., and Sutton, J. (1978) *J. Chem. Phys.* **69,** 2258.
Bhatia, A. B., and Singh, R. N. (1982) *Phys. Chem. Liquids* **11,** 285.
Bhatia, A. B., and Singh, R. N. (1984) *Phys. Chem. Liquids* **13**, 177.
Bhatia, A. B., March, N. H., and Tosi, M. P. (1980) *Phys. Chem. Liquids* **9,** 229.
Bhatia, A. B., and Thornton, D. E. (1970) *Phys. Rev.* **B2,** 3004.
Bhatia, A. B., Thornton, D. E., and March, N. H. (1974) *Phys. Chem. Liquids* **4,** 93.
Blandin, A., and Friedel, J. (1959) *J. Phys. Radium,* Paris **20,** 160.
Boos, A., and Steeb, S. (1977) *Phys. Lett.* **63A,** 333.
Borchi, E., and de Gennaro, S. (1972) *Phys. Rev.* **B5,** 4761.
Borland, R. E. (1960) *Proc. Phys. Soc.* **52,** 129.
Bottyan, L., Dupree, R., and Freyland, W. (1983) *J. Phys.* **F.13,** L173.
Bratby, P., Gaskell, T., and March, N. H. (1970) *Phys. Chem. Liquids* **2,** 53.
Briane, M. (1973a) *C. r. hebd. Seanc. Acad. Sci. Paris* **276,** 139.
Briane, M. (1973b) *C. r. hebd. Seanc. Acad. Sci. Paris* **277,** 695.
Brindley, G. W., and Hoare, F. E. (1937) *Trans. Far. Soc.* **33,** 268; *Proc. Phys. Soc.* **49,** 619.
Brinkman, W. F., and Rice, T. M. (1970) *Phys. Rev.* **B2,** 4302.
Brown, J. S. (1981) *J. Phys.* **F11,** 2099.
Brown, J. M., and McQueen, R. G. (1980) *J. Geophys. Res.* **7,** 533.
Brown, J. M., and McQueen, R. G. (1982) in *High-pressure research in geophysics*, Akimoto, S. and Manghnani, M. H. (eds.) (Centre for Academic Publications: Tokyo) p. 611.
Brown, R. C., and March, N. H. (1968) *Phys. Chem. Liquids* **1,** 141.
Brown, R. C., and March, N. H. (1973) *J. Phys.* **C6,** L363.
Brown, R. C., and March, N. H. (1976) *Phys. Repts.* **24,** 77.
Buff, F. P., and Lovett, R. (1968) in *Simple dense fluids*, Frisch, H. L., and Salsburg, Z. W. (eds.) (Academic: New York).
Busch, G., Güntherodt, H. J., Kunzi, H. U., and Meier, H. A. (1973) *The properties of liquid metals*, Takeuchi, S. (ed.) (Taylor and Francis: London) p. 263.
Bushman, A. V., and Fortov, V. E. (1983) *Sov. Phys. Usp.* **26,** 465.
Cahn, J. W., and Hilliard, J. E. (1958) *J. Chem. Phys.* **28,** 258.
Cahn, R. W. (1980) *Contemp. Phys.* **21,** 43.
Callaway, J. (1976) *Quantum theory of the solid state* (Academic: New York).
Callaway, J., and March, N. H. (1984) *Solid State Phys.* **38,** 135.
Callaway, J., Zou, X., and Bagayoko, D. (1983) *Phys. Rev.* **B27,** 631.

Campbell, C. E. (1972) *Annals Phys.* **74,** 43.
Campbell, C. E., and Zabolitzky, J. G. (1984) *Phys. Rev.* **B29,** 123.
Campbell, C. E., Kürten, K. E., and Krotscheck, E. (1982) *Phys. Rev.* **B25,** 1633.
Carnahan, N. F., and Starling, K. E. (1969) *J. Chem. Phys.* **51,** 635.
Castellani, C., and DiCastro, C. (1985) in *Localization and metal-insulator transitions,* Fritzsche, H. and Adler, D. (eds.) (Plenum: New York); (1986) *Phys. Rev.* **B34,** 5935.
Castellani, C., et al. (1984) *Phys. Rev.* **B30,** 527.
Castellani, C., et al. (1986) *Phys. Rev.* **B33,** 6169.
Castellani, C., DiCastro, C., Kotliar, G., and Lee, P. A. (1987) *Phys. Rev. Lett.* **59,** 477.
Ceperley, D. M. (1989) private communication and to be published.
Ceperley, D. M., and Alder, B. J. (1980) *Phys. Rev. Lett.* **45,** 566.
Ceperley, D. M., and Alder, B. J. (1987) *Phys. Rev.* **B36,** 2092.
Chakraborty, T. et al. (1983) *Phys. Rev.* **B27,** 3061.
Chakravarty and Ashcroft, N. W. (1978) *Phys. Rev.* **18,** 5488.
Chapman, S., and Cowling, T. G. (1960) *Theory of non-uniform gases* (Cambridge University Press).
Chapman, R. G., and March, N. H. (1986) *Phys. Chem. Liquids* **16,** 77.
Chapman, R. G., and March, N. H. (1987) *Phys. Chem. Liquids* **17,** 165.
Chapman, R. G., and March, N. H. (1988) *Phys. Rev.* **B38,** 792.
Chapman, R. G., and March, N. H. (1989) to be published.
Chaturvedi, D. K., Rovere, M., Senatore, G., and Tosi, M. P. (1981) *Physica* **111B,** 11.
Chen, H. S. (1980) *Rept. Prog. Phys.* **43,** 353.
Chester, G. V., and Thellung, A. (1961) *Proc. Phys. Soc.* **77,** 1005.
Chieux, P., and Ruppersberg, H. (1980) *J. Phys.* (Las Illis, Fr.) **41,** C8-145.
Chihara, J. (1973) 2nd Int. Conf. Liquid Metals (Taylor and Francis: London) p. 137.
Chihara, J. (1984) *J. Phys.* **C17,** 1633; and private communication.
Choy, T. R., and Mayer, J. E. (1967) *J. Chem. Phys.* **46,** 110.
Christiansen, J. A. (1936) *Z. Phys. Chem.* **B33,** 145.
Cohen, C., Sutherland, J. W. H., and Deutch, J. M. (1971) *Phys. Chem. Liquids* **2,** 213.
Cohen, M. H., and Grest, G. S. (1979) *Phys. Rev.* **B20,** 1077.
Cohen, M. H., and Turnbull, D. (1959) *J. Chem. Phys.* **31,** 1164.
Collins, F. C., and Raffel, H. (1954) *J. Chem. Phys.* **22,** 1728.
Copestake, A. P., et al. (1983) *J. Phys.* **F13,** 1993.
Copley, J. R. D., and Rowe, J. M. (1974) *Phys. Rev. Lett.* **32,** 49.
Corless, G. K., and March, N. H. (1961) *Phil. Mag.* **6,** 1285.
Cowan, R. D., and Kirkwood, J. G. (1958) *J. Chem. Phys.* **29,** 264.
Cowley, J. M. (1950) *Phys. Rev.* **77,** 667.
Croxton, C. A., and Ferrier, R. P. (1971) *Phil. Mag.* **24,** 489.
Cummings, P. T., and Morriss, G. P. (1987) *J. Phys.* **F.17,** 593.
Curtin, W., and Ashcroft, N. W. (1985) *Phys. Rev.* **A32,** 2909; (1986) *Phys. Rev. Lett.* **56,** 2775
Cusack, N. E. (1963) *Repts. Prog. Phys.* **26,** 361.
Cusack, N. E. (1987) *The physics of structurally disordered matter* (Adam Hilger: Bristol).
Cyrot-Lackmann, F. (1964) *Phys. Kondens. Mater.* **3,** 75.
Dagens, L., Rasolt, M., and Taylor, R. (1975) *Phys. Rev.* **B11,** 2726.
Darken, L. S. (1967) *Trans. Metall. Soc. AIME* **239,** 80.
Davison, L., and Graham, R. A. (1979) *Phys. Rep.* **55,** 255.
de Angelis, U., and March, N. H. (1976) *Phys. Lett.* **56A,** 287.
DeGroot, S., and Mazur, P. (1963) *Non-equilibrium thermodynamics* (North-Holland: Amsterdam).
Delley, B., and Beck, H. (1979) *J. Phys.* **F9,** 517, 2231.
DeMarcus, W. C. (1958) *Astron. J.* **63,** 2.
Devillers, M. A. C., and Ross, R. G. (1975) *J. Phys.* **F5,** 73.

DeWitt, H. E. (1978) *Phys. Rev.* **A14,** 1290; also in *Strongly coupled plasmas,* Kalman, G., and Carini, P. (eds.) (Plenum; New York).
Dharma-wardana, M. W. C., and Aers, G. C. (1983) *Phys. Rev.* **B28,** 1701.
Dharma-wardana, M. W. C., and Aers, G. C. (1986) *Phys. Rev. Lett.* **56,** 1211.
Dharma-wardana, M. W. C., and Perrot, F. (1982) *Phys. Rev.* **A26,** 2096.
Dharma-wardana, M. W. C., Perrot, F., and Aers, G. C. (1983) *Phys. Rev.* **A28,** 344.
Dobson, P. J. (1978) *J. Phys.* **C11,** L295.
Doniach, S. (1967) *Proc. Phys. Soc.* **91,** 86.
Donovan, B. (1967) *Elementary theory of metals* (Pergamon: Oxford)
Dupree, R., and Geldart, D. J. W. (1971) *Solid State Commun.* **9,** 145.
Dupree, R., and Seymour, E. F. W. (1970) *Phys. Kondens. Mater.* **12,** 97.
Dupree, R., and Sholl, C. A. (1975) *Z. Phys.* **B20,** 275.
Duvall, G. E., and Fowles, G. R. (1963) *High pressure physics and chemistry,* Vol. 2, Bradley, B. S. (ed.) (Academic: New York) p. 209.
Ebbsjö, I., Robinson, G., and March, N. H. (1983) *Phys. Chem. Liquids* **13,** 65.
Edwards, S. F. (1958) *Phil. Mag.* **3,** 1020.
Edwards, S. F. (1962) *Proc. Roy. Soc.* **A267,** 518.
Edwards, D. J., and Jarzynski, J. (1972) *J. Phys.* **C5,** 1745.
Egelstaff, P. A. (1967) *Introduction to the liquid state* (Academic: New York).
Egelstaff, P. A. (1987) *Phys. Chem. Liquids* **16,** 293.
Egelstaff, P. A., March, N. H., and McGill, N. C. (1974) *Can. J. Phys.* **52,** 1651.
Egelstaff, P. A., Page, D. I., and Heard, C. R. T. (1971) *J. Phys.* **C4,** 1453.
Egelstaff, P. A., Suck, J. B., Gläser, W., McPherson, R., and Teitsma, A. J. (1980) *J. Physique Coll.* **41,** C8-222.
Egelstaff, P. A., and Widom, B. (1970) *J. Chem. Phys.* **53,** 2667.
Eisenschitz, R., and Wilford, M. J. (1962) *Proc. Phys. Soc.* **80,** 1078.
El-Hanany, U., and Warren, W. W. (1975) *Phys. Rev.* **B12,** 861: *Phys. Rev. Lett.* **34,** 1276.
Enderby, J. E., Gaskell, T., and March, N. H. (1965) *Proc. Phys. Soc.* **85,** 217.
Enderby, J. E., North, D. M., and Egelstaff, P. A. (1966) *Phil. Mag.* **14,** 961.
Endo, H., Tamura, K., and Yao, M. (1987) *Can. J. Phys.* **65,** 266.
Ernst, M. H., Hauge, E. H., and Van Leeuwen, J. M. J. (1970) *Phys. Rev. Lett.* **25,** 1254.
Ernst, M. H., Hauge, E. H., and Van Leeuwen, J. M. J. (1971) *Phys. Rev.* **A4,** 2055.
Evans, M. G., and Polanyi, M. (1935) *Trans. Far. Soc.* **31,** 875.
Faber, T. E. (1966) *Adv. Phys.* **15,** 547.
Faber, T. E. (1967) *Adv. Phys.* **16,** 637.
Faber, T. E. (1972) *Theory of liquid metals* (Cambridge University Press).
Feenberg, E. (1969) *Theory of quantum fluids* (Academic: New York).
Ferraz, A., and March, N. H. (1979) *Phys. Chem. Liquids* **8,** 271.
Ferraz, A., and March, N. H. (1980) *Solid State Commun.* **36,** 977.
Finkelstein, A. M. (1983) *Zh. Eksp. Teor. Fiz.* **84,** 168; *Sov. Phys. JETP* **57,** 97.
Finnis, M. W. (1974) *J. Phys.* **F4,** 1645.
Fischer, J., and Lago, S. (1983) *J. Chem. Phys.* **78,** 5750.
Fisher, M. E. (1967) *Repts. Prog. Phys.* **30,** 615.
Fisher, R. A., and Watts, R. O. (1972) *Aust. J. Phys.* **25,** 21, 529.
Fleming, P. D., Yang, A. J. M., and Gibbs, J. H. (1976) *J. Chem. Phys.* **65,** 7.
Flores, F., and March, N. H. (1981) *J. Phys. Chem. Solids* **42,** 439.
Flory, P. J. (1942) *J. Chem. Phys.* **10,** 51.
Flynn, C. P., Peters, J. J., and Wert, C. A. (1971) *Phys. Lett.* **A35,** 1571.
Flynn, C. P., Rigney, D. A., and Gardner, J. A. (1967) *Phil. Mag.* **15,** 1255.
Fortov, V. E. et al. (1974) *JETP Lett.* **20,** 13.
Fowler, R. H. (1937) *Proc. Roy. Soc.* **A159,** 229.
Franz, G. (1980), thesis, University of Marburg.

Franz, G. et al. (1980) *J. Phys. (Paris) Colloq.* **41,** C8-194.
Franz, J. R. (1984) *Phys. Rev.* **B29,** 1565.
Freeman, K. S. C., and McDonald, I. R. (1973) *Mol. Phys.* **26,** 529.
Freyland, W. (1979) *Phys. Rev.* **B20,** 5104.
Freyland, W. (1980) *J. Phys. (Paris) Colloq.* **41,** C8-74.
Friedel, J. (1956) *Can. J. Phys.* **34,** 1190.
Friedel, J. (1958) Suppl. to *Nuovo Cim.* **7,** 287.
Friedel, J. (1962) *J. Phys. Radium,* Paris **23,** 692.
Fritzson, P., and Berggren, K.-F. (1976) *Solid State Commun.* **19,** 385.
Fröhlich, H. (1967) *Physica* **37,** 215.
Fulde, P. (1988) *J. Phys. F.* **18,** 601.
Fulde, P., Keller, J., and Zwicknagl, G. (1988) *Solid State Phys.* **41,** 1.
Fumi, F. G. (1955) *Phil. Mag.* **46,** 1007.
Furukawa, K. (1960) *Sci. Rep. Res. Inst. Tohoku Univ.* **A12,** 368.
Galam, S., and Hansen, J. P. (1976) *Phys. Rev.* **A14,** 816.
Gardner, J. A., and Ardary, C. (1976) *Solid State Commun.* **19,** 143.
Gardner, J. A. and Flynn, C. P. (1967) *Phil. Mag.* **15,** 1233.
Gaskell, T. (1982) *J. Phys.* **C15,** 1601.
Gaskell, T. (1988) *J. Phys.* **C21,** 1.
Gaskell, T., and March, N. H. (1970) *Phys. Lett.* **33A,** 460.
Gaskell, T., and Miller, S. (1978) *J. Phys.* **C11,** 3749.
Gaskell, T., Balucani, U., and Vallauri, R. (1989) *Phys. Chem. Liquids* **19,** 193.
Gaspari, G. D., and Gyorffy, B. L. (1972) *Phys. Rev. Lett.* **28,** 801.
Gathers, G. R. (1983) *Int. J. Therm. Phys.* **4,** 209.
Gathers, G. R., and Ross, M. (1984) *J. Non-Cryst Solids* **61,** 59.
Geertsma, W. (1985) *Physica* **132B,** 337.
Geldart, D. J. W., and Rasolt, M. (1987): see March, N. H., and Deb, B. M. (1987).
Geldart, D. J. W., and Taylor, R. (1970) *Can. J. Phys.* **48,** 167.
Gell-Mann, M., and Low, F. E., (1951) *Phys. Rev.* **84,** 350.
Gerjuoy, E. (1965) *J. Math. Phys.* **6,** 993.
Gerl, M. (1971) *Zeits für Naturforschung* **26A,** 1.
Gerl, M. (1985) in *Amorphous solids and the liquid state,* March, N. H., Street, R. A., and Tosi, M. P. (eds.) (Plenum: New York).
Gerling, U., Pool, M. J., and Predel, R. (1983) *Z. Metallkd.* **74,** 616.
Gillan, M. J. (1989) *J. Phys. Cond. Mat.* **1,** 689.
Gingrich, N. S., and Heaton, L. (1961) *J. Chem. Phys.* **34,** 873.
Ginoza, M., and March, N. H. (1985) *Phys. Chem. Liquids* **15,** 75.
Goldstein, R. E., and Ashcroft, N. W. (1985) *Phys. Rev. Lett.* **55,** 2164.
Gonzalez, O. D., and Oriani, R. A. (1965) *Trans. Metall. Soc. A.I.M.E.* **233,** 1878.
Goodman, B., and Sjölander, A. (1975) *Phys. Rev.* **B11,** 113.
Götze, W. (1979) *J. Phys.* **C12,** 1279; (1981) *Phil. Mag.* **43,** 219.
Gray, P. (1973) *J. Phys.* **F3,** L43.
Gray, P., and Young, W. H. (1983) *Phys. Chem. Liquids* **13,** 159.
Gray, P., Yokoyama, I., and Young, W. H. (1980) *J. Phys.* **F10,** 197.
Green, H. S. (1952) *Molecular theory of fluids* (North-Holland: Amsterdam).
Greenfield, A. J., Wellendorf, J., and Wiser, N. (1971) *Phys. Rev.* **A4,** 1607.
Greenwood, D. A. (1958) *Proc. Phys. Soc.* **71,** 585.
Grew, K. E. (1969) in *Transport of fluids,* Hanley, H. J. M. (ed.) (Dekker: New York).
Griffiths, R. B. (1965) *J. Chem. Phys.* **43,** 1958.
Gschneider, K. A. (1964) *Solid State Phys.* **16,** 275.
Guggenheim, E. A. (1949) *Thermodynamics* (North-Holland: Amsterdam); (1952) *Mixtures* (Oxford University Press).

Gunnarsson, O., and Lundqvist, B. (1980) *Phys. Rev.* **B22,** 3710.
Gutzwiller, M. C. (1965) *Phys. Rev.* **137,** A1726.
Gyorffy, B. L., and March, N. H. (1969) *Phys. Chem. Liquids* **1,** 253.
Gyorffy, B. L., and March, N. H. (1971) *Phys. Chem. Liquids* **2,** 197.
Haeffner, E. (1953) *Nature* **172,** 775.
Hafner, J. (1985) in *Amorphous solids and the liquid state,* March N. H., Street, R. A., and Tosi, M. P. (eds.) (Plenum: New York).
Hafner, J., and Heine, V. (1983) *J. Phys.* **F13,** 2479.
Hafner, J., and Kahl, G. (1984) *J. Phys.* **F14,** 2259.
Halperin, B. I. (1973) in *Collective properties of physical systems,* Lundqvist, B., and Lundqvist, S. (eds.) (Nobel Foundation: Stockholm).
Halperin, B. I., and Hohenberg, P. C. (1967) *Phys. Rev. Lett.* **19,** 700.
Halperin, B. I., Hohenberg, P. C., and Ma, S. (1972) *Phys. Rev. Lett.* **29,** 1548.
Ham, F. S. (1962) *Phys. Rev.* **128,** 82.
Hamann, D. R., and Overhauser, A. W. (1966) *Phys. Rev.* **143,** 183.
Hanabusa, M., Kushida, T., and Murphy, J. C. (1976) *Phys. Rev.* **B13,** 5179.
Hansen, J. P., Torrie, G. M., and Viellefosse, P. (1977) *Phys. Rev.* **A16,** 2153.
Hansen, J. P., and Verlet, L. (1970) *Phys. Rev.* **184,** 151.
Harris, R. (1972) *J. Phys.* **C5,** L56.
Harrowell, et al. (1989) to appear.
Hasegawa, M., and Watabe, M. (1972) *J. Phys. Soc. Jpn.* **32,** 14.
Hasegawa, M., and Watabe, M. (1987) *Can. J. Phys.* **65,** 348.
Hasegawa, M., and Young, W. H. (1981) *J. Phys.* **F11,** 977.
Haymet, A. D. J. (1989) to appear.
Haymet, A. D. J., and Oxtoby, D. W. (1981) *J. Chem. Phys.* **74,** 2559.
Hebborn, J. E., and March, N. H. (1970) *Adv. Phys.* **19,** 175.
Hemmer, P. C. (1964) *J. Math. Phys.* **5,** 75.
Hensel, F. (1981) *Proc. 8th symp. on thermophysical properties,* Sengers, J. V. (ed.) (Am. Soc. Mech. Engrs: New York) p. 151.
Hernandez, J. P. (1986a) *Phys. Rev.* **A34,** A316.
Hernandez, J. P. (1986b) *Phys. Rev. Lett.* **57,** 3183.
Hernando, J. A. (1986) *J. Chem. Phys.* **84,** 2853; (1986) *Phys. Rev.* **A33,** 1338.
Herring, C. (1966) *Magnetism,* Vol. 4, Rado, G., and Suhl, H. (eds.) (Academic: New York).
Heyes, D. M. (1986) *Can. J. Phys.* **64,** 773.
Heyes, D. M. (1989) *Phys. Chem. Liquids* **19,** 125.
Hill, T. L. (1956) *Statistical mechanics* (McGraw-Hill: New York).
Hill, H. H., and Kmetko, E. A. (1975) *J. Phys.* **F5,** 1119.
Hilton, D., March, N. H., and Curtis, A. R. (1967) *Proc. Roy. Soc.* **A300,** 391.
Hinkelmann, H. (1970) *Phys. Lett.* **33A,** 479.
Howe, R. A., Rigney, D. A., and Flynn, C. P. (1972) *Phys. Rev.* **B6,** 3358.
Huang, K. (1948) *Proc. Phys. Soc.* **60,** 161.
Hubbard, J., and Beeby, J. L. (1969) *J. Phys.* **C2,** 556.
Huberman, M., and Chester, G. V. (1975) *Adv. Phys.* **24,** 489.
Huijben, M. J., and van der Lugt, W. (1979) *Acta Cryst.* **A35,** 431.
Huntington, H. B. (1973) in American Society for Metals, Metals Park, Ohio, p. 155.
Hynes, J. T. (1977) *Ann. Rev. Phys. Chem.* **28,** 301.
Hynes, J. T. (1985) *Ann. Rev. Phys. Chem.* **36,** 573.
Hynes, J. T., Kapral, R., and Weinberg, M. (1979) *J. Chem. Phys.* **70,** 1456.
Ichikawa, K., Granstaff, S. M., and Thompson, J. C. (1974) *J. Chem. Phys.* **61,** 4059.
Ichimaru, S., and Totsuji, H. (1973) *Prog. Theor. Phys.* **50,** 753; **52,** 42.
Iglói, F., and Hafner, J. (1986) *J. Phys.* **C19,** 5799.

Inagaki, T., Arakawa, E. T., Birkhoff, R. D., and Williams, M. W. (1976) *Phys. Rev.* **B13,** 5610.
Ishida, Y., and Yonezawa, F. (1973) *Prog. Theor. Phys.* **49,** 731.
Ishihara, A. (1968) *J. Phys.* **A1,** 539.
Jacobeus, P., Madsen, J. U., Kragh, F., and Cotterill, R. M. J. (1980) *Phil. Mag.* **B41,** 11.
Jacobs, R. E., and Andersen, H. C. (1975) *Chem. Phys.* **10,** 73.
Johnson, M. D., and March, N. H. (1963) *Phys. Lett.* **3,** 313.
Johnson, M. D., Hutchinson, P., and March, N. H. (1964) *Proc. Roy. Soc.* **A282,** 283.
Johnson, M. W. (1990) private communication.
Johnson, M. W. et al. (1974) *J. Phys.* **C8,** 751.
Johnson, M. W., McCoy, B., March, N. H., and Page, D. I. (1977) *Phys. Chem. Liquids* **6,** 243.
Jones, W. (1973) *J. Phys.* **F3,** 1577.
Jones, W. (1980) *Phys. Chem. Liquids* **10,** 99.
Jones, W., and Barker, G. C. (1987) *Phys. Chem. Liquids* **17,** 105.
Jones, W., and Dunleavy, H. N. (1979) *J. Phys.* **F9,** 1541.
Jones, W., and March, N. H. (1973) *Theoretical solid state physics,* Vols. 1 and 2 (Wiley: Interscience: London); (1985) Dover (Reprint Series): New York.
Jüngst, S., Knuth, B., and Hensel, F. (1985) *Phys. Rev. Lett.* **55,** 2160.
Kadanoff, L. (1966) *Physics* **2,** 263.
Kadanoff, L. (1970) *Comments on solids* **2.**
Kadanoff, L., and Baym, G. (1962) *Ann. Phys.* **24,** 419.
Kadanoff, L., and Houghton, A. (1975) *Phys. Rev.* **B11,** 377.
Kadanoff, L., and Swift, J. (1968) *Phys. Rev.* **166,** 89.
Kaeck, J. A. (1972) *Phys. Rev.* **B5,** 1659.
Kanazawa, H., and Matsudaira, N. (1960) *Prog. Theor. Phys.* **23,** 433.
Kanazawa, H., and Matsudaira, N. (1960) *Prog. Theor. Phys.* **22,** 463.
Kasowski, R. V. (1969) *Phys. Rev.* **187,** 891.
Katz, I., and Rice, S. A. (1972) *J. Am. Chem. Soc.* **94,** 4874.
Kawasaki, K. (1976) in *Phase transitions and critical phenomena,* Domb, C., and Green, M. S. (eds.) (Academic: New York) Vol. 5A, p. 165.
Kayser, R. F., and Raveché, H. J. (1982) *Phys. Rev.* **A26,** 2123.
Kelly, P. J., and Glötzel, D. (1986) *Phys. Rev.* **B33,** 5284.
Kerr, W. (1968) *Phys. Rev.* **174,** 316.
Kimura, M., and Yonezawa, F. (1983) in *Topological disorder in condensed matter.* Yonezawa, F., and Ninomiya, T. (eds.) (Springer: Berlin).
Kirkwood, J. G. (1933) *Phys. Rev.* **44,** 749.
Kirkwood, J. G. (1935) *J. Chem. Phys.* **3,** 300.
Kirkwood, J. G. (1946) *J. Chem. Phys.* **14,** 180.
Kirkwood, J. G., and Buff, F. P. (1949) *J. Chem. Phys.* **17,** 338.
Kirkwood, J. G., and Buff, F. P. (1951) *J. Chem. Phys.* **19,** 775.
Kirkwood, J. G., and Monroe, E. (1941) *J. Chem. Phys.* **9,** 514.
Klein, A. P., and Heeger, A. J. (1966) *Phys. Rev.* **144,** 458.
Kleppa, O. J. (1950) *J. Am. Chem. Soc.* **72,** 3346.
Kohn, W. (1959) *Phys. Rev. Lett.* **2,** 393.
Kohn, W., and Sham, L. J. (1965) *Phys. Rev.* **B140,** 1133.
Kohn, W., and Vosko, S. H. (1960) *Phys. Rev.* **119,** 912.
Korringa, J. (1950) *Physica* **16,** 60.
Kramers, H. C. (1940) *Physica* **7,** 284.
Kreuzer, H. J. (1981) *Non-equilibrium thermodynamics and its statistical foundations* (Oxford Univ. Press).

Kreuzer, H. J., Chapman, R. G., and March, N. H. (1988) *Phys. Rev.* **A37,** 582.
Krotscheck, E., Qian, G.-X., and Kohn, W. (1985) *Phys. Rev.* **B31,** 4245.
Kubo, R. (1965) *Statistical mechanics* (North Holland: Amsterdam).
Kubo, R. (1966) *Rep. Prog. Phys.* **29,** 265.
Kumar, N., March, N. H., and Wasserman, A. (1982) *Phys. Chem. Liquids* **11,** 271.
Kunz, H., and Souillard, G. (1979) *Helv. Phys. Acta* **50,** 111.
Landau, L. D., and Lifshitz, E. M. (1959) *Fluid mechanics* (Pergamon: Oxford).
Landau, L. D., and Zeldovich, Ya. B. (1943) *Acta Phys. Chem. USSR* **18,** 194.
Landauer, R., and Woo, J. W. F. (1974) *Phys. Rev.* **B10,** 1266.
Lang, N. D. (1983) in *Theory of the inhomogeneous electron gas,* Lundqvist, S., and March, N. H. (eds.) (Plenum: New York) p. 309.
Lang, N. D., and Kohn, W. (1970) *Phys. Rev.* **B1,** 4555.
Lang, N. D., and Kohn, W. (1971) *Phys. Rev.* **B3,** 1215.
Langer, J. S., and Neal, F. (1966) *Phys. Rev. Lett.* **16,** 984.
Lantto, L. J. (1980) *Phys. Rev.* **B22,** 1380.
Lasocka, M. (1975) *Phys. Lett.* **51A,** 137.
Leavens, C. R., MacDonald, A. H., Taylor, R., Ferraz, A., and March, N. H. (1981) *Phys. Chem. Liquids* **11,** 115.
Lebowitz, J. L. (1964) *Phys. Rev.* **133,** A895.
Leung, C. H., and March, N. H. (1977) *Plasma Phys.* (GB) **11,** 277.
Levesque, D., Weis, J. J., and Reatto, L. (1985) *Phys. Rev. Lett.* **54,** 451.
Li, D. H., Moore, R. A., and Wang, S. (1988a) *J. Chem. Phys.* **88,** 2700.
Li, D. H., Moore, R. A., and Wang, S. (1988b) *J. Chem. Phys.* **89,** 4309.
Liboff, R. L. (1986) *Phys. Rev.* **A31,** 1883.
Lighthill, M. J. (1958) *An introduction to Fourier analysis and generalized functions* (Cambridge University Press).
Lindhard, J. (1954) *Mat.-Fys. Medd.-K. Dan. Vidensk. Selsk.* **28** (8).
Lloyd, P. (1967a) *Proc. Phys. Soc.* **90,** 207.
Lloyd, P. (1976b) *Proc. Phys. Soc.* **90,** 217.
Lloyd, P., and Smith, P. V. (1972) *Adv. Phys.* **21,** 69.
Longuet-Higgins, H. C. (1951) *Proc. Roy. Soc.* **A205,** 247.
Longuet-Higgins, H. C., and Pople, J. A. (1956) *J. Chem. Phys.* **25,** 884.
Loubeyre, P., Le Toullec, R., and Pinceaux, J. P. (1987) *Phys. Rev.* **B36,** 3723.
Lovett, R. (1977) *J. Chem. Phys.* **66,** 1225.
Lovett, R., and Buff, F. P. (1980) *J. Chem. Phys.* **72,** 2425.
Lovett, R., et al. (1973) *J. Chem. Phys.* **58,** 1880.
Lovett, R., Buff, F. P., and Mou, C. Y. (1976) *J. Chem. Phys.* **65,** 570.
Lukes, T., and Jones, R. (1968) *J. Phys.* **A1,** 29.
Lundqvist, S., and March, N. H. (1983) *Theory of the inhomogeneous electron gas* (Plenum: New York).
Ma, S. (1973) *Phys. Lett.* **43A,** 475.
Mansoori, G. A., and Canfield, F. B. (1969) *J. Chem. Phys.* **51,** 4958.
March, N. H. (1966) *Phys. Lett.* **20,** 231.
March, N. H. (1968) in *Theory of condensed matter: (IAEA, Vienna): Liquid metals* (Pergamon: Oxford)
March, N. H. (1974) in *Nobel Symposium, 'Collective properties of physical systems',* Lundqvist, B., and Lundqvist, S. (eds.) (Academic: New York) p. 230.
March, N. H. (1975) *Phil. Mag.* **32,** 497.
March, N. H. (1976) in *Linear and non-linear electron transport in solids,* Devreese, J. T., and van Doren, V. E. (eds.) (Plenum: New York) p. 131.
March, N. H. (1977) *Can. J. Chem.* **55,** 2165.

March, N. H. (1977) *Physics and contemporary needs*, Vol. I, Riazuddin (ed.) (Plenum: New York) p. 53.

March, N. H. (1984) *J. Chem. Phys.* **80,** 5345.

March, N. H. (1984) *Phys. Chem. Liquids* **14,** 79.

March, N. H. (1985) *Proc. Int. School Physics E. Fermi Highlights of Condensed Matter Theory* (North-Holland: Amsterdam) p. 64.

March, N. H. (1986) *Phys. Chem. Liquids* **16,** 117.

March, N. H. (1986) *Chemical bonds outside metal surfaces* (Plenum: New York).

March, N. H. (1987a) *Phys. Chem. Liquids* **16,** 205.

March, N. H. (1987b) *Phys. Chem. Liquids* **16,** 209.

March, N. H. (1987c) *Phys. Chem. Liquids* **17,** 1.

March, N. H. (1987d) *Can. J. Phys.* **65,** 219.

March, N. H. (1987e) *Solid State Commun.* **63,** 1075.

March, N. H. (1989) *Phys. Chem. Liquids* **19,** 59; (1989) *Phys. Rev. B* **B40,** 3356.

March, N. H., and Deb, B. M. (1987) *The single-particle density in physics and chemistry* (Academic: New York).

March, N. H., and Donovan, B. (1954) *Proc. Phys. Soc.* **A67,** 464.

March, N. H., and Murray, A. M. (1960) *Phys. Rev.* **120,** 831.

March, N. H., and Murray, A. M. (1961) *Proc. Roy. Soc.* **A261,** 119.

March, N. H., and Paranjape, B. V. (1987b) *Phys. Chem. Liquids* **17,** 55; (1987a) *Phys. Rev.* **A35,** 5285.

March, N. H., and Sayers, C. M. (1979) *Adv. Phys.* **28,** 1.

March, N. H., and Senatore, G. (1988) *Phys. Chem. Liquids* **17,** 331, and unpublished work.

March, N. H., Suzuki, M., and Parrinello, M. (1979) *Phys. Rev.* **B19,** 2027.

March, N. H., and Tosi, M. P. (1972) *Proc. Roy. Soc.* **A330,** 373.

March, N. H., and Tosi, M. P. (1973) *Ann. Phys.* **81,** 414.

March, N. H., and Tosi, M. P. (1974) *Nuovo Cim.* **B15,** 308.

March, N. H., and Tosi, M. P. (1976) *Atomic dynamics in liquids* (Macmillan: London).

March, N. H., and Tosi, M. P. (1980) *Phys. Chem. Liquids* **10,** 113.

March, N. H., and Tosi, M. P. (1981a,b,c) *Phys. Chem. Liquids* **11,** 79, 89, 129.

March, N. H., and Tosi, M. P. (1984) *Coulomb liquids* (Academic: New York).

March, N. H., and Tosi, M. P. (1986) *J. Phys. Chem. Solids* **47,** 999.

March, N. H., and Tosi, M. P. (1988) *Phys. Chem. Liquids* **18,** 269.

March, N. H., Tosi, M. P., and Bhatia, A. B. (1973) *J. Phys.* **C6,** L59.

March, N. H., Tosi, M. P., and Chapman, R. G. (1988) *Phys. Chem. Liquids* **18,** 195.

March, N. H., Wilkins, S., and Tibballs, J. F. (1976) *Cryst. Lattice Defects* **6,** 253.

Matsunaga, S., Ishigura, T., and Tamaki, S. (1983) *J. Phys.* **F13,** 587.

Matthai, C. C., and March, N. H. (1982) *Phys. Chem. Liquids* **11,** 207.

Mattheiss, L. F., and Warren, W. W. (1977) *Phys. Rev.* **B16,** 624.

Mazenko, G. F. (1973) *Phys. Rev.* **A7,** 209.

Mazenko, G. F. (1974) *Phys. Rev.* **A9,** 360.

McAlister, S. P., and Turner, R. (1972) *J. Phys.* **F2,** L51.

McCaskill, J. S., and March, N. H. (1982) *Phys. Chem. Liquids* **12,** 1.

McCaskill, J. S., and March, N. H. (1984) *J. Phys. Chem. Solids* **45,** 215.

McDonald, I. R., and O'Gorman, S. P. (1978) *Phys. Chem. Liquids* **8,** 57.

McGreevy, R. L., and Mitchell, E. W. J. (1985) *Phys. Rev. Lett.* **55,** 398.

McLaughlin, I. L., and Young, W. H. (1984) *J. Phys.* **F14,** 1.

McMillan, W. (1985) *Phys. Rev.* **B31,** 2750.

Mermin, N. D. (1965) *Phys. Rev.* **A137,** 1441.

Mermin, N. D. (1970) *Phys. Rev.* **B1,** 2362..

Meyer, A. et al. (1981) *Phys. Chem. Liquids* **10,** 279.

Meyer, A. et al. (1984) *Phys. Chem. Liquids* **13,** 293.
Minchin, P., Meyer, A., and Young, W. H. (1974) *J. Phys.* **F4,** 2117.
Mo, K. C., Gubbins, K. E., Jacucci, G., and McDonald, I. R. (1974) *Mol. Phys.* **27,** 1173.
Mon, K. K., Clester, G. V., and Ashcroft, N. W. (1980) *Phys. Rev.* **B21,** 2641.
Mon, K. K., Gann, R., and Stroud, D. (1981) *Phys. Rev.* **A24,** 2145.
Mori, H. (1965a) *Prog. Theor. Phys.* **33,** 423.
Mori, H. (1965b) *Prog. Theor. Phys.* **34,** 399.
Mori, H. (1975) *Prog. Theor. Phys.* **53,** 1617.
Moriya, T. (1963) *J. Phys. Soc. Jpn.* **18,** 516.
Mott, N. F. (1936) *Proc. Camb. Phil. Soc.* **32,** 281.
Mott, N. F., and Jones, H. (1936) *Properties of metals and alloys* (Oxford University Press).
Mountain, R. D. (1982) *Phys. Rev.* **A26,** 2859.
Mountain, R. D., and Thirumalai, D. (1987) *Phys. Rev.* **A36,** 3300; (1983) *Phys. Rev.* **A27,** 2767(E).
Mukherjee, K. (1965) *Phil. Mag.* **12,** 915.
Munakata, T. (1983) *J. Chem. Phys.* **78,** 7290.
Munakata, T., and Igarashi, A. (1977) *Prog. Theor. Phys.* **58,** 1345.
Munakata, T., and Igarashi, A. (1978) *Prog. Theor. Phys.* **60,** 45.
Muto, T., et al. (1962) *J. Phys. Chem. Solids* **23,** 1303.
Nakagawa, Y. (1956) *J. Phys. Soc. Jpn.* **11,** 855.
Narath, A., and Weaver, H. T. (1968) *Phys. Rev.* **175,** 373.
Neale, F. E., and Cusack, N. E. (1982) *J. Phys.* **F12,** 2839.
Nellis, W. J. (1983) *High pressure measurement techniques*, Reggs, G. N. (ed.) (Applied Sciences Publ.: London) p. 68.
Nellis, W. J., and Mitchell, A. C. (1980) *J. Chem. Phys.* **73,** 6137.
Niemeyer, Th., and Van Leeuwen, J. M. J. (1974) *Physica* **71,** 17.
Niemeyer, Th., and Van Leeuwen, J. M. J. (1976) in *Phase transitions and critical phenomena*, Domb, C., and Green, M. S. (eds.) Vol. 6 (Academic: New York).
Nijboer, B. R. A., and Van Hove, L. (1952) *Phys. Rev.* **85,** 777.
Norskov, J. K. (1979) *Phys. Rev.* **B20,** 446.
Olchowy, G. A., and Sengers, J. V. (1988) *Phys. Rev. Lett.* **61,** 15.
Omini, M. (1986) *Phil. Mag.* **54A,** 561.
Ornstein, L. S., and Zernike, F. (1918) *Phys. Z.* **19,** 134.
Ott, A., and Lunden, A. (1964) *Zeits. für Naturforschung* **19A,** 822.
Overhof, H., Uchtmann, H., and Hensel, F. (1976) *J. Phys.* **F6,** 523.
Owen, J. et al. (1957) *J. Phys. Chem. Solids* **2,** 85.
Page, D. I., de Angelis, U., and March, N. H. (1982) *Phys. Chem. Liquids* **12,** 53.
Pant, M. M., Das, M. P., and Joshi, S. K. (1971) *Phys. Rev.* **B4,** 4379.
Pant, M. M., Das, M. P., and Joshi, S. K. (1973) *Phys. Rev.* **B7,** 4741.
Parrinello, M., March, N. H., and Tosi, M. P. (1977) *Nuovo Cimento* **39B,** 233.
Parrinello, M., and Tosi, M. P. (1972) *Nuovo Cimento* **12B,** 155.
Parrinello, M., Tosi, M. P., and March, N. H. (1974) *Proc. Roy. Soc.* **A341,** 91.
Parrinello, M., Tosi, M. P., and March, N. H. (1975) *Lett. Nuovo Cimento* **12,** 605.
Paskin, A. (1967) in *Properties of liquid metals*, Adams, P. D., and Davies H. H. (eds.) p. 223.
Pathak, K. N., and Singwi, K. S. (1970) *Phys. Rev.* **A2,** 2427.
Pearson, W. B. (1964) *A handbook of lattice spacings and structures of metals and alloys* (Pergamon: Oxford).
Pearson, F. J., and Rushbrooke, G. S. (1957) *Proc. Roy. Soc. Edinburgh* **A64,** 305.
Pechukas, P. (1981) *Ann. Rev. Phys. Chem.* **32,** 159.
Peierls, R. E. (1936) *Proc. Camb. Phil. Soc.* **32,** 471.
Pelzer, H., and Wigner, E. P. (1932) *Z. Phys. Chem.* **B15,** 445.

Percus, J., and Yevick, G. (1958) *Phys. Rev.* **110,** 1.
Perdew, J. P., and Wilkins, J. W. (1970) *Solid State Commun.* **8,** 2041; (1973) *Phys. Rev.* **B7,** 2461.
Perrot, F., and Rasolt, M. (1983) *Phys. Rev.* **B27,** 3273.
Perrot, F., Rasolt, M., Kahn, L. M. (1983) *Phys. Rev.* **B27,** 5110.
Peters, J. J., and Flynn, C. P. (1972) *Phys. Rev.* **B6,** 3343.
Peterson, H. K. et al. (1976) in *Structure and excitation in amorphous solids* (*AIR Conf. Proc.* **31:** New York) p. 378.
Pethick, C. J. (1970) *Phys. Rev.* **B2,** 1789.
Pettifor, D. G. (1970) *J. Phys.* **C3,** 367.
Pfeifer, H. P., Freyland, W., and Hensel, F. (1979) *Ber. Bunsenges. Phys. Chem.* **83,** 204.
Pfeuty, P., and Toulouse, G. (1977) *Introduction to the renormalization group and to critical phenomena* (Wiley: New York).
Pines, D., and Nozières, P. (1966) *The theory of quantum liquids* (Benjamin: New York).
Pippard, A. B. (1960) *Rep. Prog. Phys.* **23,** 176.
Pomeau, Y. (1972) *J. Chem. Phys.* **57,** 2800.
Pomeau, Y., and Résibois, P. (1975) *Phys. Rep.* **19C,** 63.
Price, D. L., Singwi, K. S., and Tosi, M. P. (1976) *Phys. Rev.* **B2,** 2983.
Pugachevich, P. P., and Timofeevicheva, O. A. (1951) *Dokl. Akad. Nauk SSSR* **79,** 831.
Pugachevich, P. P., and Timofeevicheva, O. A. (1954) *Dokl. Akad. Nauk SSSR* **94,** 285.
Rahman, A. (1974a) *Phys. Rev. Lett.* **32,** 52.
Rahman, A. (1974b) *Phys. Rev.* **A9,** 1667.
Ramakrishnan, T. V., and Yussouff, M. (1977) *Solid State Commun.* **21,** 389; (1979) *Phys. Rev.* **B19,** 2775.
Ramchandra Rao, P., Singh, R. N., and Lele, S. (1984) *J. Non-Cryst. Solids* **64,** 387.
Randolph, P. D. (1964) *Phys. Rev.* **134,** A1238.
Rasaiah, J., and Stell, G. (1970) *Mol. Phys.* **18,** 249.
Rashid, R. I. M. A., and March, N. H. (1989) *Phys. Chem. Liquids* **19,** 41.
Rashid, R. I. M. A., Senatore, G., and March, N. H. (1986) *Phys. Chem. Liquids* **16,** 1.
Ratti, V. K., and Bhatia, A. B. (1977) *J. Phys.* **F7,** 647.
Ratti, V. K., and Bhatia, A. B. (1978) *Nuovo Cimento* **43B,** 1.
Raveche, H. J., Mountain, R. D., and Streett, W. B. (1974) *J. Chem. Phys.* **61,** 1970.
Reatto, L. (1988) *Phil. Mag.* **A58,** 37.
Reatto, L., and Chester, G. V. (1970) *Phys. Rev.;* (1988) *Phil. Mag.*
Reatto, L., Levesque, D., and Weis, J. J. (1986) *Phys. Rev.* **A33,** 3451.
Ree, F. H. (1983) *J. Chem. Phys.* **78,** 409; **87,** 2846.
Ree, F. H., and Bender, C. F. (1979) *J. Chem. Phys.* **71,** 5362.
Ree, F. H., and Bender, C. F. (1980) *J. Chem. Phys.* **73,** 4712.
Renkert, H., Hensel, F., and Franck, E. U. (1969) *Phys. Lett.* **30A,** 494; (1970), *J. Non-Cryst. Solids* **4,** 180.
Rice, S. A. et al. (1974) *Adv. Chem. Phys.* **27,** 543.
Rice, T. M., Ueda, K., Ott, H. R., and Rudigier, H. (1985) *Phys. Rev.* **B31,** 594.
Rigney, D. A. (1977) *Liquid metals* (Inst. Phys. Conf. series No. 30: Bristol) p. 619.
Root, L. J., Stillinger, F. H., and Washington, G. E. (1988) *J. Chem. Phys.* **88,** 7791.
Root, L. J., Stillinger, F. H., and Weber, T. A. (1981) *J. Chem. Phys.* **74,** 4015, 4020.
Rosenfeld, Y., and Ashcroft, N. W. (1979) *Phys. Rev.* **A20,** 1208.
Ross, M. (1979) *J. Chem. Phys.* **71,** 1567.
Ross, M. (1985) *Rept. Prog. Phys.* **48,** 1.
Ross, M., deWitt, H. E., and Hubbard, W. B. (1981) *Phys. Rev.* **A24,** 1016.
Ross, M., Pee, F. H., and Young, D. A. (1983) *J. Chem. Phys.* **79,** 1487.
Roth, L. M. (1969) *Phys. Rev.* **184,** 451.

Roth, L. M. (1972) *Phys. Rev. Lett.* **28,** 1570.
Roth, L. M. (1973) *Phys. Rev.* **B7,** 4321.
Roth, L. M. (1974) *Phys. Rev.* **B9,** 2476.
Rousseau, J. S. (1971) *J. Phys.* **C4,** L351.
Rousseau, J. S., Stoddart, J. C., and March, N. H. (1970) *Proc. Roy. Soc.* **A317,** 211.
Rousseau, J. S., Stoddart, J. C., and March, N. H. (1971) *J. Phys.* **C4,** L59.
Rousseau, J. S., Stoddart, J. C., and March, N. H. (1972) *J. Phys.* **C5,** L175.
Rovere, M., and Tosi, M. P. (1988) *Z. Phys. Chem. Neue. Folge.* **156,** pt. 2, 411.
Rowland, T. J. (1960) *Phys. Rev.* **119,** 900.
Rowlinson, J. S. (1969) *Mixtures* (Cambridge University Press).
Ruppersberg, H., and Egger, H. (1975) *J. Chem. Phys.* **63,** 4095.
Rushbrooke, G. S. (1960) *Physica* **26,** 259.
Saboungi, M. L., Herron, S. J., and Kumar, R. (1985) *Ber. Bunsen-Ges. Phys. Chem.* **89,** 375.
Sampson, J. B., and Seitz, F. (1940) *Phys. Rev.* **58,** 633.
Sceats, M. G. (1986) *J. Chem. Phys.* **84,** 5206.
Sceats, M. G. (1988a) *J. Chem. Phys.* **88,** 7811.
Sceats, M. G. (1988b) *Adv. Chem. Phys.* **70,** 357.
Sceats, M. G. (1988c) *J. Colloid and Interface Sci.* (in press).
Schofield, P. (1966) *Proc. Phys. Soc.* **88,** 149.
Schofield, P. (1979) *Chem. Phys. Lett.* **62,** 413.
Schwartz, L., and Ehrenreich, H. (1971) *Ann. Phys.* **64,** 100.
Schwartz, L. et al. (1975) *Phys. Rev.* **B12,** 313.
Senatore, G. S., and March, N. H. (1984) *J. Chem. Phys.* **80,** 5242.
Senatore, G. S., and March, N. H. (1986) *Phys. Chem. Liquids* **16,** 131.
Senatore, G., Rashid, R. I. M. A., and March, N. H. (1986) *Phys. Chem. Liquids* **16,** 1.
Seymour, E. (1974) *Pure Appl. Chem.* **40,** 41.
Shaner, J. W., Brown, J. M., and McQueen, R. G. (1984) *High pressure science and technology,* Vol. 3., Homan, C., MacCrone, R., and Whalley, F. (eds.) (Elsevier: Amsterdam) p. 137.
Sharma, S. K., Mao, H. K., and Bell, P. M. (1980) *Phys. Rev. Lett.* **44,** 886.
Shaw, R. W., and Warren, W. W. (1971) *Phys. Rev.* **B3,** 1562.
Silbert, M., and Young, W. H. (1976) *Phys. Lett.* **58A,** 469.
Silvera, I. F., and Wijngaarden, R. J. (1981) *Phys. Rev. Lett.* **47,** 39.
Singh, R. N. (1987) *Can. J. Phys.* **65,** 309.
Singh, R. N., and Bhatia, A. B. (1984) *J. Phys.* **F14,** 2309.
Singwi, K. S., and Tosi, M. P. (1981) *Solid State Physics* **36,** 177.
Singwi, K. S., Sköld, K., and Tosi, M. P. (1970) *Phys. Rev.* **A1,** 454.
Singwi, K. S., Sjölander, A., Tosi, M. P., and Land, R. H. (1970) *Phys. Rev.* **B1,** 1044.
Siringo, F. et al. (1989) *Phys. Rev.* **B38,** 9567.
Sköld, K. (1967) *Phys. Rev. Lett.* **19,** 1023.
Slater, J. C. (1951) *Phys. Rev.* **81,** 385.
Slater, J. C., and Koster, G. F. (1954) *Phys. Rev.* **94,** 1498.
Smith, R. T., Webber, G. M. B., Young, F. R., and Stephens, R. W. B. (1967) in *Properties of liquid metals,* Adams, P. D., Davies, H. H., and Epstein, S. G. (eds.) p. 515.
Smith, W. R., and Henderson, D. (1970) *Mol. Phys.* **19,** 411.
Sorbello, R. S. (1978) *Phys. Stat. Sol.* **B86,** 671.
Sorbello, R. S., and Rimbey, P. R. (1988) *Phys. Rev.* **B38,** 1095.
Stanley, H. E. (1971) *Phase transitions and critical phenomena* (Oxford University Press); (1974) *Cooperative phenomena near phase transitions* (M.I.T. Press: Cambridge, Mass).
Steeb, S., Falch, S., and Lamparter, P. (1984) *Z. Metallkd.* **75,** 599.
Stevenson, D. J., and Salpeter, E. E. (1976) *Jupiter,* Gehrels, T. (ed.) (Univ. of Arizona Press: Tucson) p. 85.

Stillinger, F. H. (1976) *J. Chem. Phys.* **65,** 3968; (1979) *J. Chem. Phys.* **70,** 4067; *Phys. Rev.* **B20,** 299.
Stillinger, F. H., and Weber, T. A. (1978) *J. Chem. Phys.* **68,** 3837.
Stillinger, F. H., and Weber, T. A. (1979) *J. Chem. Phys.* **70,** 1074E.
Stillinger, F. H., and Weber, T. A. (1980) *Phys. Rev.* **B22,** 3790.
Stillinger, F. H., and Weber, T. A. (1981) *J. Chem. Phys.* **74,** 4020.
Stinchcombe, R. B. (1984) in *Polymers, liquid crystals and low-dimensional solids,* March, N. H., and Tosi, M. P. (eds.) (Plenum: New York) Ch. 13.
Stishov, S. M., Makarenko, I. N., Ivanov, V. A., and Nikolaenko, A. M. (1973) *Phys. Lett.* **45A,** 18.
Stone, J. P. et al. (1966) *J. Chem. Eng. Data* **11,** 309.
Stott, M. J., and Young, W. H. (1981) *Phys. Chem. Liquids* **11,** 95.
Stroud, D. (1976) *Phys. Rev.* **B13,** 4221.
Stroud, D., and Ashcroft, N. W. (1972) *Phys. Rev.* **B5,** 371.
Sundström, L. J. (1965) *Phil. Mag.* **11,** 657.
Swamy, K. N. (1986) *Phys. Chem. Liquids* **15,** 309.
Szabo, N. (1972) *J. Phys.* **C5,** L241.
Takahashi, Y., and Shimizu, M. (1973) *J. Phys. Soc. Jpn.* **34,** 942; **35,** 1046.
Tallon, J. L. (1982) *Phys. Lett.* **87A,** 362.
Tamaki, S. (1967) *J. Phys. Soc. Jpn.* **22,** 865.
Tamaki, S. (1968) *J. Phys. Soc. Jpn.* **25,** 374, 1596, 1602.
Tamaki, S. (1987) *Can. J. Phys.* **65,** 286.
Tarazona, P. (1984) *Mol. Phys.* **52,** 81; (1985) *Phys. Rev.* **A31,** 2672.
Taylor, R. (1978) *J. Phys.* **F8,** 1699.
Taylor, R., and Watts, R. O. (1981) *Solid State Commun.* **38,** 965.
Telo de Gama, M. M., and Evans, R. (1980) *Mol. Phys.* **41,** 1091.
Thiele, E. (1963) *J. Chem. Phys.* **39,** 474.
Thouless, D. J. (1979) in *Ill-condensed matter,* Balian, R., Maynard, R., and Toulouse, G. (eds.) (North-Holland: Amsterdam).
Timbie, J. P., and White, R. M. (1970) *Phys. Rev.* **B1,** 2409.
Tosi, M. P. (1985) in *Amorphous solids and the liquid state,* March, N. H., Street, R. A., and Tosi, M. P. (eds.) (Plenum: New York).
Tosi, M. P., and March, N. H. (1973a) *Nuovo Cimento* **B15,** 308.
Tosi, M. P., and March, N. H. (1973b): see March, N. H. (1974).
Tosi, M. P., Parrinello, M., and March, N. H. (1974) *Nuovo Cimento* **B23,** 135.
Tosi, M. P., and Rovere, M. (1988) *Z. Physik Chem.: liquid metals conf.,* to appear.
Triezenberg, D. G., and Zwanzig, R. (1972) *Phys. Rev. Lett.* **28,** 1183.
Turnbull, D., and Cohen, M. H. (1961) *J. Chem. Phys.* **34,** 120.
Tyrrell, H. J. V., and Harris, K. R. (1984) *Diffusion in liquids* (Butterworths: London).
Van den Bergh, L. C., Schouten, J. A., and Trappeniers, N. J. (1987) *Physica* **A141,** 524.
Van den Bergh, L. C., and Schouten, J. A. (1988) *J. Chem. Phys.* **89,** 2336.
Van Hove, L. (1954) *Phys. Rev.* **95,** 249; *Phys. Rev.* **93,** 1374.
Van der Lugt, W., and Geertsma, W. (1987) *Can. J. Phys.* **65,** 326.
Vashishta, P., and Singwi, K. S. (1972) *Phys. Rev.* **B6,** 875, 4883.
Vashishta, P., Bhattacharya, P., and Singwi, K. S. (1973) *Phys. Rev. Lett.* **30,** 1248; see also (1974) *Phys. Rev.* **B10,** 5108.
Verhoeven, J. D. (1969) *Phys. Fluids* **12,** 1783.
Verlet, L. (1968) *Phys. Rev.* **165,** 201.
Verlet, L., and Weis, J. J. (1972) *Phys. Rev.* **A5,** 939.
Vineyard, G. H. (1958) *Phys. Rev.* **110,** 999.
Visscher, P. B. (1976) *Phys. Rev.* **B13,** 3272.
Vosko, S. H., Wilk, L., and Nasair, M. (1980) *Can. J. Phys.* **58,** 1200.

Warren, B. E. (1969) *X-ray diffraction* (Addison-Wesley: Reading, Mass.).
Warren, W. W. (1984) *Phys. Rev.* **B29,** 7012.
Warren, W. W. (1987) in *Amorphous and liquid metals*, Lücher, E., Fritsch, G., and Jacucci, G. (eds.) NATO Advanced Study Institute, Series E18 (Martinus Nijhoff: Dordrecht) p. 304.
Warren, W. W., Brennert, G. F., and El-Hanany, U. (1989) *Phys. Rev.* **B39,** 4038.
Warren, W. W., and Mattheiss, L. F. (1984) *Phys. Rev.* **B30,** 3103.
Waseda, Y., and Suzuki, K. (1970) *Phys. Status Solidi* **40,** 183.
Waseda, Y., Yokoyama, K., and Suzuki, K. (1974) *Phil. Mag.* **29,** 1427.
Watabe, M. (1977) *3rd Int. Conf. Liq. Metals* (Inst. Phys.: Bristol) p. 288
Watabe, M., and Hasegawa, M. (1972) *2nd. Int. Conf. Liquid Metals, Tokyo* (Taylor and Francis: London) p. 133.
Watabe, M., and Hasegawa, M. (1972) *J. Phys. Soc. Jpn.* **32,** 14; (1983) *J. Phys.* **C16,** L699.
Watabe, M., and Young, W. H. (1974) *J. Phys.* **F4,** L29.
Watabe, M. et al. (1965) *Phil. Mag.* **12,** 347.
Weeks, J. D., Chandler, D., and Andersen, H. C. (1971) *J. Chem. Phys.* **54,** 5237; **55,** 5422.
Wertheim, M. S. (1963) *Phys. Rev. Lett.* **10,** 321.
Widom, B. (1965) *J. Chem. Phys.* **43,** 3892, 3898.
Wigner, E. P. (1932) *Phys. Rev.* **40,** 749.
Wilson, J. R. (1965) *Metall. Rev.* **10,** 381.
Wilson, K. G. (1972) *Phys. Rev. Lett.* **28,** 458.
Wilson, K. G. (1974) *Physica* **73,** 119.
Wilson, K. G., and Fisher, M. E. (1972) *Phys. Rev. Lett.* **28,** 240.
Wilson, K. G., and Kogut, J. (1974) *Phys. Rept.* **12C,** 75.
Winter, J. (1971) *Magnetic resonance in metals* (Oxford Univ. Press: Oxford).
Winter, R., Bodensteiner, T., and Hensel, F. (1989) *High Pressure Research* **1,** 1.
Winter, R., Hensel, F., Bodensteiner, T., and Glaser, W. (1987) *Ber. Bunsengere Phys. Chem.* **91,** 1327.
Wong, Y. J., and Chester, G. V. (1987) *Phys. Rev.* **B35,** 3506.
Woodhead-Galloway, J., Gaskell, T., and March, N. H. (1968) *J. Phys.* **C1,** 271.
Worster, J., and March, N. H. (1964) *Solid State Commun.* **2,** 245.
Worster, J., and March, N. H. (1964) *J. Phys. Chem. Solids* **24,** 1305.
Yang, A. J. M., Fleming, P. D., and Gibbs, J. H. (1976) *J. Chem. Phys.* **67,** 74.
Yao, M. et al. (1980): see Endo et al. (1987).
Yarnell, J. L., Katz, M. J., Wenzel, R. G., and Koenig, S. H. (1973) *Phys. Rev.* **A7,** 2130.
Yonezawa, F., and Martino, F. (1976) *Solid State Commun.* **18,** 1471.
Yonezawa, F., Martino, F., and Asano, S. (1977) in *Liquid metals*, Evans, R., and Greenwood, D. A. (eds.) (Inst. Phys.: London).
Yonezawa, F., and Ogawa, T. (1982) *Prog. Theor. Phys. Suppl.* **72,** 1.
Young, D. A., McMahan, A. K., and Ross, M. (1981) *Phys. Rev.* **B24,** 5119.
Young, W. H. (1987) *Can. J. Phys.* **65,** 241: see also (1977) *Conf. Ser. Inst. Phys.* **30,** 1.
Zabolitzky, J. G. (1980) *Phys. Rev.* **B22,** 2353.
Zabolitzky, J. G. (1981) in *Advances in nuclear physics*, Negele, J., and Vogt, E. (eds.) (Plenum: New York), Ch. 1.
Zeldovitch, Ya B., and Raizer, Yu P. (1966) *Physics by shock waves and high temperature hydrodynamic phenomena*, Vols. 1 and 2 (Academic: New York).
Zemansky, M. W. (1951) *Heat and thermodynamics* (McGraw-Hill: New York).
Ziman, J. M. (1960) *Electrons and phonons* (Oxford University Press).
Ziman, J. M. (1961) *Phil. Mag.* **6,** 1013.
Ziman, J. M. (1964) *Adv. Phys.* **13,** 89.
Zwanzig, R. (1960) *J. Chem. Phys.* **33,** 1338.
Zwanzig, R. (1983) *J. Chem. Phys.* **79,** 4507.

Notes added in proof

Since the manuscript was completed, some further noteworthy studies have come to the writer's attention.

Chapter 3

J. F. Lutsko and M. Baus (*Phys. Rev. Lett.* **64,** 761, 1990) raise the question as to whether the thermodynamic properties of a solid can be mapped on to those of a liquid.

Chapter 5

F. Perrot and N. H. March (*Phys. Rev.* **A,** 1990) have made an all-electron calculation of the pair potential in liquid Na near freezing. All of the main features of the potential shown in Figure 5.2, obtained by inversion of the structure factor $S(k)$, are evident in the electron theory results.

Experimental evidence for many-body forces in liquids has been reviewed by P. A. Egelstaff (*Phys. Scripta* **T29,** 288, 1989).

Chapter 6

M. P. Tosi (*Phys. Scripta* **T29,** 277, 1989) has discussed the freezing of Coulomb liquids.

Chapter 7

G. Srinati et al. (*Phys. Scripta* **T29,** 130, 1989) review the theory of transport in disordered many-body systems.

Chapter 10

A dynamical simulated annealing (DSA) approach to the electronic structure of liquid metals is reported by J. Hafner and M. C. Payne (*J. Phys. Cond. Mat.* **2,** 221, 1990) following the seminal work on DSA by R. Car and M. Parrinello (*Phys. Rev. Lett.* **55,** 2471, 1985).

Structural and electronic properties of the group IV elements Si, Ge, Sn and Pb have been studied extensively by W. Jank and J. Hafner (*Phys. Rev.* **B41,** 1497, 1990). It is also relevant to mention work on liquid Te (J. Hafner, *J. Phys. Cond. Mat.* **2,** 1271, 1990) with low coordination number as well as on Cs near the critical point (N. H. March, *J. Math. Chem., in press, 1990).*

Chapter 13

Glass transitions are treated by L. Sjögren (*Phys. Scripta* **T29,** 282, 1989). An excellent review of the theory of disordered alloys is that of J. S. Faulkner (*Prog. Mater. Sci. (GB)* **27,** 3, 1982: see also references to B. L. Gyorffy there).

Thermodynamic properties of liquid Rb-Pb alloys have been deduced from EMF measurements by P. J. Tumidajski et al. (*J. Phys. Cond. Mat.* **2,** 209, 1990) and related to the theory of concentration fluctuations.

Chapter 14

S. Takeda et al. (*J. Phys. Soc. Japan,* **58,** 3999, 1990) have extracted electronic correlation functions for liquid Na from diffraction data.

Chapter 16

Electron-ion correlations in H and He plasmas have been examined further by F. Perrot et al. (*Phys. Rev.* **A41,** 1096, 1990).

Helium-hydrogen mixtures are reviewed by J. A. Schouten (*Phys. Reps.* **172,** 33, 1989).

Index